STUDENT SOLUTIONS MANUAL

Loren Argabrig

Drexel University

to accompany

Contemporary Linear Algebra

Howard Anton
Drexel University
Robert C. Busby
Drexel University

WILEY

ISBN 0-471-17059-3

10 9 8 7 6 5 4 3 2 1

CONTENTS

CHAPTER 1
Vectors

1. **(a)** $\mathbf{v}_1 = (3, 6)$ **(b)** $\mathbf{v}_2 = (-4, 8)$

   (c) $\mathbf{v}_3 = (3, 3, 0)$ **(d)** $\mathbf{v}_4 = (0, 0, -3)$

3. **(a)** $2\mathbf{u} = (2, 2)$ **(b)** $\mathbf{u} + \mathbf{v} = (0, 2)$ **(c)** $2\mathbf{u} + 2\mathbf{v} = (0, 4)$

   (d) $\mathbf{u} - \mathbf{v} = (2, 0)$ **(e)** $\mathbf{u} + 2\mathbf{v} = (-1, 3)$

5. **(a)** $\mathbf{v} = (4 - 1, 1 - 5) = (3, -4)$ **(b)** $\mathbf{v} = (0 - 2, 0 - 3, 4 - 0) = (-2, -3, 4)$

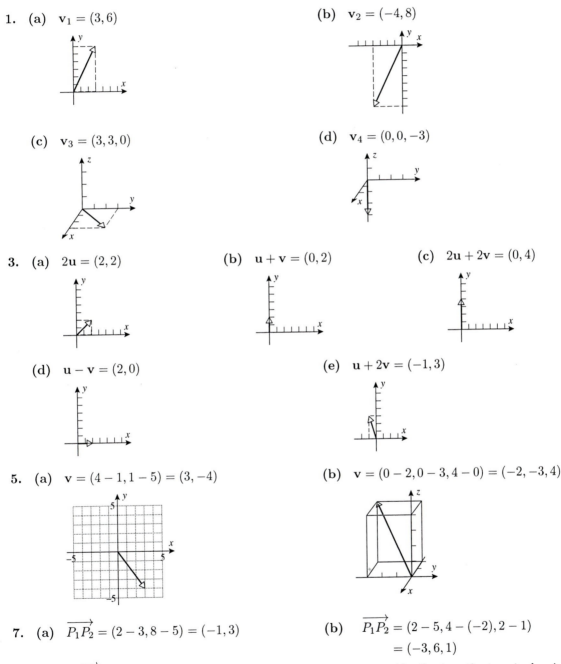

7. **(a)** $\overrightarrow{P_1P_2} = (2 - 3, 8 - 5) = (-1, 3)$ **(b)** $\overrightarrow{P_1P_2} = (2 - 5, 4 - (-2), 2 - 1)$
   $$= (-3, 6, 1)$$

9. **(a)** If $\overrightarrow{AB}$ is equivalent to $\mathbf{u} = (1, 2)$, and the initial point is $A(1, 1)$, then the terminal point is $B(1 + 1, 1 + 2) = B(2, 3)$.

   (b) If $\overrightarrow{AB}$ is equivalent to $\mathbf{u} = (1, 1, 3)$ and the terminal point is $B(-1, -1, 2)$, then the initial point is $A(-1 - 1, -1 - 1, 2 - 3) = A(-2, -2, -1)$.

1

11. **(a)** $\mathbf{v} - \mathbf{w} = (4 - 6, 0 - (-1), -8 - (-4), 1 - 3, 2 - (-5)) = (-2, 1, -4, -2, 7)$

(b) $6\mathbf{u} + 2\mathbf{v} = (-18, 6, 12, 24, 24) + (8, 0, -16, 2, 4) = (-10, 6, -4, 26, 28)$

(c) $(2\mathbf{u} - 7\mathbf{w}) - (8\mathbf{v} + \mathbf{u}) = (-48, 9, 32, -13, 43) - (29, 1, -62, 12, 20) = (-77, 8, 94, -25, 23)$

13. If $2\mathbf{u} - \mathbf{v} + \mathbf{x} = 7\mathbf{x} + \mathbf{w}$, then $2\mathbf{u} - \mathbf{v} - \mathbf{w} = 6\mathbf{x}$ and so $\mathbf{x} = \frac{1}{6}(2\mathbf{u} - \mathbf{v} - \mathbf{w}) = \frac{1}{6}(-16, 3, 16, 4, 11) = (-\frac{8}{3}, \frac{1}{2}, \frac{8}{3}, \frac{2}{3}, \frac{11}{6})$.

15. **(a)** The given vector is not equal to $k\mathbf{u}$ for any scalar k, and thus is not parallel to $\mathbf{u}$.

(b) The given vector is equal to $-2\mathbf{u}$, and thus is parallel to $\mathbf{u}$.

(c) The zero vector is equal to $0\mathbf{u}$, and thus is parallel to $\mathbf{u}$.

17. **(a)** $\mathbf{u} + \mathbf{v} + \mathbf{w} = (-2, 5)$

(b) $\mathbf{u} + \mathbf{v} + \mathbf{w} = (-2 - (-5), -4 - 4)$
$= (3, -8)$

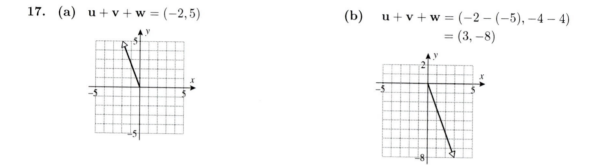

19. If $\mathbf{u} = (1, -1, 3, 5)$ and $\mathbf{v} = (2, 1, 0, -3)$, then $a\mathbf{u} + b\mathbf{v} = (1, -4, 9, 18)$ if and only if $a + 2b = 1$, $-a + b = -4$, $3a = 9$, and $5a - 3b = 18$. The third equation requires that $a = 3$ and, substituting this into the second equation, it follows that $b = -1$. Finally, it is easy to check that these values also satisfy the first and fourth equations.

21. Three parallelograms having the points $A(0, -1)$, $B(-1, 3)$ and $C(1, 2)$ as vertices.

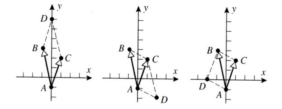

23. The figure indicates three forces, represented by $\mathbf{F}_1 = (3, -2)$, $\mathbf{F}_2 = (0, 5)$, and $\mathbf{F}_3 = (-4, 0)$, acting on a particle at the origin. The sum of these three forces is $\mathbf{F}_1 + \mathbf{F}_2 + \mathbf{F}_3 = (-1, 3)$. Thus the force $\mathbf{F} = -(\mathbf{F}_1 + \mathbf{F}_2 + \mathbf{F}_3) = (1, -3)$ must be applied in order to produce static equilibrium.

25. **(a)**

To

$$\text{From} \begin{bmatrix} 0 & 1 & 0 & 0 & 0 \\ 1 & 0 & 1 & 0 & 0 \\ 0 & 0 & 0 & 1 & 0 \\ 0 & 0 & 0 & 0 & 1 \\ 1 & 0 & 0 & 0 & 0 \end{bmatrix}$$

(b)

To

$$\text{From} \begin{bmatrix} 0 & 0 & 0 & 0 & 0 \\ 1 & 0 & 0 & 1 & 1 \\ 1 & 0 & 0 & 1 & 1 \\ 0 & 0 & 0 & 0 & 0 \\ 0 & 0 & 0 & 0 & 0 \end{bmatrix}$$

27.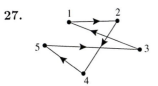

DISCUSSION AND DISCOVERY

D1. One obvious example is (x, y, z, t), where $P(x, y, z)$ is spatial position and t is time.

D2. Time is a scalar.

D3. Yes. If $u_1 + u_2 + u_3 = 0$, then $u_1 = -(u_2 + u_3)$ and so u_1 lies in the same plane as u_2 and u_3.

D4. The total displacement is zero. The sum of the two displacement vectors (going up and coming back) is zero, so one is the negative of the other.

D5. (a) The sum of the six radial vectors is zero. There are three pairs of vectors, each pair consisting of a vector and its negative.

    (b) The sum is still zero. Same reason.

    (c) If a is removed, then the sum of the remaining five vectors is $-a$.

    (d) If any vector is removed, then the sum of the remaining five vectors is the negative of the one that was removed.

D6. The sum of all of the radial vectors is zero for the same reason as in D5.

D7. (a) The sum of the twelve vectors (unmodified) in the figure is zero. [There are six pairs of vectors, each pair consisting of a vector and its negative.] If the vector terminating at 12 is doubled in length, then the new sum will be equal to that vector. [Reason: $2v_{12} = v_{12} + v_{12}$; thus the new sum is equal to $v_{12} +$ the old sum $= v_{12} + 0 = v_{12}$.]

    (b) The sum will still be zero. [Reason: v_3 and v_9 are one of the opposite pairs; thus when both are tripled they (and the total sum) still add to zero.]

    (c) The new sum will be equal to v_2. [Reason: v_5 and v_{11} are one of the opposite pairs, $v_5 + v_{11} = 0$, and so nothing changes if both are removed. On the other hand, if v_8 is removed, then there is nothing left to cancel v_2.]

D8. This picture illustrates four nonzero vectors in the plane, one of which is the sum of the other three.

$$v = v_1 + v_2 + v_3.$$

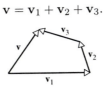

D9. (a) True. Subtract x from both sides of the equation.

    (b) False. If $u + v = 0$, then $v = -u$ and so $au + bv = (a - b)u = 0$ if and only if $a = b$ or $u = 0$.

(c) False. Any vector $\mathbf{v}$ and its negative $-\mathbf{v}$ are parallel and have the same length.

(d) True. If $a\mathbf{x} = 0$ and $a \neq 0$, then $\mathbf{x} = 1\mathbf{x} = (\frac{1}{a})(a\mathbf{x}) = (\frac{1}{a})\mathbf{0} = \mathbf{0}$.

(e) False. If $a = b = 0$, then $a\mathbf{u} + b\mathbf{v} = \mathbf{0}$ for any $\mathbf{u}$ and $\mathbf{v}$.

(f) False. The given vectors have the same direction but different lengths ($\mathbf{u} = 2\mathbf{v}$).

WORKING WITH PROOFS

P1. Let $\mathbf{u} = (u_1, u_2, \ldots, u_n)$. Then

$$(k+l)\mathbf{u} = ((k+l)u_1, (k+l)u_2, \ldots, (k+l)u_n) = (ku_1 + lu_1, ku_2 + lu_2, \ldots, ku_n + lu_n)$$
$$= (ku_1, ku_2, \ldots, ku_n) + (lu_1, lu_2, \ldots, lu_n)$$
$$= k(u_1, u_2, \ldots, u_n) + l(u_1, u_2, \ldots, u_n) = k\mathbf{u} + l\mathbf{u}$$

P2. Let $\mathbf{u} = (u_1, u_2, \ldots, u_n)$ and $\mathbf{v} = (v_1, v_2, \ldots, v_n)$. Then

$$k(\mathbf{u} + \mathbf{v}) = k(u_1 + v_1, u_2 + v_2, \ldots, u_n + v_n) = (k(u_1 + v_1), k(u_2 + v_2), \ldots, k(u_n + v_n))$$
$$= (ku_1 + kv_1, ku_2 + kv_2, \ldots, ku_n + kv_n)$$
$$= (ku_1, ku_2, \ldots, ku_n) + (kv_1, kv_2, \ldots, kv_n) = k\mathbf{u} + k\mathbf{v}$$

P3. These proofs refer to properties already established in Theorem 1.1.5.

(a) Using properties (c) and (e), we have $0\mathbf{v} + \mathbf{0} = 0\mathbf{v} = (0+0)\mathbf{v} = 0\mathbf{v} + 0\mathbf{v}$. Thus, on adding $-0\mathbf{v}$ to both sides of this equation (and regrouping using properties (a) and (b)), it follows that $\mathbf{0} = 0\mathbf{v}$.

(b) Using properties (c) and (f), we have $k\mathbf{0} + \mathbf{0} = k\mathbf{0} = k(\mathbf{0} + \mathbf{0}) = k\mathbf{0} + k\mathbf{0}$. Thus, on adding $-k\mathbf{0}$ to both sides of this equation (and regrouping), it follows that $\mathbf{0} = k\mathbf{0}$.

(c) Using properties (h) and (e), we have $\mathbf{v} + (-1)\mathbf{v} = 1\mathbf{v} + (-1)\mathbf{v} = (1 + (-1))\mathbf{v} = 0\mathbf{v} = \mathbf{0}$, where the last equality follows from part (a) above. Thus, on adding $-\mathbf{v}$ to both sides of this equation, it follows that $(-1)\mathbf{v} = -\mathbf{v}$.

EXERCISE SET 1.2

1. (a) The norm of $\mathbf{v}$ is $\|\mathbf{v}\| = \sqrt{4^2 + (-3)^2} = \sqrt{25} = 5$. The vector $\frac{\mathbf{v}}{\|\mathbf{v}\|} = (\frac{4}{5}, -\frac{3}{5})$ is a unit vector having the same direction as $\mathbf{v}$, and $-\frac{\mathbf{v}}{\|\mathbf{v}\|} = (-\frac{4}{5}, \frac{3}{5})$ is a unit vector in the opposite direction.

(b) $\|\mathbf{v}\| = \sqrt{2^2 + 2^2 + 2^2} = \sqrt{12} = 2\sqrt{3}$ $\pm\frac{\mathbf{v}}{\|\mathbf{v}\|} = \pm\frac{1}{2\sqrt{3}}(2, 2, 2) = \pm\left(\frac{1}{\sqrt{3}}, \frac{1}{\sqrt{3}}, \frac{1}{\sqrt{3}}\right)$

(c) $\|\mathbf{v}\| = \sqrt{1^2 + 0^2 + 2^2 + 1^2 + 3^3} = \sqrt{15}$ $\pm\frac{\mathbf{v}}{\|\mathbf{v}\|} = \pm\frac{1}{\sqrt{15}}(1, 0, 2, 1, 3)$

3. (a) $\mathbf{u} + \mathbf{v} = (3, -5, 7)$ $\|\mathbf{u} + \mathbf{v}\| = \sqrt{3^2 + (-5)^5 + 7^2} = \sqrt{9 + 25 + 49} = \sqrt{83}$

(b) $\|\mathbf{u}\| + \|\mathbf{v}\| = \sqrt{4 + 4 + 9} + \sqrt{1 + 9 + 16} = \sqrt{17} + \sqrt{26}$

(c) $-2\mathbf{u} + 2\mathbf{v} = (-2 - 2, 2)$ $\|-2\mathbf{u} + 2\mathbf{v}\| = \sqrt{(-2)^2 + (-2)^2 + 2^2} = \sqrt{12} = 2\sqrt{3}$

(d) $3\mathbf{u} - 5\mathbf{v} + \mathbf{w} = (4, 15, -15)$ $\|3\mathbf{u} - 5\mathbf{v} + \mathbf{w}\| = \sqrt{16 + 225 + 225} = \sqrt{466}$

5. (a) $3\mathbf{u} - 5\mathbf{v} + \mathbf{w} = (-27, -6, 38, -19)$ $\|3\mathbf{u} - 5\mathbf{v} + \mathbf{w}\| = \sqrt{729 + 36 + 1444 + 361} = \sqrt{2570}$

   (b) $\|3\mathbf{u}\| - 5\|\mathbf{v}\| + \|\mathbf{u}\| = 3\|\mathbf{u}\| - 5\|\mathbf{v}\| + \|\mathbf{u}\| = 4\|\mathbf{u}\| - 5\|\mathbf{v}\| = 4\sqrt{46} - 5\sqrt{84}$

   (c) $\|\mathbf{u}\| = \sqrt{46}$ $\| - \|\mathbf{u}\|\mathbf{v}\| = \| - \sqrt{46}\mathbf{v}\| = \sqrt{46}\|\mathbf{v}\| = \sqrt{46}\sqrt{84} = \sqrt{3864} = 2\sqrt{966}$

7. $\|k\mathbf{v}\| = |k|\,\|\mathbf{v}\| = |k|\sqrt{49} = 7|k|$; thus $\|k\mathbf{v}\| = 5$ if and only if $k = \pm\frac{5}{7}$.

9. (a) $\mathbf{u} \cdot \mathbf{v} = (3)(2) + (1)(2) + (4)(-4) = -8$ $\mathbf{u} \cdot \mathbf{u} = \|\mathbf{u}\|^2 = 26$ $\mathbf{v} \cdot \mathbf{v} = \|\mathbf{v}\|^2 = 24$

   (b) $\mathbf{u} \cdot \mathbf{v} = (1)(2) + (1)(-2) + (4)(3) + (6)(-2) = 0$ $\mathbf{u} \cdot \mathbf{u} = \|\mathbf{u}\|^2 = 1 + 1 + 16 + 36 = 54$

      $\mathbf{v} \cdot \mathbf{v} = \|\mathbf{v}\|^2 = 4 + 4 + 9 + 4 = 21$

11. (a) $\|\mathbf{u} - \mathbf{v}\| = \sqrt{(3-1)^2 + (3-0)^2 + (3-4)^2} = \sqrt{4 + 9 + 1} = \sqrt{14}$

   (b) $\|\mathbf{u} - \mathbf{v}\| = \sqrt{(0-(-3))^2 + (-2-2)^2 + (-1-4)^2 + (1-4)^2} = \sqrt{9 + 16 + 25 + 9} = \sqrt{59}$

   (c) $\|\mathbf{u} - \mathbf{v}\| = \sqrt{(3-(-4))^2 + (-2-1)^2 + (-2-(-1))^2 + (0-5)^2 + (-3-0)^2 + (13-(-11))^2 + (5-4)^2}$

      $= \sqrt{49 + 16 + 1 + 25 + 9 + 576 + 1} = \sqrt{677}$

13. (a) $\cos\theta = \dfrac{\mathbf{u} \cdot \mathbf{v}}{\|\mathbf{u}\|\|\mathbf{v}\|} = \dfrac{(3)(1) + (3)(0) + (3)(4)}{\sqrt{9 + 9 + 9}\sqrt{1 + 0 + 16}} = \dfrac{15}{\sqrt{27}\sqrt{17}}$ θ is acute

   (b) $\cos\theta = \dfrac{\mathbf{u} \cdot \mathbf{v}}{\|\mathbf{u}\|\|\mathbf{v}\|} = \dfrac{(0)(-3) + (-2)(2) + (-1)(4) + (1)(4)}{\sqrt{0 + 4 + 1 + 1}\sqrt{9 + 4 + 16 + 16}} = \dfrac{-4}{\sqrt{6}\sqrt{45}}$ θ is obtuse

   (c) $\cos\theta = \dfrac{\mathbf{u} \cdot \mathbf{v}}{\|\mathbf{u}\|\|\mathbf{v}\|} = \dfrac{(3)(-4) + (-3)(1) + (-2)(-1) + (0)(5) + (-3)(0) + (13)(-11) + (5)(4)}{\sqrt{9 + 9 + 4 + 0 + 9 + 169 + 25}\sqrt{16 + 1 + 1 + 25 + 0 + 121 + 16}}$

      $= \dfrac{-136}{\sqrt{225}\sqrt{180}}$ θ is obtuse

15. If θ is the angle between two vectors $\mathbf{a}$ and $\mathbf{b}$, then $\mathbf{a} \cdot \mathbf{b} = \|\mathbf{a}\|\|\mathbf{b}\| \cos\theta$. In this problem we have $\|\mathbf{a}\| = 9$, $\|\mathbf{b}\| = 5$, and $\theta = 30°$; thus $\mathbf{a} \cdot \mathbf{b} = (9)(5)\cos 30° = 45\frac{\sqrt{3}}{2}$.

17. If $5\mathbf{x} - 2\mathbf{v} = 2(\mathbf{w} - 5\mathbf{x})$, then $15\mathbf{x} = 2\mathbf{v} + 2\mathbf{w} = 2(1, 2, -4, 0) + 2(-3, 5, 1, 1) = (-4, 14, -6, 2)$ and so $\mathbf{x} = \frac{1}{15}(-4, 14, -6, 2) = \frac{2}{15}(-2, 7, -3, 1) = (-\frac{4}{15}, \frac{14}{15}, -\frac{2}{5}, \frac{2}{15})$.

19. (a) $\mathbf{u} \cdot (\mathbf{v} \cdot \mathbf{w})$ doesn't make sense, since the factor $\mathbf{v} \cdot \mathbf{w}$ is a scalar rather than a vector.

   (b) $\mathbf{u} \cdot (\mathbf{v} + \mathbf{w})$ is a valid mathematical expression.

   (c) $\|\mathbf{u} \cdot \mathbf{v}\|$ is not a valid expression since the quantity inside the norm is a scalar rather than a vector.

   (d) $(\mathbf{u} \cdot \mathbf{v}) - \|\mathbf{u}\|$ is a valid expression (both terms are scalars).

21. (a) $\mathbf{u} \cdot \mathbf{v} = 10$, $\|\mathbf{u}\| = \sqrt{13}$, and $\|\mathbf{v}\| = \sqrt{17}$. Thus $|\mathbf{u} \cdot \mathbf{v}| = 10$, and $\|\mathbf{u}\|\|\mathbf{v}\| = \sqrt{13}\sqrt{17} \approx 20.025$.

   (b) $\mathbf{u} \cdot \mathbf{v} = -7$, $\|\mathbf{u}\| = \sqrt{10}$, and $\|\mathbf{v}\| = \sqrt{14}$. Thus $|\mathbf{u} \cdot \mathbf{v}| = 7$, and $\|\mathbf{u}\|\|\mathbf{v}\| = \sqrt{10}\sqrt{14} \approx 11.832$.

   (c) $\mathbf{u} \cdot \mathbf{v} = 5$, $\|\mathbf{u}\| = 3$, and $\|\mathbf{v}\| = 2$. Thus $|\mathbf{u} \cdot \mathbf{v}| = 5$, and $\|\mathbf{u}\|\|\mathbf{v}\| = (3)(2) = 6$.

23. To show that $\mathbf{v}_1$, $\mathbf{v}_2$, $\mathbf{v}_3$, $\mathbf{v}_4$ form an orthonormal set we must verify that $\mathbf{v}_i \cdot \mathbf{v}_j = 0$ for $i \neq j$, and $\mathbf{v}_i \cdot \mathbf{v}_i = \|\mathbf{v}_i\|^2 = 1$ for $i = 1, 2, 3, 4$. Here are the required calculations:

$$\mathbf{v}_1 \cdot \mathbf{v}_2 = \tfrac{1}{4} - \tfrac{5}{12} + \tfrac{1}{12} + \tfrac{1}{12} = 0 \qquad \mathbf{v}_1 \cdot \mathbf{v}_3 = \tfrac{1}{4} + \tfrac{1}{12} + \tfrac{1}{12} - \tfrac{5}{12} = 0$$

$$\mathbf{v}_1 \cdot \mathbf{v}_4 = \tfrac{1}{4} + \tfrac{1}{12} - \tfrac{5}{12} + \tfrac{1}{12} = 0 \qquad \mathbf{v}_2 \cdot \mathbf{v}_3 = \tfrac{1}{4} - \tfrac{5}{36} + \tfrac{1}{36} - \tfrac{5}{36} = 0$$

$$\mathbf{v}_2 \cdot \mathbf{v}_4 = \tfrac{1}{4} - \tfrac{5}{36} - \tfrac{5}{36} + \tfrac{1}{36} = 0 \qquad \mathbf{v}_3 \cdot \mathbf{v}_4 = \tfrac{1}{4} + \tfrac{1}{36} - \tfrac{5}{36} - \tfrac{5}{36} = 0$$

$$\mathbf{v}_1 \cdot \mathbf{v}_1 = \tfrac{1}{4} + \tfrac{1}{4} + \tfrac{1}{4} + \tfrac{1}{4} = 1 \qquad \mathbf{v}_2 \cdot \mathbf{v}_2 = \tfrac{1}{4} + \tfrac{25}{36} + \tfrac{1}{36} + \tfrac{1}{36} = 1$$

$$\mathbf{v}_3 \cdot \mathbf{v}_3 = \tfrac{1}{4} + \tfrac{1}{36} + \tfrac{1}{36} + \tfrac{25}{36} = 1 \qquad \mathbf{v}_4 \cdot \mathbf{v}_4 = \tfrac{1}{4} + \tfrac{1}{36} + \tfrac{25}{36} + \tfrac{1}{36} = 1$$

25. We begin with the observation that $\mathbf{v} = (b, -a)$ and $-\mathbf{v} = (-b, a)$ are both orthogonal to the vector $\mathbf{u} = (a, b)$. This follows from $\mathbf{v} \cdot \mathbf{u} = ba - ab = 0$ and $(-\mathbf{v}) \cdot \mathbf{u} = -ba + ab = 0$. To get unit vectors with this property we only need to normalize. The vectors $\pm \frac{\mathbf{v}}{\|\mathbf{v}\|} = \pm(\frac{b}{\sqrt{a^2+b^2}}, -\frac{a}{\sqrt{a^2+b^2}})$ are unit vectors which are orthogonal to $\mathbf{u} = (a, b)$.

27. (a) $\mathbf{u} \cdot \mathbf{v} = k + 7 + 3k = 4k + 7$; thus $\mathbf{u}$ and $\mathbf{v}$ are orthogonal if and only if $k = -\frac{7}{4}$.

(b) $\mathbf{u} \cdot \mathbf{v} = -2k + 5k + k^2 = k^2 + 3k = k(k + 3)$; thus $\mathbf{u}$ and $\mathbf{v}$ are orthogonal if and only if $k = 0$ or $k = -3$.

29. Let α, β, γ be the angles at A, B, C respectively. Then

$$\cos \beta = \frac{\overrightarrow{BA} \cdot \overrightarrow{BC}}{\|\overrightarrow{BA}\|\|\overrightarrow{BC}\|} = \frac{(-1, -3, 2) \cdot (4, -2, -1)}{\|(-1, -3, 2)\|\|(4, -2, -1)\|} = \frac{-4 + 6 - 2}{\sqrt{14}\sqrt{21}} = 0$$

and so $\beta = 90°$, i.e. there is a right angle at the vertex B.

31. (a) The dot product of $\mathbf{a} = (10, 9, 8, 7, 6, 5, 4, 3, 2)$ and $\mathbf{b} = (0, 4, 7, 1, 0, 6, 3, 6, 8)$ is 175. Dividing 175 by 11 produces a quotient of 15 and a remainder of 10, so the check digit is $c = 11 - 10 = 1$. Thus $0-471-06368-1$ is a valid ISBN number.

(b) The dot product of $\mathbf{a} = (10, 9, 8, 7, 6, 5, 4, 3, 2)$ and $\mathbf{b} = (0, 1, 3, 9, 4, 7, 7, 5, 2)$ is 202. Dividing 202 by 11 produces a quotient of 18 and a remainder of 4, so the check digit is $c = 11 - 4 = 7$. Thus $0-13-947752-3$ is not a valid ISBN number.

33. $d = \|\mathbf{v} - \mathbf{w}\| = \sqrt{\sum_{k=1}^{n}(v_k - w_k)^2} = \left(\sum_{k=1}^{n}(v_k - w_k)^2\right)^{1/2}$

35. (a) $\sum_{k=1}^{n}(a_k + b_k) = (a_1 + b_1) + \cdots + (a_n + b_n) = (a_1 + \cdots + a_n) + (b_1 + \cdots b_n) = \sum_{k=1}^{n} a_k + \sum_{k=1}^{n} b_k$

(b) $\sum_{k=1}^{n}(a_k - b_k) = (a_1 - b_1) + \cdots + (a_n - b_n) = (a_1 + \cdots + a_n) - (b_1 + \cdots b_n) = \sum_{k=1}^{n} a_k - \sum_{k=1}^{n} b_k$

(c) $\sum_{k=1}^{n} ca_k = ca_1 + ca_2 + \cdots + ca_n = c(a_1 + a_2 + \cdots + a_n) = c\sum_{k=1}^{n} a_k$

DISCUSSION AND DISCOVERY

D2. If $\|k\mathbf{v}\| = k\|\mathbf{v}\|$ then k must be positive or zero $(k \geq 0)$.

D3. (a) A line through the origin, perpendicular to the given vector.

(b) A plane through the origin, perpendicular to the given vector.

(c) The only vector in R^2 that is orthogonal to (both of) two noncollinear vectors is the zero vector. Thus the set of vectors described is just $\{0\}$, i.e. a single point (the origin).

(d) A line through the origin, perpendicular to the plane containing the two noncollinear vectors.

D4. The given vectors $\mathbf{v}_1$ and $\mathbf{v}_2$ are normalized (scaled) versions of $\mathbf{w}_1 = (2, 1, 2)$ and $\mathbf{w}_2 = (1, 2, -2)$. Note that $\mathbf{w}_1 \cdot \mathbf{w}_2 = (2)(1) + (1)(2) + (2)(-2) = 2 + 2 - 4 = 0$; thus $\mathbf{w}_1$ and $\mathbf{w}_2$ (and the corresponding scaled versions) are orthogonal. It is easy to check that the vector $\mathbf{w}_3 = (-2, 2, 1)$ is orthogonal to both $\mathbf{w}_1$ and $\mathbf{w}_2$. Thus the vectors are $\mathbf{w}_1$, $\mathbf{w}_2$ and $\mathbf{w}_3$ are mutually orthogonal, and the normalized versions $\mathbf{v}_1 = (\frac{2}{3}, \frac{1}{3}, \frac{2}{3})$, $\mathbf{v}_2 = (\frac{1}{3}, \frac{2}{3}, -\frac{2}{3})$, and $\mathbf{v}_3 = (-\frac{2}{3}, \frac{2}{3}, \frac{1}{3})$ form an orthonormal set.

D5. The expression $(\mathbf{u} \cdot \mathbf{v}) + \mathbf{w}$ makes no sense since the first term is a scalar and the second term is a vector.

D6. (a) $\|\mathbf{x} - \mathbf{x}_0\| = 1$ (b) $\|\mathbf{x} - \mathbf{x}_0\| < 1$ (c) $\|\mathbf{x} - \mathbf{x}_0\| > 1$

D7. If $\|\mathbf{u}\| = \|\mathbf{v}\| = 1$, and if $\mathbf{u} \cdot \mathbf{v} = 0$ (i.e. if $\mathbf{u}$ and $\mathbf{v}$ are orthogonal), then

$$\|\mathbf{u} - \mathbf{v}\|^2 = (\mathbf{u} - \mathbf{v}) \cdot (\mathbf{u} - \mathbf{v}) = \mathbf{u} \cdot \mathbf{u} - \mathbf{u} \cdot \mathbf{v} - \mathbf{v} \cdot \mathbf{u} + \mathbf{v} \cdot \mathbf{v} = 1 - 0 - 0 + 1 = 2$$

and so $d(\mathbf{u}, \mathbf{v}) = \|\mathbf{u} - \mathbf{v}\| = \sqrt{2}$. Here is a picture illustrating this result in R^2.

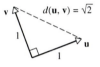

D8. The formulas alluded to are $\sum_{k=1}^{n} k = \frac{n(n+1)}{2}$ and $\sum_{k=1}^{n} k^2 = \frac{n(n+1)(2n+1)}{6}$.

(a) If $\mathbf{u} = (1, \sqrt{2}, \sqrt{3}, \ldots, \sqrt{n})$, then $\|\mathbf{u}\|^2 = 1 + 2 + 3 + \cdots + n = \frac{n(n+1)}{2}$ and so $\|\mathbf{u}\| = \sqrt{\frac{n(n+1)}{2}}$.

(b) If $\mathbf{u} = (1, 2, 3, \ldots, n)$, then $\|\mathbf{u}\|^2 = 1^2 + 2^2 + 3^2 + \cdots + n^2 = \frac{n(n+1)(2n+1)}{6}$, so $\|\mathbf{u}\| = \sqrt{\frac{n(n+1)(2n+1)}{6}}$.

D9. (a) True. We have $\|\mathbf{u} + \mathbf{v}\|^2 = (\mathbf{u} + \mathbf{v}, \mathbf{u} + \mathbf{v}) = \|\mathbf{u}\|^2 + 2(\mathbf{u} \cdot \mathbf{v}) + \|\mathbf{v}\|^2$. Thus from $\|\mathbf{u} + \mathbf{v}\|^2 = \|\mathbf{u}\|^2 + \|\mathbf{v}\|^2$ it follows that $\mathbf{u} \cdot \mathbf{v} = 0$.

(b) True. If $\mathbf{u} \cdot \mathbf{v} = \mathbf{u} \cdot \mathbf{w} = 0$, then $\mathbf{u} \cdot (\mathbf{v} + \mathbf{w}) = \mathbf{u} \cdot \mathbf{v} + \mathbf{u} \cdot \mathbf{w} = 0 + 0 = 0$.

(c) False. For example, suppose $\mathbf{u} = \mathbf{v} \neq \mathbf{0}$ and $\mathbf{w} = -\mathbf{v}$. Then $\mathbf{u} \cdot (\mathbf{v} + \mathbf{w}) = \mathbf{u} \cdot (\mathbf{v} - \mathbf{v}) = \mathbf{u} \cdot \mathbf{0} = 0$; thus $\mathbf{u}$ is orthogonal to $\mathbf{v} + \mathbf{w}$, but $\mathbf{u}$ is not orthogonal to $\mathbf{v}$ or $\mathbf{w}$.

(d) False. If $\mathbf{a}$ is orthogonal to $\mathbf{b} - \mathbf{c}$, then $\mathbf{a} \cdot \mathbf{b} - \mathbf{a} \cdot \mathbf{c} = \mathbf{a} \cdot (\mathbf{b} - \mathbf{c}) = 0$ and so $\mathbf{a} \cdot \mathbf{b} = \mathbf{a} \cdot \mathbf{c}$.

(e) True. If $\|\mathbf{u} + \mathbf{v}\| = 0$, then $\mathbf{u} + \mathbf{v} = \mathbf{0}$ and so $\mathbf{u} = -\mathbf{v}$.

(f) True. An orthonormal set is an orthogonal set of vectors all of which have length 1.

D10. (a) True. If $k\mathbf{u} = \mathbf{0}$, and $k \neq 0$, then $\mathbf{u} = (\frac{1}{k})(k\mathbf{u}) = (\frac{1}{k})(\mathbf{0}) = \mathbf{0}$.

(b) True (assuming $\mathbf{u}$ and $\mathbf{v}$ are nonzero). For example, if $\mathbf{u} = (c, d)$ is orthogonal to $\mathbf{w} = (a, b)$ and if $b \neq 0$, then from $ca + db = 0$ it follows that $d = -\frac{ca}{b}$ and so $\mathbf{u} = (c, -\frac{ca}{b}) = c(1, -\frac{a}{b})$ where $c \neq 0$. Similarly, $\mathbf{v} = c'(1, -\frac{a}{b})$ where $c' \neq 0$. Thus $\mathbf{u}$ and $\mathbf{v}$ are scalar multiples of each other.

(c) True. The vector $\mathbf{u} = (0, 0, 0)$ is such a vector.

(d) True. If $\mathbf{u} = (a, b, c)$ is orthogonal to $(1, 0, 0)$, then $a = \mathbf{u} \cdot (1, 0, 0) = 0$. Similarly, if $\mathbf{u}$ is orthogonal to $(0, 1, 0)$ and to $(0, 0, 1)$, then $b = 0$ and $c = 0$.

(e) False. For example, $\mathbf{u} = (1, 0, 0)$ is orthogonal to $\mathbf{v} = (0, 1, 0)$ and $\mathbf{v}$ is orthogonal to $\mathbf{w} = (1, 0, 1)$, but $\mathbf{u}$ is not orthogonal to $\mathbf{w}$.

(f) False. For example, if $\mathbf{v} = -\mathbf{u}$ where $\mathbf{u} \neq \mathbf{0}$, then $\|\mathbf{u} + \mathbf{v}\| = 0$ but $\|\mathbf{u}\| + \|\mathbf{v}\| = 2\|\mathbf{u}\| \neq 0$.

WORKING WITH PROOFS

P1. If $\mathbf{u}_1, \mathbf{u}_2, \ldots, \mathbf{u}_n$ are pairwise orthogonal, then $\mathbf{u}_i \cdot \mathbf{u}_j = 0$ for $i \neq j$. Thus

$$\|\mathbf{u}_1 + \mathbf{u}_2 + \cdots + \mathbf{u}_n\|^2 = (\mathbf{u}_1 + \mathbf{u}_2 + \cdots + \mathbf{u}_n) \cdot (\mathbf{u}_1 + \mathbf{u}_2 + \cdots + \mathbf{u}_n) = \sum_{i=1}^{n} \sum_{j=1}^{n} \mathbf{u}_i \cdot \mathbf{u}_j$$

$$= \sum_{i=1}^{n} \mathbf{u}_i \cdot \mathbf{u}_i = \|\mathbf{u}_1\|^2 + \|\mathbf{u}_2\|^2 + \cdots + \|\mathbf{u}_n\|^2$$

P2. If $\mathbf{u} = (\sqrt{a_1}, \sqrt{a_2})$ and $\mathbf{v} = (\sqrt{a_2}, \sqrt{a_1})$, then $\mathbf{u} \cdot \mathbf{v} = \sqrt{a_1}\sqrt{a_2} + \sqrt{a_2}\sqrt{a_1} = 2\sqrt{a_1 a_2}$ and $\|\mathbf{u}\| = \|\mathbf{v}\| = \sqrt{(\sqrt{a_1})^2 + (\sqrt{a_2})^2} = \sqrt{a_1 + a_2}$. Thus, from the Cauchy–Schwarz inequality, we have

$$2\sqrt{a_1 a_2} \leq a_1 + a_2 \quad \text{or} \quad \sqrt{a_1 a_2} \leq \frac{a_1 + a_2}{2}$$

P3. If $\mathbf{u} = (a_1, a_2, \ldots, a_n)$ and $\mathbf{v} = (b_1, b_2, \ldots, b_n)$, then from $(\mathbf{u} \cdot \mathbf{v})^2 \leq \|\mathbf{u}\|^2 \|\mathbf{v}\|^2$, we have

$$(a_1 b_1 + a_2 b_2 + \cdots + a_n b_n)^2 \leq (a_1^2 + a_2^2 + \cdots + a_n^2)(b_1^2 + b_2^2 + \cdots + b_n^2)$$

P4. (a) $\|\mathbf{u} + \mathbf{v}\|^2 - \|\mathbf{u} - \mathbf{v}\|^2 = (\mathbf{u} + \mathbf{v}) \cdot (\mathbf{u} + \mathbf{v}) - (\mathbf{u} - \mathbf{v}) \cdot (\mathbf{u} - \mathbf{v})$

$$= (\mathbf{u} \cdot \mathbf{u} + \mathbf{u} \cdot \mathbf{v} + \mathbf{v} \cdot \mathbf{u} + \mathbf{v} \cdot \mathbf{v}) - (\mathbf{u} \cdot \mathbf{u} - \mathbf{u} \cdot \mathbf{v} - \mathbf{v} \cdot \mathbf{u} + \mathbf{v} \cdot \mathbf{v})$$

$$= (\|\mathbf{u}\|^2 + 2(\mathbf{u} \cdot \mathbf{v}) + \|\mathbf{v}\|^2) - (\|\mathbf{u}\|^2 - 2(\mathbf{u} \cdot \mathbf{v}) + \|\mathbf{v}\|^2) = 4(\mathbf{u} \cdot \mathbf{v})$$

From this it follows that $\mathbf{u} \cdot \mathbf{v} = \dfrac{\|\mathbf{u} + \mathbf{v}\|^2 - \|\mathbf{u} - \mathbf{v}\|^2}{4} = \dfrac{1}{4}\|\mathbf{u} + \mathbf{v}\|^2 - \dfrac{1}{4}\|\mathbf{u} - \mathbf{v}\|^2.$

(b) If $\|\mathbf{u} + \mathbf{v}\| = 1$ and $\|\mathbf{u} - \mathbf{v}\| = 5$, then from the formula obtained in part (a) we have

$$\mathbf{u} \cdot \mathbf{v} = \frac{1}{4}\|\mathbf{u} + \mathbf{v}\|^2 - \frac{1}{4}\|\mathbf{u} - \mathbf{v}\|^2 = \frac{1}{4}(1)^2 - \frac{1}{4}(5)^2 = \frac{1}{4} - \frac{25}{4} = -6$$

P5. If L_1 and L_2 are nonvertical lines which are parallel to the vectors $\mathbf{u} = (a, b)$ and $\mathbf{v} = (c, d)$, then $a \neq 0, c \neq 0$, $m_1 = \text{slope}(L_1) = \frac{b}{a}$, and $m_2 = \text{slope}(L_2) = \frac{d}{c}$. The lines L_1 and L_2 are perpendicular if and only if $\mathbf{u} \cdot \mathbf{v} = 0$, i.e. if and only if $ac + bd = 0$. But, since $a \neq 0$ and $c \neq 0$, the equality $ac + bd = 0$ is equivalent to $1 + m_1 m_1 = 1 + (\frac{b}{a})(\frac{d}{c}) = 1 + \frac{bd}{ac} = 0$. Thus the lines L_1 and L_2 are perpendicular if and only if $m_1 m_2 = -1$.

P6. (a) $d(\mathbf{u}, \mathbf{v}) = \sqrt{(u_1 - v_1)^2 + (u_2 - v_2)^2 + \cdots + (u_n - v_n)^2} \geq 0$ for any $\mathbf{u}$ and $\mathbf{v}$.

(b) $d(\mathbf{u}, \mathbf{v}) = 0$ if and only if $(u_1 - v_1)^2 + (u_2 - v_2)^2 + \cdots + (u_n - v_n)^2 = 0$, and this is true if and only if $u_1 = v_1, u_2 = v_2, \ldots$, and $u_n = v_n$, i.e. if and only if $\mathbf{u} = \mathbf{v}$.

(c) $d(\mathbf{u}, \mathbf{v}) = \sqrt{(u_1 - v_1)^2 + \cdots + (u_n - v_n)^2} = \sqrt{(v_1 - u_1)^2 + \cdots + (v_n - u_n)^2} = d(\mathbf{v}, \mathbf{u})$

P7. (a) $\mathbf{u} \cdot \mathbf{v} = u_1 v_1 + u_2 v_2 + \cdots + u_n v_n = v_1 u_1 + v_2 u_2 + \cdots + v_n u_n = \mathbf{v} \cdot \mathbf{u}$

(b) $\mathbf{u} \cdot (\mathbf{v} + \mathbf{w}) = u_1(v_1 + w_1) + u_2(v_2 + w_2) + \cdots + u_n(v_n + w_n)$

$$= (u_1 v_1 + u_1 w_1) + (u_2 v_2 + u_2 w_2) + \cdots + (u_n v_n + u_n w_n)$$

$$= (u_1 v_1 + u_2 v_2 + \cdots + u_n v_n) + (u_1 w_1 + u_2 w_2 + \cdots + u_n w_n)$$

$$= \mathbf{u} \cdot \mathbf{v} + \mathbf{u} \cdot \mathbf{w}$$

P8. **(a)** Using parts (a) and (c) of Theorem 1.2.6, we have:

$$k(\mathbf{u} \cdot \mathbf{v}) = k(\mathbf{v} \cdot \mathbf{u}) = (k\mathbf{v}) \cdot \mathbf{u} = \mathbf{u} \cdot (k\mathbf{v})$$

(b) Using parts (c) and (a) of Theorem 1.2.6, we have:

$$\mathbf{0} \cdot \mathbf{v} = ((0)\mathbf{0}) \cdot \mathbf{v} = (0)(\mathbf{0} \cdot \mathbf{v}) = 0 \text{ and } \mathbf{v} \cdot \mathbf{0} = \mathbf{0} \cdot \mathbf{v} = 0$$

P9. From the accompanying figure (see below) we have $\overrightarrow{AX} = -\mathbf{a} + \mathbf{x}$ and $\overrightarrow{BX} = \mathbf{a} + \mathbf{x}$; thus

$$\overrightarrow{AX} \cdot \overrightarrow{BX} = (-\mathbf{a} + \mathbf{x}) \cdot (\mathbf{a} + \mathbf{x}) = -\mathbf{a} \cdot \mathbf{a} - \mathbf{a} \cdot \mathbf{x} + \mathbf{x} \cdot \mathbf{a} + \mathbf{x} \cdot \mathbf{x} = -\|\mathbf{a}\|^2 + \|\mathbf{x}\|^2$$

Since $\|\mathbf{a}\|^2 = \|\mathbf{x}\|^2$ (both $\|\mathbf{a}\|$ and $\|\mathbf{x}\|$ are equal to the radius of the circle) it follows that $\overrightarrow{AX} \cdot \overrightarrow{BX} = 0$, i.e. the angle at X is a right angle.

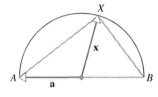

EXERCISE SET 1.3

1. **(a)**
$$L_1: \quad x = 1, \qquad y = t \quad -\infty < t < \infty$$
$$L_2: \quad x = t, \qquad y = 1$$
$$L_3: \quad x = t, \qquad y = t$$
$$L_4: \quad x = 1 - t, \quad y = t$$

Note. These equations are just one possibility. There are many different parametric representations for a given line.

(b)
$$L_1: \quad x = 1, \quad y = 1, \quad z = t \quad -\infty < t < \infty$$
$$L_2: \quad x = t, \quad y = 1, \quad z = 1$$
$$L_3: \quad x = 1, \quad y = t, \quad z = t$$
$$L_4: \quad x = t, \quad y = t, \quad z = t$$

3. **(a)** $(x, y) = t(2, 3)$ **(b)** $(x, y) = (1, 1) + t(1, -1)$

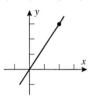   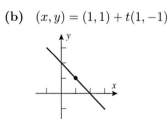

5. **(a)** Vector equation: $(x, y) = (0, 0) + t(3, 5) = t(3, 5)$
Parametric equations: $x = 3t, \; y = 5t$

(b) Vector equation: $(x, y, z) = (1, 1, 1) + t((0, 0, 0) - (1, 1, 1)) = (1, 1, 1) - t(1, 1, 1)$
Parametric equations: $x = 1 - t, \; y = 1 - t, \; z = 1 - t$

(c) Vector equation: $(x, y, z) = (1, -1, 1) + t((2, 1, 1) - (1, -1, 1)) = (1, -1, 1) + t(1, 2, 0)$
Parametric equations: $x = 1 + t, \; y = -1 + 2t, \; z = 1$
Note. Once again these equations represent just one of the many possibilities.

7. (a) Vector equation: $(x, y) = (1, 1) + t(1, 2)$
  Parametric equations: $x = 1 + t, y = 1 + 2t$
  Points other than P_0: $Q(2, 3)$ corresponds to $t = 1$; $R(3, 5)$ corresponds to $t = 2$

(b) Vector equation: $(x, y, z) = (2, 0, 3) + t(1, -1, 1)$
  Parametric equations: $x = 2 + t, y = -t, z = 3 + t$
  Points other than P_0: $Q(3, -1, 4)$ corresponds to $t = 1$; $R(1, 1, 2)$ corresponds to $t = -1$

(c) Vector equation: $(x, y, z) = (0, 0, 0) + t(3, 2, -3) = t(3, 2, -3)$
  Parametric equations: $x = 3t, y = 2t, z = -3t$
  Points other than P_0: $Q(-3, -2, 3)$ corresponds to $t = -1$; $R(6, 4, -6)$ corresponds to $t = 2$

9. A point-normal form is $3(x + 1) + 2(y + 1) + (z + 1) = 0$. The corresponding general equation is $3x + 2y + z = -6$.

11. If P, Q, R are the points $P(1, 1, 4)$, $Q(2, -3, 1)$, $R(3, 5, -2)$, then the vectors $\overrightarrow{PQ} = (1, -4, -3)$ and $\overrightarrow{PR} = (2, 4, -6)$ are parallel to the plane through P, Q and R. Thus a vector equation for the plane, expressed in component form, is

$$(x, y, z) = (1, 1, 4) + t_1(1, -4, -3) + t_2(2, 4, -6)$$

where $-\infty < t_1, t_2 < \infty$. The corresponding parametric equations are

$$x = 1 + t_1 + 2t_2, \quad y = 1 - 4t_1 + 4t_2, \quad z = 4 - 3t_1 - 6t_2$$

Some other points on the plane are:

$$
\begin{array}{lll}
S_1(4, 1, -5), & \text{corresponding to} & t_1 = 1, \ t_2 = 1 \\
S_2(0, -7, 7), & \text{corresponding to} & t_1 = 1, \ t_2 = -1 \\
S_3(2, 9, 1), & \text{corresponding to} & t_1 = -1, \ t_2 = 1
\end{array}
$$

13. (a) The parametric equations can be written in vector form as

$$(x, y, z) = (2 + 4t, -1 + t, t) = (2, -1, 0) + t(4, 1, 1)$$

which is the vector equation of the line which passes through the point $P(2, -1, 0)$ and is parallel to the vector $\mathbf{v} = (4, 1, 1)$.

(b) The parametric equations can be written in vector form as

$$(x, y, z) = (1 + 2t_1 + t_2, -2 - t_1 + 5t_2, 4t_1 - t_2) = (1, -2, 0) + t_1(2, -1, 4) + t_2(1, 5, -1)$$

which is the vector equation of the plane which passes through the point $P(1, -2, 0)$ and is parallel to the vectors $\mathbf{v}_1 = (2, -1, 4)$ and $\mathbf{v}_2 = (1, 5, -1)$.

(c) One way to do this is to solve the equation $3x + 4y - 2z = 4$ for x in terms of y and z, and then make y and z into parameters. This leads to the parametric equations

$$x = \tfrac{4}{3} - \tfrac{4}{3}t_1 + \tfrac{2}{3}t_2, \quad y = t_1, \quad z = t_2$$

where $-\infty < t_1, t_2 < \infty$. There are of course other ways as well. For example, solving the equation for z in terms of x and y, then making x and y into parameters, leads to

$$x = s_1, \quad y = s_2, \quad z = -2 + \tfrac{3}{2}s_1 + 2s_2$$

15. First note that the points $P(1,2,4)$, $Q(1,-1,6)$, $R(1,4,8)$, all lie on the vertical plane $x = 1$. This is the general equation of the plane. To find a vector equation, we observe that the points which are on this plane are exactly those points which are of the form $(x,y,z) = (1,t_1,t_2)$ where $-\infty < t_1, t_2 < \infty$. This corresponds to the parametric equations $x = 1$, $y = t_1$, $z = t_2$, or to the vector equation

$$(x,y,z) = (1,0,0) + t_1(0,1,0) + t_2(0,0,1)$$

17. (a) This is a vector equation for the line that passes through the origin and the point $P(1,-2,5,7)$; it is represented by the parametric equations $x_1 = t$, $x_2 = -2t$, $x_3 = 5t$, $x_4 = 7t$.

(b) This is a vector equation for the line that passes through the point $P(4,5,-6,1)$ and is parallel to the vector $\mathbf{v} = (1,1,1,1)$; it is represented by the parametric equations

$$x_1 = 4+t, \quad x_2 = 5+t, \quad x_3 = -6+t, \quad x_4 = 1+t$$

(c) This is a vector equation for the plane that passes through the point $P(-1,0,4,2)$ and is parallel to the vectors $\mathbf{v}_1 = (-3,5,-7,4)$ and $\mathbf{v}_2 = (6,3,-1,2)$; it is represented by the parametric equations

$$x_1 = -1 - 3t_1 + 6t_2, \quad x_2 = 5t_1 + 3t_2, \quad x_3 = 4 - 7t_1 - t_2, \quad x_4 = 2 + 4t_1 + 2t_2$$

19. (a) These are parametric equations for the line in R^5 that passes through the origin and is parallel to the vector $\mathbf{u} = (3,4,7,1,9)$.

(b) These are parametric equations for the plane in R^4 that passes through the point $P(3,4,-2,1)$ and is parallel to the vectors $\mathbf{v}_1 = (-2,-3,-2,-2)$ and $\mathbf{v}_2 = (5,6,7,-1)$.

21. Any plane with general equation of the form $3x + 2y - z = k$ has $\mathbf{n} = (3,2,-1)$ as a normal vector, and thus is parallel to $3x + 2y - z = 1$. In order for such a plane to pass though the point $P(1,1,1)$ we must have $k = 4$. Thus the general equation of the plane is $3x + 2y - z = 4$. Finally, we can get parametric equations for the plane by (for example) solving the equation for z in terms of x and y, and then making x and y into parameters. This leads to:

$$x = t_1, \quad y = t_2, \quad z = 3t_1 + 2t_2 - 4$$

23. A plane is parallel to $3x + y - 2z = 5$ if and only if it has a normal vector which is parallel to (i.e. a scalar multiple of) the vector $\mathbf{n} = (3,1,-2)$.

(a) Normal vector $(1,1,-1)$. Not parallel.

(b) Normal vector $(3,1,-2)$. Parallel.

(c) Normal vector $(1,\frac{1}{3},-\frac{2}{3}) = \frac{1}{3}(3,1,-2)$. Parallel.

25. The plane $x + y + z = 0$ has normal vector $\mathbf{n} = (1,1,1)$; thus any line perpendicular to the plane must be parallel to $\mathbf{n}$. Parametric equations for the line that passes through the point $P(2,0,1)$ and is parallel to $\mathbf{n} = (1,1,1)$ are

$$x = 2+t, \quad y = t, \quad z = 1+t$$

27. If the plane passes through the origin $O(0,0,0)$ and contains the points $P(5,4,3)$ and $Q(1,-1-2)$, then it must be parallel to the vectors $\overrightarrow{OP} = (5,4,3)$ and $\overrightarrow{OQ} = (1,-1,-2)$. Thus a vector equation for the plane is given by

$$(x,y,z) = t_1(5,4,3) + t_2(1,-1,-2)$$

29. A vector equation for the given line is $(x, y, z) = (4, -2, 0) + t(2, 3, -5)$. The vector $\mathbf{n} = (2, 3, -5)$ is parallel to the line and thus serves as a normal vector for a plane perpendicular to the line. Thus a point-normal equation for the plane through the point $P(-2, 1, 7)$ and perpendicular to the line is $2(x + 2) + 3(y - 1) - 5(z - 7) = 0$. This reduces to $2x + 3y - 5z + 36 = 0$, and from this we can obtain parametric equations by (for example) solving for y in terms of x and z, and then making x and z into parameters. This leads to the parametric equations $x = t_1$, $y = -\frac{2}{3}t_1 + \frac{5}{3}t_2 - 12$, $z = t_2$. Note. This is just one possibility for a parametric representation of the plane.

31. (a) The line has vector equation $(x, y, z) = (-5, 1, 3) + t(-4, -1, 2)$ and thus is parallel to the vector $\mathbf{v} = (-4, -1, 2)$. The plane has general equation $x + 2y + 3z = 9$ and thus is perpendicular to the vector $\mathbf{n} = (1, 2, 3)$. It follows that the line and plane are parallel if and only if $\mathbf{v}$ is perpendicular to $\mathbf{n}$. Since $\mathbf{v} \cdot \mathbf{n} = (-4)(1) + (-1)(2) + (2)(3) = 0$ we see that this is in fact the case, so the line and plane are parallel.

    (b) In this case (same reasoning), we have $\mathbf{v} = (3, 2, -1)$ and $\mathbf{n} = (4, 1, 2)$. Since $\mathbf{v} \cdot \mathbf{n} \neq 0$ the line and plane are not parallel.

33. (a) The plane $3x - y + z - 4 = 0$ has normal vector $\mathbf{n}_1 = (3, -1, 1)$ and the plane $x + 2z = -1$ has normal vector $\mathbf{n}_2 = (1, 0, 2)$. The two planes are perpendicular if and only if these two vectors are perpendicular. Since $\mathbf{n}_1 \cdot \mathbf{n}_2 = (3)(1) + (-1)(0) + (1)(2) \neq 0$ we see that this is not the case, so the planes are not perpendicular.

    (b) In this case (same reasoning), we have $\mathbf{n}_1 = (1, -2, 3)$ and $\mathbf{n}_2 = (-2, 5, 4)$. Since $\mathbf{n}_1 \cdot \mathbf{n}_2 = 0$, the planes are perpendicular.

35. (a) The line passes through the origin and is parallel to the vector $\mathbf{v} = (0, 1, 1)$; it has the vector equation $(x, y, z) = t(0, 1, 1)$. It is easy to check that any point of the form $(x, y, z) = (0, t, t)$ satisfies the equation $6x + 4y - 4z = 0$ and thus lies on the plane. Another way to see this is to observe that the plane has normal vector $\mathbf{n} = (6, 4, -4)$, and that $\mathbf{v}$ is perpendicular to $\mathbf{n}$ (since $\mathbf{v} \cdot \mathbf{n} = 0$). Thus the line is parallel to the plane, and since both pass through the origin it follows that the line is contained in the plane.

    (b) This plane has normal vector $\mathbf{n} = (5, -3, 3)$, and since $\mathbf{v} \cdot \mathbf{n} = 0$ it follows (as above) that the line is parallel to the plane. On the other hand, the line goes through the origin whereas the plane intersects the z-axis at the point $(0, 0, \frac{1}{3})$. Thus the line is below the plane.

    (c) This plane has normal vector $\mathbf{n} = (6, 2, -2)$, and since $\mathbf{v} \cdot \mathbf{n} = 0$ it follows (as above) that the line is parallel to the plane. On the other hand, the line goes through the origin whereas the plane intersects the z-axis at the point $(0, 0, \frac{3}{2})$. Thus the line is below the plane.

37. (a) One way to do this is to solve the two equations for x and y in terms of z and then make z into a parameter. Here are the calculations:

$$\left.\begin{array}{r} 7x - 2y = -2 - 3z \\ -3x + y = -5 - 2z \end{array}\right\} \Rightarrow \left.\begin{array}{r} 7x - 2y = -2 - 3z \\ -6x + 2y = -10 - 4z \end{array}\right\} \Rightarrow x = -12 - 7z$$

From this it follows that $y = -5 - 2z + 3x = -5 - 2z + 3(-12 - 7x) = -41 - 23z$, and this leads to the following parametric equations for the line of intersection:

$$x = -12 - 7t, \quad y = -41 - 23t, \quad z = t$$

    (b) The plane $2x + 3y - 5z = 0$ has normal vector $\mathbf{n}_1 = (2, 3, -5)$ and $4x + 6y - 10z = 8$ has normal vector $\mathbf{n}_2 = (4, 6, -10) = 2\mathbf{n}_1$; thus these two planes are parallel. Furthermore, they do not coincide, since the first passes through the origin whereas the second intersects the z-axis at the point $(0, 0, -\frac{4}{5})$. Thus these planes do not intersect (the first plane is above the second).

39. **(a)** The points on the line are of the form $(x, y, z) = (9 - 5t, -1 - t, 3 + t)$ where $-\infty < t < \infty$. In order that such a point satisfy the equation $2x - 3y + 4z + 7 = 0$ we must have

$$2(9 - 5t) - 3(-1 - t) + 4(3 + t) + 7 = 0$$

from which it follows that $-3t + 40 = 0$ or $t = \frac{40}{3}$. For this value of t we have

$$(x, y, z) = (9 - \tfrac{200}{3}, -1 - \tfrac{40}{3}, 3 + \tfrac{40}{3}) = (-\tfrac{173}{3}, -\tfrac{43}{3}, \tfrac{49}{3})$$

(b) The points on the line are of the form $(x, y, z) = (t, t, t)$ where $-\infty < t < \infty$. None of these points satisfy the equation $x + y - 2z = 3$ since $t + t - 2t = 0$ for every value of t. There is no point of intersection; the line is parallel to the plane.

41. **(a)**   **(b)**

43. The line segment from $P(-2, 4, 1)$ to $Q(0, 4, 7)$ is represented by the vector equation

$$(x, y, z) = (1 - t)(-2, 4, 1) + t(0, 4, 7)$$

45. **(a)** A vector equation for the line segment from P to Q is $(x, y, z) = (1 - t)(2, 3, -2) + t(7, -4, 1)$. The midpoint of the segment is obtained by taking $t = \frac{1}{2}$:

$$(x, y, z) = \tfrac{1}{2}(2, 3, -2) + \tfrac{1}{2}(7, -4, 1) = (\tfrac{9}{2}, -\tfrac{1}{2}, -\tfrac{1}{2}).$$

(b) The point that is $\frac{3}{4}$ of the way from P to Q is obtained by taking $t = \frac{3}{4}$:

$$(x, y, z) = \tfrac{1}{4}(2, 3, -2) + \tfrac{3}{4}(7, -4, 1) = (\tfrac{23}{4}, -\tfrac{9}{4}, \tfrac{1}{4})$$

DISCUSSION AND DISCOVERY

D1. The lines L_1: $x = x_0 + at$, $y = y_0 + bt$, $z = z_0 + ct$, and L_2: $x = x_0 + dt$, $y = y_0 + et$, $z = z_0 + ft$, will be perpendicular if and only if the vectors $\mathbf{v}_1 = (a, b, c)$ and $\mathbf{v}_2 = (d, e, f)$ are perpendicular, i.e. if and only if $\mathbf{v}_1 \cdot \mathbf{v}_2 = ad + be + cf = 0$. For example, the line

$$L_2: x = x_0 + bt, \quad y = y_0 - at, \quad z = z_0$$

passes through the point (x_0, y_0, z_0) and is perpendicular to L_1.

D2. **(a)** The line $\mathbf{x} = \mathbf{x}_0 + \mathbf{v}t$ is parallel to the plane $\mathbf{x} = \mathbf{x}_0 + t_1\mathbf{v}_1 + t_2\mathbf{v}_2$ if and only if $\mathbf{v} = a\mathbf{v}_1 + b\mathbf{v}_2$ for some scalars a and b.

(b) The statement in part (a) serves as an appropriate definition in R^n as well.

D3. **(a)** If $\mathbf{v}$ is orthogonal to both $\mathbf{w}_1$ and $\mathbf{w}_2$, then

$$\mathbf{v} \cdot (k_1\mathbf{w}_1 + k_2\mathbf{w}_2) = k_1(\mathbf{v} \cdot \mathbf{w}_1) + k_2(\mathbf{v} \cdot \mathbf{w}_2) = k_1(0) + k_2(0) = 0$$

and so $\mathbf{v}$ is orthogonal to any vector of the form $\mathbf{x} = k_1\mathbf{w}_1 + k_2\mathbf{w}_2$.

(b) If $\mathbf{v}$ is orthogonal to both $\mathbf{w}_1$ and $\mathbf{w}_2$, then $\mathbf{v}$ is orthogonal to any vector that lies in the plane spanned by $\mathbf{w}_1$ and $\mathbf{w}_2$.

D4. (a) When interpreted in R^3, the equation $Ax + By = 0$ represents a plane through the origin that contains the z-axis.

   (b) No. When interpreted in R^4, the equation $Ax + By + Cz = 0$ will represent a three-dimensional object, not a plane.

D5. (a) True. The line is parallel to $\mathbf{n} = (a, b, c)$, and this is a normal vector for the plane.

   (b) False. For example, the lines with vector equations $(x, y, z) = (t, 0, 0)$ and $(x, y, z) = (0, s, 1)$ are not parallel and do not intersect each other.

   (c) True. If $\mathbf{u} + \mathbf{v} + \mathbf{w} = \mathbf{0}$, then $\mathbf{w} = -(\mathbf{u} + \mathbf{v})$; thus $\mathbf{w}$ lies in the plane determined by $\mathbf{u}$ and $\mathbf{v}$.

   (d) False. If $\mathbf{v} = \mathbf{0}$, then the equation represents a point (the origin); otherwise it will represent a line through the origin.

CHAPTER 2
Systems of Linear Equations

EXERCISE SET 2.1

1. (a) and (c) are linear. (b) is not linear due to the $x_1 x_3$ term. (d) is not linear due to the x_1^{-2} term.

3. (a) is linear. (b) is linear if $k \neq 0$. (c) is linear only if $k = 1$.

5. (a), (d), and (e) are solutions; these sets of values satisfy all three equations. (b) and (c) are not solutions.

7. The three lines intersect at the point $(1, 0)$ (see figure). The values $x = 1$, $y = 0$ satisfy all three equations and this is the unique solution of the system.

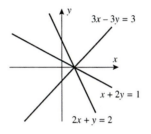

The augmented matrix of the system is

$$\begin{bmatrix} 1 & 2 & \vdots & 1 \\ 2 & 1 & \vdots & 2 \\ 3 & -3 & \vdots & 3 \end{bmatrix}.$$

Add -2 times row 1 to row 2 and add -3 times row 1 to row 3:

$$\begin{bmatrix} 1 & 2 & \vdots & 1 \\ 0 & -3 & \vdots & 0 \\ 0 & -9 & \vdots & 0 \end{bmatrix}$$

Multiply row 2 by $-\frac{1}{3}$ and add 9 times the new row 2 to row 3:

$$\begin{bmatrix} 1 & 2 & \vdots & 1 \\ 0 & 1 & \vdots & 0 \\ 0 & 0 & \vdots & 0 \end{bmatrix}$$

From the last row we see that the system is redundant (reduces to only two equations). From the second row we see that $y = 0$ and, from back substitution, it follows that $x = 1 - 2y = 1$.

9. (a) The solution set of the equation $7x - 5y = 3$ can be described parametrically by (for example) solving the equation for x in terms of y and then making y into a parameter. This leads to $x = \frac{3 + 5t}{7}$, $y = t$, where $-\infty < t < \infty$.

(b) The solution set of $3x_1 - 5x_2 + 4x_3 = 7$ can be described by solving the equation for x_1 in terms of x_2 and x_3, then making x_2 and x_3 into parameters. This leads to $x_1 = \frac{7 + 5s - 4t}{3}$, $x_2 = s$, $x_3 = t$, where $-\infty < s, t < \infty$.

(c) The solution set of $-8x_1 + 2x_2 - 5x_3 + 6x_4 = 1$ can be described by (for example) solving the equation for x_2 in terms of x_1, x_3, and x_4, then making x_1, x_3, and x_4 into parameters. This leads to $x_1 = r$, $x_2 = \frac{1 + 8r + 5s - 6t}{2}$, $x_3 = s$, $x_4 = t$, where $-\infty < r, s, t < \infty$.

(d) The solution set of $3v - 8w + 2x - y + 4z = 0$ can be described by (for example) solving the equation for y in terms of the other variables, and then making those variables into parameters. This leads to $v = t_1$, $w = t_2$, $x = t_3$, $y = 3t_1 - 8t_2 + 2t_3 + 4t_4$, $z = t_4$, where $-\infty < t_1, t_2, t_3, t_4 < \infty$.

11. (a) If the solution set is described by the equations $x = 5 + 2t$, $y = t$, then on replacing t by y in the first equation we have $x = 5 + 2y$ or $x - 2y = 5$. This is a linear equation with the given solution set.

(b) The solution set can also be described by solving the equation for y in terms of x, and then making x into a parameter. This leads to the equations $x = t$, $y = \frac{1}{2}t - \frac{5}{2}$.

13. We can find parametric equations for the line of intersection by (for example) solving the given equations for x and y in terms of z, then making z into a parameter:

$$\left. \begin{array}{r} x + y = 3 + z \\ 2x + y = 4 - 3z \end{array} \right\} \quad \Rightarrow \quad \left. \begin{array}{r} 2x + y = 4 - 3z \\ -x - y = -3 - z \end{array} \right\} \quad \Rightarrow \quad x = 1 - 4z$$

From the above it follows that $y = 3 + z - x = 3 + x - 1 + 4z = 2 + 5z$, and this leads to the parametric equations

$$x = 1 - 4t, \quad y = 2 + 5t, \quad z = t$$

for the line of intersection. The corresponding vector equation is

$$(x, y, z) = (1, 2, 0) + t(-4, 5, 1)$$

15. If $k \neq 6$, then the equations $x - y = 3$, $2x - 2y = k$ represent nonintersecting parallel lines and so the system of equations has no solution. If $k = 6$, the two lines coincide and so there are infinitely many solutions: $x = 3 + t$, $y = t$, where $-\infty < t < \infty$.

17. The augmented matrix of the system is $\begin{bmatrix} 3 & -2 & | & -1 \\ 4 & 5 & | & 3 \\ 7 & 3 & | & 2 \end{bmatrix}$.

19. The augmented matrix of the system is $\begin{bmatrix} 1 & 2 & 0 & -1 & 1 & | & 1 \\ 0 & 3 & 2 & 0 & -1 & | & 2 \\ 0 & 3 & 1 & 7 & 0 & | & 1 \end{bmatrix}$.

21. A system of equations corresponding to the given augmented matrix is:

$$\begin{array}{rcl} 2x_1 & = & 0 \\ 3x_1 - 4x_2 & = & 0 \\ x_2 & = & 1 \end{array}$$

23. A system of equations corresponding to the given augmented matrix is:

$$\begin{array}{rcl} 7x_1 + 2x_2 + x_3 - 3x_4 & = & 5 \\ x_1 + 2x_2 + 4x_3 & = & 1 \end{array}$$

25. **(a)** B is obtained from A by adding 2 times the first row to the second row. A is obtained from B by adding -2 times the first row to the second row.

(b) B is obtained from A by multiplying the first row by $\frac{1}{2}$. A is obtained from B by multiplying the first row by 2.

27.
$$2x + 3y + z = 7$$
$$2x + y + 3z = 9$$
$$4x + 2y + 5z = 16$$

29.
$$x + y + z = 12$$
$$2x + y + 2z = 5$$
$$x - z = -1$$

31. **(a)**
$$3c_1 + c_2 + 2c_3 - c_4 = 5$$
$$c_2 + 3c_3 + 2c_4 = 6$$
$$-c_1 + c_2 + 5c_4 = 5$$
$$2c_1 + c_2 + 2c_3 = 5$$

(b)
$$3c_1 + c_2 + 2c_3 - c_4 = 8$$
$$c_2 + 3c_3 + 2c_4 = 3$$
$$-c_1 + c_2 + 5c_4 = -2$$
$$2c_1 + c_2 + 2c_3 = 6$$

(c)
$$3c_1 + c_2 + 2c_3 - c_4 = 4$$
$$c_2 + 3c_3 + 2c_4 = 4$$
$$-c_1 + c_2 + 5c_4 = 6$$
$$2c_1 + c_2 + 2c_3 = 2$$

DISCUSSION AND DISCOVERY

D1. **(a)** There is no common intersection point.
(b) There is exactly one common point of intersection.
(c) The three lines coincide.

D2. A consistent system has at least one solution; moreover, it either has exactly one solution or it has infinitely many solutions.

If the system has exactly one solution, then there are two possibilities. If the three lines are all distinct but have a common point of intersection, then any one of the three equations can be discarded without altering the solution set. On the other hand, if two of the lines coincide, then one of the corresponding equations can be discarded without altering the solution set.

If the system has infinitely many solutions, then the three lines coincide. In this case any one (in fact any two) of the equations can be discarded without altering the solution set.

D3. Yes. If B can be obtained from A by multiplying a row by a nonzero constant, then A can be obtained from B by multiplying the same row by the reciprocal of that constant. If B can be obtained from A by interchanging two rows, then A can be obtained from B by interchanging the same two rows. Finally, if B can be obtained from A by adding a multiple of a row to another row, then A can be obtained from B by subtracting the same multiple of that row from the other row.

D4. If $k = l = m = 0$, then $x = y = 0$ is a solution of all three equations and so the system is consistent. If the system has exactly one solution then the three lines intersect at the origin.

D5. The parabola $y = ax^2 + bx + c$ will pass through the points $(1, 1)$, $(2, 4)$, and $(-1, 1)$ if and only if

$$a + b + c = 1$$
$$4a + 2b + c = 4$$
$$a - b + c = 1$$

Since there is a unique parabola passing through any three non-collinear points, one would expect this system to have exactly one solution.

D6. The parabola $y = ax^2 + bx + c$ passes through the points (x_1, y_1), (x_2, y_2), and (x_3, y_3) if and only if

$$ax_1^2 + bx_1 + c = y_1$$
$$4ax_2^2 + 2bx_2 + c = y_2$$
$$ax_3^2 - bx_3 + c = y_3$$

i.e. if and only if a, b, and c satisfy the linear system whose augmented matrix is

$$\begin{bmatrix} x_1^2 & x_1 & 1 & y_1 \\ x_2^2 & x_2 & 1 & y_2 \\ x_3^2 & x_3 & 1 & y_3 \end{bmatrix}$$

D7. To say that the equations have the same solution set is the same thing as to say that they represent the same line. From the first equation the x_1-intercept of the line is $x_1 = c$, and from the second equation the x_1-intercept is $x_1 = d$; thus $c = d$. If the line is vertical then $k = l = 0$. If the line is not vertical then from the first equation the slope is $m = \frac{1}{k}$, and from the second equation the slope is $m = \frac{1}{l}$; thus $k = l$. In summary, we conclude that $c = d$ and $k = l$; thus the two equations are identical.

D8. **(a)** True. If there are $n \geq 2$ columns, then the first $n - 1$ columns correspond to the coefficients of the variables that appear in the equations and the last column corresponds to the constants that appear on the right-hand side of the equal sign.

(b) False. Referring to Example 6: The sequence of linear systems appearing in the left-hand column all have the same solution set, but the corresponding augmented matrices appearing in the right-hand column are all different.

(c) False. Multiplying a row of the augmented matrix by zero corresponds to multiplying both sides of the corresponding equation by zero. But this is equivalent to discarding one of the equations!

(d) True. If the system is consistent, one can solve for two of the variables in terms of the third or (if further redundancy is present) for one of the variables in terms of the other two. In any case, there is at least one "free" variable that can be made into a parameter in describing the solution set of the system. Thus if the system is consistent, it will have infinitely many solutions.

D9. **(a)** True. A plane in 3-space corresponds to a linear equation in three variables. Thus a set of four planes corresponds to a system of four linear equations in three variables. If there is enough redundancy in the equations so that the system reduces to a system of two independent equations, then the solution set will be a line. For example, four vertical planes each containing the z-axis and intersecting the xy-plane in four distinct lines.

(b) False. Interchanging the first two columns corresponds to interchanging the coefficients of the first two variables. This results in a different system with a different solution set. [It is okay to interchange rows since this corresponds to interchanging equations and therefore does not alter the solution set.]

(c) False. If there is enough redundancy so that the system reduces to a system of only two (or fewer) equations, and if these equations are consistent, then the original system will be consistent.

(d) True. Such a system will always have the trivial solution $x_1 = x_2 = \cdots = x_n = 0$.

EXERCISE SET 2.2

1. The matrices (a), (c), and (d) are in reduced row echelon form. The matrix (b) does not satisfy property 4 of the definition, and the matrix (e) does not satisfy property 2.

3. The matrices (a) and (b) are in row echelon form. The matrix (c) does not satisfy property 1 or property 3 of the definition.

5. The matrices (a) and (c) are in reduced row echelon form. The matrix (b) does not satisfy property 3 and thus is not in row echelon form or reduced row echelon form.

7. The possible 2 by 2 reduced row echelon forms are $\begin{bmatrix} 0 & 0 \\ 0 & 0 \end{bmatrix}$, $\begin{bmatrix} 0 & 1 \\ 0 & 0 \end{bmatrix}$, $\begin{bmatrix} 1 & 0 \\ 0 & 1 \end{bmatrix}$, and $\begin{bmatrix} 1 & * \\ 0 & 0 \end{bmatrix}$ with any real number substituted for the *.

9. The given matrix corresponds to the system

$$
\begin{aligned}
x_1 && &= -3 \\
& x_2 & &= 0 \\
&& x_3 &= 7
\end{aligned}
$$

which clearly has the unique solution $x_1 = -3$, $x_2 = 0$, $x_3 = 7$.

11. The given matrix corresponds to the system

$$
\begin{aligned}
x_1 - 6x_2 && + 3x_5 &= -2 \\
& x_3 & + 4x_5 &= 7 \\
&& x_4 + 5x_5 &= 8
\end{aligned}
$$

where the equation corresponding to the zero row has been omitted. Solving these equations for the leading variables (x_1, x_3, and x_4) in terms of the free variables (x_2 and x_5) results in $x_1 = -2 + 6x_2 - 3x_5$, $x_3 = 7 - 4x_5$ and $x_4 = 8 - 5x_5$. Thus, assigning arbitrary values to x_2 and x_5, the solution set can be represented by the parametric equations

$$
x_1 = -2 + 6s - 3t, \quad x_2 = s, \quad x_3 = 7 - 4t, \quad x_4 = 8 - 5t, \quad x_5 = t
$$

where $-\infty < s, t < \infty$. The corresponding vector form is

$$
(x_1, x_2, x_3, x_4, x_5) = (-2, 0, 7, 8, 0) + s(6, 1, 0, 0, 0) + t(-3, 0, -4, -5, 1)
$$

13. The given matrix corresponds to the system

$$
\begin{aligned}
x_1 && - 7x_4 &= 8 \\
& x_2 & + 3x_4 &= 2 \\
&& x_3 + x_4 &= -5
\end{aligned}
$$

Solving these equations for the leading variables in terms of the free variable results in $x_1 = 8 + 7x_4$, $x_2 = 2 - 3x_4$, and $x_3 = -5 - x_4$. Thus, making x_4 into a parameter, the solution set of the system can be represented by the parametric equations

$$
x_1 = 8 + 7t, \quad x_2 = 2 - 3t, \quad x_3 = -5 - t, \quad x_4 = t
$$

where $-\infty < t < \infty$. The corresponding vector form is

$$
(x_1, x_2, x_3, x_4) = (8, 2, -5, 0) + t(7, -3, -1, 1)
$$

15. The system of equations corresponding to the given matrix is

$$x_1 - 3x_2 + 4x_3 = 7$$
$$x_2 + 2x_3 = 2$$
$$x_3 = 5$$

Starting with the last equation and working up, it follows that $x_3 = 5$, $x_2 = 2 - 2x_3 = 2 - 10 = -8$, and $x_1 = 7 + 3x_2 - 4x_3 = 7 - 24 - 20 = -37$.

Alternate solution via Gauss–Jordan (starting from the original matrix and reducing further):

$$\begin{bmatrix} 1 & -3 & 4 & 7 \\ 0 & 1 & 2 & 2 \\ 0 & 0 & 1 & 5 \end{bmatrix}$$

Add -2 times row 3 to row 2. Add -4 times row 3 to row 1.

$$\begin{bmatrix} 1 & -3 & 0 & -13 \\ 0 & 1 & 0 & -8 \\ 0 & 0 & 1 & 5 \end{bmatrix}$$

Add 3 times row 2 to row 1.

$$\begin{bmatrix} 1 & 0 & 0 & -37 \\ 0 & 1 & 0 & -8 \\ 0 & 0 & 1 & 5 \end{bmatrix}$$

From this we conclude (as before) that $x_1 = -37$, $x_2 = -8$, and $x_3 = 5$.

17. The corresponding system of equations is

$$x_1 + 7x_2 - 2x_3 \qquad - 8x_5 = -3$$
$$x_3 + x_4 + 6x_5 = 5$$
$$x_4 + 3x_5 = 9$$

Starting with the last equation and working up, it follows that $x_4 = 9 - 3x_5$, $x_3 = 5 - x_4 - 6x_5 = 5 - (9 - 3x_5) - 6x_5 = -4 - 3x_5$, and $x_1 = -3 - 7x_2 + 2x_3 + 8x_5 = -3 - 7x_2 + 2(-4 - 3x_5) + 8x_5 = -11 - 7x_2 + 2x_5$. Finally, assigning arbitrary values to x_2 and x_5, the solution set can be described by

$$x_1 = -11 - 7s + 2t, \quad x_2 = s, \quad x_3 = -4 - 3t, \quad x_4 = 9 - 3t, \quad x_5 = t$$

19. The corresponding system is

$$x_1 + x_2 - 3x_3 + 2x_4 = 1$$
$$x_2 + 4x_3 = 3$$
$$x_4 = 2$$

Starting with the last equation, we have $x_4 = 2$, $x_2 = 3 - 4x_3$, $x_1 = 1 - x_2 + 3x_3 - 2x_4 = 1 - (3 - 4x_3) + 3x_3 - 2(2) = -6 + 7x_3$. Thus, making x_3 into a parameter, the solution set can be described by the parametric equations

$$x_1 = -6 + 7t, \quad x_2 = 3 - 4t, \quad x_3 = t, \quad x_4 = 2$$

21. Starting with the first equation and working down, we have $x_1 = 2$, $x_2 = \frac{1}{3}(5 - x_1) = \frac{1}{3}(5 - 2) = 1$, and $x_3 = \frac{1}{3}(12 - 3x_1 - 2x_2) = \frac{1}{3}(12 - 6 - 2) = \frac{4}{3}$.

23. The augmented matrix of the system is

$$\begin{bmatrix} 1 & 1 & 2 & 8 \\ -1 & -2 & 3 & 1 \\ 3 & -7 & 4 & 10 \end{bmatrix}$$

Add row 1 to row 2. Add -3 times row 1 to row 3.

$$\begin{bmatrix} 1 & 1 & 2 & 8 \\ 0 & -1 & 5 & 9 \\ 0 & -10 & -2 & -14 \end{bmatrix}$$

Multiply row 2 by -1. Add 10 times the new row 2 to row 3.

$$\begin{bmatrix} 1 & 1 & 2 & 8 \\ 0 & 1 & -5 & -9 \\ 0 & 0 & -52 & -114 \end{bmatrix}$$

Multiply row 3 by $-\frac{1}{52}$.

$$\begin{bmatrix} 1 & 1 & 2 & 8 \\ 0 & 1 & -5 & -9 \\ 0 & 0 & 1 & 2 \end{bmatrix}$$

Add 5 times row 3 to row 2. Add -2 times row 3 to row 1.

$$\begin{bmatrix} 1 & 1 & 0 & 4 \\ 0 & 1 & 0 & 1 \\ 0 & 0 & 1 & 2 \end{bmatrix}$$

Add -1 times row 2 to row 1.

$$\begin{bmatrix} 1 & 0 & 0 & 3 \\ 0 & 1 & 0 & 1 \\ 0 & 0 & 1 & 2 \end{bmatrix}$$

Thus the solution is $x_1 = 3$, $x_2 = 1$, $x_3 = 2$.

25. The augmented matrix of the system is

$$\begin{bmatrix} 1 & -1 & 2 & -1 & -1 \\ 2 & 1 & -2 & -2 & -2 \\ -1 & 2 & -4 & 1 & 1 \\ 3 & 0 & 0 & -3 & -3 \end{bmatrix}$$

Add -2 times row 1 to row 2. Add row 1 to row 3. Add -3 times row 1 to row 4.

$$\begin{bmatrix} 1 & -1 & 2 & -1 & -1 \\ 0 & 3 & -6 & 0 & 0 \\ 0 & -1 & -2 & 0 & 0 \\ 0 & 3 & -6 & 0 & 0 \end{bmatrix}$$

Multiply row 2 by $\frac{1}{3}$. Add the new row 2 to row 3. Add -3 times the new row 2 to row 4.

$$\begin{bmatrix} 1 & -1 & 2 & -1 & -1 \\ 0 & 1 & -2 & 0 & 0 \\ 0 & 0 & 0 & 0 & 0 \\ 0 & 0 & 0 & 0 & 0 \end{bmatrix}$$

Add row 2 to row 1.

$$\begin{bmatrix} 1 & 0 & 0 & -1 & -1 \\ 0 & 1 & -2 & 0 & 0 \\ 0 & 0 & 0 & 0 & 0 \\ 0 & 0 & 0 & 0 & 0 \end{bmatrix}$$

Thus, setting $z = s$ and $w = t$, the solution set of the system is represented by the parametric equations

$$x = -1 + t, \quad y = 2s, \quad z = s, \quad w = t$$

27. The augmented matrix of the system is

$$\begin{bmatrix} 2 & -3 & -2 \\ 2 & 1 & 1 \\ 3 & 2 & 1 \end{bmatrix}$$

Multiply row 3 by 2, then add -1 times row 1 to row 2 and -3 times row 1 to the new row 3.

$$\begin{bmatrix} 2 & -3 & -2 \\ 0 & 4 & 3 \\ 0 & 13 & 8 \end{bmatrix}$$

The last two rows correspond to the (incompatible) equations $4x_2 = 3$ and $13x_2 = 8$; thus the system is inconsistent.

29. As an intermediate step in Exercise 23, the augmented matrix of the system was reduced to

$$\begin{bmatrix} 1 & 1 & 2 & 8 \\ 0 & 1 & -5 & -9 \\ 0 & 0 & 1 & 2 \end{bmatrix}$$

Starting with the last row and working up, it follows that $x_3 = 2$, $x_2 = -9 + 5x_3 = -9 + 10 = 1$, and $x_1 = 8 - x_1 - 2x_3 = 8 - 1 - 4 = 3$.

31. As an intermediate step in Exercise 25, the augmented matrix of the system was reduced to

$$\begin{bmatrix} 1 & -1 & 2 & -1 & -1 \\ 0 & 1 & -2 & 0 & 0 \\ 0 & 0 & 0 & 0 & 0 \\ 0 & 0 & 0 & 0 & 0 \end{bmatrix}$$

It follows that $y = 2z$ and $x = -1 + y - 2z + w = -1 + w$. Thus, setting $z = s$ and $w = t$, the solution set of the system is represented by the parametric equations $x = -1 + t$, $y = 2s$, $z = s$, $w = t$.

33. **(a)** There are more unknowns than equations in this homogeneous system. Thus, by Theorem 2.2.3, there are infinitely many nontrivial solutions.

(b) From back substitution it is clear that $x_1 = x_2 = x_3 = 0$. This system has only the trivial solution.

35. The augmented matrix of the homogeneous system is

$$\begin{bmatrix} 2 & 1 & 3 & 0 \\ 1 & 2 & 0 & 0 \\ 0 & 1 & 2 & 0 \end{bmatrix}$$

Interchange rows 1 and 2. Add -2 times the new row 1 to the new row 2.

$$\begin{bmatrix} 1 & 2 & 0 & 0 \\ 0 & -3 & 3 & 0 \\ 0 & 1 & 2 & 0 \end{bmatrix}$$

Multiply row 2 by $-\frac{1}{3}$. Add -1 times row 2 to row 3. Multiply the new row 3 by $\frac{1}{3}$.

$$\begin{bmatrix} 1 & 2 & 0 & 0 \\ 0 & 1 & -1 & 0 \\ 0 & 0 & 1 & 0 \end{bmatrix}$$

The last row of this matrix corresponds to $x_3 = 0$ and, from back substitution, it follows that $x_2 = x_3 = 0$ and $x_1 = -2x_2 = 0$. This system has only the trivial solution.

37. The augmented matrix of the homogeneous system is

$$\begin{bmatrix} 0 & 2 & 2 & 4 & 0 \\ 1 & 0 & -1 & -3 & 0 \\ -2 & 1 & 3 & -2 & 0 \end{bmatrix}$$

Interchange rows 1 and 2. Add 2 times the new row 1 to row 3.

$$\begin{bmatrix} 1 & 0 & -1 & -3 & 0 \\ 0 & 2 & 2 & 4 & 0 \\ 0 & 1 & 1 & -8 & 0 \end{bmatrix}$$

Multiply row 2 by $\frac{1}{2}$. Add -1 times the new row 2 to row 3. Multiply the new row 3 by $-\frac{1}{10}$

$$\begin{bmatrix} 1 & 0 & -1 & -3 & 0 \\ 0 & 1 & 1 & 2 & 0 \\ 0 & 0 & 0 & 1 & 0 \end{bmatrix}$$

Add -2 times row 3 to row 2. Add 3 times row 3 to row 1.

$$\begin{bmatrix} 1 & 0 & -1 & 0 & 0 \\ 0 & 1 & 1 & 0 & 0 \\ 0 & 0 & 0 & 1 & 0 \end{bmatrix}$$

This is the reduced row echelon form of the matrix. From this we see that y (the third variable) is a free variable and, on setting $y = t$, the solution set of the system can be described by the parametric equations $w = t$, $x = -t$, $y = t$, $z = 0$.

39. The augmented matrix of this homogeneous system is

$$\begin{bmatrix} 0 & 1 & 3 & -2 & 0 \\ 2 & 1 & -4 & 3 & 0 \\ 2 & 3 & 2 & -1 & 0 \\ -4 & -3 & 5 & -4 & 0 \end{bmatrix}$$

and the reduced row echelon form of this matrix is

$$\begin{bmatrix} 1 & 0 & -\frac{7}{2} & \frac{5}{2} & 0 \\ 0 & 1 & 3 & -2 & 0 \\ 0 & 0 & 0 & 0 & 0 \\ 0 & 0 & 0 & 0 & 0 \end{bmatrix}$$

Thus, setting $w = 2s$ and $x = 2t$, the solution set of the system can be described by the parametric equations $u = 7s - 5t$, $v = -6s + 4t$, $w = 2s$, $x = 2t$.

41. We will solve the system by Gaussian elimination, i.e. by reducing the augmented matrix of the system to a row-echelon form: The augmented matrix of the original system is

$$\begin{bmatrix} 2 & -1 & 3 & 4 & 0 \\ 1 & 0 & -2 & 7 & 0 \\ 3 & -3 & 1 & 5 & 0 \\ 2 & 1 & 4 & 4 & 0 \end{bmatrix}$$

Interchange rows 1 and 2. Add -2 times the new row 1 to the new row 2. Add -3 times the new row 1 to row 3. Add -2 times the new row 1 to row 4.

$$\begin{bmatrix} 1 & 0 & -2 & 7 & 0 \\ 0 & -1 & 7 & -10 & 0 \\ 0 & -3 & 7 & -16 & 0 \\ 0 & 1 & 8 & -10 & 0 \end{bmatrix}$$

Multiply row 2 by -1. Add 3 times the new row 2 to row 3. Add -1 times the new row 2 to row 4.

$$\begin{bmatrix} 1 & 0 & -2 & 7 & 0 \\ 0 & 1 & -7 & 10 & 0 \\ 0 & 0 & -14 & 14 & 0 \\ 0 & 0 & 15 & -20 & 0 \end{bmatrix}$$

Multiply row 3 by $-\frac{1}{14}$. Add -15 times the new row 3 to row 4. Multiply the new row 4 by $\frac{1}{5}$.

$$\begin{bmatrix} 1 & 0 & -2 & 7 & 0 \\ 0 & 1 & -7 & 10 & 0 \\ 0 & 0 & 1 & -1 & 0 \\ 0 & 0 & 0 & 1 & 0 \end{bmatrix}$$

This is a row-echelon form for the augmented matrix. From the last row we conclude that $I_4 = 0$, and from back substitution it follows that $I_3 = I_2 = I_1 = 0$ also. This system has only the trivial solution.

43. The augmented matrix of the system is

$$\begin{bmatrix} 1 & 2 & 3 & 4 \\ 3 & -1 & 5 & 2 \\ 4 & 1 & -14 & a+2 \end{bmatrix}$$

Add -3 times row 1 to row2. Add -4 times row 1 to row 3.

$$\begin{bmatrix} 1 & 2 & 3 & 4 \\ 0 & -7 & -4 & -10 \\ 0 & -7 & -26 & a-14 \end{bmatrix}$$

Multiply row 2 by -1. Add the new row 2 to row 3.

$$\begin{bmatrix} 1 & 2 & 3 & 4 \\ 0 & 7 & 4 & 10 \\ 0 & 0 & -22 & a-4 \end{bmatrix}$$

From the last row we conclude that $z = \frac{a-4}{-22}$ and, from back substitution, it is clear that y and x are uniquely determined as well. This system has exactly one solution for every value of a.

45. The augmented matrix of the system is

$$\begin{bmatrix} 1 & 2 & 1 \\ 2 & a^2-5 & a-1 \end{bmatrix}$$

Add -2 times row 1 to row 2.

$$\begin{bmatrix} 1 & 2 & 1 \\ 0 & a^2-9 & a-3 \end{bmatrix}$$

If $a = 3$, then the last row corresponds to $0 = 0$ and the system has infinitely many solutions. If $a = -3$, the last row corresponds to $0 = -6$ and the system is inconsistent. If $a \neq \pm 3$, then $y = \frac{a-3}{a^2-9} = \frac{1}{a+3}$ and, from back substitution, x is uniquely determined as well; the system has exactly one solution in this case.

47. (a) If $x + y + z = 1$, then $2x + 2y + 2z = 2 \neq 4$; thus the system has no solution. The planes represented by the two equations do not intersect (they are parallel).

(b) If $x + y + z = 0$, then $2x + 2y + 2z = 0$ also; thus the system is redundant and has infinitely many solutions. Any set of values of the form $x = -s - t$, $y = s$, and $z = t$ will satisfy both equations. The planes represented by the equations coincide.

49. The system is linear in the variables $x = \sin \alpha$, $y = \cos \beta$, and $z = \tan \gamma$.

$$\begin{array}{rcrcrcl} 2x & - & y & + & 3z & = & 3 \\ 4x & + & 2y & - & 2z & = & 2 \\ 6x & - & 3y & + & z & = & 9 \end{array}$$

We solve the system by performing the indicated row operations on the augmented matrix

$$\begin{bmatrix} 2 & -1 & 3 & 3 \\ 4 & 2 & -2 & 2 \\ 6 & -3 & 1 & 9 \end{bmatrix}$$

Add -2 times row 1 to row 2. Add -3 times row 1 to row 3.

$$\begin{bmatrix} 2 & -1 & 3 & 3 \\ 0 & 4 & -8 & -4 \\ 0 & 0 & -8 & 0 \end{bmatrix}$$

From this we conclude that $\tan \gamma = z = 0$ and, from back substitution, that $\cos \beta = y = -1$ and $\sin \alpha = x = 1$. Thus $\alpha = \frac{\pi}{2}$, $\beta = \pi$, and $\gamma = 0$.

51. This system is homogeneous with augmented matrix

$$\begin{bmatrix} 2-\lambda & -1 & 0 & 0 \\ 2 & 1-\lambda & 1 & 0 \\ -2 & -2 & 1-\lambda & 0 \end{bmatrix}$$

If $\lambda = 1$, the augmented matrix is $\begin{bmatrix} 1 & -1 & 0 & 0 \\ 2 & 0 & 1 & 0 \\ -2 & -2 & 0 & 0 \end{bmatrix}$, and the reduced row-echelon form of this

matrix is $\begin{bmatrix} 1 & 0 & 0 & 0 \\ 0 & 1 & 0 & 0 \\ 0 & 0 & 1 & 0 \end{bmatrix}$. Thus $x = y = z = 0$, i.e. the system has only the trivial solution.

If $\lambda = 2$, the augmented matrix is $\begin{bmatrix} 0 & -1 & 0 & 0 \\ 2 & -1 & 1 & 0 \\ -2 & -2 & -1 & 0 \end{bmatrix}$, and the reduced row-echelon form of this

matrix is $\begin{bmatrix} 1 & 0 & \frac{1}{2} & 0 \\ 0 & 1 & 0 & 0 \\ 0 & 0 & 0 & 0 \end{bmatrix}$. Thus the system has infinitely many solutions: $x = -\frac{1}{2}t$, $y = 0$, $z = t$,

where $-\infty < t < \infty$.

53. **(a)** Starting with the given system and proceeding as directed, we have

$$\begin{aligned} 0.0001x + 1.000y &= 1.000 \\ 1.000x - 1.000y &= 0.000 \end{aligned}$$

$$\begin{aligned} 1.000x + 10000y &= 10000 \\ 1.000x - 1.000y &= 0.000 \end{aligned}$$

$$\begin{aligned} 1.000x + 10000y &= 10000 \\ - 10000y &= -10000 \end{aligned}$$

which results in $y \approx 1.000$ and $x \approx 0.000$.

(b) If we first interchange rows and then proceed as directed, we have

$$\begin{aligned} 1.000x - 1.000y &= 0.000 \\ 0.0001x + 1.000y &= 1.000 \end{aligned}$$

$$\begin{aligned} 1.000x - 1.000y &= 0.000 \\ 1.000y &= 1.000 \end{aligned}$$

which results in $y \approx 1.000$ and $x \approx 1.000$.

(c) The exact solution is $x = \frac{100000}{49999}$ and $y = \frac{49997}{49999}$.

The approximate solution without using partial pivoting is

$$\begin{aligned} 0.00002x + 1.000y &= 1.000 \\ 1.000x + 1.000y &= 3.000 \end{aligned}$$

$$\begin{aligned} 1.000x + 50000y &= 50000 \\ 1.000x + 1.000y &= 3.000 \end{aligned}$$

$$\begin{aligned} 1.000x + 50000y &= 50000 \\ - 50000y &= -50000 \end{aligned}$$

which results in $y \approx 1.000$ and $x \approx 0.000$.

The approximate solution using partial pivoting is

$$0.00002x + 1.000y = 1.000$$
$$1.000x + 1.000y = 3.000$$

$$1.000x + 1.000y = 3.000$$
$$0.00002x + 1.000y = 1.000$$

$$1.000x + 1.000y = 3.000$$
$$1.000y = 1.000$$

which results in $y \approx 1.000$ and $x \approx 2.000$.

DISCUSSION AND DISCOVERY

D1. If the homogeneous system has only the trivial solution, then the non-homogeneous system will either be inconsistent or have exactly one solution.

D2. (a) All three lines pass through the origin and at least two of them do not coincide.
  (b) If the system has nontrivial solutions then the lines must coincide and pass through the origin.

D3. (a) Yes. If $ax_0 + by_0 = 0$ then $a(kx_0) + b(ky_0) = k(ax_0 + by_0) = 0$. Similarly for the other equation.
  (b) Yes. If $ax_0 + by_0 = 0$ and $ax_1 + by_1 = 0$ then

$$a(x_0 + x_1) + b(y_0 + y_1) = (ax_0 + by_0) + (ax_1 + by_1) = 0$$

  and similarly for the other equation.
  (c) Yes in both cases. These statements are not true for non-homogenous systems.

D4. The first system may be inconsistent, but the second system always has (at least) the trivial solution. If the first system is consistent then the solution sets will be parallel objects (points, lines, or the entire plane) with the second containing the origin.

D5. (a) At most three (the number of rows in the matrix).
  (b) At most five (if B is the zero matrix). If B is not the zero matrix, then there are at most 4 free variables ($5 - r$ where r is the number of non-zero rows in a row echelon form).
  (c) At most three (the number of rows in the matrix).

D6. (a) At most three (the number of columns).
  (b) At most three (if B is the zero matrix). If B is not the zero matrix, then there are at most 2 free variables ($3 - r$ where r is the number of non-zero rows in a row echelon form).
  (c) At most three (the number of columns).

D7. (a) False. For example, $x + y + z = 0$ and $x + y + z = 1$ are inconsistent.
  (b) False. If there is more than one solution then there are infinitely many solutions.
  (c) False. If the system is consistent then, since there is at least one free variable, there will be infinitely many solutions.
  (d) True. A homogeneous system always has (at least) the trivial solution.

D8. **(a)** True. For example $\begin{bmatrix} 1 & 1 \\ -1 & 1 \end{bmatrix}$ can be reduced to either $\begin{bmatrix} 1 & 1 \\ 0 & 1 \end{bmatrix}$ or $\begin{bmatrix} 1 & 0 \\ 0 & 1 \end{bmatrix}$.

(b) False. The reduced row echelon form of a matrix is unique.

(c) False. The appearance of a row of zeros means that there was some redundancy in the system. But the remaining equations may be inconsistent, have exactly one solution, or have infinitely many solutions. All of these are possible.

(d) False. There may be redundancy in the system. For example, the system consisting of the equations $x + y = 1$, $2x + 2y = 2$, and $3x + 3y = 3$ has infinitely many solutions.

D9. The system is linear in the variables $x = \sin \alpha$, $y = \cos \beta$, $z = \tan \gamma$, and this system has only the trivial solution $x = y = z = 0$. Thus $\sin \alpha = 0$, $\cos \beta = 0$, $\tan \gamma = 0$. It follows that $\alpha = 0$, π, or 2π; $\beta = \frac{\pi}{2}$ or $\frac{3\pi}{2}$; and $\gamma = 0$, π, or 2π. There are eighteen possible combinations in all. This does not contradict Theorem 2.1.1 since the equations are not linear in the variables α, β, γ.

WORKING WITH PROOFS

P1. **(a)** If $a \neq 0$, then the reduction can be accomplished as follows:

$$\begin{bmatrix} a & b \\ c & d \end{bmatrix} \to \begin{bmatrix} 1 & \frac{b}{a} \\ c & d \end{bmatrix} \to \begin{bmatrix} 1 & \frac{b}{a} \\ 0 & \frac{ad-bc}{a} \end{bmatrix} \to \begin{bmatrix} 1 & \frac{b}{a} \\ 0 & 1 \end{bmatrix} \to \begin{bmatrix} 1 & 0 \\ 0 & 1 \end{bmatrix}$$

If $a = 0$, then $b \neq 0$ and $c \neq 0$, so the reduction can be carried out as follows:

$$\begin{bmatrix} 0 & b \\ c & d \end{bmatrix} \to \begin{bmatrix} c & d \\ 0 & b \end{bmatrix} \to \begin{bmatrix} 1 & \frac{d}{c} \\ 0 & b \end{bmatrix} \to \begin{bmatrix} 1 & \frac{d}{c} \\ 0 & 1 \end{bmatrix} \to \begin{bmatrix} 1 & 0 \\ 0 & 1 \end{bmatrix}$$

(b) If $\begin{bmatrix} a & b \\ c & d \end{bmatrix}$ can be reduced to $\begin{bmatrix} 1 & 0 \\ 0 & 1 \end{bmatrix}$, then the corresponding row operations on the augmented matrix $\begin{bmatrix} a & b & k \\ c & d & l \end{bmatrix}$ will reduce it to a matrix of the form $\begin{bmatrix} 1 & 0 & K \\ 0 & 1 & L \end{bmatrix}$ and from this it follows that the system has the unique solution $x = K$, $y = L$.

EXERCISE SET 2.3

1. This problem is concerned with the network

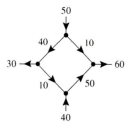

in which the known and unknown flow rates (and assumed directions) are as indicated. There are four nodes. Starting with the top node and working clockwise, application of the principle of conservation of flow leads to the system of equations

$$x_1 + x_2 \qquad\quad = \quad 50$$
$$x_2 \qquad\quad = \quad 10$$
$$x_3 \quad\; = \; -10$$
$$x_1 \qquad + x_3 \; = \quad 30$$

which has the unique solution $x_1 = 40$, $x_2 = 10$, $x_3 = -10$. Thus the flow rates and directions in the network are as indicated below:

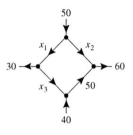

3. This problem is concerned with the one-way traffic network

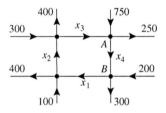

in which the known traffic flow rates, and directions of the unknown rates, are as indicated.

(a) At each intersection the flow in must be equal to the flow out. Starting with intersection A and working clockwise, this leads to the following system of equations:

$$x_3 + 750 = x_4 + 250, \quad x_4 + 200 = x_1 + 300, \quad x_1 + 100 = x_2 + 400, \quad x_2 + 300 = x_3 + 400$$

or,

$$x_3 - x_4 = -500$$
$$-x_1 \qquad\quad + x_4 = \quad 100$$
$$x_1 - x_2 \qquad\quad = \quad 300$$
$$x_2 - x_3 \qquad\quad = \quad 100$$

(b) The augmented matrix of the system is

$$\begin{bmatrix} 0 & 0 & 1 & -1 & -500 \\ -1 & 0 & 0 & 1 & 100 \\ 1 & -1 & 0 & 0 & 300 \\ 0 & 1 & -1 & 0 & 100 \end{bmatrix}$$

Interchange rows 1 and 3. Add the new row 1 to row 2.

$$\begin{bmatrix} 1 & -1 & 0 & 0 & 300 \\ 0 & -1 & 0 & 1 & 400 \\ 0 & 0 & 1 & -1 & -500 \\ 0 & 1 & -1 & 0 & 100 \end{bmatrix}$$

Multiply row 2 by -1. Add -1 times the new row 2 to row 4. Add row 3 to the new row 4.

$$\begin{bmatrix} 1 & -1 & 0 & 0 & 300 \\ 0 & 1 & 0 & -1 & -400 \\ 0 & 0 & 1 & -1 & -500 \\ 0 & 0 & 0 & 0 & 0 \end{bmatrix}$$

Add row 2 to row 1.

$$\begin{bmatrix} 1 & 0 & 0 & -1 & -100 \\ 0 & 1 & 0 & -1 & -400 \\ 0 & 0 & 1 & -1 & -500 \\ 0 & 0 & 0 & 0 & 0 \end{bmatrix}$$

This is the reduced row echelon form of the augmented matrix. From it we conclude that x_4 is a free variable, and that the general solution of the system is

$$x_1 = -100 + t, \quad x_2 = -400 + t, \quad x_3 = -500 + t, \quad x_4 = t$$

(c) In the context of this problem, all of the variables must be nonnegative since they represent flow rates (cars/hour) on one-way streets. The flow rate along the road from A to B is represented by the variable $x_4 = t$. Thus, in order to satisfy the requirement that all flow rates be nonnegative, the minimum permissible value of x_4 is 500 cars/hour.

5. From Kirchoff's current law, applied at either of the nodes, we have $I_1 + I_2 = I_3$. From Kirchoff's voltage law, applied to the left and right loops, we have $2I_1 = 2I_2 + 6$ and $2I_2 + 4I_3 = 8$. Thus the currents I_1, I_2, and I_3 satisfy the following system of equations:

$$\begin{array}{rrrrl} I_1 + & I_2 - & I_3 & = 0 \\ 2I_1 - & 2I_2 & & = 6 \\ & 2I_2 + & 4I_3 & = 8 \end{array}$$

The augmented matrix of this system is:

$$\begin{bmatrix} 1 & 1 & -1 & 0 \\ 2 & -2 & 0 & 6 \\ 0 & 2 & 4 & 8 \end{bmatrix}$$

Multiply rows 2 and 3 by $\frac{1}{2}$.

$$\begin{bmatrix} 1 & 1 & -1 & 0 \\ 1 & -1 & 0 & 3 \\ 0 & 1 & 2 & 4 \end{bmatrix}$$

Add -1 times row 1 to row 2.

$$\begin{bmatrix} 1 & 1 & -1 & 0 \\ 0 & -2 & 1 & 3 \\ 0 & 1 & 2 & 4 \end{bmatrix}$$

Interchange rows 2 and 3. Add 2 times the new row 2 to the new row 3.

$$\begin{bmatrix} 1 & 1 & -1 & 0 \\ 0 & 1 & 2 & 4 \\ 0 & 0 & 5 & 11 \end{bmatrix}$$

From the last row we conclude that $I_3 = \frac{11}{5} = 2.2$ A, and from back substitution it then follows that $I_2 = 4 - 2I_3 = 4 - \frac{22}{5} = -\frac{2}{5} = -0.4$ A, and $I_1 = I_3 - I_2 = \frac{11}{5} + \frac{2}{5} = \frac{13}{5} = 2.6$ A. Since I_2 is negative, its direction is opposite to that indicated in the figure.

7. From application of the current law at each of the four nodes (clockwise from upper left) we have $I_1 = I_2 + I_4$, $I_4 = I_3 + I_5$, $I_6 = I_3 + I_5$, and $I_1 = I_2 + I_6$. These four equations can be reduced to the following:
$$I_1 = I_2 + I_4, \quad I_4 = I_3 + I_5, \quad I_4 = I_6$$

From application of the voltage law to the three inner loops (left to right) we have $10 = 20I_1 + 20I_2$, $20I_2 = 20I_3$, and $10 + 20I_3 = 20I_5$. These equations can be simplified to:
$$2I_1 + 2I_2 = 1, \quad I_3 = I_2, \quad 2I_5 - 2I_3 = 1$$

This gives us six equations in the unknown currents $I_1, \ldots, I_6$. But, since $I_3 = I_2$ and $I_6 = I_4$, this can be simplified to a system of only four equations in the variables I_1, I_2, I_4, and I_5:
$$
\begin{aligned}
I_1 - I_2 - I_4 &= 0 \\
I_2 - I_4 + I_5 &= 0 \\
2I_1 + 2I_2 &= 1 \\
- 2I_2 + 2I_5 &= 1
\end{aligned}
$$

The augmented matrix of the system is
$$
\begin{bmatrix}
1 & -1 & -1 & 0 & 0 \\
0 & 1 & -1 & 1 & 0 \\
2 & 2 & 0 & 0 & 1 \\
0 & -2 & 0 & 2 & 1
\end{bmatrix}
$$

Add -2 times row 1 to row 3.
$$
\begin{bmatrix}
1 & -1 & -1 & 0 & 0 \\
0 & 1 & -1 & 1 & 0 \\
0 & 4 & 2 & 0 & 1 \\
0 & -2 & 0 & 2 & 1
\end{bmatrix}
$$

Add -4 times row 2 to row 3. Add 2 times row 2 to row 4.
$$
\begin{bmatrix}
1 & -1 & -1 & 0 & 0 \\
0 & 1 & -1 & 1 & 0 \\
0 & 0 & 6 & -4 & 1 \\
0 & 0 & -2 & 4 & 1
\end{bmatrix}
$$

Add 3 times row 4 to row 3. Interchange the new rows 3 and 4, then multiply the new row 3 by $-\frac{1}{2}$ and multiply the new row 4 by $\frac{1}{8}$.
$$
\begin{bmatrix}
1 & -1 & -1 & 0 & 0 \\
0 & 1 & -1 & 1 & 0 \\
0 & 0 & 1 & -2 & -\frac{1}{2} \\
0 & 0 & 0 & 1 & \frac{1}{2}
\end{bmatrix}
$$

From this we conclude that $I_5 = \frac{1}{2}$, $I_6 = I_4 = \frac{1}{2} + 2x_5 = -\frac{1}{2} - 1 = \frac{1}{2}$, $I_3 = I_2 = I_4 - I_5 = \frac{1}{2} - \frac{1}{2} = 0$, and $I_1 = I_2 + I_4 = 0 + \frac{1}{2} = \frac{1}{2}$.

9. We seek positive integers x_1, x_2, x_3, and x_4 that will balance the chemical equation

$$x_1(C_3H_8) + x_2(O_2) \rightarrow x_3(CO_2) + x_4(H_2O)$$

For each of the atoms in the equation (C, H, and O) the number of atoms on the left must be equal to the number of atoms on the right. This leads to the equations $3x_1 = x_3$, $8x_1 = 2x_4$, and $2x_2 = 2x_3 + x_4$. These equations form the homogeneous linear system

$$
\begin{aligned}
3x_1 \quad\quad - x_3 \quad\quad &= 0 \\
8x_1 \quad\quad\quad\quad - 2x_4 &= 0 \\
2x_2 - 2x_3 - x_4 &= 0
\end{aligned}
$$

represented by the augmented matrix

$$
\begin{bmatrix}
3 & 0 & -1 & 0 & 0 \\
8 & 0 & 0 & -2 & 0 \\
0 & 2 & -2 & -1 & 0
\end{bmatrix}
$$

Multiply row 1 by $\frac{1}{3}$. Add -8 times the new row 1 to row 2.

$$
\begin{bmatrix}
1 & 0 & -\frac{1}{3} & 0 & 0 \\
0 & 0 & \frac{8}{3} & -2 & 0 \\
0 & 2 & -2 & -1 & 0
\end{bmatrix}
$$

Interchange rows 2 and 3. Multiply the new row 2 by $\frac{1}{2}$. Multiply the new row 3 by $\frac{3}{8}$.

$$
\begin{bmatrix}
1 & 0 & -\frac{1}{3} & 0 & 0 \\
0 & 1 & -1 & -\frac{1}{2} & 0 \\
0 & 0 & 1 & -\frac{3}{4} & 0
\end{bmatrix}
$$

From this we conclude that x_4 is a free variable, and that the general solution of the system is given by $x_4 = t$, $x_3 = \frac{3}{4}x_4 = \frac{3}{4}t$, $x_2 = x_3 + \frac{1}{2}x_4 = \frac{3}{4}t + \frac{1}{2}t = \frac{5}{4}t$, and $x_1 = \frac{1}{3}x_3 = \frac{1}{4}t$. The smallest positive integer solutions are obtained by taking $t = 4$, in which case we obtain $x_1 = 1$, $x_2 = 5$, $x_3 = 3$, and $x_4 = 4$. Thus the balanced equation is $C_3H_8 + 5O_2 \rightarrow 3CO_2 + 4H_2O$.

11. We must find positive integers x_1, x_2, x_3, and x_4 that balance the chemical equation

$$x_1(CH_3COF) + x_2(H_2O) \rightarrow x_3(CH_3COOH) + x_4(HF)$$

For each of the elements in the equation (C, H, O, and F) the number of atoms on the left must equal the number of atoms on the right. This leads to the equations $x_1 = x_3$, $3x_1 + 2x_2 = 4x_3 + x_4$, $x_1 + x_2 = 2x_3$, and $x_1 = x_4$. This is a very simple system of equations having (by inspection) the general solution

$$x_1 = x_2 = x_3 = x_4 = t$$

Thus, taking $t = 1$, the balanced equation is

$$CH_3COF + H_2O \rightarrow CH_3COOH + HF$$

13. The graph of the polynomial $p(x) = a_0 + a_1x + a_2x^2$ passes through the points $(1, 1)$, $(2, 2)$, and $(3, 5)$ if and only if the coefficients a_0, a_1, and a_2 satisfy the equations

$$
\begin{aligned}
a_0 + a_1 + a_2 &= 1 \\
a_0 + 2a_1 + 4a_2 &= 2 \\
a_0 + 3a_1 + 9a_2 &= 5
\end{aligned}
$$

The augmented matrix of this system is

$$\begin{bmatrix} 1 & 1 & 1 & 1 \\ 1 & 2 & 4 & 2 \\ 1 & 3 & 9 & 5 \end{bmatrix}$$

Add -1 times row 1 to rows 2 and 3.

$$\begin{bmatrix} 1 & 1 & 1 & 1 \\ 0 & 1 & 3 & 1 \\ 0 & 2 & 8 & 4 \end{bmatrix}$$

Add -2 times row 2 to row 3.

$$\begin{bmatrix} 1 & 1 & 1 & 1 \\ 0 & 1 & 3 & 1 \\ 0 & 0 & 2 & 2 \end{bmatrix}$$

We conclude that $a_2 = 1$, and from back substitution it follows that $a_1 = 1 - 3a_2 = -2$, and $a_0 = 1 - a_1 - a_2 = 2$. Thus the interpolating quadratic polynomial is $p(x) = 2 - 2x + x^2$.

15. The graph of the polynomial $p(x) = a_0 + a_1x + a_2x^2 + a_3x^3$ passes through the points $(-1, -1)$, $(0, 1)$, $(1, 3)$, and $(4, -1)$, if and only if the coefficients a_0, a_1, and a_2 satisfy the equations

$$
\begin{array}{rcrcrcrcr}
a_0 & - & a_1 & + & a_2 & - & a_3 & = & -1 \\
a_0 & & & & & & & = & 1 \\
a_0 & + & a_1 & + & a_2 & + & a_3 & = & 3 \\
a_0 & + & 4a_1 & + & 16a_2 & + & 64a_3 & = & -1
\end{array}
$$

The augmented matrix of this system is

$$\begin{bmatrix} 1 & -1 & 1 & -1 & -1 \\ 1 & 0 & 0 & 0 & 1 \\ 1 & 1 & 1 & 1 & 3 \\ 1 & 4 & 16 & 64 & -1 \end{bmatrix}$$

Interchange rows 1 and 2. Add -1 times the new row 1 to each of the other rows.

$$\begin{bmatrix} 1 & 0 & 0 & 0 & 1 \\ 0 & -1 & 1 & -1 & -2 \\ 0 & 1 & 1 & 1 & 2 \\ 0 & 4 & 16 & 64 & -2 \end{bmatrix}$$

Multiply row 2 by -1. Add -1 times the new row 2 to row 3. Add -4 times the new row 2 to row 4.

$$\begin{bmatrix} 1 & 0 & 0 & 0 & 1 \\ 0 & 1 & -1 & 1 & 2 \\ 0 & 0 & 2 & 0 & 0 \\ 0 & 0 & 20 & 60 & -10 \end{bmatrix}$$

Multiply row 2 by $\frac{1}{2}$. Add -20 times the new row 3 to row 4.

$$\begin{bmatrix} 1 & 0 & 0 & 0 & 1 \\ 0 & 1 & -1 & 1 & 2 \\ 0 & 0 & 1 & 0 & 0 \\ 0 & 0 & 0 & 60 & -10 \end{bmatrix}$$

From the first row we see that $a_0 = 1$, and from last two rows we conclude that $a_2 = 0$ and $a_3 = -\frac{1}{6}$. Finally, from back substitution, it follows that $a_1 = 2 + a_2 - a_3 = \frac{13}{6}$. Thus the interpolating polynomial is $p(x) = 1 + \frac{13}{6}x - \frac{1}{6}x^3$.

DISCUSSION AND DISCOVERY

D1. **(a)** The quadratic polynomial $p(x) = a_0 + a_1 x + a_2 x^2$ passes through the points $(0, 1)$ and $(1, 2)$ if and only if the coefficients a_0, a_1, a_2 satisfy the equations

$$\begin{aligned} a_0 & = 1 \\ a_0 + a_1 + a_2 & = 2 \end{aligned}$$

The general solution of this system, using a_1 as a parameter, is $a_0 = 1$, $a_1 = k$, $a_2 = 1 - k$. Thus this family of polynomials is represented by the equation $p(x) = 1 + kx + (1 - k)x^2$ where $-\infty < k < \infty$.

(b) Below are graphs of four curves in the family: $y = 1 + x^2$ $(k = 0)$, $y = 1 + x$ $(k = 1)$, $y = 1 + 2x - x^2$ $(k = 2)$, and $y = 1 + 3x - 2x^2$ $(k = 3)$.

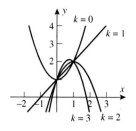

Matrices and Matrix Algebra

EXERCISE SET 3.1

1. Since two matrices are equal if and only if their corresponding entries are equal, we have $a - b = 8$, $b + a = 1$, $3d + c = 7$, and $2d - c = 6$. Adding the first two equations we see that $2a = 9$; thus $a = \frac{9}{2}$ and it follows that $b = 1 - a = -\frac{7}{2}$. Adding the second two equations we see that $5d = 13$; thus $d = \frac{13}{5}$ and $c = 7 - 3d = -\frac{4}{5}$.

3. **(a)** A has size 3×4, B^T has size 4×2.

   (b) $a_{32} = 3$, $a_{23} = 11$

   (c) $a_{ij} = 3$ if and only if $(i, j) = (1, 1), (3, 1)$, or $(3, 2)$

   (d) $\mathbf{c}_3(A^T) = \begin{bmatrix} 3 \\ 3 \\ 9 \\ -4 \end{bmatrix}$

   (e) $\mathbf{r}_2(2B^T) = \begin{bmatrix} 1 & 2 \end{bmatrix}$

5. **(a)** $A + 2B = \begin{bmatrix} 3 & 0 \\ -1 & 2 \\ 1 & 1 \end{bmatrix} + \begin{bmatrix} 4 & 2 \\ -6 & 2 \\ 8 & 0 \end{bmatrix} = \begin{bmatrix} 7 & 2 \\ -7 & 4 \\ 9 & 1 \end{bmatrix}$

   (b) $A - B^T$ is not defined

   (c) $4\hat{D} - 3C^T = \begin{bmatrix} 4 & 4 \\ -12 & 12 \end{bmatrix} - \begin{bmatrix} 3 & 9 \\ 0 & -3 \end{bmatrix} = \begin{bmatrix} 1 & -5 \\ -12 & 15 \end{bmatrix}$

   (d) $D - D^T = \begin{bmatrix} 1 & 1 \\ -3 & 3 \end{bmatrix} - \begin{bmatrix} 1 & -3 \\ 1 & 3 \end{bmatrix} = \begin{bmatrix} 0 & 4 \\ -4 & 0 \end{bmatrix}$

   (e) $G + (2F^T) = \begin{bmatrix} 6 & 1 & 3 \\ -1 & 1 & 2 \\ 4 & 1 & 3 \end{bmatrix} + \begin{bmatrix} 2 & -2 & 6 \\ 10 & 0 & 4 \\ 4 & 2 & 8 \end{bmatrix} = \begin{bmatrix} 8 & -1 & 9 \\ 9 & 1 & 6 \\ 8 & 3 & 11 \end{bmatrix}$

   (f) $(7A - B) + E$ is not defined

7. **(a)** $CD = \begin{bmatrix} 1 & 0 \\ 3 & -1 \end{bmatrix} \begin{bmatrix} 1 & 1 \\ -3 & 3 \end{bmatrix} = \begin{bmatrix} 1 & 1 \\ 6 & 0 \end{bmatrix}$

   (b) $AE = \begin{bmatrix} 3 & 0 \\ -1 & 2 \\ 1 & 1 \end{bmatrix} \begin{bmatrix} 1 & 4 & 2 \\ 3 & 1 & 5 \end{bmatrix} = \begin{bmatrix} 3 & 12 & 6 \\ 5 & -2 & 8 \\ 4 & 5 & 7 \end{bmatrix}$

   (c) $FG = \begin{bmatrix} 1 & 5 & 2 \\ -1 & 0 & 1 \\ 3 & 2 & 4 \end{bmatrix} \begin{bmatrix} 6 & 1 & 3 \\ -1 & 1 & 2 \\ 4 & 1 & 3 \end{bmatrix} = \begin{bmatrix} 9 & 8 & 19 \\ -2 & 0 & 0 \\ 32 & 9 & 25 \end{bmatrix}$

   (d) $B^T F = \begin{bmatrix} 2 & -3 & 4 \\ 1 & 1 & 0 \end{bmatrix} \begin{bmatrix} 1 & 5 & 2 \\ -1 & 0 & 1 \\ 3 & 2 & 4 \end{bmatrix} = \begin{bmatrix} 17 & 18 & 17 \\ 0 & 5 & 3 \end{bmatrix}$

(e) $BB^T = \begin{bmatrix} 2 & 1 \\ -3 & 1 \\ 4 & 0 \end{bmatrix} \begin{bmatrix} 2 & -3 & 4 \\ 1 & 1 & 0 \end{bmatrix} = \begin{bmatrix} 5 & -5 & 8 \\ -5 & 10 & -12 \\ 8 & -12 & 16 \end{bmatrix}$

(f) GE is not defined

9. $A\mathbf{x} = \begin{bmatrix} 1 & 5 & 2 \\ -4 & 9 & 1 \\ 2 & 0 & 3 \end{bmatrix} \begin{bmatrix} 2 \\ -1 \\ 3 \end{bmatrix} = 2 \begin{bmatrix} 1 \\ -4 \\ 2 \end{bmatrix} - 1 \begin{bmatrix} 5 \\ 9 \\ 0 \end{bmatrix} + 3 \begin{bmatrix} 2 \\ 1 \\ 3 \end{bmatrix} = \begin{bmatrix} 3 \\ -14 \\ 13 \end{bmatrix}$

11. (a) $\begin{bmatrix} 2 & -3 & 5 \\ 9 & -1 & 1 \end{bmatrix} \begin{bmatrix} x_1 \\ x_2 \\ x_3 \end{bmatrix} = \begin{bmatrix} 7 \\ -1 \end{bmatrix}$ **(b)** $\begin{bmatrix} 1 & 1 & 1 \\ 2 & -3 & 4 \\ 1 & 5 & -2 \end{bmatrix} \begin{bmatrix} x_1 \\ x_2 \\ x_3 \end{bmatrix} = \begin{bmatrix} 4 \\ 3 \\ -2 \end{bmatrix}$

13. $5x_1 + 6x_2 - 7x_3 = 2$
  $-x_1 - 2x_2 + 3x_3 = 0$
  $4x_2 - x_3 = 3$

15. $(AB)_{23} = \mathbf{r}_2(A) \cdot \mathbf{c}_3(B) = (6)(4) + (5)(3) + (4)(5) = 59$

17. (a) $\mathbf{r}_1(AB) = \mathbf{r}_1(A)B = \begin{bmatrix} 3 & -2 & 7 \end{bmatrix} \begin{bmatrix} 6 & -2 & 4 \\ 0 & 1 & 3 \\ 7 & 7 & 5 \end{bmatrix} = \begin{bmatrix} 67 & 41 & 41 \end{bmatrix}$

(b) $\mathbf{r}_3(AB) = \mathbf{r}_3(A)B = \begin{bmatrix} 0 & 4 & 9 \end{bmatrix} \begin{bmatrix} 6 & -2 & 4 \\ 0 & 1 & 3 \\ 7 & 7 & 5 \end{bmatrix} = \begin{bmatrix} 63 & 67 & 57 \end{bmatrix}$

(c) $\mathbf{c}_2(AB) = A\mathbf{c}_2(B) = \begin{bmatrix} 3 & -2 & 7 \\ 6 & 5 & 4 \\ 0 & 4 & 9 \end{bmatrix} \begin{bmatrix} -2 \\ 1 \\ 7 \end{bmatrix} = \begin{bmatrix} 41 \\ 21 \\ 67 \end{bmatrix}$

19. (a) $\mathrm{tr}(A) = (A)_{11} + (A)_{22} + (A)_{33} = 3 + 5 + 9 = 17$

(b) $\mathrm{tr}(A^T) = (A^T)_{11} + (A^T)_{22} + (A^T)_{33} = 3 + 5 + 9 = 17$

(c) $\mathrm{tr}(AB) = (AB)_{11} + (AB)_{22} + (AB)_{33} = 67 + 21 + 57 = 145$, $\mathrm{tr}(B) = 6 + 1 + 5 = 12$; thus

$$\mathrm{tr}(AB) - \mathrm{tr}(A)\mathrm{tr}(B) = 145 - (17)(12) = 145 - 204 = -59$$

21. (a) $\mathbf{u}^T\mathbf{v} = \begin{bmatrix} -2 & 3 \end{bmatrix} \begin{bmatrix} 4 \\ 5 \end{bmatrix} = -8 + 15 = 7$ **(b)** $\mathbf{u}\mathbf{v}^T = \begin{bmatrix} -2 \\ 3 \end{bmatrix} \begin{bmatrix} 4 & 5 \end{bmatrix} = \begin{bmatrix} -8 & -10 \\ 12 & 15 \end{bmatrix}$

(c) $\mathrm{tr}(\mathbf{u}\mathbf{v}^T) = -8 + 15 = 7 = \mathbf{u}^T\mathbf{v}$ **(d)** $\mathbf{v}^T\mathbf{u} = \begin{bmatrix} 4 & 5 \end{bmatrix} \begin{bmatrix} -2 \\ 3 \end{bmatrix} = -8 + 15 = 7 = \mathbf{u}^T\mathbf{v}$

(e) $\mathrm{tr}(\mathbf{u}\mathbf{v}^T) = \mathrm{tr}(\mathbf{v}\mathbf{u}^T) = \mathbf{u} \cdot \mathbf{v} = \mathbf{v} \cdot \mathbf{u} = \mathbf{u}^T\mathbf{v} = \mathbf{v}^T\mathbf{u} = 7$

23. $\begin{bmatrix} k & 1 & 1 \end{bmatrix} \begin{bmatrix} 1 & 1 & 0 \\ 1 & 0 & 2 \\ 0 & 2 & -3 \end{bmatrix} \begin{bmatrix} k \\ 1 \\ 1 \end{bmatrix} = \begin{bmatrix} k & 1 & 1 \end{bmatrix} \begin{bmatrix} k+1 \\ k+2 \\ -1 \end{bmatrix} = k(k+1) + (k+2) - 1 = k^2 + 2k + 1 =$
$(k+1)^2 = 0$ if and only if $k = -1$.

25. Let $F = (C(DE)$. Then, from Theorem 3.1.7, the entry in the ith row and jth column of F is the dot product of the ith row of C and the jth column of DE. Thus $(F)_{23}$ can be computed as follows:

$$(F)_{23} = \mathbf{r}_2(C) \cdot \mathbf{c}_3(DE) = \mathbf{r}_2(C) \cdot D\mathbf{c}_3(E) = \begin{bmatrix} 3 & -1 \end{bmatrix} \left(\begin{bmatrix} 1 & 1 \\ -3 & 3 \end{bmatrix} \begin{bmatrix} 2 \\ 5 \end{bmatrix} \right) = \begin{bmatrix} 3 & -1 \end{bmatrix} \begin{bmatrix} 7 \\ 9 \end{bmatrix} = 21 - 9 = 12$$

27. Suppose that A is $m \times n$ and B is $r \times s$. If AB is defined, then $n = r$. On the other hand, if BA is defined, then $s = m$. Thus A is $m \times n$ and B is $n \times m$. It follows that AB is $m \times m$ and BA is $n \times n$.

29. (a) If the ith row of A is a row of zeros, then (using the row rule) $\mathbf{r}_i(AB) = \mathbf{r}_i(A)B = \mathbf{0}B = \mathbf{0}$ and so the ith row of AB is a row of zeros.

(b) If the jth column of B is a column of zeros, then (using the column rule) $\mathbf{c}_j(AB) = A\mathbf{c}_j(B) = A\mathbf{0} = \mathbf{0}$ and so the jth column of AB is a column of zeros.

31. (a) If $i \neq j$, then a_{ij} has unequal row and column numbers; that is, it is off (above or below) the main diagonal of the matrix $[a_{ij}]$; thus the matrix has zeros in all of the positions that are above or below the main diagonal.

$$[a_{ij}] = \begin{bmatrix} a_{11} & 0 & 0 & 0 & 0 & 0 \\ 0 & a_{22} & 0 & 0 & 0 & 0 \\ 0 & 0 & a_{33} & 0 & 0 & 0 \\ 0 & 0 & 0 & a_{44} & 0 & 0 \\ 0 & 0 & 0 & 0 & a_{55} & 0 \\ 0 & 0 & 0 & 0 & 0 & a_{66} \end{bmatrix}$$

(b) If $i > j$, then the entry a_{ij} has row number larger than column number; that is, it lies below the main diagonal. Thus $[a_{ij}]$ has zeros in all of the positions below the main diagonal.

(c) If $i < j$, then the entry a_{ij} has row number smaller than column number; that is, it lies above the main diagonal. Thus $[a_{ij}]$ has zeros in all of the positions above the main diagonal.

(d) If $|i - j| > 1$, then either $i - j > 1$ or $i - j < -1$; that is, either $i > j + 1$ or $j > i + 1$. The first of these inequalities says that the entry a_{ij} lies below the main diagonal and also below the "subdiagonal" consisting of entries immediately below the diagonal entries. The second inequality says that the entry a_{ij} lies above the diagonal and also above the entries immediately above the diagonal entries. Thus the matrix A has the following form:

$$A = [a_{ij}] = \begin{bmatrix} a_{11} & a_{12} & 0 & 0 & 0 & 0 \\ a_{21} & a_{22} & a_{23} & 0 & 0 & 0 \\ 0 & a_{32} & a_{33} & a_{34} & 0 & 0 \\ 0 & 0 & a_{43} & a_{44} & a_{45} & 0 \\ 0 & 0 & 0 & a_{54} & a_{55} & a_{56} \\ 0 & 0 & 0 & 0 & a_{65} & a_{66} \end{bmatrix}$$

33. The components of the matrix product $\begin{bmatrix} 3 & 4 & 3 \\ 5 & 6 & 0 \\ 2 & 9 & 4 \\ 1 & 1 & 7 \end{bmatrix} \begin{bmatrix} 1 \\ 2 \\ 3 \end{bmatrix} = \begin{bmatrix} 20 \\ 17 \\ 32 \\ 24 \end{bmatrix}$ represent the total expenditures for purchases during each of the first four months of the year. For example, the February expenditures were $(5)(\$1) + (6)(\$2) + (0)(\$3) = \$17$.

DISCUSSION AND DISCOVERY

D1. If AB has 6 rows and 8 columns, then A must have size $6 \times k$ and B must have six $k \times 8$; thus A has 6 rows and B has 8 columns.

D2. If $A = \begin{bmatrix} 0 & 0 \\ 1 & 0 \end{bmatrix}$, then $AA = \begin{bmatrix} 0 & 0 \\ 1 & 0 \end{bmatrix} \begin{bmatrix} 0 & 0 \\ 1 & 0 \end{bmatrix} = \begin{bmatrix} 0 & 0 \\ 0 & 0 \end{bmatrix}$.

D3. Let $A = \begin{bmatrix} 1 & 2 \\ 1 & 1 \end{bmatrix}$ and $B = \begin{bmatrix} 1 & 1 \\ 3 & 2 \end{bmatrix}$. In the following we illustrate three different methods for computing the product AB.

Method 1. Using Definition 3.1.6. This is the same as what is later referred to as the column rule. Since

$$\mathbf{c}_1(AB) = A\mathbf{c}_1(B) = \begin{bmatrix} 1 & 2 \\ 1 & 1 \end{bmatrix} \begin{bmatrix} 1 \\ 3 \end{bmatrix} = \begin{bmatrix} 7 \\ 4 \end{bmatrix} \quad \text{and} \quad \mathbf{c}_2(AB) = A\mathbf{c}_2(B) = \begin{bmatrix} 1 & 2 \\ 1 & 1 \end{bmatrix} \begin{bmatrix} 1 \\ 2 \end{bmatrix} = \begin{bmatrix} 5 \\ 3 \end{bmatrix}$$

we have $AB = \begin{bmatrix} 7 & 5 \\ 4 & 3 \end{bmatrix}$.

Method 2. Using Theorem 3.1.7 (the dot product rule), we have

$$AB = \begin{bmatrix} \mathbf{r}_1(A) \cdot \mathbf{c}_1(B) & \mathbf{r}_1(A) \cdot \mathbf{c}_2(B) \\ \mathbf{r}_2(A) \cdot \mathbf{c}_1(B) & \mathbf{r}_2(A) \cdot \mathbf{c}_2(B) \end{bmatrix} = \begin{bmatrix} 1+6 & 1+4 \\ 1+3 & 1+2 \end{bmatrix} = \begin{bmatrix} 7 & 5 \\ 4 & 3 \end{bmatrix}$$

Method 3. Using the row rule. Since

$$\mathbf{r}_1(AB) = \mathbf{r}_1(A)B = \begin{bmatrix} 1 & 2 \end{bmatrix} \begin{bmatrix} 1 & 1 \\ 3 & 2 \end{bmatrix} = \begin{bmatrix} 7 & 5 \end{bmatrix} \text{ and } \mathbf{r}_2(AB) = \mathbf{r}_2(A)B = \begin{bmatrix} 1 & 1 \end{bmatrix} \begin{bmatrix} 1 & 1 \\ 3 & 2 \end{bmatrix} = \begin{bmatrix} 4 & 3 \end{bmatrix}$$

we have $AB = \begin{bmatrix} 7 & 5 \\ 4 & 3 \end{bmatrix}$.

D4. The matrix $A = \begin{bmatrix} 1 & 1 & 0 \\ 1 & -1 & 0 \\ 0 & 0 & 0 \end{bmatrix}$ is the only 3×3 with this property. Here is a proof:

Since $x\mathbf{c}_1(A) + y\mathbf{c}_2(A) + z\mathbf{c}_3(A) = A \begin{bmatrix} x \\ y \\ z \end{bmatrix}$, we must have $x\mathbf{c}_1(A) + y\mathbf{c}_2(A) + z\mathbf{c}_3(A) = \begin{bmatrix} x+y \\ x-y \\ 0 \end{bmatrix}$ for

all x, y, and z. Taking $x = 1$, $y = 0$, $z = 0$, it follows that $\mathbf{c}_1(A) = \begin{bmatrix} 1 \\ 1 \\ 0 \end{bmatrix}$. Similarly, $\mathbf{c}_2(A) = \begin{bmatrix} 1 \\ -1 \\ 0 \end{bmatrix}$

and $\mathbf{c}_3(A) = \begin{bmatrix} 0 \\ 0 \\ 0 \end{bmatrix}$.

D5. There is no such matrix. Here is a proof (by contradiction):

Suppose A is a 3×3 for which $A \begin{bmatrix} x \\ y \\ z \end{bmatrix} = \begin{bmatrix} xy \\ 0 \\ 0 \end{bmatrix}$ for all x, y, and z. Then $A \begin{bmatrix} 1 \\ 1 \\ 1 \end{bmatrix} = \begin{bmatrix} 1 \\ 0 \\ 0 \end{bmatrix}$ and, on the other

hand, we must have $A \begin{bmatrix} 1 \\ 1 \\ 1 \end{bmatrix} = -A \begin{bmatrix} -1 \\ -1 \\ -1 \end{bmatrix} = -\begin{bmatrix} 1 \\ 0 \\ 0 \end{bmatrix}$. Thus there is no such matrix.

D6. **(a)** $S_1 = \begin{bmatrix} 1 & 1 \\ 1 & 1 \end{bmatrix}$ and $S_2 = \begin{bmatrix} -1 & -1 \\ -1 & -1 \end{bmatrix}$ are two square roots of the matrix $A = \begin{bmatrix} 2 & 2 \\ 2 & 2 \end{bmatrix}$.

(b) The matrices $S = \begin{bmatrix} \pm\sqrt{5} & 0 \\ 0 & 3 \end{bmatrix}$ and $S = \begin{bmatrix} \pm\sqrt{5} & 0 \\ 0 & -3 \end{bmatrix}$ are four different square roots of $B = \begin{bmatrix} 5 & 0 \\ 0 & 9 \end{bmatrix}$.

(c) Not all matrices have square roots. For example, it is easy to check that the matrix $\begin{bmatrix} -1 & 0 \\ 0 & 0 \end{bmatrix}$ has no square root.

D7. Yes, the zero matrix $A = \begin{bmatrix} 0 & 0 & 0 \\ 0 & 0 & 0 \\ 0 & 0 & 0 \end{bmatrix}$ has the property that AB has three equal rows (rows of zeros) for every 3×3 matrix B.

D8. Yes, the matrix $A = \begin{bmatrix} 2 & 0 & 0 \\ 0 & 2 & 0 \\ 0 & 0 & 2 \end{bmatrix}$ has the property that $AB = 2B$ for every 3×3 matrix B.

D9. **(a)** False. For example, if A is 2×3 and B is 3×2, then AB and BA are both defined.

(b) True. If AB and BA are both defined and A is $m \times n$, then B must be $n \times m$; thus AB is $m \times m$ and BA is $n \times n$. If, in addition, $AB + BA$ is defined then AB and BA must have the same size, i.e. $m = n$.

(c) True. From the column rule, $c_j(AB) = Ac_j(B)$. Thus if B has a column of zeros, then AB will have a column of zeros.

(d) False. For example, $\begin{bmatrix} 0 & 1 \\ 0 & 1 \end{bmatrix}\begin{bmatrix} 0 & 0 \\ 1 & 1 \end{bmatrix} = \begin{bmatrix} 1 & 1 \\ 1 & 1 \end{bmatrix}$.

(e) True. If A is $n \times m$, then A^T is $m \times n$, AA^T is $n \times n$, and A^TA is $m \times m$. Thus A^TA and AA^T both are square matrices, and so $\mathrm{tr}(A^TA)$ and $\mathrm{tr}(AA^T)$ are both defined.

(f) False. If $\mathbf{u}$ and $\mathbf{v}$ are $1 \times n$ row vectors, then $\mathbf{u}^T\mathbf{v}$ is an $n \times n$ matrix.

D10. The second column of AB is also the sum of the first and third columns:
$$\mathbf{c}_2(AB) = A\mathbf{c}_2(B) = A(\mathbf{c}_1(B) + \mathbf{c}_3(B)) = A\mathbf{c}_1(B) + A\mathbf{c}_3(B) = \mathbf{c}_1(AB) + \mathbf{c}_3(AB)$$

D11. **(a)** $\displaystyle\sum_{k=1}^{s}(a_{ik}b_{kj}) = a_{i1}b_{1j} + a_{i2}b_{2j} + \cdots + a_{is}b_{sj}$

(b) This sum represents $(AB)_{ij}$, the ijth entry of the matrix AB.

WORKING WITH PROOFS

P1. Suppose $B = [b_{ij}]$ is an $s \times n$ matrix and $\mathbf{y} = [y_1 \quad y_2 \quad \cdots \quad y_s]$. Then $\mathbf{y}B$ is the $1 \times n$ row vector
$$\mathbf{y}B = [\mathbf{y} \cdot \mathbf{c}_1(B) \quad \mathbf{y} \cdot \mathbf{c}_2(B) \quad \cdots \quad \mathbf{y} \cdot \mathbf{c}_n(B)]$$

whose jth component is $\mathbf{y} \cdot \mathbf{c}_j(B)$. On the other hand, the jth component of the vector
$$y_1\mathbf{r}_1(B) + y_2\mathbf{r}_2(B) + \cdots + y_s\mathbf{r}_s(B)$$

is $y_1b_{1j} + y_2b_{2j} + \cdots + y_sb_{sj} = \mathbf{y} \cdot \mathbf{c}_j(B)$. Thus Formula (21) is valid.

P2. Since $A\mathbf{x} = x_1\mathbf{c}_1(A) + x_2\mathbf{c}_2(A) + \cdots + x_n\mathbf{c}_n(A)$, the linear system $A\mathbf{x} = \mathbf{b}$ is equivalent to
$$x_1\mathbf{c}_1(A) + x_2\mathbf{c}_2(A) + \cdots + x_n\mathbf{c}_n(A) = \mathbf{b}$$

Thus the system $A\mathbf{x} = \mathbf{b}$ is consistent if and only if the vector $\mathbf{b}$ can be expressed as a linear combination of the column vectors of A.

EXERCISE SET 3.2

1. **(a)** $(A+B)+C = \begin{bmatrix} 10 & -4 & -2 \\ 0 & 5 & 7 \\ 2 & -6 & 10 \end{bmatrix} + \begin{bmatrix} 0 & -2 & 3 \\ 1 & 7 & 4 \\ 3 & 5 & 9 \end{bmatrix} = \begin{bmatrix} 10 & -6 & 1 \\ 1 & 12 & 11 \\ 5 & -1 & 19 \end{bmatrix}$

$A+(B+C) = \begin{bmatrix} 2 & -1 & 3 \\ 0 & 4 & 5 \\ -2 & 1 & 4 \end{bmatrix} + \begin{bmatrix} 8 & -5 & -2 \\ 1 & 8 & 6 \\ 7 & -2 & 15 \end{bmatrix} = \begin{bmatrix} 10 & -6 & 1 \\ 1 & 12 & 11 \\ 5 & -1 & 19 \end{bmatrix}$

(b) $(AB)C = \begin{bmatrix} 28 & -28 & 6 \\ 20 & -31 & 38 \\ 0 & -21 & 36 \end{bmatrix} \begin{bmatrix} 0 & -2 & 3 \\ 1 & 7 & 4 \\ 3 & 5 & 9 \end{bmatrix} = \begin{bmatrix} -10 & -222 & 26 \\ 83 & -67 & 278 \\ 87 & 33 & 240 \end{bmatrix}$

$A(BC) = \begin{bmatrix} 2 & -1 & 3 \\ 0 & 4 & 5 \\ -2 & 1 & 4 \end{bmatrix} \begin{bmatrix} -18 & -62 & -33 \\ 7 & 17 & 22 \\ 11 & -27 & 38 \end{bmatrix} = \begin{bmatrix} -10 & -222 & 26 \\ 83 & -67 & 278 \\ 87 & 33 & 240 \end{bmatrix}$

(c) $(a+b)C = (-3)\begin{bmatrix} 0 & -2 & 3 \\ 1 & 7 & 4 \\ 3 & 5 & 9 \end{bmatrix} = \begin{bmatrix} 0 & 6 & -9 \\ -3 & -21 & -12 \\ -9 & -15 & -27 \end{bmatrix}$

$aC+bC = \begin{bmatrix} 0 & -8 & 12 \\ 4 & 28 & 16 \\ 12 & 20 & 36 \end{bmatrix} + \begin{bmatrix} 0 & 14 & -21 \\ -7 & -49 & -28 \\ -21 & -35 & -63 \end{bmatrix} = \begin{bmatrix} 0 & 6 & -9 \\ -3 & -21 & -12 \\ -9 & -15 & -27 \end{bmatrix}$

(d) $a(B-C) = (4)\begin{bmatrix} 8 & -1 & -8 \\ -1 & -6 & -2 \\ 1 & -12 & -3 \end{bmatrix} = \begin{bmatrix} 32 & -4 & -32 \\ -4 & -24 & -8 \\ 4 & -48 & -12 \end{bmatrix}$

$aB-aC = \begin{bmatrix} 32 & -12 & -20 \\ 0 & 4 & 8 \\ 16 & -28 & 24 \end{bmatrix} - \begin{bmatrix} 0 & -8 & 12 \\ 4 & 28 & 16 \\ 12 & 20 & 36 \end{bmatrix} = \begin{bmatrix} 32 & -4 & -32 \\ -4 & -24 & -8 \\ 4 & -48 & -12 \end{bmatrix}$

3. **(a)** $(A^T)^T = \begin{bmatrix} 2 & 0 & -2 \\ -1 & 4 & 1 \\ 3 & 5 & 4 \end{bmatrix}^T = \begin{bmatrix} 2 & -1 & 3 \\ 0 & 4 & 5 \\ -2 & 1 & 4 \end{bmatrix} = A$

(b) $(A+B)^T = \begin{bmatrix} 10 & -4 & -2 \\ 0 & 5 & 7 \\ 2 & -6 & 10 \end{bmatrix}^T = \begin{bmatrix} 10 & 0 & 2 \\ -4 & 5 & -6 \\ -2 & 7 & 10 \end{bmatrix}$

$A^T+B^T = \begin{bmatrix} 2 & 0 & -2 \\ -1 & 4 & 1 \\ 3 & 5 & 4 \end{bmatrix} + \begin{bmatrix} 8 & 0 & 4 \\ -3 & 1 & -7 \\ -5 & 2 & 6 \end{bmatrix} = \begin{bmatrix} 10 & 0 & 2 \\ -4 & 5 & -6 \\ -2 & 7 & 10 \end{bmatrix}$

(c) $(3C)^T = \begin{bmatrix} 0 & -6 & 9 \\ 3 & 21 & 12 \\ 9 & 15 & 27 \end{bmatrix}^T = \begin{bmatrix} 0 & 3 & 9 \\ -6 & 21 & 15 \\ 9 & 12 & 27 \end{bmatrix} = (3)\begin{bmatrix} 0 & 1 & 3 \\ -2 & 7 & 5 \\ 3 & 4 & 9 \end{bmatrix} = 3C^T$

(d) $(AB)^T = \begin{bmatrix} 28 & -28 & 6 \\ 20 & -31 & 38 \\ 0 & -21 & 36 \end{bmatrix}^T = \begin{bmatrix} 28 & 20 & 0 \\ -28 & -31 & -21 \\ 6 & 38 & 36 \end{bmatrix}$

$B^T A^T = \begin{bmatrix} 8 & 0 & 4 \\ -3 & 1 & -7 \\ -5 & 2 & 6 \end{bmatrix} \begin{bmatrix} 2 & 0 & -2 \\ -1 & 4 & 1 \\ 3 & 5 & 4 \end{bmatrix} = \begin{bmatrix} 28 & 20 & 0 \\ -28 & -31 & -21 \\ 6 & 38 & 36 \end{bmatrix}$

5. (a) $\mathrm{tr}(A) = \mathrm{tr}\left(\begin{bmatrix} 2 & -1 & 3 \\ 0 & 4 & 5 \\ -2 & 1 & 4 \end{bmatrix} \right) = 2 + 4 + 4 = 10$, and

$\mathrm{tr}(A^T) = \mathrm{tr}\left(\begin{bmatrix} 2 & 0 & -2 \\ -1 & 4 & 1 \\ 3 & 5 & 4 \end{bmatrix} \right) = 2 + 4 + 4 = 10.$

(b) $\mathrm{tr}(3A) = \mathrm{tr}\left(\begin{bmatrix} 6 & -3 & 9 \\ 0 & 12 & 15 \\ -6 & 3 & 12 \end{bmatrix} \right) = 6 + 12 + 12 = 30 = 3\mathrm{tr}(A).$

(c) $\mathrm{tr}(A) = 10, \mathrm{tr}(B) = \mathrm{tr}\left(\begin{bmatrix} 8 & -3 & -5 \\ 0 & 1 & 2 \\ 4 & -7 & 6 \end{bmatrix} \right) = 8 + 1 + 6 = 15,$

and $\mathrm{tr}(A + B) = \mathrm{tr}\left(\begin{bmatrix} 10 & -4 & -2 \\ 0 & 5 & 7 \\ 2 & -6 & 10 \end{bmatrix} \right) = 25$; thus $\mathrm{tr}(A + B) = \mathrm{tr}(A) + \mathrm{tr}(B).$

(d) $\mathrm{tr}(AB) = \mathrm{tr}\left(\begin{bmatrix} 28 & -28 & 6 \\ 20 & -31 & 38 \\ 0 & -21 & 36 \end{bmatrix} \right) = 28 - 31 + 36 = 33,$

and $\mathrm{tr}(BA) = \mathrm{tr}\left(\begin{bmatrix} 26 & -25 & -11 \\ -4 & 6 & 13 \\ -4 & -26 & 1 \end{bmatrix} \right) = 26 + 6 + 1 = 33$; thus $\mathrm{tr}(AB) = \mathrm{tr}(BA).$

7. (a) A matrix X satisfies the equation $\mathrm{tr}(B)A + 3X = BC$ if and only if $3X = BC - \mathrm{tr}(B)A$, i.e.

$3X = \begin{bmatrix} 8 & -3 & -5 \\ 0 & 1 & 2 \\ 4 & -7 & 6 \end{bmatrix} \begin{bmatrix} 0 & -2 & 3 \\ 1 & 7 & 4 \\ 3 & 5 & 9 \end{bmatrix} - 15 \begin{bmatrix} 2 & -1 & 3 \\ 0 & 4 & 5 \\ -2 & 1 & 4 \end{bmatrix}$

$= \begin{bmatrix} -18 & -62 & -33 \\ 7 & 17 & 22 \\ 11 & -42 & 38 \end{bmatrix} - \begin{bmatrix} 30 & -15 & 45 \\ 0 & 60 & 75 \\ -30 & 15 & 60 \end{bmatrix} = \begin{bmatrix} -48 & -47 & -78 \\ 7 & -47 & -53 \\ 41 & -42 & -22 \end{bmatrix}$

in which case we have $X = \dfrac{1}{3} \begin{bmatrix} -48 & -47 & -78 \\ 7 & -47 & -53 \\ 41 & -42 & -22 \end{bmatrix}$

(b) A matrix X satisfies the equation $B + (A + X)^T = C$ if and only if

$$(A + X)^T = C - B$$
$$A + X = ((A + X)^T)^T = (C - B)^T$$
$$X = (C - B)^T - A = C^T - B^T - A$$

Thus $X = \begin{bmatrix} 0 & 1 & 3 \\ -2 & 7 & 5 \\ 3 & 4 & 9 \end{bmatrix} - \begin{bmatrix} 8 & 0 & 4 \\ -3 & 1 & -7 \\ -5 & 2 & 6 \end{bmatrix} - \begin{bmatrix} 2 & -1 & 3 \\ 0 & 4 & 5 \\ -2 & 1 & 4 \end{bmatrix} = \begin{bmatrix} -10 & 2 & -4 \\ 1 & 2 & 7 \\ 10 & 1 & -1 \end{bmatrix}.$

9. (a) $\det(A) = 6 - 5 = 1$ $\qquad$ $A^{-1} = \begin{bmatrix} 2 & -1 \\ -5 & 3 \end{bmatrix}$

(b) $\det(A^{-1}) = 1$ $\qquad$ $(A^{-1})^{-1} = \begin{bmatrix} 3 & 1 \\ 5 & 2 \end{bmatrix} = A$

(c) $A^T = \begin{bmatrix} 3 & 5 \\ 1 & 2 \end{bmatrix}, \det(A^T) = 1$ $\qquad$ $(A^T)^{-1} = \begin{bmatrix} 2 & -5 \\ -1 & 3 \end{bmatrix} = (A^{-1})^T$

(d) $2A = \begin{bmatrix} 6 & 2 \\ 10 & 4 \end{bmatrix}, \det(2A) = 4$ $\qquad$ $(2A)^{-1} = \frac{1}{4}\begin{bmatrix} 4 & -2 \\ -10 & 6 \end{bmatrix} = \frac{1}{2}\begin{bmatrix} 2 & -1 \\ -5 & 3 \end{bmatrix} = \frac{1}{2}A^{-1}$

11. (a) $(AB)^{-1} = \begin{bmatrix} 10 & -5 \\ 18 & -7 \end{bmatrix}^{-1} = \frac{1}{20}\begin{bmatrix} -7 & 5 \\ -18 & 10 \end{bmatrix}$

$B^{-1}A^{-1} = \frac{1}{20}\begin{bmatrix} 4 & 3 \\ -4 & 2 \end{bmatrix}\begin{bmatrix} 2 & -1 \\ -5 & 3 \end{bmatrix} = \frac{1}{20}\begin{bmatrix} -7 & 5 \\ -18 & 10 \end{bmatrix}$

(b) $(ABC)^{-1} = \begin{bmatrix} 70 & 45 \\ 122 & 79 \end{bmatrix}^{-1} = \frac{1}{40}\begin{bmatrix} 79 & -45 \\ -122 & 70 \end{bmatrix}$

$C^{-1}B^{-1}A^{-1} = \frac{1}{2}\begin{bmatrix} -1 & -4 \\ 2 & 6 \end{bmatrix}\frac{1}{20}\begin{bmatrix} 4 & 3 \\ -4 & 2 \end{bmatrix}\begin{bmatrix} 2 & -1 \\ -5 & 3 \end{bmatrix} = \frac{1}{40}\begin{bmatrix} 79 & -45 \\ -122 & 70 \end{bmatrix}$

13. $X = A^{-1}(BC - B) = \begin{bmatrix} 2 & -1 \\ -5 & 3 \end{bmatrix}\left(\begin{bmatrix} 18 & 11 \\ 16 & 12 \end{bmatrix} - \begin{bmatrix} 2 & -3 \\ 4 & 4 \end{bmatrix}\right) = \begin{bmatrix} 2 & -1 \\ -5 & 3 \end{bmatrix}\begin{bmatrix} 16 & 14 \\ 12 & 8 \end{bmatrix} = \begin{bmatrix} 20 & 20 \\ -44 & -46 \end{bmatrix}$

15. (a) $A^{-2} = (A^{-1})^2 = \begin{bmatrix} 2 & -1 \\ -5 & 3 \end{bmatrix}^2 = \begin{bmatrix} 9 & -5 \\ -25 & 14 \end{bmatrix}$

(b) $p(A) = A + 2I = \begin{bmatrix} 3 & 1 \\ 5 & 2 \end{bmatrix} + 2\begin{bmatrix} 1 & 0 \\ 0 & 1 \end{bmatrix} = \begin{bmatrix} 5 & 1 \\ 5 & 4 \end{bmatrix}$

(c) $p(A) = A^2 - 2A + I = \begin{bmatrix} 14 & 5 \\ 25 & 9 \end{bmatrix} - 2\begin{bmatrix} 3 & 1 \\ 5 & 2 \end{bmatrix} + \begin{bmatrix} 1 & 0 \\ 0 & 1 \end{bmatrix} = \begin{bmatrix} 9 & 3 \\ 15 & 6 \end{bmatrix}$

17. $(AB)^{-1}(AC^{-1})(D^{-1}C)^{-1}D^{-1} = B^{-1}A^{-1}AC^{-1}C^{-1}DD^{-1} = B^{-1} = \frac{1}{20}\begin{bmatrix} 4 & 3 \\ -4 & 2 \end{bmatrix}$

19. (a) Given that $A^{-1} = \begin{bmatrix} 2 & -1 \\ 3 & 5 \end{bmatrix}$, and noting that $\det(A^{-1}) = 13$, we have $A = (A^{-1})^{-1} = \frac{1}{13}\begin{bmatrix} 5 & 1 \\ -3 & 2 \end{bmatrix}$.

(b) Given that $(7A)^{-1} = \begin{bmatrix} 2 & -1 \\ 3 & 5 \end{bmatrix}$, we have $A = \frac{1}{7} \begin{bmatrix} -3 & 7 \\ 1 & -2 \end{bmatrix}^{-1} = \frac{1}{7} \begin{bmatrix} 2 & 7 \\ 1 & 3 \end{bmatrix}$.

21. The matrix $A = \begin{bmatrix} c & 1 \\ c & c \end{bmatrix}$ is invertible if and only if $\det(A) = c^2 - c \neq 0$, i.e. if and only if $c \neq 0, 1$.

23. One such example is $A = \begin{bmatrix} 1 & 2 & 3 \\ 2 & 2 & 2 \\ 3 & 2 & 3 \end{bmatrix}$. In general, any matrix of the form $\begin{bmatrix} a & d & e \\ d & b & f \\ e & f & c \end{bmatrix}$.

25. Let $X = [x_{ij}]_{3\times 3}$. Then $AX = I$ if and only if

$$\begin{bmatrix} 1 & 0 & 1 \\ 1 & 1 & 0 \\ 0 & 1 & 1 \end{bmatrix} \begin{bmatrix} x_{11} & x_{12} & x_{13} \\ x_{21} & x_{22} & x_{23} \\ x_{31} & x_{32} & x_{33} \end{bmatrix} = \begin{bmatrix} 1 & 0 & 0 \\ 0 & 1 & 0 \\ 0 & 0 & 1 \end{bmatrix}$$

i.e. if and only if the entries of X satisfy the following system of equations:

$$
\begin{array}{rcl}
x_{11} \phantom{+ x_{12}} + x_{31} & = & 1 \\
x_{12} \phantom{+ x_{21}} + x_{32} & = & 0 \\
x_{13} \phantom{+ x_{21}} + x_{33} & = & 0 \\
x_{11} + x_{21} & = & 0 \\
x_{12} + x_{22} & = & 1 \\
x_{13} + x_{23} & = & 0 \\
x_{21} + x_{31} & = & 0 \\
x_{22} + x_{32} & = & 0 \\
x_{23} + x_{33} & = & 1
\end{array}
$$

This system has the unique solution $x_{11} = x_{12} = x_{22} = x_{23} = x_{31} = x_{33} = \frac{1}{2}$ and $x_{13} = x_{21} = x_{32} = -\frac{1}{2}$. Thus A is invertible and $A^{-1} = \begin{bmatrix} \frac{1}{2} & \frac{1}{2} & -\frac{1}{2} \\ -\frac{1}{2} & \frac{1}{2} & \frac{1}{2} \\ \frac{1}{2} & -\frac{1}{2} & \frac{1}{2} \end{bmatrix}$.

27. **(a)** $\mathbf{rc} = \begin{bmatrix} 0 & -2 & 3 \end{bmatrix} \begin{bmatrix} 8 \\ 0 \\ 4 \end{bmatrix} = 12$ $\operatorname{tr}(\mathbf{cr}) = \operatorname{tr}\left(\begin{bmatrix} 0 & -16 & 24 \\ 0 & 0 & 0 \\ 0 & -8 & 12 \end{bmatrix} \right) = 12$

(b) $A\mathbf{u} \cdot \mathbf{v} = \left(\begin{bmatrix} 3 & 2 & -1 \\ 1 & 5 & 0 \\ -2 & 4 & 6 \end{bmatrix} \begin{bmatrix} 3 \\ -2 \\ 4 \end{bmatrix} \right) \cdot \begin{bmatrix} 1 \\ -1 \\ 3 \end{bmatrix} = \begin{bmatrix} 1 \\ -7 \\ 10 \end{bmatrix} \cdot \begin{bmatrix} 1 \\ -1 \\ 3 \end{bmatrix} = 1 + 7 + 30 = 38$

$\mathbf{u} \cdot A^T \mathbf{v} = \begin{bmatrix} 3 \\ -2 \\ 4 \end{bmatrix} \cdot \left(\begin{bmatrix} 3 & 1 & -2 \\ 2 & 5 & 4 \\ -1 & 0 & 6 \end{bmatrix} \begin{bmatrix} 1 \\ -1 \\ 3 \end{bmatrix} \right) = \begin{bmatrix} 3 \\ -2 \\ 4 \end{bmatrix} \cdot \begin{bmatrix} -4 \\ 9 \\ 17 \end{bmatrix} = -12 - 18 + 68 = 38$

29. If A is invertible and $AB = AC$ then, using the Associative law (Theorem 3.2.2(a)), we have

$$B = IB = (A^{-1}A)B = A^{-1}(AB) = A^{-1}(AC) = (A^{-1}A)C = IC = C$$

31. (a) If $A = \begin{bmatrix} \cos\theta & \sin\theta \\ -\sin\theta & \cos\theta \end{bmatrix}$, then $\det(A) = \cos^2\theta + \sin^2\theta = 1$. Thus A is invertible and

$$A^{-1} = \begin{bmatrix} \cos\theta & -\sin\theta \\ \sin\theta & \cos\theta \end{bmatrix}$$

for every value of θ.

(b) The given system can be written as $\begin{bmatrix} x \\ y \end{bmatrix} = \begin{bmatrix} \cos\theta & \sin\theta \\ -\sin\theta & \cos\theta \end{bmatrix}\begin{bmatrix} x' \\ y' \end{bmatrix}$, and so

$$\begin{bmatrix} x' \\ y' \end{bmatrix} = \begin{bmatrix} \cos\theta & \sin\theta \\ -\sin\theta & \cos\theta \end{bmatrix}^{-1}\begin{bmatrix} x \\ y \end{bmatrix} = \begin{bmatrix} \cos\theta & -\sin\theta \\ \sin\theta & \cos\theta \end{bmatrix}\begin{bmatrix} x \\ y \end{bmatrix}$$

i.e. $x' = x\cos\theta - y\sin\theta$ and $y' = x\sin\theta + y\cos\theta$.

33. (a) If A and B are invertible square matrices of the same size, and if $A + B$ is also invertible, then

$$A(A^{-1} + B^{-1})B(A + B)^{-1} = (I + AB^{-1})B(A + B)^{-1} = (B + A)(A + B)^{-1} = I$$

(b) From part (a) it follows that $A(A^{-1} + B^{-1})B = A + B$, and so $A^{-1} + B^{-1} = A^{-1}(A + B)B^{-1}$. Thus the matrix $A^{-1} + B^{-1}$ is invertible and $(A^{-1} + B^{-1})^{-1} = B(A + B)^{-1}A$.

35. If $A = \begin{bmatrix} a & b \\ c & d \end{bmatrix}$ and $p(x) = x^2 - (a + b)x + (ad - bc)$, then

$$p(A) = A^2 - (a + b)A + (ad - bc)I = \begin{bmatrix} a & b \\ c & d \end{bmatrix}^2 - (a + b)\begin{bmatrix} a & b \\ c & d \end{bmatrix} + (ad - bc)\begin{bmatrix} 1 & 0 \\ 0 & 1 \end{bmatrix}$$

$$= \begin{bmatrix} a^2 + bc & ab + bd \\ ca + dc & cb + d^2 \end{bmatrix} - \begin{bmatrix} a^2 + da & ab + db \\ ac + dc & ad + d^2 \end{bmatrix} + (ad - bc)\begin{bmatrix} 1 & 0 \\ 0 & 1 \end{bmatrix}$$

$$= \begin{bmatrix} bc - da & 0 \\ 0 & cb - ad \end{bmatrix} + (ad - bc)\begin{bmatrix} 1 & 0 \\ 0 & 1 \end{bmatrix} = \begin{bmatrix} 0 & 0 \\ 0 & 0 \end{bmatrix}$$

37. The adjacency matrix A and its square (computed using the row-column rule) are as follows:

$$A = \begin{bmatrix} 0 & 1 & 1 & 1 & 0 & 0 & 0 \\ 0 & 0 & 0 & 0 & 1 & 1 & 0 \\ 0 & 0 & 0 & 0 & 0 & 1 & 0 \\ 0 & 0 & 0 & 0 & 0 & 1 & 1 \\ 0 & 0 & 0 & 0 & 0 & 0 & 1 \\ 0 & 0 & 0 & 0 & 0 & 0 & 1 \\ 0 & 0 & 0 & 0 & 0 & 0 & 0 \end{bmatrix} \qquad A^2 = \begin{bmatrix} 0 & 0 & 0 & 0 & 1 & 3 & 1 \\ 0 & 0 & 0 & 0 & 0 & 0 & 2 \\ 0 & 0 & 0 & 0 & 0 & 0 & 1 \\ 0 & 0 & 0 & 0 & 0 & 0 & 1 \\ 0 & 0 & 0 & 0 & 0 & 0 & 0 \\ 0 & 0 & 0 & 0 & 0 & 0 & 0 \\ 0 & 0 & 0 & 0 & 0 & 0 & 0 \end{bmatrix}$$

The entry in the ijth position of the matrix A^2 represents the number of ways of traveling from i to j with one intermediate stop. For example, there are three such ways of traveling from 1 to 6 (through 2, 3 or 4) and two such ways of traveling from 2 to 7 (through 5 or 6).

In general the entry in the ijth position of the matrix A^n represents the number of ways of traveling from i to j with exactly $n - 1$ intermediate stops.

DISCUSSION AND DISCOVERY

D1. (a) Let $A = \begin{bmatrix} 1 & 1 \\ 0 & 1 \end{bmatrix}$ and $B = \begin{bmatrix} 1 & 0 \\ 1 & 1 \end{bmatrix}$. Then $A^2 = \begin{bmatrix} 1 & 2 \\ 0 & 1 \end{bmatrix}$, $B^2 = \begin{bmatrix} 1 & 0 \\ 2 & 1 \end{bmatrix}$, and $A^2 - B^2 = \begin{bmatrix} 0 & 2 \\ -2 & 0 \end{bmatrix}$. On the other hand, $(A + B)(A - B) = \begin{bmatrix} 2 & 1 \\ 1 & 2 \end{bmatrix}\begin{bmatrix} 0 & 1 \\ -1 & 0 \end{bmatrix} = \begin{bmatrix} -1 & 2 \\ -2 & 1 \end{bmatrix}$.

(b) $(A + B)(A - B) = A^2 - AB + BA - B^2$

(c) $(A + B)(A - B) = A^2 - B^2$ if and only if $AB = BA$.

D2. If A is any one of the eight matrices $A = \begin{bmatrix} \pm 1 & 0 & 0 \\ 0 & \pm 1 & 0 \\ 0 & 0 & \pm 1 \end{bmatrix}$, then $A^2 = I_3$.

D3. **(a)** If $A^2 + 2A + I = 0$, then $I = -A^2 - 2A = A(-A - 2I)$; thus A is invertible and we have $A^{-1} = -A - 2I$.

(b) If $p(A) = a_n A^n + a_{n-1} A^{n-1} + \cdots + a_1 A + a_0 I = 0$, where $a_0 \neq 0$, then we have

$$A\left(-\frac{a_n}{a_0} A^{n-1} - \frac{a_{n-1}}{a_0} A^{n-2} - \cdots - \frac{a_1}{a_0} I\right) = I$$

Thus A is invertible, with $A^{-1} = -\frac{a_n}{a_0} A^{n-1} - \frac{a_{n-1}}{a_0} A^{n-2} - \cdots - \frac{a_1}{a_0} I$.

D4. No. First note that if A^3 is defined then A must be square. Thus if $A^3 = AA^2 = I$, it follows that A is invertible with $A^{-1} = A^2$.

D5. **(a)** False. $(AB)^2 = (AB)(AB) = A(BA)B$. If A and B commute then $(AB)^2 = A^2 B^2$, but if $BA \neq AB$ then this will not in general be true.

(b) True. Expanding both expressions, we have $(A - B)^2 = A^2 - AB - BA + B^2$ and $(B - A)^2 = B^2 - BA - AB + A^2$; thus $(A - B)^2 = (B - A)^2$.

(c) True. The basic fact (from Theorem 3.2.11) is that $(A^T)^{-1} = (A^{-1})^T$, and from this it follows that $(A^{-n})^T = ((A^n)^{-1})^T = ((A^n)^T)^{-1} = ((A^T)^n)^{-1} = (A^T)^{-n}$.

(d) False. For example, if $A = \begin{bmatrix} 1 & 1 \\ 0 & 1 \end{bmatrix}$ and $B = \begin{bmatrix} 1 & 0 \\ 1 & 1 \end{bmatrix}$ then $\operatorname{tr}(AB) = \operatorname{tr}\left(\begin{bmatrix} 2 & 1 \\ 1 & 1 \end{bmatrix}\right) = 3$, whereas $\operatorname{tr}(A)\operatorname{tr}(B) = (2)(2) = 4$.

(e) False. For example, if $B = -A$ then $A + B = 0$ is not invertible (whether A is invertible or not).

D6. **(a)** If A is invertible, then the system $A\mathbf{x} = \mathbf{b}$ has a unique solution for every vector $\mathbf{b}$ in R^3, namely $\mathbf{x} = A^{-1}\mathbf{b}$. Let $\mathbf{x}_1$, $\mathbf{x}_2$, and $\mathbf{x}_3$ be the solutions of $A\mathbf{x} = \mathbf{e}_1$, $A\mathbf{x} = \mathbf{e}_2$, and $A\mathbf{x} = \mathbf{e}_3$ respectively, and let B be the matrix having these vectors as its columns; $B = [\mathbf{x}_1 \ \ \mathbf{x}_2 \ \ \mathbf{x}_3]$. Then we have $AB = A[\mathbf{x}_1 \ \ \mathbf{x}_2 \ \ \mathbf{x}_3] = [A\mathbf{x}_1 \ \ A\mathbf{x}_2 \ \ A\mathbf{x}_3] = [\mathbf{e}_1 \ \ \mathbf{e}_2 \ \ \mathbf{e}_3] = I$. Thus $A^{-1} = B = [\mathbf{x}_1 \ \ \mathbf{x}_2 \ \ \mathbf{x}_3]$.

(b) From part (a), the columns of the matrix A^{-1} are the solutions of $A\mathbf{x} = \mathbf{e}_1$, $A\mathbf{x} = \mathbf{e}_2$, and $A\mathbf{x} = \mathbf{e}_3$. The augmented matrix of $A\mathbf{x} = \mathbf{e}_1$ is

$$\begin{bmatrix} 1 & 2 & 0 & 1 \\ 1 & 3 & 0 & 0 \\ 0 & 0 & 4 & 0 \end{bmatrix}$$

and the reduced row echelon form of this matrix is

$$\begin{bmatrix} 1 & 0 & 0 & 3 \\ 0 & 1 & 0 & -1 \\ 0 & 0 & 1 & 0 \end{bmatrix}$$

Thus $\mathbf{x}_1 = \begin{bmatrix} 3 \\ -1 \\ 0 \end{bmatrix}$. Similarly, $\mathbf{x}_2 = \begin{bmatrix} -2 \\ 1 \\ 0 \end{bmatrix}$ and $\mathbf{x}_3 = \begin{bmatrix} 0 \\ 0 \\ \frac{1}{4} \end{bmatrix}$. Thus $A^{-1} = \begin{bmatrix} 3 & -2 & 0 \\ -1 & 1 & 0 \\ 0 & 0 & \frac{1}{4} \end{bmatrix}$.

D7. The matrices $\begin{bmatrix} 1 & 0 \\ 0 & 1 \end{bmatrix}$, $\begin{bmatrix} 1 & 0 \\ 1 & 1 \end{bmatrix}$, and $\begin{bmatrix} 1 & 1 \\ 0 & 1 \end{bmatrix}$ have determinant equal to 1. The matrices $\begin{bmatrix} 0 & 1 \\ 1 & 0 \end{bmatrix}$, $\begin{bmatrix} 0 & 1 \\ 1 & 1 \end{bmatrix}$, and $\begin{bmatrix} 1 & 1 \\ 1 & 0 \end{bmatrix}$ have determinant equal to -1. These six matrices are invertible. The other ten matrices all have determinant equal to 0, and thus are not invertible.

D8. True. If $AB = BA$, then $B^{-1}A^{-1} = (AB)^{-1} = (BA)^{-1} = A^{-1}B^{-1}$.

WORKING WITH PROOFS

P1. We proceed as in the proof of part (b) given in the text. It is clear that the matrices $(ab)A$ and $a(bA)$ have the same size. The following shows that corresponding entries are equal:

$$[(ab)A]_{ij} = (ab)a_{ij} = a(ba_{ij}) = a[bA]_{ij} = [a(bA)]_{ij}$$

P2. The following shows that corresponding entries on the two sides are equal:

$$[a(A + B)]_{ij} = a(a_{ij} + b_{ij}) = aa_{ij} + ab_{ij} = [aA]_{ij} + [aB]_{ij} = [aA + aB]_{ij}$$

P3. The argument that the matrices $A(B - C)$ and $AB - AC$ must have the same size is the same as in the proof of part (b) given in the text. The following shows that corresponding column vectors are equal:

$$\mathbf{c}_j[A(B - C)] = A\mathbf{c}_j(B - C) = A(\mathbf{c}_j(B) - \mathbf{c}_j(C))$$
$$= A\mathbf{c}_j(B) - A\mathbf{c}_j(C) = \mathbf{c}_j(AB) - \mathbf{c}_j(AC) = \mathbf{c}_j(AB - AC)$$

P4. These three matrices clearly have the same size. The following shows that corresponding column vectors are equal:

$$\mathbf{c}_j[a(BC)] = a\mathbf{c}_j(BC) = a(B\mathbf{c}_j(C)) = (aB)\mathbf{c}_j(C) = \mathbf{c}_j[(aB)C]$$
$$\mathbf{c}_j[a(BC)] = a\mathbf{c}_j(BC) = a(B\mathbf{c}_j(C)) = B(a\mathbf{c}_j(C)) = B(\mathbf{c}_j(aC)) = \mathbf{c}_j[B(aC)]$$

P5. If $cA = 0$ and $c \neq 0$ then, using Theorem 3.2.1(c)), we have $A = 1A = \left(\left(\frac{1}{c}\right)c\right)A = \frac{1}{c}(cA) = \frac{1}{c}0 = 0$.

P6. (a) If A is invertible, then $AA^{-1} = I$ and $A^{-1}A = I$; thus A^{-1} is invertible and $(A^{-1})^{-1} = A$.

(b) If A is invertible then, from Theorem 3.2.8, A^2 is invertible and $(A^2)^{-1} = A^{-1}A^{-1} = (A^{-1})^2$. It then follows that A^3 is invertible and $(A^3)^{-1} = (A^2)^{-1}A^{-1} = (A^{-1})^2A^{-1} = (A^{-1})^3$, etc. In general, from the remark following Theorem 3.2.8, $(A^n)^{-1} = A^{-1} \cdots A^{-1}A^{-1} = (A^{-1})^n$.

P7. (a) $\quad AA^{-1} = \begin{bmatrix} a & b \\ c & d \end{bmatrix} \left(\dfrac{1}{ad - bc} \begin{bmatrix} d & -b \\ -c & a \end{bmatrix} \right) = \dfrac{1}{ad - bc} \begin{bmatrix} ad - bc & 0 \\ 0 & -cb + da \end{bmatrix} = \begin{bmatrix} 1 & 0 \\ 0 & 1 \end{bmatrix}$

$\quad A^{-1}A = \left(\dfrac{1}{ad - bc} \begin{bmatrix} d & -b \\ -c & a \end{bmatrix} \right) \begin{bmatrix} a & b \\ c & d \end{bmatrix} = \dfrac{1}{ad - bc} \begin{bmatrix} da - bc & 0 \\ 0 & -cb + ad \end{bmatrix} = \begin{bmatrix} 1 & 0 \\ 0 & 1 \end{bmatrix}$

(b) Let $A = \begin{bmatrix} a & b \\ c & d \end{bmatrix}$ where $ad - bc = 0$. Then A is invertible if and only if there are scalars e, f, g, and h, such that

$$\begin{bmatrix} a & b \\ c & d \end{bmatrix} \begin{bmatrix} e & f \\ g & h \end{bmatrix} = \begin{bmatrix} 1 & 0 \\ 0 & 1 \end{bmatrix}$$

i.e. if and only if the following system of equations is consistent:

$$ae + bg \qquad\qquad\quad = 1$$
$$ce + dg \qquad\qquad\quad = 0$$
$$af + bh = 0$$
$$cf + dh = 1$$

Multiply the first equation by d, multiply the second equation by b, and subtract. This leads to

$$(da - bc)e = d$$

and from this we conclude that $d = 0$ is a necessary condition in order for the system to be consistent. It then follows (since $ad - bc = 0$) that $bc = 0$ and so either $b = 0$ or $c = 0$. Let us assume that $b = 0$ (the case $c = 0$ can be handled similarly). Then the equations reduce to

$$ae \qquad\quad = 1$$
$$ce \qquad\quad = 0$$
$$af = 0$$
$$cf = 1$$

and these equations are easily seen to be inconsistent. Why? From $ae = 1$ we conclude that $e \neq 0$; and from $ce = 0$ we conclude that $c = 0$ or $e = 0$. From this it follows that c must be equal to 0. But this is inconsistent with $cf = 1$! In summary, we have shown that if $ad - bc = 0$, then the system of equations has no solution and so the matrix A is not invertible.

P8. If $A = [a_{ij}]$ and $B = [b_{ij}]$, then the kth diagonal entry of AB is $(AB)_{kk} = \sum_{l=1}^{n} a_{kl}b_{lk}$. Thus

$$\operatorname{tr}(AB) = \sum_{k=1}^{n} (AB)_{kk} = \sum_{k=1}^{n}\sum_{l=1}^{n} a_{kl}b_{lk} = \sum_{l=1}^{n}\sum_{k=1}^{n} b_{lk}a_{kl} = \sum_{l=1}^{n}(BA)_{ll} = \operatorname{tr}(BA)$$

EXERCISE SET 3.3

1. **(a)** This matrix is elementary; it is obtained from I_2 by adding -5 times the first row to the second row.
   (b) This matrix is not elementary since two row operations are needed to obtain it from I_2 (follow the one in part (a) by interchanging the rows).
   (c) This matrix is elementary; it is obtained from I_3 by interchanging the first and third rows.
   (d) This matrix is not invertible, and therefore not elementary, since it has a row of zeros.

3. **(a)** Add 3 times the first row to the second row.
   (b) Multiply the third row by $\frac{1}{3}$.
   (c) Interchange the first and fourth rows.
   (d) Add $\frac{1}{7}$ times the third row to the first row.

5. **(a)** $\begin{bmatrix} 1 & 0 \\ -3 & 1 \end{bmatrix}^{-1} = \begin{bmatrix} 1 & 0 \\ 3 & 1 \end{bmatrix}$

   (b) $\begin{bmatrix} 1 & 0 & 0 \\ 0 & 1 & 0 \\ 0 & 0 & 3 \end{bmatrix}^{-1} = \begin{bmatrix} 1 & 0 & 0 \\ 0 & 1 & 0 \\ 0 & 0 & \frac{1}{3} \end{bmatrix}$

   (c) $\begin{bmatrix} 0 & 0 & 0 & 1 \\ 0 & 1 & 0 & 0 \\ 0 & 0 & 1 & 0 \\ 1 & 0 & 0 & 0 \end{bmatrix}^{-1} = \begin{bmatrix} 0 & 0 & 0 & 1 \\ 0 & 1 & 0 & 0 \\ 0 & 0 & 1 & 0 \\ 1 & 0 & 0 & 0 \end{bmatrix}$

   (d) $\begin{bmatrix} 1 & 0 & -\frac{1}{7} & 0 \\ 0 & 1 & 0 & 0 \\ 0 & 0 & 1 & 0 \\ 0 & 0 & 0 & 1 \end{bmatrix}^{-1} = \begin{bmatrix} 1 & 0 & \frac{1}{7} & 0 \\ 0 & 1 & 0 & 0 \\ 0 & 0 & 1 & 0 \\ 0 & 0 & 0 & 1 \end{bmatrix}$

7. (a) B is obtained from A by interchanging the first and third rows; thus $EA = B$ where

$$E = \begin{bmatrix} 0 & 0 & 1 \\ 0 & 1 & 0 \\ 1 & 0 & 0 \end{bmatrix}.$$

(b) $EB = A$ where E is the same as in part (a).

(c) C is obtained from A by adding -2 times the first row to the third row; thus $EA = C$ where

$$E = \begin{bmatrix} 1 & 0 & 0 \\ 0 & 1 & 0 \\ -2 & 0 & 1 \end{bmatrix}.$$

(d) $EC = A$ where E is the inverse of the matrix in part (c), i.e. $E = \begin{bmatrix} 1 & 0 & 0 \\ 0 & 1 & 0 \\ 2 & 0 & 1 \end{bmatrix}.$

9. Using the method of Example 3, we start with the partitioned matrix $[A \mid I]$ and perform row operations aimed at reducing the left side to I; these same row operations performed simultaneously on the right side will produce the matrix A^{-1}.

$$\begin{bmatrix} 1 & 5 & \vdots & 1 & 0 \\ 2 & 20 & \vdots & 0 & 1 \end{bmatrix}$$

Add -2 times the first row to the second row.

$$\begin{bmatrix} 1 & 5 & \vdots & 1 & 0 \\ 0 & 10 & \vdots & -2 & 1 \end{bmatrix}$$

Multiply the second row by $\frac{1}{10}$, then add -5 times the new second row to the first row.

$$\begin{bmatrix} 1 & 0 & \vdots & 2 & -\frac{1}{2} \\ 0 & 1 & \vdots & -\frac{1}{5} & \frac{1}{10} \end{bmatrix}$$

Thus $A^{-1} = \begin{bmatrix} 2 & -\frac{1}{2} \\ -\frac{1}{5} & \frac{1}{10} \end{bmatrix}$. On the other hand, using the formula from Theorem 3.2.7, and the fact that $\det(A) = 10$, we have $A^{-1} = \frac{1}{10}\begin{bmatrix} 20 & -5 \\ -2 & 1 \end{bmatrix} = \begin{bmatrix} 2 & -\frac{1}{2} \\ -\frac{1}{5} & \frac{1}{10} \end{bmatrix}.$

11. (a) Start with the partitioned matrix $[A \mid I]$.

$$\begin{bmatrix} 3 & 4 & -1 & \vdots & 1 & 0 & 0 \\ 1 & 0 & 3 & \vdots & 0 & 1 & 0 \\ 2 & 5 & -4 & \vdots & 0 & 0 & 1 \end{bmatrix}$$

Interchange rows 1 and 2.

$$\begin{bmatrix} 1 & 0 & 3 & \vdots & 0 & 1 & 0 \\ 3 & 4 & -1 & \vdots & 1 & 0 & 0 \\ 2 & 5 & -4 & \vdots & 0 & 0 & 1 \end{bmatrix}$$

Add -3 times row 1 to row 2. Add -2 times row 1 to row 3.

$$\begin{bmatrix} 1 & 0 & 3 & \vdots & 0 & 1 & 0 \\ 0 & 4 & -10 & \vdots & 1 & -3 & 0 \\ 0 & 5 & -10 & \vdots & 0 & -2 & 1 \end{bmatrix}$$

Add -1 times row 2 to row 3.

$$\begin{bmatrix} 1 & 0 & 3 & | & 0 & 1 & 0 \\ 0 & 4 & -10 & | & 1 & -3 & 0 \\ 0 & 1 & 0 & | & -1 & 1 & 1 \end{bmatrix}$$

Add -4 times row 3 to row 2, then interchange rows 2 and 3.

$$\begin{bmatrix} 1 & 0 & 3 & | & 0 & 1 & 0 \\ 0 & 1 & 0 & | & -1 & 1 & 1 \\ 0 & 0 & -10 & | & 5 & -7 & -4 \end{bmatrix}$$

Multiply row 3 by $-\frac{1}{10}$, then add -3 times the new row 3 to row 1.

$$\begin{bmatrix} 1 & 0 & 0 & | & \frac{3}{2} & -\frac{11}{10} & -\frac{6}{5} \\ 0 & 1 & 0 & | & -1 & 1 & 1 \\ 0 & 1 & 1 & | & -\frac{1}{2} & \frac{7}{10} & \frac{2}{5} \end{bmatrix}$$

From this we conclude that A is invertible, and that $A^{-1} = \begin{bmatrix} \frac{3}{2} & -\frac{11}{10} & -\frac{6}{5} \\ -1 & 1 & 1 \\ -\frac{1}{2} & \frac{7}{10} & \frac{2}{5} \end{bmatrix}$.

(b) Start with the partitioned matrix $[A \mid I]$.

$$\begin{bmatrix} -1 & 3 & -4 & | & 1 & 0 & 0 \\ 2 & 4 & 1 & | & 0 & 1 & 0 \\ -4 & 2 & -9 & | & 0 & 0 & 1 \end{bmatrix}$$

Multiply row 1 by -1. Add -2 times the new row 1 to row 2; add 4 times the new row 1 to row 3.

$$\begin{bmatrix} 1 & -3 & 4 & | & -1 & 0 & 0 \\ 0 & 10 & -7 & | & 2 & 1 & 0 \\ 0 & -10 & 7 & | & -4 & 0 & 1 \end{bmatrix}$$

Add row 2 to row 3.

$$\begin{bmatrix} 1 & -3 & 4 & | & -1 & 0 & 0 \\ 0 & 10 & -7 & | & 2 & 1 & 0 \\ 0 & 0 & 0 & | & -2 & 1 & 1 \end{bmatrix}$$

At this point, since we have obtained a row of zeros on the left side, we conclude that the matrix A is not invertible.

(c) Start with the partitioned matrix $[A \mid I]$.

$$\begin{bmatrix} 1 & 0 & 1 & | & 1 & 0 & 0 \\ 0 & 1 & 1 & | & 0 & 1 & 0 \\ 1 & 1 & 0 & | & 0 & 0 & 1 \end{bmatrix}$$

Add -1 times row 1 to row 3.

$$\begin{bmatrix} 1 & 0 & 1 & | & 1 & 0 & 0 \\ 0 & 1 & 1 & | & 0 & 1 & 0 \\ 0 & 1 & -1 & | & -1 & 0 & 1 \end{bmatrix}$$

Add -1 times row 2 to row 3, then multiply the new row 3 by $-\frac{1}{2}$.

$$\left[\begin{array}{ccc|ccc} 1 & 0 & 1 & 1 & 0 & 0 \\ 0 & 1 & 1 & 0 & 1 & 0 \\ 0 & 0 & 1 & \frac{1}{2} & \frac{1}{2} & -\frac{1}{2} \end{array}\right]$$

Add -1 times row 3 to rows 2 and 1.

$$\left[\begin{array}{ccc|ccc} 1 & 0 & 0 & \frac{1}{2} & -\frac{1}{2} & \frac{1}{2} \\ 0 & 1 & 0 & -\frac{1}{2} & \frac{1}{2} & \frac{1}{2} \\ 0 & 0 & 1 & \frac{1}{2} & \frac{1}{2} & -\frac{1}{2} \end{array}\right]$$

From this we conclude that A is invertible, and that $A^{-1} = \begin{bmatrix} \frac{1}{2} & -\frac{1}{2} & \frac{1}{2} \\ -\frac{1}{2} & \frac{1}{2} & \frac{1}{2} \\ \frac{1}{2} & \frac{1}{2} & -\frac{1}{2} \end{bmatrix}$.

13. As in the inversion algorithm, we start with the partitioned matrix $[A \mid I]$ and perform row operations aimed at reducing the left side to its reduced row echelon form R. If A is invertible, then R will be the identity matrix and the matrix produced on the right side will be A^{-1}. In the more general situation the reduced matrix will have the form $[R \mid B] = [BA \mid BI]$ where the matrix B on the right has the property that $BA = R$. [Note that B is the product of elementary matrices and thus is always an invertible matrix, whether A is invertible or not.]

$$\left[\begin{array}{ccc|ccc} 1 & 2 & 3 & 1 & 0 & 0 \\ 0 & 0 & 1 & 0 & 1 & 0 \\ 1 & 2 & 4 & 0 & 0 & 1 \end{array}\right]$$

Add -1 times row 1 to row 3.

$$\left[\begin{array}{ccc|ccc} 1 & 2 & 3 & 1 & 0 & 0 \\ 0 & 0 & 1 & 0 & 1 & 0 \\ 0 & 0 & 1 & -1 & 0 & 1 \end{array}\right]$$

Add -1 times row 2 to row 3.

$$\left[\begin{array}{ccc|ccc} 1 & 2 & 3 & 1 & 0 & 0 \\ 0 & 0 & 1 & 0 & 1 & 0 \\ 0 & 0 & 0 & -1 & -1 & 1 \end{array}\right]$$

Add -3 times row 2 to row 1.

$$\left[\begin{array}{ccc|ccc} 1 & 2 & 0 & 1 & -3 & 0 \\ 0 & 0 & 1 & 0 & 1 & 0 \\ 0 & 0 & 0 & -1 & -1 & 1 \end{array}\right]$$

The reduced row echelon form of A is $R = \begin{bmatrix} 1 & 2 & 0 \\ 0 & 0 & 1 \\ 0 & 0 & 0 \end{bmatrix}$, and the matrix $B = \begin{bmatrix} 1 & -3 & 0 \\ 0 & 1 & 0 \\ -1 & -1 & 1 \end{bmatrix}$ has the property that $BA = R$.

15. If $c = 0$, then the first row is a row of zeros, so the matrix is not invertible. If $c \neq 0$, then after multiplying the first row by $1/c$, we have:

$$\begin{bmatrix} 1 & 1 & 1 \\ 1 & c & c \\ 1 & 1 & c \end{bmatrix}$$

Add -1 times the first row to the second row and to the third row.

$$\begin{bmatrix} 1 & 1 & 1 \\ 0 & c-1 & c-1 \\ 0 & 0 & c-1 \end{bmatrix}$$

If $c = 1$, then the second and third rows are rows of zeros, and so the matrix is not invertible. If $c \neq 1$, then we can divide the second and third rows by $c - 1$ obtaining

$$\begin{bmatrix} 1 & 1 & 1 \\ 0 & 1 & 1 \\ 0 & 0 & 1 \end{bmatrix}$$

and from this it is clear that the reduced row echelon form is the identity matrix. Thus we conclude that the matrix is invertible if and only if $c \neq 0, 1$.

17. The matrix B is obtained by starting with the identity matrix and performing the same sequence of row operations; thus $B = \begin{bmatrix} 0 & 1 & 0 \\ 1 & 0 & 0 \\ 0 & 0 & 6 \end{bmatrix} = \begin{bmatrix} 1 & 0 & 0 \\ 0 & 1 & 0 \\ 0 & 0 & 6 \end{bmatrix} \begin{bmatrix} 0 & 1 & 0 \\ 1 & 0 & 0 \\ 0 & 0 & 1 \end{bmatrix}$.

19. If any one of the k_i's is 0, then the matrix A has a zero row and thus is not invertible. If the k_i's are all nonzero, then multiplying the ith row of the matrix $[A \mid I]$ by $1/k_i$ for $i = 1, 2, 3, 4$ and then reversing the order of the rows yields

$$\left[\begin{array}{cccc|cccc} 1 & 0 & 0 & 0 & 0 & 0 & 0 & \frac{1}{k_4} \\ 0 & 1 & 0 & 0 & 0 & 0 & \frac{1}{k_3} & 0 \\ 0 & 0 & 1 & 0 & 0 & \frac{1}{k_2} & 0 & 0 \\ 0 & 0 & 0 & 1 & \frac{1}{k_1} & 0 & 0 & 0 \end{array}\right]$$

thus A is invertible and A^{-1} is the matrix in the right-hand block above.

21. (a) The identity matrix can be obtained from A by first adding 5 times row 1 to row 3, and then multiplying row 3 by $\frac{1}{2}$. Thus if $E_1 = \begin{bmatrix} 1 & 0 \\ 5 & 1 \end{bmatrix}$ and $E_2 = \begin{bmatrix} 1 & 0 \\ 0 & \frac{1}{2} \end{bmatrix}$, then $E_2 E_1 A = I$.

(b) $A^{-1} = E_2 E_1$ where E_1 and E_2 are as in part (a).

(c) $A = E_1^{-1} E_2^{-1} = \begin{bmatrix} 1 & 0 \\ -5 & 1 \end{bmatrix} \begin{bmatrix} 1 & 0 \\ 0 & 2 \end{bmatrix}$.

23. The identity matrix can be obtained from A by the following sequence of row operations. The corresponding elementary matrices and their inverses are as indicated.

(1) Interchange rows 1 and 3. $E_1 = \begin{bmatrix} 0 & 0 & 1 \\ 0 & 1 & 0 \\ 1 & 0 & 0 \end{bmatrix}$ $E_1^{-1} = \begin{bmatrix} 0 & 0 & 1 \\ 0 & 1 & 0 \\ 1 & 0 & 0 \end{bmatrix}$

(2) Add -1 times row 1 to row 2. $E_2 = \begin{bmatrix} 1 & 0 & 0 \\ -1 & 1 & 0 \\ 0 & 0 & 1 \end{bmatrix}$ $E_2^{-1} = \begin{bmatrix} 1 & 0 & 0 \\ 1 & 1 & 0 \\ 0 & 0 & 1 \end{bmatrix}$

(3) Add -2 times row 1 to row 3. $E_3 = \begin{bmatrix} 1 & 0 & 0 \\ 0 & 1 & 0 \\ -2 & 0 & 1 \end{bmatrix}$ $E_3^{-1} = \begin{bmatrix} 1 & 0 & 0 \\ 0 & 1 & 0 \\ 2 & 0 & 1 \end{bmatrix}$

(4) Add row 2 to row 3.
$$E_4 = \begin{bmatrix} 1 & 0 & 0 \\ 0 & 1 & 0 \\ 0 & 1 & 1 \end{bmatrix} \qquad E_4^{-1} = \begin{bmatrix} 1 & 0 & 0 \\ 0 & 1 & 0 \\ 0 & -1 & 1 \end{bmatrix}$$

(5) Multiply row 3 by $-\frac{1}{4}$.
$$E_5 = \begin{bmatrix} 1 & 0 & 0 \\ 0 & 1 & 0 \\ 0 & 0 & -\frac{1}{4} \end{bmatrix} \qquad E_5^{-1} = \begin{bmatrix} 1 & 0 & 0 \\ 0 & 1 & 0 \\ 0 & 0 & -4 \end{bmatrix}$$

(6) Add row 3 to row 2.
$$E_6 = \begin{bmatrix} 1 & 0 & 0 \\ 0 & 1 & 1 \\ 0 & 0 & 1 \end{bmatrix} \qquad E_6^{-1} = \begin{bmatrix} 1 & 0 & 0 \\ 0 & 1 & -1 \\ 0 & 0 & 1 \end{bmatrix}$$

(7) Add -2 times row 3 to row 1.
$$E_7 = \begin{bmatrix} 1 & 0 & -2 \\ 0 & 1 & 0 \\ 0 & 0 & 1 \end{bmatrix} \qquad E_7^{-1} = \begin{bmatrix} 1 & 0 & 2 \\ 0 & 1 & 0 \\ 0 & 0 & 1 \end{bmatrix}$$

(8) Add -1 times row 2 to row 1.
$$E_8 = \begin{bmatrix} 1 & -1 & 0 \\ 0 & 1 & 0 \\ 0 & 0 & 1 \end{bmatrix} \qquad E_8^{-1} = \begin{bmatrix} 1 & 1 & 0 \\ 0 & 1 & 0 \\ 0 & 0 & 1 \end{bmatrix}$$

It follows that $A^{-1} = E_8 E_7 E_6 E_5 E_4 E_3 E_2 E_1$ and $A = E_1^{-1} E_2^{-1} E_3^{-1} E_4^{-1} E_5^{-1} E_6^{-1} E_7^{-1} E_8^{-1}$.

25. The two systems have the same coefficient matrix. If we augment this matrix with the two columns of constants on the right sides, we obtain

$$\begin{bmatrix} 1 & 2 & 1 & -1 & 0 \\ 1 & 3 & 2 & 3 & 0 \\ 0 & 1 & 2 & 4 & 4 \end{bmatrix}$$

and the reduced row echelon form of this matrix is

$$\begin{bmatrix} 1 & 0 & 0 & -9 & 4 \\ 0 & 1 & 0 & 4 & -4 \\ 0 & 0 & 1 & 0 & 4 \end{bmatrix}$$

From this we conclude that the first system has the solution $x_1 = -9$, $x_2 = 4$, $x_3 = 0$; and the second system has the solution $x_1 = 4$, $x_2 = -4$, $x_3 = 4$.

27. (a) The systems can be written in matrix form as

$$\begin{bmatrix} 1 & 2 & 1 \\ 1 & 3 & 2 \\ 0 & 1 & 2 \end{bmatrix} \begin{bmatrix} x_1 \\ x_2 \\ x_3 \end{bmatrix} = \begin{bmatrix} -1 \\ 3 \\ 4 \end{bmatrix} \quad \text{and} \quad \begin{bmatrix} 1 & 2 & 1 \\ 1 & 3 & 2 \\ 0 & 1 & 2 \end{bmatrix} \begin{bmatrix} x_1 \\ x_2 \\ x_3 \end{bmatrix} = \begin{bmatrix} 0 \\ 0 \\ 4 \end{bmatrix}$$

The inverse of the coefficient matrix $A = \begin{bmatrix} 1 & 2 & 1 \\ 1 & 3 & 2 \\ 0 & 1 & 2 \end{bmatrix}$ is $A^{-1} = \begin{bmatrix} 4 & -3 & 1 \\ -2 & 2 & -1 \\ 1 & -1 & 1 \end{bmatrix}$. Thus the solutions are given by

$$\begin{bmatrix} 4 & -3 & 1 \\ -2 & 2 & -1 \\ 1 & -1 & 1 \end{bmatrix} \begin{bmatrix} -1 \\ 3 \\ 4 \end{bmatrix} = \begin{bmatrix} -9 \\ 4 \\ 0 \end{bmatrix} \quad \text{and} \quad \begin{bmatrix} 4 & -3 & 1 \\ -2 & 2 & -1 \\ 1 & -1 & 1 \end{bmatrix} \begin{bmatrix} 0 \\ 0 \\ 4 \end{bmatrix} = \begin{bmatrix} 4 \\ -4 \\ 4 \end{bmatrix}$$

(b) $A^{-1}[\mathbf{b}_1 \quad \mathbf{b}_2] = \begin{bmatrix} 4 & -3 & 1 \\ -2 & 2 & -1 \\ 1 & -1 & 1 \end{bmatrix} \begin{bmatrix} -1 & 0 \\ 3 & 0 \\ 4 & 4 \end{bmatrix} = \begin{bmatrix} -9 & 4 \\ 4 & -4 \\ 0 & 4 \end{bmatrix}.$

29. The augmented matrix $\begin{bmatrix} 6 & -4 & | & b_1 \\ 3 & -2 & | & b_2 \end{bmatrix}$ can be row reduced to $\begin{bmatrix} 3 & -2 & | & b_2 \\ 0 & 0 & | & b_1 - 2b_2 \end{bmatrix}$. Thus the system is consistent if and only if $b_1 - 2b_2 = 0$, i.e. if and only if $b_1 = 2b_2$.

31. The augmented matrix $\begin{bmatrix} 1 & -2 & -1 & | & b_1 \\ -2 & 3 & 2 & | & b_2 \\ -4 & 7 & 4 & | & b_3 \end{bmatrix}$ can be row reduced to $\begin{bmatrix} 1 & -2 & -1 & | & b_1 \\ 0 & -1 & 0 & | & b_2 + 2b_1 \\ 0 & 0 & 0 & | & 2b_1 - b_2 + b_3 \end{bmatrix}$. Thus the system is consistent if and only if $2b_1 - b_2 + b_3 = 0$.

33. The matrix A can be reduced to a row echelon from R by the following sequence of row operations:

$$A = \begin{bmatrix} 0 & 1 & 7 & 8 \\ 1 & 3 & 3 & 8 \\ -2 & -5 & 1 & -8 \end{bmatrix}$$

(1) Interchange rows 1 and 2.

$$\begin{bmatrix} 1 & 3 & 3 & 8 \\ 0 & 1 & 7 & 8 \\ -2 & -5 & 1 & -8 \end{bmatrix}$$

(2) Add 2 times row 1 to row 3.

$$\begin{bmatrix} 1 & 3 & 3 & 8 \\ 0 & 1 & 7 & 8 \\ 0 & 1 & 7 & 8 \end{bmatrix}$$

(3) Add -1 times row 2 to row 3.

$$R = \begin{bmatrix} 1 & 3 & 3 & 8 \\ 0 & 1 & 7 & 8 \\ 0 & 0 & 0 & 0 \end{bmatrix}$$

It follows from this that $R = E_3 E_2 E_1 A$ where E_1, E_2, E_3 are the elementary matrices corresponding to the row operations indicated above. Finally, we have the factorization

$$A = EFGR = \begin{bmatrix} 0 & 1 & 0 \\ 1 & 0 & 0 \\ 0 & 0 & 1 \end{bmatrix} \begin{bmatrix} 1 & 0 & 0 \\ 0 & 1 & 0 \\ -2 & 0 & 1 \end{bmatrix} \begin{bmatrix} 1 & 0 & 0 \\ 0 & 1 & 0 \\ 0 & 1 & 1 \end{bmatrix} \begin{bmatrix} 1 & 3 & 3 & 8 \\ 0 & 1 & 7 & 8 \\ 0 & 0 & 0 & 0 \end{bmatrix}$$

where $E = E_1^{-1}$, $F = E_2^{-1}$, and $G = E_3^{-1}$.

DISCUSSION AND DISCOVERY

D1. Suppose the matrix A can be reduced to the identity matrix by a known sequence of elementary row operations, and let $E_1, E_2, \ldots, E_k$ be the corresponding sequence of elementary matrices. Then we have $E_k \cdots E_2 E_1 A = I$, and so $A = E_1^{-1} E_2^{-1} \cdots E_k^{-1} I$. Thus we can find A by applying the inverse row operations to I in the reverse order.

D2. There is not. For example, let $b = 1$ and $a = c = d = 0$. Then $A \begin{bmatrix} 0 & 1 \\ 0 & 0 \end{bmatrix} = \begin{bmatrix} 0 & a_{11} \\ 0 & a_{21} \end{bmatrix} \neq \begin{bmatrix} 1 & 0 \\ 0 & 0 \end{bmatrix}.$

D3. There is no nontrivial solution. From the last equation we see that $x_4 = 0$ and, from back substitution, it follows immediately that $x_3 = x_2 = x_1 = 0$ also. The coefficient matrix is invertible.

D4. (a) The matrix $B = \begin{bmatrix} 1 & 1 & 0 \\ 0 & -\frac{1}{2} & 0 \end{bmatrix}$ has the required property.

(b) Yes, the matrix $B = \begin{bmatrix} 0 & 2 & 1 \\ 0 & -\frac{1}{2} & 0 \end{bmatrix}$ works just as well.

(c) The matrix B must be of size 3×2; it is not square and therefore not invertible.

D5. (a) False. Only invertible matrices can be expressed as a product of elementary matrices.

(b) False. For example the product $\begin{bmatrix} 0 & 1 \\ 2 & 0 \end{bmatrix} = \begin{bmatrix} 0 & 1 \\ 1 & 0 \end{bmatrix} \begin{bmatrix} 2 & 0 \\ 0 & 1 \end{bmatrix}$ cannot be obtained from the identity by a single elementary row operation.

(c) True. This row operation is equivalent to multiplying the given matrix by an elementary matrix and, since any elementary matrix is invertible, the product is still invertible.

(d) True. If A is invertible and $AB = 0$, then $B = IB = (A^{-1}A)B = A^{-1}(AB) = A^{-1}0 = 0$.

(e) True. If A is invertible then the homogeneous system $A\mathbf{x} = \mathbf{0}$ has only the trivial solution; otherwise (if A is singular) there are infinitely many solutions.

D6. All of these statements are true.

D7. No. An invertible matrix cannot have a row of zeros. Thus, for A to be invertible we must have $a \neq 0$ and $h \neq 0$. But (assuming this) if we add $-d/a$ times row 1 to row 3, and add $-e/h$ times row 5 to row 3; we have a matrix with a row of zeros in the third row.

WORKING WITH PROOFS

P1. If AB is invertible then, by Theorem 3.3.8, both A and B are invertible. Thus, if either A or B is singular, then the product AB must be singular.

P2. Suppose that $A = BC$ where B is invertible, and that B is reduced to I by a sequence of elementary row operations corresponding to the elementary matrices $E_1, E_2, \ldots, E_k$. Then $I = E_k \cdots E_2 E_1 B$, and so $B^{-1} = E_k \cdots E_2 E_1$. From this it follows that $E_k \cdots E_2 E_1 A = E_k \cdots E_2 BC = C$; thus the same sequence of row operations will reduce A to C.

P3. Suppose $A\mathbf{x} = \mathbf{0}$ iff $\mathbf{x} = \mathbf{0}$. We wish to prove that, for any positive integer k, $A^k\mathbf{x} = \mathbf{0}$ iff $\mathbf{x} = \mathbf{0}$. Our proof is by induction on the exponent k.

Step 1. If $k = 1$, then $A^1\mathbf{x} = A\mathbf{x} = \mathbf{0}$ iff $\mathbf{x} = \mathbf{0}$. Thus the statement is true for $k = 1$.

Step 2 (induction step): Suppose the statement is true for $k = j$, where j is any fixed integer ≥ 1. Then $A^{j+1}\mathbf{x} = A^j(A\mathbf{x}) = \mathbf{0}$ iff $A\mathbf{x} = \mathbf{0}$, and this is true iff $\mathbf{x} = \mathbf{0}$. This shows that if the statement is true for $k = j$ it is also true for $k = j + 1$.

These two steps complete the proof by induction.

P4. From Theorem 3.3.7, the system $A\mathbf{x} = \mathbf{0}$ has only the trivial solution if and only if A is invertible. But, since B is invertible, A is invertible iff BA is invertible. Thus $A\mathbf{x} = \mathbf{0}$ has only the trivial solution iff $(BA)\mathbf{x} = \mathbf{0}$ has only the trivial solution.

P5. Let $e_1, e_2, \ldots, e_m$ be the standard unit vectors in R^m. Note that $e_1, e_2, \ldots, e_m$ are the rows of the identity matrix $I = I_m$. Thus, for any $m \times n$ matrix A, we have

$$r_i(A) = r_i(IA) = r_i(I)A = e_iA$$

for $i = 1, 2, \ldots, m$. Suppose now that E is an elementary matrix that is obtained by performing a single elementary row operation on I. We consider the three types of row operations separately.

Row Interchange. Suppose E is obtained from I by interchanging rows i and j. Then

$$r_i(EA) = r_i(E)A = e_jA = r_j(A)$$
$$r_j(EA) = r_j(E)A = e_iA = r_i(A)$$

and $r_k(EA) = r_k(E)A = e_kA = r_k(A)$ for $k \neq i, j$. Thus EA is the matrix that is obtained from A by interchanging rows i and j.

Row Scaling. Suppose E is obtained from I by multiplying row i by a nonzero scalar c. Then

$$r_i(EA) = r_i(E)A = ce_iA = cr_i(A)$$

and $r_k(EA) = r_k(E)A = e_kA = r_k(A)$ for all $k \neq i$. Thus EA is the matrix that is obtained from A by multiplying row i by the scalar c.

Row Replacement. Suppose E is obtained from I by adding c times row i to row j $(i \neq j)$. Then

$$r_j(EA) = r_j(E)A = (e_j + ce_i)A = e_jA + ce_iA = r_j(A) + cr_i(A)$$

and $r_k(EA) = r_k(E)A = e_kA = r_k(A)$ for all $k \neq j$. Thus EA is the matrix that is obtained from A by adding c times row i to row j.

P6. Let $A = \begin{bmatrix} a & b \\ c & d \end{bmatrix}$, and let $B = \begin{bmatrix} 0 & 1 \\ 1 & 0 \end{bmatrix}$. Then $AB = BA$ if and only if $\begin{bmatrix} b & a \\ d & c \end{bmatrix} = \begin{bmatrix} c & d \\ a & b \end{bmatrix}$, i.e. if and only if $a = d$ and $b = c$. Thus the only matrices that commute with B are those of the form $A = \begin{bmatrix} a & b \\ b & a \end{bmatrix}$.

Suppose now that A is a matrix of this type, and let $C = \begin{bmatrix} 1 & 0 \\ 0 & 2 \end{bmatrix}$. Then $AC = CA$ if and only if $\begin{bmatrix} a & 2b \\ b & 2a \end{bmatrix} = \begin{bmatrix} a & b \\ 2b & 2a \end{bmatrix}$, and this is true if and only if $b = 0$. Thus the only 2×2 matrices that commute with both B and C are those of the form $A = \begin{bmatrix} a & 0 \\ 0 & a \end{bmatrix} = a\begin{bmatrix} 1 & 0 \\ 0 & 1 \end{bmatrix}$ where $-\infty < a < \infty$. It is easy to see that such a matrix will commute with all other 2×2 matrices as well.

P7. Every $m \times n$ matrix A can be transformed to reduced row echelon form B by a sequence of elementary row operations. Let $E_1, E_2, \ldots, E_k$ be the corresponding sequence of elementary matrices. Then we have $B = E_k \cdots E_2E_1A = CA$ where $C = E_k \cdots E_2E_1$ is an invertible matrix.

P8. Suppose that A is reduced to $I = I_n$ by a sequence of elementary row operations, and that $E_1, E_2, \ldots, E_k$ are the corresponding elementary matrices. Then $E_k \cdots E_2E_1A = I$ and so $E_k \cdots E_2E_1 = A^{-1}$. It follows that the same sequence of elementary row operations will reduce the matrix $[A \mid B]$ to the matrix $[E_k \cdots E_2E_1A \mid E_k \cdots E_2E_1B] = [I \mid A^{-1}B]$.

EXERCISE SET 3.4

1. **(a)** $(x_1, x_2) = t(1, -1);$ $x_1 = t, x_2 = -t$
   (b) $(x_1, x_2, x_3) = t(2, 1, -4);$ $x_1 = 2t, x_2 = t, x_3 = -4t$
   (c) $(x_1, x_2, x_3, x_4) = t(1, 1, -2, 3);$ $x_1 = t, x_2 = t, x_3 = -2t, x_4 = 3t$

3. (a) $(x_1, x_2, x_3) = s(4, -4, 2) + t(-3, 5, 7);$ $\quad x_1 = 4s - 3t, \; x_2 = -4s + 5t, \; x_3 = 2s + 7t$

(b) $(x_1, x_2, x_3, x_4) = s(1, 2, 1, -3) + t(3, 4, 5, 0);$ $\quad x_1 = s + 3t, \; x_2 = 2s + 4t, \; x_3 = s + 5t, \; x_4 = -3s$

5. (a) $\mathbf{u} = -2\mathbf{v}$; thus $\mathbf{u}$ is in the subspace span$\{\mathbf{v}\}$.

(b) $\mathbf{u} \neq k\mathbf{v}$ for any scalar k; thus $\mathbf{u}$ is not in the subspace span$\{\mathbf{v}\}$.

7. (a) Two vectors are linearly dependent if and only if one is a scalar multiple of the other; thus these vectors are linearly dependent.

(b) These vectors are linearly independent.

9. $\mathbf{u} = 2\mathbf{v} - \mathbf{w}$; thus the vectors $\mathbf{u}$, $\mathbf{v}$, $\mathbf{w}$ are linearly dependent.

11. (a) A line (a 1-dimensional subspace) in R^4 that passes through the origin and is parallel to the vector $\mathbf{u} = (2, -3, 1, 4) = \frac{1}{2}(4, -6, 2, 8)$.

(b) A plane (a 2-dimensional subspace) in R^4 that passes through the origin and is parallel to the vectors $\mathbf{u} = (3, -2, 2, 5)$ and $\mathbf{v} = (6, -4, 4, 0)$.

13. The augmented matrix of the homogenous system is

$$\begin{bmatrix} 1 & 6 & 2 & -5 & 0 \\ -1 & -6 & -1 & -3 & 0 \\ 2 & 12 & 5 & -18 & 0 \end{bmatrix}$$

and the reduced row-echelon form of this matrix is

$$\begin{bmatrix} 1 & 6 & 0 & 11 & 0 \\ 0 & 0 & 1 & -8 & 0 \\ 0 & 0 & 0 & 0 & 0 \end{bmatrix}$$

Thus a general solution of the system can be written in parametric form as

$$x_1 = -6s - 11t, \quad x_2 = s, \quad x_3 = 8t, \quad x_4 = t$$

or in vector form as

$$(x_1, x_2, x_3, x_4) = s(-6, 1, 0, 0) + t(-11, 0, 8, 1)$$

This shows that the solution space is span$\{\mathbf{v}_1, \mathbf{v}_2\}$ where $\mathbf{v}_1 = (-6, 1, 0, 0)$ and $\mathbf{v}_2 = (-11, 0, 8, 1)$.

15. The augmented matrix of the homogeneous system is

$$\begin{bmatrix} 1 & 2 & 1 & 1 & 1 & 0 \\ 2 & 4 & -1 & 0 & 1 & 0 \end{bmatrix}$$

and the reduced row echelon form of this matrix is

$$\begin{bmatrix} 1 & 2 & 0 & \frac{1}{3} & \frac{2}{3} & 0 \\ 0 & 0 & 1 & \frac{2}{3} & \frac{1}{3} & 0 \end{bmatrix}$$

Thus a general solution of the system can be written in parametric form as

$$x_1 = -2r - \tfrac{1}{3}s - \tfrac{2}{3}t, \quad x_2 = r, \quad x_3 = -\tfrac{2}{3}s - \tfrac{1}{3}t, \quad x_4 = s, \quad x_5 = t$$

or in vector form as

$$(x_1, x_2, x_3, x_4, x_5) = r(-2, 1, 0, 0, 0) + s(-\tfrac{1}{3}, 0, -\tfrac{2}{3}, 1, 0) + t(-\tfrac{2}{3}, 0, -\tfrac{1}{3}, 0, 1)$$

The vectors $\mathbf{v}_1 = (-2, 1, 0, 0, 0)$, $\mathbf{v}_2 = (-\tfrac{1}{3}, 0, -\tfrac{2}{3}, 1, 0)$, and $\mathbf{v}_3 = (-\tfrac{2}{3}, 0, -\tfrac{1}{3}, 0, 1)$ span the solution space.

17. (a) v_2 is a scalar multiple of v_1($v_2 = -5v_1$); thus these two vectors are linearly dependent.

(b) Any set of more than 2 vectors in R^2 is linearly dependent (Theorem 3.4.8).

19. (a) These two vectors are linearly independent since neither is a scalar multiple of the other.

(b) v_1 is a scalar multiple of v_2($v_1 = -3v_2$); thus these two vectors are linearly dependent.

(c) These three vectors are linearly independent since the system $\begin{bmatrix} -3 & 5 & 1 \\ 0 & -1 & 1 \\ 4 & 2 & 3 \end{bmatrix} \begin{bmatrix} c_1 \\ c_2 \\ c_3 \end{bmatrix} = \begin{bmatrix} 0 \\ 0 \\ 0 \end{bmatrix}$ has only the trivial solution. The coefficient matrix, which is the matrix having the given vectors as its columns, is invertible.

(d) These four vectors are linearly dependent since any set of more than 3 vectors in R^3 is linearly dependent.

21. (a) The matrix having these vectors as its columns is invertible. Thus the vectors are linearly independent; they do not lie in a plane. *will have 2 rows of zero in GJ matrix*

(b) These vectors are linearly dependent ($v_1 = 2v_2 - 3v_3$); they lie in a plane but not on a line.

(c) These vectors line on a line; $v_1 = 2v_2$ and $v_3 = 3v_2$. *will have only 1 row of zero in GJ matrix*

23. (a) This set of vectors is a subspace; it is closed under scalar multiplication and addition: $k(a, 0, 0) = (ka, 0, 0)$ and $(a_1, 0, 0) + (a_2, 0, 0) = (a_1 + a_2, 0, 0)$.

(b) This set of vectors is not a subspace; it is not closed under scalar multiplication.

(c) This set of vectors is a subspace. If $b = a + c$, then $kb = ka + kc$. If $b_1 = a_1 + c_1$, and $b_2 = a_2 + c_2$, then $(b_1 + b_2) = (a_1 + a_2) + (c_1 + c_2)$.

(d) This set of vectors is not a subspace; it is not closed under addition or scalar multiplication.

25. The set W consists of all vectors of the form $x = a(1, 0, 1, 0)$; thus $W = \text{span}\{v\}$ where $v = (1, 0, 1, 0)$. This corresponds to a line (i.e. a 1-dimensional subspace) through the origin in R^4.

27. (a) $7v_1 - 2v_2 + 3v_3 = 0$

(b) $v_1 = \frac{2}{7}v_2 - \frac{3}{7}v_3$, $v_2 = \frac{7}{2}v_1 + \frac{3}{2}v_3$, $v_3 = -\frac{7}{3}v_1 + \frac{2}{3}v_2$

29. (a) Suppose $S = \{v_1, v_2, v_3\}$ is a linearly independent set. Note first that none of these vectors can be equal to 0 (otherwise S would be linearly dependent), and so each of the sets $\{v_1\}$, $\{v_2\}$, and $\{v_3\}$ is linearly independent. Suppose then that T is a 2-element subset of S, e.g. $T = \{v_1, v_2\}$. If T is linearly dependent then there are scalars c_1 and c_2, not both zero, such that $c_1v_1 + c_2v_2 = 0$. But, if this were true, then $c_1v_1 + c_2v_2 + 0v_3 = 0$ would be a nontrivial linear relationship among the vectors v_1, v_2, v_3, and so S would be linearly dependent. Thus $T = \{v_1, v_2\}$ is linearly independent. The same argument applies to any 2-element subset of S. Thus if S is linearly independent, then each of its nonempty subsets is linearly independent.

(b) If $S = \{v_1, v_2, v_3\}$ is linearly dependent, then there are scalars c_1, c_2, and c_3, not all zero, such that $c_1v_1 + c_2v_2 + c_3v_3 = 0$. Thus, for any vector v in R^n, we have $c_1v_1 + c_2v_2 + c_3v_3 + 0v = 0$ and this is a nontrivial linear relationship among the vectors v_1, v_2, v_3, v. This shows that if $S = \{v_1, v_2, v_3\}$ is linearly dependent then so is $T = \{v_1, v_2, v_3, v\}$ for any v.

31. $(u - v) + (v - w) + (w - u) = 0$; thus the vectors $u - v, v - w$, and $w - u$ form a linearly dependent set.

33. (a) No. It is not closed under either addition or scalar multiplication.

(b) $\mathbf{p}_{876} = 0.38\mathbf{c} + 0.59\mathbf{m} + 0.73\mathbf{y} + 0.07\mathbf{k}$

   $\mathbf{p}_{216} = 0.83\mathbf{m} + 0.34\mathbf{y} + 0.47\mathbf{k}$

   $\mathbf{p}_{328} = \mathbf{c} + 0.47\mathbf{y} + 0.30\mathbf{k}$

(c) $\frac{1}{2}(\mathbf{p}_{876} + \mathbf{p}_{216})$ corresponds to the CMYK vector $(0.19, 0.71, 0.535, 0.27)$.

35. (a) $k_1 = k_2 = k_3 = k_4 = \frac{1}{4}$ (b) $k_1 = k_2 = k_3 = k_4 = k_5 = \frac{1}{5}$

  (c) The components of $\mathbf{x} = \frac{1}{3}\mathbf{r}_1 + \frac{1}{3}\mathbf{r}_2 + \frac{1}{3}\mathbf{r}_3$ represent the average total population of Philadelphia, Bucks, and Delaware counties in each of the sampled years.

DISCUSSION AND DISCOVERY

D1. (a) Two nonzero vectors will span R^2 if and only if they are do not lie on a line.

  (b) Three nonzero vectors will span R^3 if and only if they do not lie in a plane.

D2. (a) Two vectors in R^n will span a plane if and only if they are nonzero and not scalar multiples of one another.

  (b) Two vectors in R^n will span a line if and only if they are not both zero and one is a scalar multiple of the other.

  (c) span$\{\mathbf{u}\}$ = span$\{\mathbf{v}\}$ if and only if one of the vectors $\mathbf{u}$ and $\mathbf{v}$ is a scalar multiple of the other.

D3. (a) Yes. If three nonzero vectors are mutually orthogonal then none of them lies in the plane spanned by the other two; thus the three are linearly independent.

  (b) Suppose the vectors $\mathbf{v}_1$, $\mathbf{v}_2$, and $\mathbf{v}_3$ are nonzero and mutually orthogonal; thus $\mathbf{v}_i \cdot \mathbf{v}_i = \|\mathbf{v}_i\|^2 > 0$ for $i = 1, 2, 3$ and $\mathbf{v}_i \cdot \mathbf{v}_j = 0$ for $i \neq j$. To prove they are linearly independent we must show that if

$$c_1\mathbf{v}_1 + c_2\mathbf{v}_2 + c_3\mathbf{v}_3 = \mathbf{0}$$

then $c_1 = c_2 = c_3 = 0$. This follows from the fact that if $c_1\mathbf{v}_1 + c_2\mathbf{v}_2 + c_3\mathbf{v}_3 = \mathbf{0}$, then

$$c_i \|\mathbf{v}_i\|^2 = \mathbf{v}_i \cdot (c_1\mathbf{v}_1 + c_2\mathbf{v}_2 + c_3\mathbf{v}_3) = \mathbf{v}_i \cdot \mathbf{0} = 0$$

for $i = 1, 2, 3$.

D4. The vectors in the first figure are linearly independent since none of them lies in the plane spanned by the other two (none of them can be expressed as a linear combination of the other two). The vectors in the second figure are linearly dependent since $\mathbf{v}_3 = \mathbf{v}_1 + \mathbf{v}_2$.

D5. This set is closed under scalar multiplication, but not under addition. For example, the vectors $\mathbf{u} = (1, 2)$ and $\mathbf{v} = (-2, -1)$ correspond to points in the set, but $\mathbf{u} + \mathbf{v} = (-1, 1)$ does not.

D6. (a) False. For example, two of the vectors may lie on a line (so one is a scalar multiple of the other), but the third vector may not lie on this same line and therefore cannot be expressed as a linear combination of the other two.

  (b) False. The set of all linear combinations of two vectors can be $\{\mathbf{0}\}$ (if both are $\mathbf{0}$), a line (if one is a scalar multiple of the other), or a plane (if they are linearly independent).

  (c) False. For example, $\mathbf{v}$ and $\mathbf{w}$ might be linearly dependent (scalar multiples of each other). [But it is true that if $\{\mathbf{v}, \mathbf{w}\}$ is a linearly independent set, and if $\mathbf{u}$ cannot be expressed as a linear combination of $\mathbf{v}$ and $\mathbf{w}$, then $\{\mathbf{u}, \mathbf{v}, \mathbf{w}\}$ is a linearly independent set.]

  (d) True. See Example 9.

(e) True. If $c_1(k\mathbf{v}_1) + c_2(k\mathbf{v}_2) + c_2(k\mathbf{v}_2) = \mathbf{0}$, then $k(c_1\mathbf{v}_1 + c_2\mathbf{v}_2 + c_3\mathbf{v}_2) = \mathbf{0}$. Thus, since $k \neq 0$, it follows that $c_1\mathbf{v}_1 + c_2\mathbf{v}_2 + c_3\mathbf{v}_2 = \mathbf{0}$ and so $c_1 = c_2 = c_3 = 0$.

D7. (a) False. The set $\{\mathbf{u}, k\mathbf{u}\}$ is always a linearly dependent set.

(b) False. This statement is true for a homogeneous system ($\mathbf{b} = \mathbf{0}$), but not for a non-homogeneous system. [The solution space of a non-homogeneous linear system is a translated subspace.]

(c) True. If W is a subspace, then W is already closed under scalar multiplication and addition, and so $\text{span}(W) = W$.

(d) False. For example, if $S_1 = \{(1,0), (0,1)\}$ and $S_2 = \{(1,1), (0,1)\}$, then $\text{span}(S_1) = \text{span}(S_2) = R^2$, but $S_1 \neq S_2$.

D8. Since $\text{span}(S)$ is a subspace (already closed under scalar multiplication and addition), we have $\text{span}(\text{span}(S)) = \text{span}(S)$.

WORKING WITH PROOFS

P1. Let θ be the angle between $\mathbf{u}$ and $\mathbf{w}$, and let ϕ be the angle between $\mathbf{v}$ and $\mathbf{w}$. We will show that $\theta = \phi$. First recall that $\mathbf{u} \cdot \mathbf{w} = \|\mathbf{u}\|\|\mathbf{w}\| \cos\theta$, so $\mathbf{u} \cdot \mathbf{w} = k\|\mathbf{w}\| \cos\theta$. Similarly, $\mathbf{v} \cdot \mathbf{w} = l\|\mathbf{w}\| \cos\phi$. On the other hand we have

$$\mathbf{u} \cdot \mathbf{w} = \mathbf{u} \cdot (l\mathbf{u} + k\mathbf{v}) = l(\mathbf{u} \cdot \mathbf{u}) + k(\mathbf{u} \cdot \mathbf{v}) = lk^2 + k(\mathbf{u} \cdot \mathbf{v})$$

and so $k\|\mathbf{w}\| \cos\theta = \mathbf{u} \cdot \mathbf{w} = lk^2 + k(\mathbf{u} \cdot \mathbf{v})$, i.e. $\|\mathbf{w}\| \cos\theta = lk + (\mathbf{u} \cdot \mathbf{v})$. Similar calculations show that $\|\mathbf{w}\| \cos\phi = (\mathbf{v} \cdot \mathbf{u}) + kl$; thus $\|\mathbf{w}\| \cos\theta = \|\mathbf{w}\| \cos\phi$. It follows that $\cos\theta = \cos\phi$ and $\theta = \phi$.

P2. If $\mathbf{x}$ belongs to $W_1 \cap W_2$ and k is a scalar, then $k\mathbf{x}$ also belongs to $W_1 \cap W_2$ since both W_1 and W_2 are subspaces. Similarly, if $\mathbf{x}_1$ and $\mathbf{x}_2$ belong to $W_1 \cap W_2$, then $\mathbf{x}_1 + \mathbf{x}_2$ belongs to $W_1 \cap W_2$. Thus $W_1 \cap W_2$ is closed under scalar multiplication and addition, i.e. $W_1 \cap W_2$ is a subspace.

P3. First we show that $W_1 + W_2$ is closed under scalar multiplication: Suppose $\mathbf{z} = \mathbf{x} + \mathbf{y}$ where $\mathbf{x}$ is in W_1 and $\mathbf{y}$ is in W_2. Then, for any scalar k, we have $k\mathbf{z} = k(\mathbf{x} + \mathbf{y}) = k\mathbf{x} + k\mathbf{y}$, where $k\mathbf{x}$ is in W_1 and $k\mathbf{y}$ is in W_2 (since W_1 and W_2 are subspaces); thus $k\mathbf{z}$ is in $W_1 + W_2$. Finally we show that $W_1 + W_2$ is closed under addition: Suppose $\mathbf{z}_1 = \mathbf{x}_1 + \mathbf{y}_1$ and $\mathbf{z}_2 = \mathbf{x}_2 + \mathbf{y}_2$, where $\mathbf{x}_1$ and $\mathbf{x}_2$ are in W_1 and $\mathbf{y}_1$ and $\mathbf{y}_2$ are in W_2. Then $\mathbf{z}_1 + \mathbf{z}_2 = (\mathbf{x}_1 + \mathbf{y}_1) + (\mathbf{x}_2 + \mathbf{y}_2) = (\mathbf{x}_1 + \mathbf{x}_2) + (\mathbf{y}_1 + \mathbf{y}_2)$, where $\mathbf{x}_1 + \mathbf{x}_2$ is in W_1 and $\mathbf{y}_1 + \mathbf{y}_2$ is in W_2 (since W_1 and W_2 are subspaces); thus $\mathbf{z}_1 + \mathbf{z}_2$ is in $W_1 + W_2$.

EXERCISE SET 3.5

1. (a) The reduced row echelon form of the augmented matrix of the homogeneous system is

$$\begin{bmatrix} 1 & \frac{2}{3} & -\frac{1}{3} & 0 \\ 0 & 0 & 0 & 0 \\ 0 & 0 & 0 & 0 \end{bmatrix}$$

thus a general solution is $x_1 = -\frac{2}{3}s + \frac{1}{3}t$, $x_2 = s$, $x_3 = t$; or (in column vector form)

$$\begin{bmatrix} x_1 \\ x_2 \\ x_3 \end{bmatrix} = s \begin{bmatrix} -\frac{2}{3} \\ 1 \\ 0 \end{bmatrix} + t \begin{bmatrix} \frac{1}{3} \\ 0 \\ 1 \end{bmatrix}$$

(b) $\begin{bmatrix} 3 & 2 & -1 \\ 6 & 4 & -2 \\ -3 & -2 & 1 \end{bmatrix} \begin{bmatrix} 1 \\ 0 \\ 1 \end{bmatrix} = \begin{bmatrix} 2 \\ 4 \\ -2 \end{bmatrix}$

(c) From (a) and (b), a general solution of the nonhomogeneous system is given by

$$\begin{bmatrix} x_1 \\ x_2 \\ x_3 \end{bmatrix} = \begin{bmatrix} 1 \\ 0 \\ 1 \end{bmatrix} + s \begin{bmatrix} -\frac{2}{3} \\ 1 \\ 0 \end{bmatrix} + t \begin{bmatrix} \frac{1}{3} \\ 0 \\ 1 \end{bmatrix}$$

(d) The reduced row echelon form of the augmented matrix of the nonhomogeneous system is

$$\begin{bmatrix} 1 & \frac{2}{3} & -\frac{1}{3} & \frac{2}{3} \\ 0 & 0 & 0 & 0 \\ 0 & 0 & 0 & 0 \end{bmatrix}$$

thus a general solution is given by $x_1 = \frac{2}{3} - \frac{2}{3}s' + \frac{1}{3}t'$, $x_2 = s'$, $x_3 = t'$; or

$$\begin{bmatrix} x_1 \\ x_2 \\ x_3 \end{bmatrix} = \begin{bmatrix} \frac{2}{3} \\ 0 \\ 0 \end{bmatrix} + s' \begin{bmatrix} -\frac{2}{3} \\ 1 \\ 0 \end{bmatrix} + t' \begin{bmatrix} \frac{1}{3} \\ 0 \\ 1 \end{bmatrix}$$

This solution is related to the one in part (c) by the change of variable $s' = s$, $t' = t + 1$.

3. (a) The reduced row echelon form of the augmented matrix of the given system is

$$\begin{bmatrix} 1 & \frac{4}{3} & \frac{1}{3} & 0 & \frac{1}{3} \\ 0 & 0 & 0 & 1 & 1 \\ 0 & 0 & 0 & 0 & 0 \end{bmatrix}$$

thus a general solution is given by $x_1 = \frac{1}{3} - \frac{4}{3}s - \frac{1}{3}t$, $x_2 = s$, $x_3 = t$, $x_4 = 1$; or

$$\begin{bmatrix} x_1 \\ x_2 \\ x_3 \\ x_4 \end{bmatrix} = \begin{bmatrix} \frac{1}{3} \\ 0 \\ 0 \\ 1 \end{bmatrix} + s \begin{bmatrix} -\frac{4}{3} \\ 1 \\ 0 \\ 0 \end{bmatrix} + t \begin{bmatrix} -\frac{1}{3} \\ 0 \\ 1 \\ 0 \end{bmatrix}$$

(b) A general solution of the associated homogeneous system is

$$\begin{bmatrix} x_1 \\ x_2 \\ x_3 \\ x_4 \end{bmatrix} = s \begin{bmatrix} -\frac{4}{3} \\ 1 \\ 0 \\ 0 \end{bmatrix} + t \begin{bmatrix} -\frac{1}{3} \\ 0 \\ 1 \\ 0 \end{bmatrix}$$

and a particular solution of the given nonhomogenous system is $x_1 = \frac{1}{3}$, $x_2 = 0$, $x_3 = 0$, $x_4 = 1$.

5. The vector $\mathbf{w}$ can be expressed as a linear combination of $\mathbf{v}_1$, $\mathbf{v}_2$, and $\mathbf{v}_3$, if and only if the system

$$\begin{bmatrix} 2 & 4 & -10 \\ 3 & 9 & -21 \\ 1 & 5 & -12 \end{bmatrix} \begin{bmatrix} c_1 \\ c_2 \\ c_3 \end{bmatrix} = \begin{bmatrix} -2 \\ 0 \\ 1 \end{bmatrix}$$

is consistent. The reduced row echelon form of the augmented matrix of this system is

$$\begin{bmatrix} 1 & 0 & 0 & -2 \\ 0 & 1 & 0 & 3 \\ 0 & 0 & 1 & 1 \end{bmatrix}$$

From this we conclude that the system has a unique solution, and that $\mathbf{w} = -2\mathbf{v}_1 + 3\mathbf{v}_2 + \mathbf{v}_3$.

7. The vector $\mathbf{w}$ is in span$\{\mathbf{v}_1, \mathbf{v}_2, \mathbf{v}_3, \mathbf{v}_4\}$ if and only if the system

$$\begin{bmatrix} 1 & 1 & 1 & 1 \\ -1 & 1 & -3 & 3 \\ 1 & 0 & 2 & -1 \end{bmatrix} \begin{bmatrix} c_1 \\ c_2 \\ c_3 \\ c_4 \end{bmatrix} = \begin{bmatrix} 1 \\ 5 \\ -2 \end{bmatrix}$$

is consistent. The row reduced echelon form of the augmented matrix of this system is

$$\begin{bmatrix} 1 & 0 & 2 & -1 & -2 \\ 0 & 1 & -1 & 2 & 3 \\ 0 & 0 & 0 & 0 & 0 \end{bmatrix}$$

From this we conclude that the system has infinitely many solutions; thus $\mathbf{w}$ is in span$\{\mathbf{v}_1, \mathbf{v}_2, \mathbf{v}_3, \mathbf{v}_4\}$.

9. (a) The hyperplane $\mathbf{a}^{\perp}$ consists of all vectors $\mathbf{x} = (x, y)$ such that $\mathbf{a} \cdot \mathbf{x} = 0$, i.e. $-2x + 3y = 0$. This corresponds to the line through the origin with parametric equations $x = \frac{3}{2}t$, $y = t$.

(b) The hyperplane $\mathbf{a}^{\perp}$ consists of all vectors $\mathbf{x} = (x, y, z)$ such that $\mathbf{a} \cdot \mathbf{x} = 0$, i.e. $4x - 5z = 0$. This corresponds to the plane through the origin with parametric equations $x = \frac{5}{4}t$, $y = s$, $z = t$.

(c) $\mathbf{a}^{\perp}$ consists of all vectors $\mathbf{x} = (x_1, x_2, x_3, x_4)$ such that $\mathbf{a} \cdot \mathbf{x} = 0$, i.e. $x_1 + 2x_2 - 3x_3 + 7x_4 = 0$. This is a hyperplane in R^4 with parametric equations $x_1 = -2r + 3s - 7t$, $x_2 = r$, $x_3 = s$, $x_4 = t$.

11. This system reduces to a single equation, $x_1 + x_2 + x_3 = 0$. Thus a general solution is given by $x_1 = -s - t, x_2 = s, x_3 = t$; or (in vector form)

$$\begin{bmatrix} x_1 \\ x_2 \\ x_3 \end{bmatrix} = s \begin{bmatrix} -1 \\ 1 \\ 0 \end{bmatrix} + t \begin{bmatrix} -1 \\ 0 \\ 1 \end{bmatrix}$$

The solution space is two-dimensional.

13. The reduced row echelon form of the augmented matrix of the system is

$$\begin{bmatrix} 1 & 0 & -\frac{3}{7} & \frac{19}{7} & \frac{8}{7} & 0 \\ 0 & 1 & \frac{2}{7} & -\frac{1}{7} & -\frac{3}{7} & 0 \end{bmatrix}$$

Thus a general solution is $x_1 = \frac{3}{7}r - \frac{19}{7}s - \frac{8}{7}t$, $x_2 = -\frac{2}{7}r + \frac{1}{7}s + \frac{3}{7}t$, $x_3 = r$, $x_4 = s$, $x_5 = t$; or

$$\begin{bmatrix} x_1 \\ x_2 \\ x_3 \\ x_4 \\ x_5 \end{bmatrix} = r \begin{bmatrix} \frac{3}{7} \\ -\frac{2}{7} \\ 1 \\ 0 \\ 0 \end{bmatrix} + s \begin{bmatrix} -\frac{19}{7} \\ \frac{1}{7} \\ 0 \\ 1 \\ 0 \end{bmatrix} + t \begin{bmatrix} -\frac{8}{7} \\ \frac{3}{7} \\ 0 \\ 0 \\ 1 \end{bmatrix}$$

The solution space is three-dimensional.

15. (a) A general solution is $x = 1 - s - t$, $y = s$, $z = t$; or $\begin{bmatrix} x \\ y \\ z \end{bmatrix} = \begin{bmatrix} 1 \\ 0 \\ 0 \end{bmatrix} + s \begin{bmatrix} -1 \\ 1 \\ 0 \end{bmatrix} + t \begin{bmatrix} -1 \\ 0 \\ 1 \end{bmatrix}$.

(b) The solution space corresponds to the plane which passes through the point $P(1, 0, 0)$ and is parallel to the vectors $\mathbf{v}_1 = (-1, 1, 0)$ and $\mathbf{v}_2 = (-1, 0, 1)$.

17. (a) A vector $\mathbf{x} = (x, y, z)$ is orthogonal to $\mathbf{a} = (1, 1, 1)$ and $\mathbf{b} = (-2, 3, 0)$ if and only if

$$x + y + z = 0$$
$$-2x + 3y = 0$$

(b) The solution space is the line through the origin that is perpendicular to the vectors $\mathbf{a}$ and $\mathbf{b}$.

(c) The reduced row echelon form of the augmented matrix of the system is

$$\begin{bmatrix} 1 & 0 & \frac{3}{5} & 0 \\ 0 & 1 & \frac{2}{5} & 0 \end{bmatrix}$$

and so a general solution is given by $x = -\frac{3}{5}t$, $y = -\frac{2}{5}t$, $z = t$; or $(x, y, z) = t(-\frac{3}{5}, -\frac{2}{5}, 1)$. Note that the vector $\mathbf{v} = (-\frac{3}{5}, -\frac{2}{5}, 1)$ is orthogonal to both $\mathbf{a}$ and $\mathbf{b}$.

19. (a) A vector $\mathbf{x} = (x_1, x_2, x_3, x_4)$ is orthogonal to $\mathbf{v}_1 = (1, 1, 2, 2)$ and $\mathbf{v}_2 = (5, 4, 3, 4)$ if and only if

$$x_1 + x_2 + 2x_3 + 2x_4 = 0$$
$$5x_1 + 4x_2 + 3x_3 + 4x_4 = 0$$

(b) The solution space is the plane (2 dimensional subspace) in R^4 that passes through the origin and is perpendicular to the vectors $\mathbf{v}_1$ and $\mathbf{v}_2$.

(c) The reduced row echelon form of the augmented matrix of the system is

$$\begin{bmatrix} 1 & 0 & -5 & -4 & 0 \\ 0 & 1 & 7 & 6 & 0 \end{bmatrix}$$

and so a general solution of the system is given by $(x_1, x_2, x_3, x_4) = s(5, -7, 1, 0) + t(4, -6, 0, 1)$. Note that the vectors $(5, -7, 1, 0)$ and $(4, -6, 0, 1)$ are orthogonal to both $\mathbf{v}_1$ and $\mathbf{v}_2$.

DISCUSSION AND DISCOVERY

D1. The solution set of $A\mathbf{x} = \mathbf{b}$ is a translated subspace $\mathbf{x}_0 + W$, where W is the solution space of $A\mathbf{x} = \mathbf{0}$.

D2. If $\mathbf{v}$ is orthogonal to every row of A, then $A\mathbf{v} = \mathbf{0}$, and so (since A is invertible) $\mathbf{v} = \mathbf{0}$.

D3. The general solution will have at least 3 free variables. Thus, assuming $A \neq 0$, the solution space will be of dimension 3, 4, 5, or 6 depending on how much redundancy there is.

D4. (a) True. The solution set of $A\mathbf{x} = \mathbf{b}$ is of the form $\mathbf{x}_0 + W$ where W is the solution space of $A\mathbf{x} = \mathbf{0}$.

(b) False. For example, the system $\begin{matrix} x - y = 0 \\ x - y = 1 \end{matrix}$ is inconsistent, but the associated homogeneous system has infinitely many solutions.

(c) True. Each hyperplane corresponds to a single homogeneous linear equation in four variables, and there must be a least four equations in order to have a unique solution.

(d) True. Every plane in R^3 corresponds to a equation of the form $ax + by + cz = d$.

(e) False. A vector $\mathbf{x}$ is orthogonal to row(A) if and only if $\mathbf{x}$ is a solution of the homogenous system $A\mathbf{x} = \mathbf{0}$.

WORKING WITH PROOFS

P1. Suppose that $A\mathbf{x} = \mathbf{0}$ has only the trivial solution and that $A\mathbf{x} = \mathbf{b}$ is consistent. If $\mathbf{x}_1$ and $\mathbf{x}_2$ are two solutions of $A\mathbf{x} = \mathbf{b}$, then $A(\mathbf{x}_1 - \mathbf{x}_2) = A\mathbf{x}_1 - A\mathbf{x}_2 = \mathbf{b} - \mathbf{b} = \mathbf{0}$ and so $\mathbf{x}_1 - \mathbf{x}_2 = \mathbf{0}$, i.e. $\mathbf{x}_1 = \mathbf{x}_2$. Thus, if $A\mathbf{x} = \mathbf{0}$ has only the trivial solution, the system $A\mathbf{x} = \mathbf{b}$ is either inconsistent or has exactly one solution.

P2. Suppose that $A\mathbf{x} = \mathbf{0}$ has infinitely many solutions and that $A\mathbf{x} = \mathbf{b}$ is consistent. Let $\mathbf{x}_0$ be any solution of $A\mathbf{x} = \mathbf{b}$. Then, for any solution $\mathbf{w}$ of $A\mathbf{x} = \mathbf{0}$, we have $A(\mathbf{x}_0 + \mathbf{w}) = A\mathbf{x}_0 + A\mathbf{w} = \mathbf{b} + \mathbf{0} = \mathbf{b}$. Thus, if $A\mathbf{x} = \mathbf{0}$ has infinitely many solutions, the system $A\mathbf{x} = \mathbf{b}$ is either inconsistent or has infinitely many solutions. Conversely, if $A\mathbf{x} = \mathbf{b}$ has at most one solution, then $A\mathbf{x} = \mathbf{0}$ has only the trivial solution.

P3. If $\mathbf{x}_1$ is a solution of $A\mathbf{x} = \mathbf{b}$ and $\mathbf{x}_2$ is a solution of $A\mathbf{x} = \mathbf{c}$, then $A(\mathbf{x}_1 + \mathbf{x}_2) = A\mathbf{x}_1 + A\mathbf{x}_2 = \mathbf{b} + \mathbf{c}$; i.e. $\mathbf{x}_1 + \mathbf{x}_2$ is a solution of $A\mathbf{x} = \mathbf{b} + \mathbf{c}$. Thus if $A\mathbf{x} = \mathbf{b}$ and $A\mathbf{x} = \mathbf{c}$ are consistent systems, then $A\mathbf{x} = \mathbf{b} + \mathbf{c}$ is also consistent. This argument can easily be adapted to prove the following:

Theorem. If $A\mathbf{x} = \mathbf{b}_j$ is consistent for each $j = 1, 2, \ldots, r$ and if $\mathbf{b} = \mathbf{b}_1 + \mathbf{b}_2 + \cdots + \mathbf{b}_r$, then $A\mathbf{x} = \mathbf{b}$ is also consistent.

P4. Since $(k\mathbf{a}) \cdot \mathbf{x} = k(\mathbf{a} \cdot \mathbf{x})$ and $k \neq 0$, it follows that $(k\mathbf{a}) \cdot x = 0$ if and only if $\mathbf{a} \cdot \mathbf{x} = 0$. This proves that $(k\mathbf{a})^\perp = \mathbf{a}^\perp$.

EXERCISE SET 3.6

1. (a) This matrix has a row of zeros; it is not invertible.

(b) A diagonal matrix with nonzero entries on the diagonal;
$$\begin{bmatrix} -1 & 0 & 0 \\ 0 & 2 & 0 \\ 0 & 0 & \frac{1}{3} \end{bmatrix}^{-1} = \begin{bmatrix} -1 & 0 & 0 \\ 0 & \frac{1}{2} & 0 \\ 0 & 0 & 3 \end{bmatrix}.$$

3. (a) $\begin{bmatrix} 3 & 0 & 0 \\ 0 & -1 & 0 \\ 0 & 0 & 2 \end{bmatrix} \begin{bmatrix} 2 & 1 \\ -4 & 1 \\ 2 & 5 \end{bmatrix} = \begin{bmatrix} 6 & 3 \\ 4 & -1 \\ 4 & 10 \end{bmatrix}$ **(b)** $\begin{bmatrix} 2 & 1 \\ -4 & 1 \\ 2 & 5 \end{bmatrix} \begin{bmatrix} 5 & 0 \\ 0 & -2 \end{bmatrix} = \begin{bmatrix} 10 & -2 \\ -20 & -2 \\ 10 & -10 \end{bmatrix}$

(c) $\begin{bmatrix} 3 & 0 & 0 \\ 0 & -1 & 0 \\ 0 & 0 & 2 \end{bmatrix} \begin{bmatrix} 2 & 1 \\ -4 & 1 \\ 2 & 5 \end{bmatrix} \begin{bmatrix} 5 & 0 \\ 0 & -2 \end{bmatrix} = \begin{bmatrix} 3 & 0 & 0 \\ 0 & -1 & 0 \\ 0 & 0 & 2 \end{bmatrix} \begin{bmatrix} 10 & -2 \\ -20 & -2 \\ 10 & -10 \end{bmatrix} = \begin{bmatrix} 30 & -6 \\ 20 & 2 \\ 20 & -20 \end{bmatrix}$

5. (a) $A^2 = \begin{bmatrix} 1 & 0 & 0 \\ 0 & 4 & 0 \\ 0 & 0 & 16 \end{bmatrix}$ **(b)** $A^{-2} = \begin{bmatrix} 1 & 0 & 0 \\ 0 & \frac{1}{4} & 0 \\ 0 & 0 & \frac{1}{16} \end{bmatrix}$ **(c)** $A^{-k} = \begin{bmatrix} 1 & 0 & 0 \\ 0 & (-\frac{1}{2})^k & 0 \\ 0 & 0 & (\frac{1}{4}^k) \end{bmatrix}$

7. A is invertible if and only if $x \neq 1, -2, 4$ (the diagonal entries must be nonzero).

9. Apply the inversion algorithm (see Section 3.3) to find the inverse of A.

$$\left[\begin{array}{ccc|ccc} 1 & 2 & 3 & 1 & 0 & 0 \\ 0 & 1 & -2 & 0 & 1 & 0 \\ 0 & 0 & 1 & 0 & 0 & 1 \end{array}\right]$$

Add 2 times row 3 to row 2. Add -3 times row 3 to row 1.

$$\left[\begin{array}{ccc|ccc} 1 & 2 & 0 & 1 & 0 & -3 \\ 0 & 1 & 0 & 0 & 1 & 2 \\ 0 & 0 & 1 & 0 & 0 & 1 \end{array}\right]$$

Add -2 times row 2 to row 1.

$$\left[\begin{array}{ccc|ccc} 1 & 0 & 0 & 1 & -2 & -7 \\ 0 & 1 & 0 & 0 & 1 & 2 \\ 0 & 0 & 1 & 0 & 0 & 1 \end{array}\right]$$

Thus the inverse of the upper triangular matrix $A = \begin{bmatrix} 1 & 2 & 3 \\ 0 & 1 & -2 \\ 0 & 0 & 1 \end{bmatrix}$ is $A^{-1} = \begin{bmatrix} 1 & -2 & -7 \\ 0 & 1 & 2 \\ 0 & 0 & 1 \end{bmatrix}$.

11. $A = \begin{bmatrix} 0 & 0 & 4 \\ 0 & 0 & 1 \\ -4 & -1 & 0 \end{bmatrix}$

13. The matrix A is symmetric if and only if a, b, and c satisfy the following equations:

$$\begin{array}{rcr} a - 2b + 2c = & 3 \\ 2a + b + c = & 0 \\ a + c = & -2 \end{array}$$

The augmented matrix of this system is

$$\begin{bmatrix} 1 & -2 & 2 & 3 \\ 2 & 1 & 1 & 0 \\ 1 & 0 & 1 & -2 \end{bmatrix}$$

and the reduced row echelon form is

$$\begin{bmatrix} 1 & 0 & 0 & 11 \\ 0 & 1 & 0 & -9 \\ 0 & 0 & 1 & -13 \end{bmatrix}$$

Thus, in order for the matrix A to be symmetric we must have $a = 11$, $b = -9$, and $c = -13$.

15. We have $A^{-1} = \begin{bmatrix} 2 & -1 \\ -1 & 3 \end{bmatrix}^{-1} = \dfrac{1}{5}\begin{bmatrix} 3 & 1 \\ 1 & 2 \end{bmatrix}$; thus A^{-1} is symmetric.

17. $AB = \begin{bmatrix} -1 & 2 & 5 \\ 0 & 1 & 3 \\ 0 & 0 & -4 \end{bmatrix}\begin{bmatrix} 2 & -8 & 0 \\ 0 & 2 & 1 \\ 0 & 0 & 3 \end{bmatrix} = \begin{bmatrix} -2 & 12 & 17 \\ 0 & 2 & 10 \\ 0 & 0 & -12 \end{bmatrix}$

19. If $A = \begin{bmatrix} 1 & 0 & 0 \\ 0 & -1 & 0 \\ 0 & 0 & -1 \end{bmatrix}$, then $A^{-5} = \begin{bmatrix} (1)^{-5} & 0 & 0 \\ 0 & (-1)^{-5} & 0 \\ 0 & 0 & (-1)^{-5} \end{bmatrix} = \begin{bmatrix} 1 & 0 & 0 \\ 0 & -1 & 0 \\ 0 & 0 & -1 \end{bmatrix}$.

21. The lower triangular matrix $A = \begin{bmatrix} 1 & 0 & 0 \\ 3 & 1 & 0 \\ 1 & 3 & 1 \end{bmatrix}$ is invertible because of the nonzero diagonal entries.
Thus, from Theorem 3.6.5, AA^T and $A^T A$ are also invertible; furthermore, since these matrices are symmetric, their inverses are also symmetric by Theorem 3.6.4. Following are the matrices AA^T, $A^T A$, and their inverses.

$$A^T A = \begin{bmatrix} 11 & 6 & 1 \\ 6 & 10 & 3 \\ 1 & 3 & 1 \end{bmatrix} \qquad (A^T A)^{-1} = \begin{bmatrix} 1 & -3 & 8 \\ -3 & 10 & -27 \\ 8 & -27 & 74 \end{bmatrix}$$

$$AA^T = \begin{bmatrix} 1 & 3 & 1 \\ 3 & 10 & 6 \\ 1 & 6 & 11 \end{bmatrix} \qquad (AA^T)^{-1} = \begin{bmatrix} 74 & -27 & 8 \\ -27 & 10 & -3 \\ 8 & -3 & 1 \end{bmatrix}$$

23. The fixed points of the matrix $A = \begin{bmatrix} 2 & 1 \\ 1 & 2 \end{bmatrix}$ are the solutions of the homogeneous system $(I - A)\mathbf{x} = \mathbf{0}$. The augmented matrix of this system is

$$\begin{bmatrix} -1 & -1 & 0 \\ -1 & -1 & 0 \end{bmatrix}$$

and from this it is easy to see that the system has the general solution $x_1 = t$, $x_2 = -t$. Thus the fixed points of A are vectors of the form $\mathbf{x} = \begin{bmatrix} t \\ -t \end{bmatrix} = t\begin{bmatrix} 1 \\ -1 \end{bmatrix}$, where $-\infty < t < \infty$.

25. (a) If $A = \begin{bmatrix} 0 & 1 \\ 0 & 0 \end{bmatrix}$, then $A^2 = \begin{bmatrix} 0 & 1 \\ 0 & 0 \end{bmatrix}\begin{bmatrix} 0 & 1 \\ 0 & 0 \end{bmatrix} = \begin{bmatrix} 0 & 0 \\ 0 & 0 \end{bmatrix}$. Thus A is nilpotent with nilpotency index 2, and the inverse of $I - A = \begin{bmatrix} 1 & -1 \\ 0 & 1 \end{bmatrix}$ is $(I - A)^{-1} = I + A = \begin{bmatrix} 1 & 1 \\ 0 & 1 \end{bmatrix}$.

(b) The nilpotency index of $A = \begin{bmatrix} 0 & 0 & 0 \\ 1 & 0 & 0 \\ 8 & 1 & 0 \end{bmatrix}$ is 3; thus the inverse of $I - A = \begin{bmatrix} 1 & 0 & 0 \\ -1 & 1 & 0 \\ -8 & -1 & 1 \end{bmatrix}$ is

$$(I - A)^{-1} = I + A + A^2 = \begin{bmatrix} 1 & 0 & 0 \\ 0 & 1 & 0 \\ 0 & 0 & 1 \end{bmatrix} + \begin{bmatrix} 0 & 0 & 0 \\ 1 & 0 & 0 \\ 8 & 1 & 0 \end{bmatrix} + \begin{bmatrix} 0 & 0 & 0 \\ 0 & 0 & 0 \\ 1 & 0 & 0 \end{bmatrix} = \begin{bmatrix} 1 & 0 & 0 \\ 1 & 1 & 0 \\ 9 & 1 & 1 \end{bmatrix}$$

27. (a) If A is invertible and skew symmetric, then $(A^{-1})^T = (A^T)^{-1} = (-A)^{-1} = -A^{-1}$; thus A^{-1} is skew symmetric.

(b) If A and B are skew symmetric, then

$$(A + B)^T = A^T + B^T = -A - B = -(A + B)$$
$$(A - B)^T = A^T - B^T = -A + B = -(A - B)$$
$$(kA)^T = kA^T = k(-A) = -kA$$

thus the matrices $A + B$, $A - B$, and kA are also skew-symmetric.

29. If $H = I_n - 2\mathbf{u}\mathbf{u}^T$ then we have $H^T = I_n^T - 2(\mathbf{u}\mathbf{u}^T)^T = I_n - 2\mathbf{u}\mathbf{u}^T = H$; thus H is symmetric.

31. Using Formula (9), we have $\operatorname{tr}(A^T A) = \|\mathbf{a}_1\|^2 + \|\mathbf{a}_2\|^2 + \cdots + \|\mathbf{a}_n\|^2 = 1 + 1 + \cdots + 1 = n$.

DISCUSSION AND DISCOVERY

D1. **(a)** $A = \begin{bmatrix} 3a_{11} & 5a_{12} & 7a_{13} \\ 3a_{21} & 5a_{22} & 7a_{23} \\ 3a_{31} & 5a_{32} & 7a_{33} \end{bmatrix} = \begin{bmatrix} a_{11} & a_{12} & a_{13} \\ a_{21} & a_{22} & a_{23} \\ a_{31} & a_{32} & a_{33} \end{bmatrix} \begin{bmatrix} 3 & 0 & 0 \\ 0 & 5 & 0 \\ 0 & 0 & 7 \end{bmatrix} = BD$

(b) There are other such factorizations as well. For example, $A = \begin{bmatrix} \frac{1}{3}a_{11} & a_{12} & 14a_{13} \\ \frac{1}{3}a_{21} & a_{22} & 14a_{23} \\ \frac{1}{3}a_{31} & a_{32} & 14a_{33} \end{bmatrix} \begin{bmatrix} 9 & 0 & 0 \\ 0 & 5 & 0 \\ 0 & 0 & \frac{1}{2} \end{bmatrix}$.

D2. **(a)** A is symmetric since $a_{ji} = j^2 + i^2 = i^2 + j^2 = a_{ij}$ for each i, j.
(b) A is not symmetric since $a_{ji} = j - i = -(i - j) = -a_{ij}$ for each i, j. In fact, A is skew-symmmetric.
(c) A is symmetric since $a_{ji} = 2j + 2i = 2i + 2j = a_{ij}$ for each i, j.
(d) A is not symmetric since $a_{ji} = 2j^2 + 2i^3 \neq 2i^2 + 2j^3 = a_{ij}$ if $i = 1$, $j = 2$.

In general, the matrix $A = [f(i, j)]$ is symmetric if and only if $f(j, i) = f(i, j)$ for each i, j.

D3. No. In fact, if A and B are commuting skew-symmetric matrices, then

$$(AB)^T = B^T A^T = (-B)(-A) = BA = AB$$

and so the product AB is symmetric rather than skew-symmetric.

D4. Using Formula (9), we have $\operatorname{tr}(A^T A) = \|\mathbf{a}_1\|^2 + \|\mathbf{a}_2\|^2 + \cdots + \|\mathbf{a}_n\|^2$. Thus, if $A^T A = 0$, then it follows that $\|\mathbf{a}_1\|^2 = \|\mathbf{a}_2\|^2 = \cdots = \|\mathbf{a}_n\|^2 = 0$ and so $A = 0$.

D5. **(a)** If $A = \begin{bmatrix} d_1 & 0 & 0 \\ 0 & d_2 & 0 \\ 0 & 0 & d_3 \end{bmatrix}$, then $A^2 = \begin{bmatrix} d_1^2 & 0 & 0 \\ 0 & d_2^2 & 0 \\ 0 & 0 & d_3^2 \end{bmatrix}$; thus $A^2 = A$ if and only if $d_i^2 = d_i$ for $i = 1, 2$, and 3, and this is true iff $d_i = 0$ or 1 for $i = 1, 2$, and 3. There are a total of eight such matrices (3×3 matrices whose diagonal entries are either 0 or 1).

(b) There are a total of 2^n such matrices ($n \times n$ matrices whose diagonal entries are either 0 or 1)

D6. If $A = \begin{bmatrix} d_1 & 0 \\ 0 & d_2 \end{bmatrix}$, then $A^2 + 5A + 6I_2 = 0$ if and only if $d_i^2 + 5d_i + 6 = 0$ for $i = 1$ and 2; i.e. if and only if $d_i = -2$ or -3 for $i = 1$ and 2. There are a total of four such matrices (any 2×2 diagonal matrix whose diagonal entries are either -2 or -3).

D7. If A is both symmetric and skew symmetric, then $A^T = A$ and $A^T = -A$; thus $A = -A$ and so $A = 0$.

D8. In a symmetric matrix, entries that are symmetrically positioned across the main diagonal are equal to each other. Thus a symmetric matrix is completely determined by the entries that lie on or above the main diagonal, and entries that appear below the main diagonal are duplicates of entries that appear above the main diagonal. An $n \times n$ matrix has n entries on the main diagonal, $n - 1$ entries on the diagonal just above the main diagonal, etc. Thus there are a total of

$$n + (n - 1) + \cdots + 2 + 1 = \frac{n(n + 1)}{2}$$

entries that lie on or above the main diagonal. For a symmetric matrix, this is the maximum number of distinct entries the matrix can have.

In a skew-symmetric matrix, the diagonal entries are 0 and entries that are symmetrically positioned across the main diagonal are the negatives of each other. The maximum number of distinct entries can be attained by selecting distinct positive entries for the $\frac{n(n-1)}{2}$ positions above the main diagonal. The entries in the $\frac{n(n-1)}{2}$ positions below the main diagonal will then automatically be distinct from each other and from the entries on or above the main diagonal. Thus the maximum number of distinct entries in a skew-symmetric matrix is

$$\frac{n(n-1)}{2} + \frac{n(n-1)}{2} + 1 = n(n-1) + 1$$

D9. If $D = \begin{bmatrix} d_1 & 0 \\ 0 & d_2 \end{bmatrix}$, then $AD = [d_1\mathbf{a}_1 \quad d_2\mathbf{a}_2]$ where $\mathbf{a}_1$ and $\mathbf{a}_2$ are the columns of A. Thus $AD = I = [\mathbf{e}_1 \quad \mathbf{e}_2]$ (where $\mathbf{e}_1$ and $\mathbf{e}_2$ are the standard unit vectors in R^2) if and only if $d_1\mathbf{a}_1 = \mathbf{e}_1$ and $d_2\mathbf{a}_2 = \mathbf{e}_2$. But this is true if and only if $d_1 \neq 0$, $d_2 \neq 0$, $\mathbf{a}_1 = \frac{1}{d_1}\mathbf{e}_1$, and $\mathbf{a}_2 = \frac{1}{d_2}\mathbf{e}_2$. Thus $A = \begin{bmatrix} \frac{1}{d_1} & 0 \\ 0 & \frac{1}{d_2} \end{bmatrix}$ where d_1, $d_2 \neq 0$. Although described here for the case $n = 2$, it should be clear that the same argument can be applied to a square matrix of any size. Thus, if $AD = I$, then the diagonal entries $d_1, d_2, \ldots, d_n$ of D must be nonzero, and $A = \begin{bmatrix} \frac{1}{d_1} & 0 & \cdots & 0 \\ 0 & \frac{1}{d_2} & \cdots & 0 \\ \vdots & \vdots & \ddots & \vdots \\ 0 & 0 & \cdots & \frac{1}{d_n} \end{bmatrix}$.

D10. (a) False. If A is not square then A is not invertible; it doesnt matter whether AA^T (which is always square) is invertible or not. [But if A is square and AA^T is invertible, then A is invertible by Theorem 3.6.5.]

(b) False. For example if $A = \begin{bmatrix} 1 & 0 \\ 1 & 1 \end{bmatrix}$ and $B = \begin{bmatrix} 1 & 1 \\ 0 & 1 \end{bmatrix}$, then $A + B = \begin{bmatrix} 1 & 2 \\ 2 & 1 \end{bmatrix}$ is symmetric.

(c) True. If A is both symmetric and triangular, then A must be a diagonal matrix. Thus $A = \begin{bmatrix} d_1 & 0 & \cdots & 0 \\ 0 & d_2 & \cdots & 0 \\ \vdots & \vdots & \ddots & \vdots \\ 0 & 0 & \cdots & d_n \end{bmatrix}$, and so $p(A) = \begin{bmatrix} p(d_1) & 0 & \cdots & 0 \\ 0 & p(d_2) & \cdots & 0 \\ \vdots & \vdots & \ddots & \vdots \\ 0 & 0 & \cdots & p(d_n) \end{bmatrix}$ is also a diagonal matrix (both symmetric and triangular).

(d) True. For example, in the 3×3 case, we have

$$\begin{bmatrix} a_{11} & a_{12} & a_{13} \\ a_{21} & a_{22} & a_{23} \\ a_{31} & a_{32} & a_{33} \end{bmatrix} = \begin{bmatrix} 0 & 0 & 0 \\ a_{21} & 0 & 0 \\ a_{31} & a_{32} & 0 \end{bmatrix} + \begin{bmatrix} 0 & a_{12} & a_{13} \\ 0 & 0 & a_{23} \\ 0 & 0 & 0 \end{bmatrix} + \begin{bmatrix} a_{11} & 0 & 0 \\ 0 & a_{22} & 0 \\ 0 & 0 & a_{33} \end{bmatrix}$$

(e) True. If $A\mathbf{x} = 0$ has only the trivial solution, then A is invertible. But if A is invertible then so is A^T (Theorem 3.2.11); thus $A^T\mathbf{x} = 0$ has only the trivial solution.

D11. (a) False. For example $\begin{bmatrix} 0 & -1 \\ 1 & 0 \end{bmatrix}\begin{bmatrix} 0 & -1 \\ 1 & 0 \end{bmatrix} = \begin{bmatrix} -1 & 0 \\ 0 & -1 \end{bmatrix}$.

(b) True. If A is invertible then A^k is invertible, and thus $A^k \neq 0$, for every $k = 1, 2, 3, \ldots$. This shows that an invertible matrix cannot be nilpotent; equivalently, a nilpotent matrix cannot be invertible.

(c) True (assuming $A \neq 0$). If $A^3 = A$, then $A^6 = A$, $A^9 = A$, $A^{12} = A, \ldots$. Thus is it not possible to have $A^k = 0$ for any positive integer k, since this would imply that $A^j = 0$ for all $j \geq k$.

(d) True. See Theorem 3.2.11.

(e) False. For example, I is invertible but $I - I = 0$ is not invertible.

WORKING WITH PROOFS

P1. If A and B are symmetric, then $A^T = A$ and $B^T = B$. It follows that

$$(A^T)^T = A^T$$

$$(A + B)^T = A^T + B^T = A + B$$

$$(A - B)^T = A^T - B^T = A - B$$

$$(kA)^T = kA^T = kA$$

thus the matrices A^T, $A + B$, $A - B$, and kA are also symmetric.

P2. Our proof is by induction on the exponent k.

Step 1. We have $D^1 = D = \begin{bmatrix} d_1 & \cdots & 0 \\ \vdots & \ddots & \vdots \\ 0 & \cdots & d_n \end{bmatrix} = \begin{bmatrix} d_1^1 & \cdots & 0 \\ \vdots & \ddots & \vdots \\ 0 & \cdots & d_n^1 \end{bmatrix}$; thus the statement is true for $k = 1$.

Step 2 (induction step). Suppose the statement is true for $k = j$, where j is an integer ≥ 1. Then

$$D^{j+1} = DD^j = \begin{bmatrix} d_1 & \cdots & 0 \\ \vdots & \ddots & \vdots \\ 0 & \cdots & d_n \end{bmatrix} \begin{bmatrix} d_1^j & \cdots & 0 \\ \vdots & \ddots & \vdots \\ 0 & \cdots & d_n^j \end{bmatrix} = \begin{bmatrix} d_1^{j+1} & \cdots & 0 \\ \vdots & \ddots & \vdots \\ 0 & \cdots & d_n^{j+1} \end{bmatrix}$$

and so the statement is also true for $k = j + 1$.

These two steps complete the proof by induction.

P3. If $d_1, d_2, \ldots, d_n$ are nonzero, then $\begin{bmatrix} d_1 & 0 & \cdots & 0 \\ 0 & d_2 & \cdots & 0 \\ \vdots & \vdots & \ddots & \vdots \\ 0 & 0 & \cdots & d_n \end{bmatrix} \begin{bmatrix} \frac{1}{d_1} & 0 & \cdots & 0 \\ 0 & \frac{1}{d_2} & \cdots & 0 \\ \vdots & \vdots & \ddots & \vdots \\ 0 & 0 & \cdots & \frac{1}{d_n} \end{bmatrix} = \begin{bmatrix} 1 & 0 & \cdots & 0 \\ 0 & 1 & \cdots & 0 \\ \vdots & \vdots & \ddots & \vdots \\ 0 & 0 & \cdots & 1 \end{bmatrix}$; thus the matrix

$D = \begin{bmatrix} d_1 & 0 & \cdots & 0 \\ 0 & d_2 & \cdots & 0 \\ \vdots & \vdots & \ddots & \vdots \\ 0 & 0 & \cdots & d_n \end{bmatrix}$ is invertible with $D^{-1} = \begin{bmatrix} \frac{1}{d_1} & 0 & \cdots & 0 \\ 0 & \frac{1}{d_2} & \cdots & 0 \\ \vdots & \vdots & & \vdots \\ 0 & 0 & \cdots & \frac{1}{d_n} \end{bmatrix}$. On the other hand if any one of

the diagonal entries is zero, then D has a row of zeros and thus is not invertible.

P4. We will show that if A is symmetric (i.e. if $A^T = A$), then $(A^n)^T = A^n$ for each positive integer n. Our proof is by induction on the exponent n.

Step 1. Since A is symmetric, we have $(A^1)^T = A^T = A = A^1$; thus the statement is true for $n = 1$.

Step 2 (induction step). Suppose the statement is true for $n = j$, where j is an integer ≥ 1. Then

$$(A^{j+1})^T = (AA^j)^T = (A^j)^T A^T = A^j A = A^{j+1}$$

and so the statement is also true for $n = j + 1$.

These two steps complete the proof by induction.

P5. If A is invertible, then Theorem 3.2.11 implies A^T is invertible; thus the products AA^T and $A^T A$ are invertible as well. On the other hand, if either AA^T or $A^T A$ is invertible, then Theorem 3.3.8 implies that A is invertible. It follows that A, AA^T, and $A^T A$ are either all invertible or all singular.

EXERCISE SET 3.7

1. We will solve $Ax = b$ by first solving $Ly = b$ for $\mathbf{y}$, and then solving $Ux = \mathbf{y}$ for $\mathbf{x}$. The system $Ly = b$ is

$$\begin{aligned} 3y_1 \quad\quad &= 0 \\ -2y_1 + y_2 &= 1 \end{aligned}$$

from which, by forward substitution, we obtain $y_1 = 0$, $y_2 = 1$. The system $Ux = \mathbf{y}$ is

$$\begin{aligned} x_1 - 2x_2 &= 0 \\ x_2 &= 1 \end{aligned}$$

from which, by back substitution, we obtain $x_1 = 2$, $x_2 = 1$. It is easy to check that this is in fact the solution of $Ax = b$.

3. We will solve $Ax = b$ by first solving $Ly = b$ for $\mathbf{y}$, and then solving $Ux = \mathbf{y}$ for $\mathbf{x}$. The system $Ly = b$ is

$$\begin{aligned} 3y_1 \quad\quad\quad\quad &= -3 \\ 2y_1 + 4y_2 \quad\quad &= -22 \\ -4y_1 - y_2 + 2y_3 &= 3 \end{aligned}$$

from which, by forward substitution, obtain $y_1 = -1$, $y_2 = -5$, $y_3 = -3$.

The system $Ux = \mathbf{y}$ is

$$\begin{aligned} x_1 - 2x_2 - x_3 &= -1 \\ x_2 + 2x_3 &= -5 \\ x_3 &= -3 \end{aligned}$$

from which, by back substitution, we obtain $x_1 = -2$, $x_2 = 1$, $x_3 = -3$. It is easy to check that this is in fact the solution of $Ax = b$.

5. The matrix A can be reduced to row echelon form by the following operations:

$$A = \begin{bmatrix} 2 & 8 \\ -1 & -1 \end{bmatrix} \rightarrow \begin{bmatrix} 1 & 4 \\ -1 & -1 \end{bmatrix} \rightarrow \begin{bmatrix} 1 & 4 \\ 0 & 3 \end{bmatrix} \rightarrow \begin{bmatrix} 1 & 4 \\ 0 & 1 \end{bmatrix} = U$$

The multipliers associated with these operations are $\frac{1}{2}$, 1, and $\frac{1}{3}$; thus

$$A = LU = \begin{bmatrix} 2 & 0 \\ -1 & 3 \end{bmatrix} \begin{bmatrix} 1 & 4 \\ 0 & 1 \end{bmatrix}$$

is an LU-factorization of A.

To solve the system $A\mathbf{x} = \mathbf{b}$ where $\mathbf{b} = \begin{bmatrix} -2 \\ -2 \end{bmatrix}$, we first solve $L\mathbf{y} = \mathbf{b}$ for $\mathbf{y}$, and then solve $U\mathbf{x} = \mathbf{y}$ for $\mathbf{x}$:

The system $L\mathbf{y} = \mathbf{b}$ is

$$\begin{aligned} 2y_1 &= -2 \\ -y_1 + 3y_2 &= -2 \end{aligned}$$

from which we obtain $y_1 = -1, y_2 = -1$.

The system $U\mathbf{x} = \mathbf{y}$ is

$$\begin{aligned} x_1 + 4x_2 &= -1 \\ x_2 &= -1 \end{aligned}$$

from which we obtain $x_1 = 3$, $x_2 = -1$. It is easy to check that this is in fact the solution of $A\mathbf{x} = \mathbf{b}$.

7. The matrix A can be reduced to row echelon form by the following sequence of operations:

$$A = \begin{bmatrix} 2 & -2 & -2 \\ 0 & -2 & 2 \\ -1 & 5 & 2 \end{bmatrix} \rightarrow \begin{bmatrix} 1 & -1 & -1 \\ 0 & -2 & 2 \\ -1 & 5 & 2 \end{bmatrix} \rightarrow \begin{bmatrix} 1 & -1 & -1 \\ 0 & -2 & 2 \\ 0 & 4 & 1 \end{bmatrix} \rightarrow$$

$$\begin{bmatrix} 1 & -1 & -1 \\ 0 & 1 & -1 \\ 0 & 4 & 1 \end{bmatrix} \rightarrow \begin{bmatrix} 1 & -1 & -1 \\ 0 & 1 & -1 \\ 0 & 0 & 5 \end{bmatrix} \rightarrow \begin{bmatrix} 1 & -1 & -1 \\ 0 & 1 & -1 \\ 0 & 0 & 1 \end{bmatrix} = U$$

The multipliers associated with these operations are $\frac{1}{2}$, 0 (for the second row), 1, $-\frac{1}{2}$, -4, and $\frac{1}{5}$; thus

$$A = LU = \begin{bmatrix} 2 & 0 & 0 \\ 0 & -2 & 0 \\ -1 & 4 & 5 \end{bmatrix} \begin{bmatrix} 1 & -1 & -1 \\ 0 & 1 & -1 \\ 0 & 0 & 1 \end{bmatrix}$$

is an LU-factorization of A.

To solve the system $A\mathbf{x} = \mathbf{b}$ where $\mathbf{b} = \begin{bmatrix} -4 \\ -2 \\ 6 \end{bmatrix}$, we first solve $L\mathbf{y} = \mathbf{b}$ for $\mathbf{y}$, and then solve $U\mathbf{x} = \mathbf{y}$ for $\mathbf{x}$:

The system $L\mathbf{y} = \mathbf{b}$ is

$$\begin{aligned} 2y_1 &= -4 \\ -2y_2 &= -2 \\ -y_1 + 4y_2 + 5y_3 &= 6 \end{aligned}$$

from which we obtain $y_1 = -2, y_2 = 1, y_3 = 0$.

The system $U\mathbf{x} = \mathbf{y}$ is

$$\begin{aligned} x_1 - x_2 - x_3 &= -2 \\ x_2 - x_3 &= 1 \\ x_3 &= 0 \end{aligned}$$

from which we obtain $x_1 = -1$, $x_2 = 1$, $x_3 = 0$. It is easy to check that this is the solution of $A\mathbf{x} = \mathbf{b}$.

9. The matrix A can be reduced to row echelon form by the following sequence of operations:

$$A = \begin{bmatrix} -1 & 0 & 1 & 0 \\ 2 & 3 & -2 & 6 \\ 0 & -1 & 2 & 0 \\ 0 & 0 & 1 & 5 \end{bmatrix} \rightarrow \begin{bmatrix} 1 & 0 & -1 & 0 \\ 2 & 3 & -2 & 6 \\ 0 & -1 & 2 & 0 \\ 0 & 0 & 1 & 5 \end{bmatrix} \rightarrow \begin{bmatrix} 1 & 0 & -1 & 0 \\ 0 & 3 & 0 & 6 \\ 0 & -1 & 2 & 0 \\ 0 & 0 & 1 & 5 \end{bmatrix} \rightarrow$$

$$\begin{bmatrix} 1 & 0 & -1 & 0 \\ 0 & 1 & 0 & 2 \\ 0 & -1 & 2 & 0 \\ 0 & 0 & 1 & 5 \end{bmatrix} \rightarrow \begin{bmatrix} 1 & 0 & -1 & 0 \\ 0 & 1 & 0 & 2 \\ 0 & 0 & 2 & 2 \\ 0 & 0 & 1 & 5 \end{bmatrix} \rightarrow \begin{bmatrix} 1 & 0 & -1 & 0 \\ 0 & 1 & 0 & 2 \\ 0 & 0 & 1 & 1 \\ 0 & 0 & 1 & 5 \end{bmatrix} \rightarrow$$

$$\begin{bmatrix} 1 & 0 & -1 & 0 \\ 0 & 1 & 0 & 2 \\ 0 & 0 & 1 & 1 \\ 0 & 0 & 0 & 4 \end{bmatrix} \rightarrow \begin{bmatrix} 1 & 0 & -1 & 0 \\ 0 & 1 & 0 & 2 \\ 0 & 0 & 1 & 1 \\ 0 & 0 & 0 & 1 \end{bmatrix} = U$$

The multipliers associated with these operations are -1, -2, 0 (for the third row), 0, $\frac{1}{3}$, 1, 0, $\frac{1}{2}$, -1, and $\frac{1}{4}$; thus

$$A = LU = \begin{bmatrix} -1 & 0 & 0 & 0 \\ 2 & 3 & 0 & 0 \\ 0 & -1 & 2 & 0 \\ 0 & 0 & 1 & 4 \end{bmatrix} \begin{bmatrix} 1 & 0 & -1 & 0 \\ 0 & 1 & 0 & 2 \\ 0 & 0 & 1 & 1 \\ 0 & 0 & 0 & 1 \end{bmatrix}$$

is an LU-decomposition of A.

To solve the system $A\mathbf{x} = \mathbf{b}$ where $\mathbf{b} = \begin{bmatrix} 5 \\ -1 \\ 3 \\ 7 \end{bmatrix}$, we first solve $L\mathbf{y} = \mathbf{b}$ for $\mathbf{y}$, and then solve $U\mathbf{x} = \mathbf{y}$

for $\mathbf{x}$:
The system $L\mathbf{y} = \mathbf{b}$ is

$$
\begin{array}{rcl}
-y_1 & = & 5 \\
2y_1 + 3y_2 & = & -1 \\
- y_2 + 2y_3 & = & 3 \\
y_3 + 4y_4 & = & 7
\end{array}
$$

from which we obtain $y_1 = -5$, $y_2 = 3$, $y_3 = 3$, $y_4 = 1$.
The system $U\mathbf{x} = \mathbf{y}$ is

$$
\begin{array}{rcl}
x_1 - x_3 & = & -5 \\
x_2 + 2x_4 & = & 3 \\
x_3 + x_4 & = & 3 \\
x_4 & = & 1
\end{array}
$$

from which we obtain $x_1 = -3$, $x_2 = 1$, $x_3 = 2$, $x_4 = 1$. It is easy to check that this is in fact the solution of $A\mathbf{x} = \mathbf{b}$.

11. Let $\mathbf{e}_1$, $\mathbf{e}_2$, $\mathbf{e}_3$ be the standard unit vectors in R^3. Then (for $j = 1, 2, 3$) the jth column $\mathbf{x}_j$ of the matrix A^{-1} is obtained by solving the system $A\mathbf{x} = \mathbf{e}_j$ for $\mathbf{x} = \mathbf{x}_j$. Using the given LU-decomposition, we will do this by first solving $L\mathbf{y} = \mathbf{e}_j$ for $\mathbf{y} = \mathbf{y}_j$, and then solving $U\mathbf{x} = \mathbf{y}_j$ for $\mathbf{x} = \mathbf{x}_j$.

(1) Computation of $\mathbf{x}_1$: The system $L\mathbf{y} = \mathbf{e}_1$ is

$$
\begin{aligned}
3y_1 &= 1 \\
2y_1 + 4y_2 &= 0 \\
-4y_1 - y_2 + 2y_3 &= 0
\end{aligned}
$$

from which we obtain $y_1 = \frac{1}{3}$, $y_2 = -\frac{1}{6}$, $y_3 = \frac{7}{12}$. Then the system $U\mathbf{x} = \mathbf{y}_1$ is

$$
\begin{aligned}
x_1 - 2x_2 - x_3 &= \tfrac{1}{3} \\
x_2 + 2x_3 &= -\tfrac{1}{6} \\
x_3 &= \tfrac{7}{12}
\end{aligned}
$$

from which we obtain $x_1 = -\frac{7}{4}$, $x_2 = -\frac{4}{3}$, $x_3 = \frac{7}{12}$. Thus $\mathbf{x}_1 = \begin{bmatrix} -\frac{7}{4} \\ -\frac{4}{3} \\ \frac{7}{12} \end{bmatrix}$

(2) Computation of $\mathbf{x}_2$: The system $L\mathbf{y} = \mathbf{e}_2$ is

$$
\begin{aligned}
3y_1 &= 0 \\
2y_1 + 4y_2 &= 1 \\
-4y_1 - y_2 + 2y_3 &= 0
\end{aligned}
$$

from which we obtain $y_1 = 0$, $y_2 = \frac{1}{4}$, $y_3 = \frac{1}{8}$. Then the system $U\mathbf{x} = \mathbf{y}_2$ is

$$
\begin{aligned}
x_1 - 2x_2 - x_3 &= 0 \\
x_2 + 2x_3 &= \tfrac{1}{4} \\
x_3 &= \tfrac{1}{8}
\end{aligned}
$$

from which we obtain $x_1 = \frac{1}{8}$, $x_2 = 0$, $x_3 = \frac{1}{8}$. Thus $\mathbf{x}_2 = \begin{bmatrix} \frac{1}{8} \\ 0 \\ \frac{1}{8} \end{bmatrix}$

(3) Computation of $\mathbf{x}_3$: The system $L\mathbf{y} = \mathbf{e}_3$ is

$$
\begin{aligned}
3y_1 &= 0 \\
2y_1 + 4y_2 &= 0 \\
-4y_1 - y_2 + 2y_3 &= 1
\end{aligned}
$$

from which we obtain $y_1 = 0$, $y_2 = 0$, $y_3 = \frac{1}{2}$. Then the system $U\mathbf{x} = \mathbf{y}_3$ is

$$
\begin{aligned}
x_1 - 2x_2 - x_3 &= 0 \\
x_2 + 2x_3 &= 0 \\
x_3 &= \tfrac{1}{2}
\end{aligned}
$$

from which we obtain $x_1 = -\frac{3}{2}$, $x_2 = -1$, $x_3 = \frac{1}{2}$. Thus $\mathbf{x}_3 = \begin{bmatrix} -\frac{3}{2} \\ -1 \\ \frac{1}{4} \end{bmatrix}$

Finally, as a result of these computations, we conclude that $A^{-1} = [\mathbf{x}_1 \quad \mathbf{x}_2 \quad \mathbf{x}_3] = \begin{bmatrix} -\frac{7}{4} & \frac{1}{8} & -\frac{3}{2} \\ -\frac{4}{3} & 0 & -1 \\ \frac{7}{12} & \frac{1}{8} & \frac{1}{2} \end{bmatrix}$.

13. The matrix A can be reduced to row echelon form by the following sequence of operations:

$$A = \begin{bmatrix} 2 & 1 & -1 \\ -2 & 0 & 2 \\ 2 & 2 & 1 \end{bmatrix} \rightarrow \begin{bmatrix} 1 & \frac{1}{2} & -\frac{1}{2} \\ -2 & 0 & 2 \\ 2 & 2 & 1 \end{bmatrix} \rightarrow \begin{bmatrix} 1 & \frac{1}{2} & -\frac{1}{2} \\ 0 & 1 & 1 \\ 2 & 2 & 1 \end{bmatrix} \rightarrow$$

$$\begin{bmatrix} 1 & \frac{1}{2} & -\frac{1}{2} \\ 0 & 1 & 1 \\ 0 & 1 & 2 \end{bmatrix} \rightarrow \begin{bmatrix} 1 & \frac{1}{2} & -\frac{1}{2} \\ 0 & 1 & 1 \\ 0 & 0 & 1 \end{bmatrix} = U$$

where the associated multipliers are $\frac{1}{2}$, 2, -2, 1 (for the leading entry in the second row), and -1. This leads to the factorization $A = LU$ where $L = \begin{bmatrix} 2 & 0 & 0 \\ -2 & 1 & 0 \\ 2 & 1 & 1 \end{bmatrix}$. If instead we prefer a lower triangular factor that has 1s on the main diagonal, this can be achieved by shifting the diagonal entries of L to a diagonal matrix D and writing the factorization as

$$A = \begin{bmatrix} 2 & 1 & -1 \\ -2 & 0 & 2 \\ 2 & 2 & 1 \end{bmatrix} = \begin{bmatrix} 1 & 0 & 0 \\ -1 & 1 & 0 \\ 1 & 1 & 1 \end{bmatrix} \begin{bmatrix} 2 & 0 & 0 \\ 0 & 1 & 0 \\ 0 & 0 & 1 \end{bmatrix} \begin{bmatrix} 1 & \frac{1}{2} & -\frac{1}{2} \\ 0 & 1 & 1 \\ 0 & 0 & 1 \end{bmatrix} = L'DU$$

15. (a) This is a permutation matrix; it is obtained by interchanging the rows of I_2.
 (b) This is not a permutation matrix; the second row is not a row of I_3.
 (c) This is a permutation matrix; it is obtained by reordering the rows of I_4 (3rd, 2nd, 4th, 1st).

17. The system $A\mathbf{x} = \mathbf{b}$ is equivalent to $P^{-1}A\mathbf{x} = P^{-1}\mathbf{b}$ where $P^{-1} = P = \begin{bmatrix} 0 & 1 & 0 \\ 1 & 0 & 0 \\ 0 & 0 & 1 \end{bmatrix}$. Using the given decomposition, the system $P^{-1}A\mathbf{x} = P^{-1}\mathbf{b}$ can be written as

$$LU\mathbf{x} = \begin{bmatrix} 1 & 0 & 0 \\ 0 & 1 & 0 \\ 3 & -5 & 1 \end{bmatrix} \begin{bmatrix} 1 & 2 & 2 \\ 0 & 1 & 4 \\ 0 & 0 & 17 \end{bmatrix} \begin{bmatrix} x_1 \\ x_2 \\ x_3 \end{bmatrix} = \begin{bmatrix} 1 \\ 2 \\ 5 \end{bmatrix} = P^{-1}\mathbf{b}$$

We solve this by first solving $L\mathbf{y} = P^{-1}\mathbf{b}$ for $\mathbf{y}$, and then solving $U\mathbf{x} = \mathbf{y}$ for $\mathbf{x}$. The system $L\mathbf{y} = P^{-1}\mathbf{b}$ is

$$\begin{aligned} y_1 &= 1 \\ y_2 &= 2 \\ 3y_1 - 5y_2 + y_3 &= 5 \end{aligned}$$

from which we obtain $y_1 = 1$, $y_2 = 2$, $y_3 = 12$. Finally, the system $U\mathbf{x} = \mathbf{y}$ is

$$\begin{aligned} x_1 + 2x_2 + 2x_3 &= 1 \\ x_2 + 4x_3 &= 2 \\ 17x_3 &= 12 \end{aligned}$$

from which we obtain $x_1 = \frac{21}{17}$, $x_2 = -\frac{14}{17}$, $x_3 = \frac{12}{17}$. This is the solution of $A\mathbf{x} = \mathbf{b}$.

19. If we interchange rows 2 and 3 of A, then the resulting matrix can be reduced to row echelon form without any further row interchanges. This is equivalent to first multiplying A on the left by the corresponding permutation matrix P:

$$\begin{bmatrix} 3 & -1 & 0 \\ 0 & 2 & 1 \\ 3 & -1 & 1 \end{bmatrix} = PA = \begin{bmatrix} 1 & 0 & 0 \\ 0 & 0 & 1 \\ 0 & 1 & 0 \end{bmatrix} \begin{bmatrix} 3 & -1 & 0 \\ 3 & -1 & 1 \\ 0 & 2 & 1 \end{bmatrix}$$

The reduction of PA to row echelon form proceeds as follows:

$$PA = \begin{bmatrix} 3 & -1 & 0 \\ 0 & 2 & 1 \\ 3 & -1 & 1 \end{bmatrix} \rightarrow \begin{bmatrix} 1 & -\frac{1}{3} & 0 \\ 0 & 2 & 1 \\ 3 & -1 & 1 \end{bmatrix} \rightarrow \begin{bmatrix} 1 & -\frac{1}{3} & 0 \\ 0 & 2 & 1 \\ 0 & 0 & 1 \end{bmatrix} \rightarrow \begin{bmatrix} 1 & -\frac{1}{3} & 0 \\ 0 & 1 & \frac{1}{2} \\ 0 & 0 & 1 \end{bmatrix} = U$$

This corresponds to the LU-decomposition

$$PA = \begin{bmatrix} 3 & -1 & 0 \\ 0 & 2 & 1 \\ 3 & -1 & 1 \end{bmatrix} = \begin{bmatrix} 3 & 0 & 0 \\ 0 & 2 & 0 \\ 3 & 0 & 1 \end{bmatrix} \begin{bmatrix} 1 & -\frac{1}{3} & 0 \\ 0 & 1 & \frac{1}{2} \\ 0 & 0 & 1 \end{bmatrix} = LU$$

of the matrix PA, or to the following PLU-decomposition of A

$$A = \begin{bmatrix} 1 & 0 & 0 \\ 0 & 0 & 1 \\ 0 & 1 & 0 \end{bmatrix} \begin{bmatrix} 3 & 0 & 0 \\ 0 & 2 & 0 \\ 3 & 0 & 1 \end{bmatrix} \begin{bmatrix} 1 & -\frac{1}{3} & 0 \\ 0 & 1 & \frac{1}{2} \\ 0 & 0 & 1 \end{bmatrix} = P^{-1}LU$$

Note that, since $P^{-1} = P$, this decomposition can also be written as $A = PLU$.

The system $A\mathbf{x} = \mathbf{b} = \begin{bmatrix} -2 \\ 1 \\ 4 \end{bmatrix}$ is equivalent to $PA\mathbf{x} = P\mathbf{b} = \begin{bmatrix} -2 \\ 4 \\ 1 \end{bmatrix}$ and, using the LU-decomposition obtained above, this can be written as

$$PA\mathbf{x} = LU\mathbf{x} = \begin{bmatrix} 3 & 0 & 0 \\ 0 & 2 & 0 \\ 3 & 0 & 1 \end{bmatrix} \begin{bmatrix} 1 & -\frac{1}{3} & 0 \\ 0 & 1 & \frac{1}{2} \\ 0 & 0 & 1 \end{bmatrix} \begin{bmatrix} x_1 \\ x_2 \\ x_3 \end{bmatrix} = \begin{bmatrix} -2 \\ 4 \\ 1 \end{bmatrix} = P\mathbf{b}$$

Finally, the solution of $L\mathbf{y} = P\mathbf{b}$ is $y_1 = -\frac{2}{3}$, $y_2 = 2$, $y_3 = 3$, and the solution of $U\mathbf{x} = \mathbf{y}$ (and of $A\mathbf{x} = \mathbf{b}$) is $x_1 = -\frac{1}{2}$, $x_2 = \frac{1}{2}$, $x_3 = 3$.

21. (a) We have $n = 10^5$ for the given system and so, from Table 3.7.1, the number of gigaflops required for the forward and backward phases are approximately

$$G_{\text{forward}} = \tfrac{2}{3}n^3 \times 10^{-9} = \tfrac{2}{3}(10^5)^3 \times 10^{-9} = \tfrac{2}{3} \times 10^6 \approx 670{,}000$$

$$G_{\text{backward}} = n^2 \times 10^{-9} = (10^5)^2 \times 10^{-9} = 10$$

Thus if a computer can execute 1 gigaflop per second it will require approximately 66.67×10^4 s for the forward phase, and approximately 10 s for the backward phase.

(b) If $n = 10^4$ then, from Example 4, the total number of gigaflops required for the forward and backward phases is approximately $\tfrac{2}{3} \times 10^3 + 10^{-1} \approx 667$. Thus, in order to complete the task in less than 0.5 s, a computer must be able to execute at least $\frac{667}{0.5} = 1334$ gigaflops per second.

DISCUSSION AND DISCOVERY

D1. The rows of P are obtained by reordering of the rows of the identity matrix I_4 (4th, 3rd, 2nd, 1st). Thus P is a permutation matrix. Multiplication of A (on the left) by P results in the corresponding reordering of the rows of A; thus

$$PA = \begin{bmatrix} 3 & -3 & 6 & -6 \\ -11 & 12 & 6 & 9 \\ 0 & 7 & 7 & 4 \\ 2 & 1 & 3 & 5 \end{bmatrix}$$

D2. If A is invertible, then the LDU-decomposition of A is unique. Furthermore, if A is symmetric, and if $A = LDU$ is its LDU-decomposition, then

$$A = A^T = (LDU)^T = U^T DL^T$$

where U^T is lower triangular and L^T is upper triangular; thus $U^T = L$ and $L^T = U$.

D3. If $A = \begin{bmatrix} 0 & 1 \\ 1 & 0 \end{bmatrix} = \begin{bmatrix} a & 0 \\ b & c \end{bmatrix} \begin{bmatrix} 1 & x \\ 0 & 1 \end{bmatrix} = LU$, then we must have $a = 0$ and $ax = 1$, which is of course not possible. Thus the matrix A has no LU decomposition.

D4. Here are two LU-decompositions of the matrix A:

$$A = \begin{bmatrix} 2 & 6 & 2 \\ -3 & -8 & 0 \\ 4 & 9 & 2 \end{bmatrix} = \begin{bmatrix} 2 & 0 & 0 \\ -3 & 1 & 0 \\ 4 & -3 & 7 \end{bmatrix} \begin{bmatrix} 1 & 3 & 1 \\ 0 & 1 & 3 \\ 0 & 0 & 1 \end{bmatrix} = LU$$

$$A = \begin{bmatrix} 2 & 6 & 2 \\ -3 & -8 & 0 \\ 4 & 9 & 2 \end{bmatrix} = \begin{bmatrix} 2 & 0 & 0 \\ -3 & 1 & 0 \\ 4 & -3 & 1 \end{bmatrix} \begin{bmatrix} 1 & 3 & 1 \\ 0 & 1 & 3 \\ 0 & 0 & 7 \end{bmatrix} = L'U'$$

EXERCISE SET 3.8

1. (a) The block version of the row-column rule for the product of partitioned matrices of this form is

$$AB = \begin{bmatrix} A_{11} & A_{12} \\ A_{21} & A_{22} \end{bmatrix} \begin{bmatrix} B_{11} & B_{12} \\ B_{21} & B_{22} \end{bmatrix} = \begin{bmatrix} A_{11}B_{11} + A_{12}B_{21} & A_{11}B_{12} + A_{12}B_{22} \\ A_{21}B_{11} + A_{22}B_{21} & A_{21}B_{12} + A_{22}B_{22} \end{bmatrix}$$

In this example, the block sizes conform and the product can be computed as follows:

$$A_{11}B_{11} + A_{12}B_{21} = \begin{bmatrix} 2 & 1 \\ 3 & 0 \end{bmatrix} \begin{bmatrix} 1 & -2 \\ 3 & 0 \end{bmatrix} + \begin{bmatrix} -1 \\ 5 \end{bmatrix} \begin{bmatrix} -1 & 3 \end{bmatrix} = \begin{bmatrix} 5 & -4 \\ 3 & -6 \end{bmatrix} + \begin{bmatrix} 1 & -3 \\ -5 & 15 \end{bmatrix} = \begin{bmatrix} 6 & -7 \\ -2 & 9 \end{bmatrix}$$

$$A_{11}B_{12} + A_{12}B_{22} = \begin{bmatrix} 2 & 1 \\ 3 & 0 \end{bmatrix} \begin{bmatrix} -1 \\ -2 \end{bmatrix} + \begin{bmatrix} -1 \\ 5 \end{bmatrix} (2) = \begin{bmatrix} -4 \\ -3 \end{bmatrix} + \begin{bmatrix} -2 \\ 10 \end{bmatrix} = \begin{bmatrix} -6 \\ 7 \end{bmatrix}$$

$$A_{21}B_{11} + A_{22}B_{21} = \begin{bmatrix} 4 & 2 \end{bmatrix} \begin{bmatrix} 1 & -2 \\ 3 & 0 \end{bmatrix} + (1)\begin{bmatrix} -1 & 3 \end{bmatrix} = \begin{bmatrix} 10 & 8 \end{bmatrix} + \begin{bmatrix} -1 & 3 \end{bmatrix} = \begin{bmatrix} 9 & -5 \end{bmatrix}$$

$$A_{21}B_{12} + A_{22}B_{22} = \begin{bmatrix} 4 & 2 \end{bmatrix} \begin{bmatrix} -1 \\ -2 \end{bmatrix} + (1)(2) = -8 + 2 = -6$$

$$AB = \begin{bmatrix} 2 & 1 & -1 \\ 3 & 0 & 5 \\ 4 & 2 & 1 \end{bmatrix} \begin{bmatrix} 1 & -2 & -1 \\ 3 & 0 & -2 \\ -1 & 3 & 2 \end{bmatrix} = \begin{bmatrix} 6 & -7 & -6 \\ -2 & 9 & 7 \\ 9 & -5 & -6 \end{bmatrix}$$

(b) In this example the block sizes do not conform. For example, A_{12} is 1×2 and B_{21} is 1×2, so the product $A_{12}B_{21}$ is not defined.

(c) In this example the block sizes do conform. In fact, this partitioning corresponds to the usual row-column rule for matrix multiplication.

(d) The block version of the row-column rule for the product of partitioned matrices of this form is

$$AB = \begin{bmatrix} A_{11} \\ \hline A_{21} \end{bmatrix} B = \begin{bmatrix} A_{11}B \\ \hline A_{21}B \end{bmatrix}$$

In this example the block sizes conform and the product can be computed as follows:

$$A_{11}B = \begin{bmatrix} 2 & 1 & -1 \\ 3 & 0 & 5 \end{bmatrix} \begin{bmatrix} 1 & -2 & -1 \\ 3 & 0 & -2 \\ -1 & 3 & 2 \end{bmatrix} = \begin{bmatrix} 6 & -7 & -6 \\ -2 & 9 & 7 \end{bmatrix}$$

$$A_{21}B = \begin{bmatrix} 4 & 2 & 1 \end{bmatrix} \begin{bmatrix} 1 & -2 & -1 \\ 3 & 0 & -2 \\ -1 & 3 & 2 \end{bmatrix} = \begin{bmatrix} 9 & -5 & -6 \end{bmatrix}$$

$$AB = \begin{bmatrix} 2 & 1 & -1 \\ 3 & 0 & 5 \\ \hline 4 & 2 & 1 \end{bmatrix} \begin{bmatrix} 1 & -2 & -1 \\ 3 & 0 & -2 \\ -1 & 3 & 2 \end{bmatrix} = \begin{bmatrix} 6 & -7 & -6 \\ -2 & 9 & 7 \\ \hline 9 & -5 & -6 \end{bmatrix}$$

3. (a) The block sizes conform and $AB = \begin{bmatrix} -1 & 2 & 1 & 5 \\ 0 & -3 & 4 & 2 \\ \hline 1 & 5 & 6 & 1 \end{bmatrix} \begin{bmatrix} -2 & 1 & 4 \\ -3 & 5 & 2 \\ \hline 7 & -1 & 5 \\ 0 & 3 & -3 \end{bmatrix} =$

$$\begin{bmatrix} \begin{bmatrix} -1 & 2 \\ 0 & -3 \end{bmatrix}\begin{bmatrix} -2 & 1 \\ -3 & 5 \end{bmatrix} + \begin{bmatrix} 1 & 5 \\ 4 & 2 \end{bmatrix}\begin{bmatrix} 7 & -1 \\ 0 & 3 \end{bmatrix} & \begin{bmatrix} -1 & 2 \\ 0 & -3 \end{bmatrix}\begin{bmatrix} 4 \\ 2 \end{bmatrix} + \begin{bmatrix} 1 & 5 \\ 4 & 2 \end{bmatrix}\begin{bmatrix} 5 \\ -3 \end{bmatrix} \\ \hline \begin{bmatrix} 1 & 5 \end{bmatrix}\begin{bmatrix} -2 & 1 \\ -3 & 5 \end{bmatrix} + \begin{bmatrix} 6 & 1 \end{bmatrix}\begin{bmatrix} 7 & -1 \\ 0 & 3 \end{bmatrix} & \begin{bmatrix} 1 & 5 \end{bmatrix}\begin{bmatrix} 4 \\ 2 \end{bmatrix} + \begin{bmatrix} 6 & 1 \end{bmatrix}\begin{bmatrix} 5 \\ -3 \end{bmatrix} \end{bmatrix} =$$

$$\begin{bmatrix} \begin{bmatrix} -4 & 9 \\ 9 & -15 \end{bmatrix} + \begin{bmatrix} 7 & 14 \\ 28 & 2 \end{bmatrix} & \begin{bmatrix} 0 \\ -6 \end{bmatrix} + \begin{bmatrix} -10 \\ 14 \end{bmatrix} \\ \hline \begin{bmatrix} -17 & 26 \end{bmatrix} + \begin{bmatrix} 42 & -3 \end{bmatrix} & 14 + 27 \end{bmatrix} = \begin{bmatrix} 3 & 23 & -10 \\ 37 & -13 & 8 \\ \hline 25 & 23 & 41 \end{bmatrix}$$

(b) The block sizes conform and $AB = \begin{bmatrix} -1 & 2 & 1 & \vdots & 5 \\ 0 & -3 & 4 & \vdots & 2 \\ 1 & 5 & 6 & \vdots & 1 \end{bmatrix} \begin{bmatrix} -2 & 1 & \vdots & 4 \\ -3 & 5 & \vdots & 2 \\ 7 & -1 & \vdots & 5 \\ \hdashline 0 & 3 & \vdots & -3 \end{bmatrix} =$

$$
\left[
\begin{array}{c:c}
[-1 \; 2 \; 1]\begin{bmatrix} -2 & 1 \\ -3 & 5 \\ 7 & -1 \end{bmatrix} + (5)[0\;3] & [-1\;2\;1]\begin{bmatrix} 4 \\ 2 \\ 5 \end{bmatrix} + (5)(-3) \\
\hdashline
\begin{bmatrix} 0 & -3 & 4 \\ 1 & 5 & 6 \end{bmatrix}\begin{bmatrix} -2 & 1 \\ -3 & 5 \\ 7 & -1 \end{bmatrix} + \begin{bmatrix} 2 \\ 1 \end{bmatrix}[0\;3] & \begin{bmatrix} 0 & -3 & 4 \\ 1 & 5 & 6 \end{bmatrix}\begin{bmatrix} 4 \\ 2 \\ 5 \end{bmatrix} + \begin{bmatrix} 2 \\ 1 \end{bmatrix}(-3)
\end{array}
\right] =
$$

$$
\left[
\begin{array}{c:c}
[3\;8] + [0\;15] & 5 - 15 \\
\hdashline
\begin{bmatrix} 37 & -19 \\ 25 & 20 \end{bmatrix} + \begin{bmatrix} 0 & 6 \\ 0 & 3 \end{bmatrix} & \begin{bmatrix} 14 \\ 44 \end{bmatrix} + \begin{bmatrix} -6 \\ -3 \end{bmatrix}
\end{array}
\right] = \begin{bmatrix} 3 & 23 & \vdots & -10 \\ 37 & -13 & \vdots & 8 \\ 25 & 23 & \vdots & 41 \end{bmatrix}
$$

5. (a) $\begin{bmatrix} 3 & -1 & 0 & \vdots & -3 \\ 2 & 1 & 4 & \vdots & 5 \end{bmatrix} \begin{bmatrix} 2 & -4 & 1 \\ 3 & 0 & 2 \\ 1 & -3 & 5 \\ \hdashline 2 & 1 & 4 \end{bmatrix} = \begin{bmatrix} 3 & -1 & 0 \\ 2 & 1 & 4 \end{bmatrix}\begin{bmatrix} 2 & -4 & 1 \\ 3 & 0 & 2 \\ 1 & -3 & 5 \end{bmatrix} + \begin{bmatrix} -3 \\ 5 \end{bmatrix}[2\;1\;4]$

$$
= \left[\begin{bmatrix} 3 & -12 & 1 \\ 11 & -20 & 24 \end{bmatrix} + \begin{bmatrix} -6 & -3 & -12 \\ 10 & 5 & 20 \end{bmatrix}\right] = \begin{bmatrix} -3 & -15 & -11 \\ 21 & -15 & 44 \end{bmatrix}
$$

(b) $\begin{bmatrix} 2 & -5 \\ 1 & 3 \\ 0 & 5 \\ \hdashline 1 & 4 \end{bmatrix} \begin{bmatrix} 2 & -1 & \vdots & 3 & -4 \\ 0 & 1 & \vdots & 5 & 7 \end{bmatrix} =$

$$
\left[
\begin{array}{c:c}
\begin{bmatrix} 2 & -5 \\ 1 & 3 \\ 0 & 5 \end{bmatrix}\begin{bmatrix} 2 & -1 \\ 0 & 1 \end{bmatrix} & \begin{bmatrix} 2 & -5 \\ 1 & 3 \\ 0 & 5 \end{bmatrix}\begin{bmatrix} 3 & -4 \\ 5 & 7 \end{bmatrix} \\
\hdashline
[1\;4]\begin{bmatrix} 2 & -1 \\ 0 & 1 \end{bmatrix} & [1\;4]\begin{bmatrix} 3 & -4 \\ 5 & 7 \end{bmatrix}
\end{array}
\right]
$$

$$
= \begin{bmatrix} 4 & -7 & \vdots & -19 & -43 \\ 2 & 2 & \vdots & 18 & 17 \\ 0 & 5 & \vdots & 25 & 35 \\ \hdashline 2 & 3 & \vdots & 23 & 24 \end{bmatrix}
$$

7. $\begin{bmatrix} 1 & 2 \\ 3 & -5 \end{bmatrix}\begin{bmatrix} -1 & 2 \\ -4 & 5 \end{bmatrix} = \begin{bmatrix} 1 \\ 3 \end{bmatrix}[-1\;2] + \begin{bmatrix} 2 \\ -5 \end{bmatrix}[-4\;5] = \begin{bmatrix} -1 & 2 \\ -3 & 6 \end{bmatrix} + \begin{bmatrix} -8 & 10 \\ 20 & -25 \end{bmatrix} = \begin{bmatrix} -9 & 12 \\ 17 & -19 \end{bmatrix}$

9. (a) $A^{-1} = \begin{bmatrix} 2 & -1 & \vdots & 0 & 0 \\ -3 & 2 & \vdots & 0 & 0 \\ \hdashline 0 & 0 & \vdots & \frac{1}{7} & \frac{4}{7} \\ 0 & 0 & \vdots & \frac{1}{7} & -\frac{3}{7} \end{bmatrix}$

(b) $A^{-1} = \begin{bmatrix} -1 & 2 & \vdots & 0 & \vdots & 0 & 0 \\ 3 & -5 & \vdots & 0 & \vdots & 0 & 0 \\ \hdashline 0 & 0 & \vdots & \frac{1}{5} & \vdots & 0 & 0 \\ \hdashline 0 & 0 & \vdots & 0 & \vdots & 4 & -7 \\ 0 & 0 & \vdots & 0 & \vdots & -1 & 2 \end{bmatrix}$

11. $A = \begin{bmatrix} 2 & 1 & | & 3 & -6 \\ 1 & 1 & | & 7 & 4 \\ \hline 0 & 0 & | & 3 & 5 \\ 0 & 0 & | & 2 & 3 \end{bmatrix} = \begin{bmatrix} A_{11} & | & A_{12} \\ \hline 0 & | & A_{22} \end{bmatrix}$ where $A_{11}^{-1} = \begin{bmatrix} 1 & -1 \\ -1 & 2 \end{bmatrix}$ and $A_{22}^{-1} = \begin{bmatrix} -3 & 5 \\ 2 & -3 \end{bmatrix}$. Thus

$$A_{11}^{-1}A_{12}A_{22}^{-1} = \begin{bmatrix} 1 & -1 \\ -1 & 2 \end{bmatrix}\begin{bmatrix} 3 & -6 \\ 7 & 4 \end{bmatrix}\begin{bmatrix} -3 & 5 \\ 2 & -3 \end{bmatrix} = \begin{bmatrix} 1 & -1 \\ -1 & 2 \end{bmatrix}\begin{bmatrix} -21 & 33 \\ -13 & 23 \end{bmatrix} = \begin{bmatrix} -8 & 10 \\ -5 & 13 \end{bmatrix}$$

$$A^{-1} = \begin{bmatrix} A_{11}^{-1} & | & -A_{11}^{-1}A_{12}A_{22}^{-1} \\ \hline 0 & | & A_{22}^{-1} \end{bmatrix} = \begin{bmatrix} 1 & -1 & | & 8 & -10 \\ -1 & 2 & | & 5 & -13 \\ \hline 0 & 0 & | & -3 & 5 \\ 0 & 0 & | & 2 & -3 \end{bmatrix}$$

13. $MN = \begin{bmatrix} A & B \\ 0 & A \end{bmatrix}\begin{bmatrix} A^{-1} & B \\ 0 & A \end{bmatrix} = \begin{bmatrix} AA^{-1} + B0 & AB + BA \\ 0A^{-1} + A0 & AA \end{bmatrix} = \begin{bmatrix} I & AB + BA \\ 0 & A^2 \end{bmatrix}$

15. Let $B_1 = \begin{bmatrix} w & x \\ y & z \end{bmatrix}$. If $\begin{bmatrix} A_1 & B_1 \\ 0 & C_1 \end{bmatrix}\begin{bmatrix} A_2 & B_2 \\ 0 & C_2 \end{bmatrix} = \begin{bmatrix} A_3 & B_3 \\ 0 & C_3 \end{bmatrix}$, then $B_3 = A_1B_2 + B_1C_2$ and so

$$B_1C_2 = B_3 - A_1B_2$$

$$\begin{bmatrix} w & x \\ y & z \end{bmatrix}\begin{bmatrix} 1 & 0 \\ 0 & 2 \end{bmatrix} = \begin{bmatrix} 2 & 1 \\ 1 & 3 \end{bmatrix} - \begin{bmatrix} 2 & 0 \\ 0 & 1 \end{bmatrix}\begin{bmatrix} 1 & 1 \\ 1 & 2 \end{bmatrix}$$

$$\begin{bmatrix} w & 2x \\ y & 2z \end{bmatrix} = \begin{bmatrix} 2 & 1 \\ 1 & 3 \end{bmatrix} - \begin{bmatrix} 2 & 2 \\ 1 & 2 \end{bmatrix} = \begin{bmatrix} 0 & -1 \\ 0 & 1 \end{bmatrix}$$

Thus $w = 0$, $x = -\frac{1}{2}$, $y = 0$, $z = \frac{1}{2}$, i.e. $B_1 = \begin{bmatrix} 0 & -\frac{1}{2} \\ 0 & \frac{1}{2} \end{bmatrix}$.

17. Let $\mathbf{u} = \begin{bmatrix} x_1 \\ x_2 \end{bmatrix}$, $\mathbf{v} = \begin{bmatrix} x_3 \\ x_4 \end{bmatrix}$. Then the given system can be written as $\begin{bmatrix} A & B \\ I & D \end{bmatrix}\begin{bmatrix} \mathbf{u} \\ \mathbf{v} \end{bmatrix} = \begin{bmatrix} \mathbf{b} \\ \mathbf{0} \end{bmatrix}$, or

$$A\mathbf{u} + B\mathbf{v} = \mathbf{b}$$
$$\mathbf{u} + D\mathbf{v} = \mathbf{0}$$

where $A = \begin{bmatrix} 5 & 2 \\ 2 & 1 \end{bmatrix}$, $B = \begin{bmatrix} 2 & 3 \\ -3 & 1 \end{bmatrix}$, $D = \begin{bmatrix} 4 & 1 \\ 0 & 2 \end{bmatrix}$, and $\mathbf{b} = \begin{bmatrix} 2 \\ 6 \end{bmatrix}$. From the second equation we have $\mathbf{u} = -D\mathbf{v}$, and substitution of this into the first equation leads to

$$(-AD + B)\mathbf{v} = \mathbf{b}$$

$$\begin{bmatrix} -18 & -6 \\ -11 & -3 \end{bmatrix}\begin{bmatrix} x_3 \\ x_4 \end{bmatrix} = \begin{bmatrix} 2 \\ 6 \end{bmatrix}$$

$$\begin{bmatrix} x_3 \\ x_4 \end{bmatrix} = \begin{bmatrix} -18 & -6 \\ -11 & -3 \end{bmatrix}^{-1}\begin{bmatrix} 2 \\ 6 \end{bmatrix} = \frac{1}{12}\begin{bmatrix} 3 & -6 \\ -11 & 18 \end{bmatrix}\begin{bmatrix} 2 \\ 6 \end{bmatrix} = \begin{bmatrix} -\frac{5}{2} \\ \frac{43}{6} \end{bmatrix}$$

$$\begin{bmatrix} x_1 \\ x_2 \end{bmatrix} = -D\mathbf{v} = -\begin{bmatrix} 4 & 1 \\ 0 & 2 \end{bmatrix}\begin{bmatrix} -\frac{5}{2} \\ \frac{43}{6} \end{bmatrix} = \begin{bmatrix} \frac{17}{6} \\ -\frac{43}{3} \end{bmatrix}$$

DISCUSSION AND DISCOVERY

D1. $M^2 = \begin{bmatrix} A & 0 \\ 0 & B \end{bmatrix} \begin{bmatrix} A & 0 \\ 0 & B \end{bmatrix} = \begin{bmatrix} A^2 & 0 \\ 0 & B^2 \end{bmatrix} = \begin{bmatrix} I_m & 0 \\ 0 & I_k \end{bmatrix} = I_{m+k}$

D2. **(a)** If A_{11} and A_{22} are invertible square matrices of size m and n respectively, then

$$\begin{bmatrix} A_{11} & A_{12} \\ 0 & A_{22} \end{bmatrix} \begin{bmatrix} A_{11}^{-1} & -A_{11}^{-1}A_{12}A_{22}^{-1} \\ 0 & A_{22}^{-1} \end{bmatrix} = \begin{bmatrix} A_{11}A_{11}^{-1} & -A_{12}A_{22}^{-1} + A_{12}A_{22}^{-1} \\ 0 & A_{22}A_{22}^{-1} \end{bmatrix} = \begin{bmatrix} I_m & 0 \\ 0 & I_n \end{bmatrix}$$

thus $A = \begin{bmatrix} A_{11} & A_{12} \\ 0 & A_{22} \end{bmatrix}$ is invertible and $A^{-1} = \begin{bmatrix} A_{11}^{-1} & -A_{11}^{-1}A_{12}A_{22}^{-1} \\ 0 & A_{22}^{-1} \end{bmatrix}$.

(b) Similarly, $\begin{bmatrix} A_{11} & 0 \\ A_{21} & A_{22} \end{bmatrix}^{-1} = \begin{bmatrix} A_{11}^{-1} & 0 \\ -A_{22}^{-1}A_{21}A_{11}^{-1} & A_{22}^{-1} \end{bmatrix}$.

WORKING WITH PROOFS

P1. Let $E_1, E_2, \ldots, E_k$ be elementary matrices corresponding to a sequence of elementary row operations which reduce A to I. Then $E_k \cdots E_2 E_1 A = I$, and so $E_k \cdots E_2 E_1 = A^{-1}$. Thus, the same sequence of row operations will reduce the matrix $M = [A \quad B]$ to

$$M' = [E_k \cdots E_2 E_1 A \quad E_k \cdots E_2 E_1 B] = [I \quad A^{-1}B]$$

CHAPTER 4

Determinants

EXERCISE SET 4.1

1. $\begin{vmatrix} 3 & 5 \\ -2 & 4 \end{vmatrix} = (3)(4) - (5)(-2) = 12 + 10 = 22$

3. $\begin{vmatrix} -5 & 7 \\ -7 & -2 \end{vmatrix} = (-5)(-2) - (7)(-7) = 10 + 49 = 59$

5. $\begin{vmatrix} a-3 & 5 \\ -3 & a-2 \end{vmatrix} = (a-3)(a-2) - (5)(-3) = (a^2 - 5a + 6) + 15 = a^2 - 5a + 21$

7. $\begin{vmatrix} -2 & 1 & 4 \\ 3 & 5 & -7 \\ 1 & 6 & 2 \end{vmatrix} = (-20 - 7 + 72) - (20 + 84 + 6) = 45 - 110 = -65$

9. $\begin{vmatrix} 3 & 0 & 0 \\ 2 & -1 & 5 \\ 1 & 9 & -4 \end{vmatrix} = (12 + 0 + 0) - (0 + 135 + 0) = -123$

11. **(a)** $\{4, 1, 3, 5, 2\}$ is an odd permutation (3 interchanges). The signed product is $-a_{14}a_{21}a_{33}a_{45}a_{52}$.
   (b) $\{5, 3, 4, 2, 1\}$ is an odd permutation (3 interchanges). The signed product is $-a_{15}a_{23}a_{34}a_{42}a_{51}$.
   (c) $\{4, 2, 5, 3, 1\}$ is an odd permutation (3 interchanges). The signed product is $-a_{14}a_{22}a_{35}a_{43}a_{51}$.
   (d) $\{5, 4, 3, 2, 1\}$ is an even permutation (2 interchanges). The signed product is $+a_{15}a_{24}a_{33}a_{42}a_{51}$.
   (e) $\{1, 2, 3, 4, 5\}$ is an even permutation (0 interchanges). The signed product is $+a_{11}a_{22}a_{33}a_{44}a_{55}$.
   (f) $\{1, 4, 2, 3, 5\}$ is an even permutation (2 interchanges). The signed product is $+a_{11}a_{24}a_{32}a_{43}a_{55}$.

13. $\det(A) = (\lambda - 2)(\lambda + 4) + 5 = \lambda^2 + 2\lambda - 3 = (\lambda - 1)(\lambda + 3)$. Thus $\det(A) = 0$ if and only if $\lambda = 1$ or $\lambda = -3$.

15. $\det(A) = (\lambda - 1)(\lambda + 1)$. Thus $\det(A) = 0$ if and only if $\lambda = 1$ or $\lambda = -1$.

17. We have $\begin{vmatrix} x & -1 \\ 3 & 1-x \end{vmatrix} = x(1-x) + 3 = -x^2 + x + 3$, and

$$\begin{vmatrix} 1 & 0 & -3 \\ 2 & x & -6 \\ 1 & 3 & x-5 \end{vmatrix} = ((x(x-5) + 0 - 18) - (-3x - 18 + 0) = x^2 - 2x$$

Thus the given equation is valid if and only if $-x^2 + x + 3 = x^2 - 2x$, i.e. if $2x^2 - 3x - 3 = 0$. The roots of this quadratic equation are $x = \frac{3 \pm \sqrt{33}}{4}$.

19. **(a)** $\begin{vmatrix} 1 & 0 & 0 \\ 0 & -1 & 0 \\ 0 & 0 & 1 \end{vmatrix} = (1)(-1)(1) = -1$ **(b)** $\begin{vmatrix} 0 & 0 & 0 & 0 \\ 1 & 2 & 0 & 0 \\ 0 & 4 & 3 & 0 \\ 1 & 2 & 3 & 8 \end{vmatrix} = 0$

(c) $\begin{vmatrix} 1 & 2 & 7 & -3 \\ 0 & 1 & -4 & 1 \\ 0 & 0 & 2 & 7 \\ 0 & 0 & 0 & 3 \end{vmatrix} = (1)(1)(2)(3) = 6$

21. $M_{11} = \begin{vmatrix} 7 & -1 \\ 1 & 4 \end{vmatrix} = 29, C_{11} = 29 \qquad M_{12} = \begin{vmatrix} 6 & -1 \\ -3 & 4 \end{vmatrix} = 21, C_{12} = -21 \quad M_{13} = 27, C_{13} = 27$

$M_{21} = \begin{vmatrix} -2 & 3 \\ 1 & 4 \end{vmatrix} = -11, C_{21} = 11 \qquad M_{22} = \begin{vmatrix} 1 & 3 \\ -3 & 4 \end{vmatrix} = 13, C_{22} = 13 \qquad M_{23} = -5, C_{23} = 5$

$M_{31} = \begin{vmatrix} -2 & 3 \\ 7 & -1 \end{vmatrix} = -19, C_{31} = -19 \quad M_{32} = \begin{vmatrix} 1 & 3 \\ 6 & -1 \end{vmatrix} = -19, C_{32} = 19 \qquad M_{33} = 19, C_{33} = 19$

23. (a) $M_{13} = \begin{vmatrix} 0 & 0 & 3 \\ 4 & 1 & 14 \\ 4 & 1 & 2 \end{vmatrix} = (0 + 0 + 12) - (12 + 0 + 0) = 0 \qquad C_{13} = 0$

(b) $M_{23} = \begin{vmatrix} 4 & -1 & 6 \\ 4 & 1 & 14 \\ 4 & 1 & 2 \end{vmatrix} = (8 - 56 + 24) - (24 + 56 - 8) = -96 \qquad C_{23} = 96$

(c) $M_{22} = \begin{vmatrix} 4 & 1 & 6 \\ 4 & 0 & 14 \\ 4 & 3 & 2 \end{vmatrix} = (0 + 56 + 72) - (0 + 8 + 168) = -48 \qquad C_{22} = -48$

(d) $M_{21} = \begin{vmatrix} -1 & 1 & 6 \\ 1 & 0 & 14 \\ 1 & 3 & 2 \end{vmatrix} = (0 + 14 + 18) - (0 + 2 - 42) = 72 \qquad C_{21} = -72$

25. (a) $\det(A) = (1)C_{11} + (-2)C_{12} + (3)C_{13} = (1)(29) + (-2)(-21) + (3)(27) = 152$

(b) $\det(A) = (1)C_{11} + (6)C_{21} + (-3)C_{31} = (1)(29) + (6)(11) + (-3)(-19) = 152$

(c) $\det(A) = (6)C_{21} + (7)C_{22} + (-1)C_{23} = (6)(11) + (7)(13) + (-1)(5) = 152$

(d) $\det(A) = (-2)C_{12} + (7)C_{22} + (1)C_{32} = (-2)(-21) + (7)(13) + (1)(19) = 152$

(e) $\det(A) = (-3)C_{31} + (1)C_{32} + (4)C_{33} = (-3)(-19) + (1)(19) + (4)(19) = 152$

(f) $\det(A) = (3)C_{13} + (-1)C_{23} + (4)C_{33} = (3)(27) + (-1)(5) + (4)(19) = 152$

27. Using column 2: $\det(A) = (5)\begin{vmatrix} -3 & 7 \\ -1 & 5 \end{vmatrix} = (5)(-15 + 7) = -40$

29. Using column 1: $\det(A) = (1)\begin{vmatrix} k & k^2 \\ k & k^2 \end{vmatrix} - (1)\begin{vmatrix} k & k^2 \\ k & k^2 \end{vmatrix} + (1)\begin{vmatrix} k & k^2 \\ k & k^2 \end{vmatrix} = (1)\begin{vmatrix} k & k^2 \\ k & k^2 \end{vmatrix} = k^3 - k^3 = 0$

31. Using column 3: $\det(A) = (-3)\begin{vmatrix} 3 & 3 & 5 \\ 2 & 2 & -2 \\ 2 & 10 & 2 \end{vmatrix} - (3)\begin{vmatrix} 3 & 3 & 5 \\ 2 & 2 & -2 \\ 4 & 1 & 0 \end{vmatrix} = (-3)(128) - (3)(-48) = -240$

33. By expanding along the third column, we have

$$\begin{vmatrix} \sin\theta & \cos\theta & 0 \\ -\cos\theta & \sin\theta & 0 \\ \sin\theta - \cos\theta & \sin\theta + \cos\theta & 1 \end{vmatrix} = (1)\begin{vmatrix} \sin\theta & \cos\theta \\ -\cos\theta & \sin\theta \end{vmatrix} = \sin^2\theta + \cos^2\theta = 1$$

for all values of θ.

35. $d_2 - d_1 = \lambda\begin{vmatrix} 1 & f \\ 0 & 1 \end{vmatrix} = \lambda$

DISCUSSION AND DISCOVERY

D1. Since the product of integers is always an integer, each elementary product is an integer and so the sum of the elementary products is an integer

D2. The signed elementary products will all be $\pm(1)(1)\cdots(1) = \pm 1$, with half of them equal to $+1$ and half equal to -1. Thus the determinant will be zero.

D3. A 3×3 matrix A can have as many as six zeros without having $\det(A) = 0$. For example, let A be a diagonal matrix with nonzero diagonal entries.

D4. If we expand along the first row, the equation $\begin{vmatrix} x & y & 1 \\ a_1 & b_1 & 1 \\ a_2 & b_2 & 1 \end{vmatrix} = 0$ becomes

$$x(b_1 - b_2) + y(a_2 - a_1) + a_1 b_2 - a_2 b_1 = 0$$

and this is an equation for the line through the points $P_1(a_1, b_1)$ and $P_2(a_2, b_2)$.

D5. If $\mathbf{u}^T = (a_1, a_2, a_3)$ and $\mathbf{v}^T = (b_1, b_2, b_3)$, then each of the six elementary products of $\mathbf{u}\mathbf{v}^T$ is of the form $(a_1 b_{j_1})(a_2 b_{j_2})(a_3 b_{j_3})$ where $\{j_1, j_2, j_3\}$ is a permutation of $\{1, 2, 3\}$; thus each of the elementary products is equal to $(a_1 a_2 a_3)(b_1 b_2 b_3)$.

WORKING WITH PROOFS

P1. If the three points lie on a vertical line then $x_1 = x_2 = x_3 = a$ and we have $\begin{vmatrix} x_1 & y_1 & 1 \\ x_2 & y_2 & 1 \\ x_3 & y_3 & 1 \end{vmatrix} = \begin{vmatrix} a & y_1 & 1 \\ a & y_2 & 1 \\ a & y_3 & 1 \end{vmatrix} = 0.$

Thus, without loss of generality, we may assume that the points do not lie on a vertical line. In this case the points are collinear if and only if $\frac{y_3 - y_1}{x_3 - x_1} = \frac{y_2 - y_1}{x_2 - x_1}$. The latter condition is equivalent to $(y_3 - y_1)(x_2 - x_1) = (y_2 - y_1)(x_3 - x_1)$ which can be written as:

$$(x_2 y_3 - x_3 y_2) - (x_1 y_3 - x_3 y_1) + (x_1 y_2 - x_2 y_1) = 0$$

On the other hand, expanding along the third column, we have:

$$\begin{vmatrix} x_1 & y_1 & 1 \\ x_2 & y_2 & 1 \\ x_3 & y_3 & 1 \end{vmatrix} = \begin{vmatrix} x_2 & y_2 \\ x_3 & y_3 \end{vmatrix} - \begin{vmatrix} x_1 & y_1 \\ x_3 & y_3 \end{vmatrix} + \begin{vmatrix} x_1 & y_1 \\ x_2 & y_2 \end{vmatrix} = (x_2 y_3 - x_3 y_2) - (x_1 y_3 - x_3 y_1) + (x_1 y_2 - x_2 y_1)$$

Thus the three points are collinear if and only if $\begin{vmatrix} x_1 & y_1 & 1 \\ x_2 & y_2 & 1 \\ x_3 & y_3 & 1 \end{vmatrix} = 0.$

P2. We wish to prove that for each positive integer n, there are $n!$ permutations of a set $\{j_1, j_2, \ldots, j_n\}$ of n distinct elements. Our proof is by induction on n.

Step 1. It is clear that there is exactly $1 = 1!$ permutation of the set $\{j_1\}$. Thus the statement is true for the case $n = 1$.

Step 2 (induction step). Suppose that the statement is true for $n = k$, where k is a fixed integer ≥ 1. Let $S = \{j_1, j_2, \ldots, j_k, j_{k+1}\}$. Then a permutation of the set S is formed by first choosing one of $k + 1$ positions for the element j_{k+1}, and then choosing a permutation for the remaining k elements in the remaining k positions. There are $k + 1$ possibilities for the first choice and, by the hypothesis, $k!$ possibilities for the second choice. Thus there are a total of $(k + 1)k! = (k + 1)!$ permutations of S. This shows that if the statement is true for $n = k$ it must also be true for $n = k + 1$. These two steps complete the proof by induction.

EXERCISE SET 4.2

1. **(a)** $\det(A) = \begin{vmatrix} -2 & 3 \\ 1 & 4 \end{vmatrix} = -8 - 3 = -11 \qquad \det(A^T) = \begin{vmatrix} -2 & 1 \\ 3 & 4 \end{vmatrix} = -8 - 3 = -11$

   (b) $\det(A) = \begin{vmatrix} 2 & -1 & 3 \\ 1 & 2 & 4 \\ 5 & -3 & 6 \end{vmatrix} = (24 - 20 - 9) - (30 - 6 - 24) = -5 - 0 = -5$

   $\det(A^T) = \begin{vmatrix} 2 & 1 & 5 \\ -1 & 2 & -3 \\ 3 & 4 & 6 \end{vmatrix} = (24 - 9 - 20) - (30 - 6 - 24) = -5 - 0 = -5$

3. **(a)** $\begin{vmatrix} 3 & 1 & 3 & 33 \\ 0 & \frac{1}{3} & 9 & 22 \\ 0 & 0 & -2 & 12 \\ 0 & 0 & 0 & 2 \end{vmatrix} = (3)\left(\frac{1}{3}\right)(-2)(2) = -4$

   (b) $\begin{vmatrix} 3 & 1 & 9 \\ -1 & 2 & -3 \\ 1 & 5 & 3 \end{vmatrix} = 0$ (first and third columns are proportional)

   (c) $\begin{vmatrix} 3 & -17 & 4 \\ 0 & 5 & 1 \\ 0 & 0 & -2 \end{vmatrix} = (3)(5)(-2) = -30$

5. **(a)** $\begin{vmatrix} d & e & f \\ g & h & i \\ a & b & c \end{vmatrix} = (-1)\begin{vmatrix} a & b & c \\ g & h & i \\ d & e & f \end{vmatrix} = (-1)(-1)\begin{vmatrix} a & b & c \\ d & e & f \\ g & h & i \end{vmatrix} = (-1)(-1)(-6) = -6$

   (b) $\begin{vmatrix} 3a & 3b & 3c \\ -d & -e & -f \\ 4g & 4h & 4i \end{vmatrix} = (3)\begin{vmatrix} a & b & c \\ -d & -e & -f \\ 4g & 4h & 4i \end{vmatrix} = (-3)\begin{vmatrix} a & b & c \\ d & e & f \\ 4g & 4h & 4i \end{vmatrix} = (-12)\begin{vmatrix} a & b & c \\ d & e & f \\ g & h & i \end{vmatrix} = (-12)(-6) = 72$

   (c) $\begin{vmatrix} a+g & b+h & c+i \\ d & e & f \\ g & h & i \end{vmatrix} = \begin{vmatrix} a & b & c \\ d & e & f \\ g & h & i \end{vmatrix} = -6$

(d) $\begin{vmatrix} -3a & -3b & -3c \\ d & e & f \\ g-4d & h-4e & i-4f \end{vmatrix} = \begin{vmatrix} -3a & -3b & -3c \\ d & e & f \\ g & h & i \end{vmatrix} = (-3)\begin{vmatrix} a & b & c \\ d & e & f \\ g & h & i \end{vmatrix} = (-3)(-6) = 18$

7. **(a)** $\det(2A) = \begin{vmatrix} -2 & 4 \\ 6 & 8 \end{vmatrix} = -16 - 24 = -40 \qquad 2^2 \det(A) = 4\begin{vmatrix} -1 & 2 \\ 3 & 4 \end{vmatrix} = 4(-4-6) = -40$

(b) $\det(-2A) = \begin{vmatrix} -4 & 2 & -6 \\ -6 & -4 & -2 \\ -2 & -8 & -10 \end{vmatrix} = (-160 + 8 - 288) - (-48 - 64 + 120) = -440 - 8 = -448$

$(-2)^3 \det(A) = (-8)\begin{vmatrix} 2 & -1 & 3 \\ 3 & 2 & 1 \\ 1 & 4 & 5 \end{vmatrix} = (-8)((20 - 1 + 36) - (6 + 8 - 15)) = (-8)(55 + 1) = -448$

9. If $x = 0$, the given matrix becomes $A = \begin{bmatrix} 0 & 0 & 2 \\ 2 & 1 & 1 \\ 0 & 0 & -5 \end{bmatrix}$ and, since the first and third rows are propor-

tional, we have $\det(A) = 0$. If $x = 2$, the given matrix becomes $B = \begin{bmatrix} 4 & 2 & 2 \\ 2 & 1 & 1 \\ 0 & 0 & -5 \end{bmatrix}$ and, since the first

and second rows are proportional, we have $\det(B) = 0$.

11. We use the properties of determinants stated in Theorem 4.2.2. Corresponding row operations are as indicated.

$\det(A) = \begin{vmatrix} 3 & 6 & -9 \\ 0 & 0 & -2 \\ -2 & 1 & 5 \end{vmatrix} = (3)\begin{vmatrix} 1 & 2 & -3 \\ 0 & 0 & -2 \\ -2 & 1 & 5 \end{vmatrix}$	A common factor of 3 was taken from the first row.

$= (3)\begin{vmatrix} 1 & 2 & -3 \\ 0 & 0 & -2 \\ 0 & 5 & -1 \end{vmatrix}$	2 times the first row was added to the third row.

$= (3)(-1)\begin{vmatrix} 1 & 2 & -3 \\ 0 & 5 & -1 \\ 0 & 0 & -2 \end{vmatrix}$	The second and third rows were interchanged.

$= (3)(-1)(-10) = 30$

13. We use the properties of determinants stated in Theorem 4.2.2. Corresponding row operations are as indicated.

$\det(A) = \begin{vmatrix} 1 & -3 & 0 \\ -2 & 4 & 1 \\ 5 & -2 & 2 \end{vmatrix} = \begin{vmatrix} 1 & -3 & 0 \\ 0 & -2 & 1 \\ 5 & -2 & 2 \end{vmatrix}$	2 times the first row was added to the second row.

$= \begin{vmatrix} 1 & -3 & 0 \\ 0 & -2 & 1 \\ 0 & 13 & 2 \end{vmatrix}$	−5 times the first row was added to the third row.

$$= (-2) \begin{vmatrix} 1 & -3 & 0 \\ 0 & 1 & -\frac{1}{2} \\ 0 & 13 & 2 \end{vmatrix}$$

A factor of -2 was taken from the second row.

$$= (-2) \begin{vmatrix} 1 & -3 & 0 \\ 0 & 1 & -\frac{1}{2} \\ 0 & 0 & \frac{17}{2} \end{vmatrix}$$

-13 times the second row was added to the third row.

$$= (-2) \left(\frac{17}{2} \right) = -17$$

15. We use the properties of determinants stated in Theorem 4.2.2. Corresponding row operations are as indicated.

$$\det(A) = \begin{vmatrix} 1 & -2 & 3 & 1 \\ 5 & -9 & 6 & 3 \\ -1 & 2 & -6 & -2 \\ 2 & 8 & 6 & 1 \end{vmatrix} = \begin{vmatrix} 1 & 2 & -3 & 1 \\ 0 & 1 & -9 & -2 \\ 0 & 0 & -3 & -1 \\ 0 & 12 & 0 & -1 \end{vmatrix}$$

-5 times row 1 was added to row 2; row 1 was added to row 3; -2 times row 1 was added to row 4.

$$= \begin{vmatrix} 1 & 2 & -3 & 1 \\ 0 & 1 & -9 & -2 \\ 0 & 0 & -3 & -1 \\ 0 & 0 & 108 & 23 \end{vmatrix}$$

-12 times row 2 was added to row 4.

$$= \begin{vmatrix} 1 & 2 & -3 & 1 \\ 0 & 1 & -9 & -2 \\ 0 & 0 & -3 & -1 \\ 0 & 0 & 0 & -13 \end{vmatrix}$$

36 times row 3 was added row 4.

$$= 39$$

17. We use the properties of determinants stated in Theorem 4.2.2. Corresponding row operations are as indicated.

$$\det(A) = \begin{vmatrix} 0 & 1 & 1 & 1 \\ \frac{1}{2} & \frac{1}{2} & 1 & \frac{1}{2} \\ \frac{2}{3} & \frac{1}{3} & \frac{1}{3} & 0 \\ -\frac{1}{3} & \frac{2}{3} & 0 & 0 \end{vmatrix} = (-1) \begin{vmatrix} \frac{1}{2} & \frac{1}{2} & 1 & \frac{1}{2} \\ 0 & 1 & 1 & 1 \\ \frac{2}{3} & \frac{1}{3} & \frac{1}{3} & 0 \\ -\frac{1}{3} & \frac{2}{3} & 0 & 0 \end{vmatrix}$$

The first and second rows were interchanged.

$$= (-1) \left(\frac{1}{2} \right) \begin{vmatrix} 1 & 1 & 2 & 1 \\ 0 & 1 & 1 & 1 \\ \frac{2}{3} & \frac{1}{3} & \frac{1}{3} & 0 \\ -\frac{1}{3} & \frac{2}{3} & 0 & 0 \end{vmatrix}$$

A factor of $\frac{1}{2}$ was taken from the first row.

$$= (-1) \left(\frac{1}{2} \right) \begin{vmatrix} 1 & 1 & 2 & 1 \\ 0 & 1 & 1 & 1 \\ 0 & -\frac{1}{3} & -1 & -\frac{2}{3} \\ 0 & 1 & \frac{2}{3} & \frac{1}{3} \end{vmatrix}$$

$-\frac{2}{3}$ times row 1 was added to row 3; $\frac{1}{3}$ times row 1 was added to row 4.

$$= (-1)\left(\frac{1}{2}\right) \begin{vmatrix} 1 & 1 & 2 & 1 \\ 0 & 1 & 1 & 1 \\ 0 & 0 & -\frac{2}{3} & -\frac{1}{3} \\ 0 & 0 & -\frac{1}{3} & -\frac{2}{3} \end{vmatrix}$$

$\frac{1}{3}$ times row 2 was added to row 3; -1 times row 2 was added to row 4.

$$= (-1)\left(\frac{1}{2}\right)\left(-\frac{2}{3}\right) \begin{vmatrix} 1 & 1 & 2 & 1 \\ 0 & 1 & 1 & 1 \\ 0 & 0 & 1 & \frac{1}{2} \\ 0 & 0 & -\frac{1}{3} & -\frac{2}{3} \end{vmatrix}$$

A factor of $-\frac{2}{3}$ was taken from row 3.

$$= (-1)\left(\frac{1}{2}\right)\left(-\frac{2}{3}\right) \begin{vmatrix} 1 & 1 & 2 & 1 \\ 0 & 1 & 1 & 1 \\ 0 & 0 & 1 & \frac{1}{2} \\ 0 & 0 & 0 & -\frac{1}{2} \end{vmatrix}$$

$\frac{1}{3}$ times row 3 was added to row 4.

$$= (-1)\left(\frac{1}{2}\right)\left(-\frac{2}{3}\right)\left(-\frac{1}{2}\right) = -\frac{1}{6}$$

19. (a)
$$\begin{vmatrix} a_1 & b_1 & a_1 + b_1 + c_1 \\ a_2 & b_2 & a_2 + b_2 + c_2 \\ a_3 & b_3 & a_3 + b_3 + c_3 \end{vmatrix} = \begin{vmatrix} a_1 & b_1 & b_1 + c_1 \\ a_2 & b_2 & b_2 + c_2 \\ a_3 & b_3 & b_3 + c_3 \end{vmatrix}$$

Add -1 times column 1 to column 3.

$$= \begin{vmatrix} a_1 & b_1 & c_1 \\ a_2 & b_2 & c_2 \\ a_3 & b_3 & c_3 \end{vmatrix}$$

Add -1 times column 2 to column 3.

(b)
$$\begin{vmatrix} a_1 + b_1 & a_1 - b_1 & c_1 \\ a_2 + b_2 & a_2 - b_2 & c_2 \\ a_3 + b_3 & a_3 - b_3 & c_3 \end{vmatrix} = \begin{vmatrix} 2a_1 & a_1 - b_1 & c_1 \\ 2a_2 & a_2 - b_2 & c_2 \\ 2a_3 & a_3 - b_3 & c_3 \end{vmatrix}$$

Add column 2 to column 1.

$$= 2 \begin{vmatrix} a_1 & a_1 - b_1 & c_1 \\ a_2 & a_2 - b_2 & c_2 \\ a_3 & a_3 - b_3 & c_3 \end{vmatrix}$$

Factor of 2 taken from column 1.

$$= 2 \begin{vmatrix} a_1 & -b_1 & c_1 \\ a_2 & -b_2 & c_2 \\ a_3 & -b_3 & c_3 \end{vmatrix}$$

Add -1 times column 1 to column 2.

$$= -2 \begin{vmatrix} a_1 & b_1 & c_1 \\ a_2 & b_2 & c_2 \\ a_3 & b_3 & c_3 \end{vmatrix}$$

Factor of -1 taken from column 2.

21. $\det(A) = \begin{vmatrix} 1 & x & x^2 \\ 1 & y & y^2 \\ 1 & z & z^2 \end{vmatrix} = \begin{vmatrix} 1 & x & x^2 \\ 0 & y-x & y^2-x^2 \\ 0 & z-x & z^2-x^2 \end{vmatrix} = (y-x)(z-x) \begin{vmatrix} 1 & x & x^2 \\ 0 & 1 & y+x \\ 0 & 1 & z+x \end{vmatrix}$

$$= (y-x)(z-x) \begin{vmatrix} 1 & x & x^2 \\ 0 & 1 & y+x \\ 0 & 0 & z-y \end{vmatrix} = (y-x)(z-x)(z-y)$$

23. If we add the first row of A to the second row, the result is a matrix B that has two identical rows; thus $\det(A) = \det(B) = 0$.

25. (a) We have $\det(A) = (k-3)(k-2) - 4 = k^2 - 5k + 2$. Thus, from Theorem 4.2.4, the matrix A is invertible if and only if $k^2 - 5k + 2 \neq 0$, i.e. $k \neq \frac{5 \pm \sqrt{17}}{2}$.

(b) We have $\det(A) = (1)\begin{vmatrix} 1 & 6 \\ 3 & 2 \end{vmatrix} - (3)\begin{vmatrix} 2 & 4 \\ 3 & 2 \end{vmatrix} + (k)\begin{vmatrix} 2 & 4 \\ 1 & 6 \end{vmatrix} = 8 + 8k$. Thus, from Theorem 4.2.4, the matrix A is invertible if and only if $8 + 8k \neq 0$, i.e. $k \neq -1$.

27. (a) $\det(3A) = 3^3 \det(A) = (27)(7) = 189$

(b) $\det(A^{-1}) = \dfrac{1}{\det(A)} = \dfrac{1}{7}$

(c) $\det(2A^{-1}) = 2^3 \det(A^{-1}) = (8)(\frac{1}{7}) = \frac{8}{7}$

(d) $\det((2A)^{-1}) = \dfrac{1}{\det(2A)} = \dfrac{1}{2^3 \det(A)} = \dfrac{1}{(8)(7)} = \dfrac{1}{56}$

29. We have $AB = \begin{bmatrix} 1 & 0 & 0 \\ 3 & 3 & 0 \\ 5 & 2 & -2 \end{bmatrix} \begin{bmatrix} 2 & 4 & -5 \\ 0 & 1 & 3 \\ 0 & 0 & 2 \end{bmatrix} = \begin{bmatrix} 2 & 4 & -5 \\ 6 & 15 & -6 \\ 10 & 22 & -23 \end{bmatrix}$; thus

$$\det(AB) = \begin{vmatrix} 2 & 4 & -5 \\ 6 & 15 & -6 \\ 10 & 22 & -23 \end{vmatrix} = \begin{vmatrix} 2 & 4 & -5 \\ 0 & 3 & 9 \\ 0 & 2 & 2 \end{vmatrix} = 2 \begin{vmatrix} 3 & 9 \\ 2 & 2 \end{vmatrix} = 2(6-18) = -24$$

On the other hand, $\det(A) = (1)(3)(-2) = -6$ and $\det(B) = (2)(1)(2) = 4$. Thus $\det(AB) = \det(A)\det(B)$.

31. The following sequence of row operations reduce A to an upper triangular form:

$$\begin{bmatrix} 1 & -2 & 3 & 1 \\ 5 & -9 & 6 & 3 \\ -1 & 2 & -6 & -2 \\ 2 & 8 & 6 & 1 \end{bmatrix} \rightarrow \begin{bmatrix} 1 & -2 & 3 & 1 \\ 0 & 1 & -9 & -2 \\ 0 & 0 & -3 & -1 \\ 0 & 12 & 0 & -1 \end{bmatrix} \rightarrow \begin{bmatrix} 1 & -2 & 3 & 1 \\ 0 & 1 & -9 & -2 \\ 0 & 0 & -3 & -1 \\ 0 & 0 & 108 & 23 \end{bmatrix} \rightarrow \begin{bmatrix} 1 & -2 & 3 & 1 \\ 0 & 1 & -9 & -2 \\ 0 & 0 & -3 & -1 \\ 0 & 0 & 0 & -13 \end{bmatrix}$$

The corresponding LU-factorization is

$$A = \begin{bmatrix} 1 & -2 & 3 & 1 \\ 5 & -9 & 6 & 3 \\ -1 & 2 & -6 & -2 \\ 2 & 8 & 6 & 1 \end{bmatrix} = \begin{bmatrix} 1 & 0 & 0 & 0 \\ 5 & 1 & 0 & 0 \\ -1 & 0 & 1 & 0 \\ 2 & 12 & -36 & 1 \end{bmatrix} \begin{bmatrix} 1 & -2 & 3 & 1 \\ 0 & 1 & -9 & -2 \\ 0 & 0 & -3 & -1 \\ 0 & 0 & 0 & -13 \end{bmatrix} = LU$$

and from this we conclude that $\det(A) = \det(L)\det(U) = (1)(39) = 39$.

33. If we add the first row of A to the second row, using the identity $\sin^2\theta + \cos^2\theta = 1$, we see that

$$\det(A) = \begin{vmatrix} \sin^2\alpha & \sin^2\beta & \sin^2\gamma \\ \cos^2\alpha & \cos^2\beta & \cos^2\gamma \\ 1 & 1 & 1 \end{vmatrix} = \begin{vmatrix} \sin^2\alpha & \sin^2\beta & \sin^2\gamma \\ 1 & 1 & 1 \\ 1 & 1 & 1 \end{vmatrix} = 0$$

since the resulting matrix has two identical rows. Thus, from Theorem 4.2.4, A is not invertible.

35. (a) Since $\det(A^T) = \det(A)$, we have $\det(A^T A) = \det(A^T)\det(A) = (\det(A))^2 = \det(A)\det(A^T) = \det(AA^T)$.

(b) Since $\det(A^T A) = (\det(A))^2$, it follows that $\det(A^T A) = 0$ if and only if $\det(A) = 0$. Thus, from Theorem 4.2.4, $A^T A$ is invertible if and only if A is invertible.

37. $\|\mathbf{x}\|^2\|\mathbf{y}\|^2 - (\mathbf{x}\cdot\mathbf{y})^2 = (x_1^2 + x_2^2 + x_3^2)(y_1^2 + y_2^2 + y_3^2) - (x_1 y_1 + x_2 y_2 + x_3 y_3)^2$

$= x_1^2 y_2^2 + x_1^2 y_3^2 + x_2^2 y_1^2 + x_2^2 y_3^2 + x_3^2 y_1^2 + x_3^2 y_2^2 - 2x_1 y_1 x_2 y_2 - 2x_1 y_1 x_3 y_3 - 2x_2 y_2 x_3 y_3$

$\begin{vmatrix} x_1 & x_2 \\ y_1 & y_2 \end{vmatrix}^2 + \begin{vmatrix} x_1 & x_3 \\ y_1 & y_3 \end{vmatrix}^2 + \begin{vmatrix} x_2 & x_3 \\ y_2 & y_3 \end{vmatrix}^2 = (x_1 y_2 - x_2 y_1)^2 + (x_1 y_3 - x_3 y_1)^2 + (x_2 y_3 - x_3 y_2)^2$

$= x_1^2 y_2^2 - 2x_1 y_2 x_2 y_1 + x_2^2 y_1^2 + x_1^2 y_3^2 - 2x_1 y_3 x_3 y_1 + x_3^2 y_1^2 + x_2^2 y_3^2 - 2x_2 y_3 x_3 y_2 + x_3^2 y_2^2$

39. (a) $\det(M) = \begin{vmatrix} 2 & -1 \\ 4 & 3 \end{vmatrix} \begin{vmatrix} 1 & 3 & 5 \\ -2 & 6 & 2 \\ 3 & 5 & 2 \end{vmatrix} = (6+4)\begin{vmatrix} 1 & 3 & 5 \\ 0 & 12 & 12 \\ 0 & -4 & -13 \end{vmatrix} = (10)(1)\begin{vmatrix} 12 & 12 \\ -4 & -13 \end{vmatrix} = -1080$

(b) $\det(M) = \begin{vmatrix} 1 & 2 & 0 \\ 0 & 1 & 2 \\ 0 & 0 & 1 \end{vmatrix} \begin{vmatrix} 1 & 2 \\ 0 & 1 \end{vmatrix} = (1)(1) = 1$

DISCUSSION AND DISCOVERY

D1. The matrices are singular if and only if the corresponding determinants are zero. This leads to the system of equations

$$\begin{vmatrix} 1 & 2 & s \\ 2 & 3 & t \\ 4 & 5 & 7 \end{vmatrix} = s\begin{vmatrix} 2 & 3 \\ 4 & 5 \end{vmatrix} - t\begin{vmatrix} 1 & 2 \\ 4 & 5 \end{vmatrix} + 7\begin{vmatrix} 1 & 2 \\ 2 & 3 \end{vmatrix} = -2s + 3t - 7 = 0$$

$$\begin{vmatrix} 4 & 5 & 8 \\ s & 2 & 3 \\ t & 1 & 8 \end{vmatrix} = 4\begin{vmatrix} 2 & 3 \\ 1 & 8 \end{vmatrix} - s\begin{vmatrix} 5 & 8 \\ 1 & 8 \end{vmatrix} + t\begin{vmatrix} 5 & 8 \\ 2 & 3 \end{vmatrix} = 52 - 32s - t = 0$$

from which it follows that $s = \frac{149}{98}$ and $t = \frac{164}{49}$.

D2. Since $\det(AB) = \det(A)\det(B) = \det(B)\det(A) = \det(BA)$, it is always true that $\det(AB) = \det(BA)$.

D3. If A or B is not invertible then either $\det(A) = 0$ or $\det(B) = 0$ (or both). It follows that $\det(AB) = \det(A)\det(B) = 0$; thus AB is not invertible.

D4. For convenience call the given matrix A_n. If $n = 2$ or 3, then A_n can be reduced to the identity matrix by interchanging the first and last rows. Thus $\det(A_n) = -1$ if $n = 2$ or 3. If $n = 4$ or 5, then two row interchanges are required to reduce A_n to the identity (interchange the first and last rows, then interchange the second and next to last rows). Thus $\det(A_n) = +1$ if $n = 4$ or 5. This pattern continues and can be summarized as follows:

$$\det(A_{2k}) = \det(A_{2k+1}) = -1 \quad \text{for } k = 1, 3, 5, \ldots$$

$$\det(A_{2k}) = \det(A_{2k+1}) = +1 \quad \text{for } k = 2, 4, 6, \ldots$$

D5. If A is skew-symmetric, then $\det(A) = \det(A^T) = \det(-A) = (-1)^n \det(A)$ where n is the size of A. It follows that if A is a skew-symmetric matrix of odd order, then $\det(A) = -\det(A)$ and so $\det(A) = 0$.

D6. Let A be an $n \times n$ matrix, and let B be the matrix that results when the rows of A are written in reverse order. Then the matrix B can be reduced to A by a series of row interchanges. If $n = 2$ or 3, then only one interchange is needed and so $\det(B) = -\det(A)$. If $n = 4$ or 5, then two interchanges are required and so $\det(B) = +\det(A)$. This pattern continues:

$$\det(B) = -\det(A) \quad \text{for } n = 2k \text{ or } 2k+1 \text{ where } k \text{ is odd}$$

$$\det(B) = +\det(A) \quad \text{for } n = 2k \text{ or } 2k+1 \text{ where } k \text{ is even}$$

D7. (a) False. For example, if $A = I = I_2$, then $\det(I + A) = \det(2I) = 4$, whereas $1 + \det(A) = 2$.

(b) True. From Theorem 4.2.5 it follows that $\det(A^n) = (\det(A))^n$ for every $n = 1, 2, 3, \ldots$.

(c) False. From Theorem 4.2.3(c), we have $\det(3A) = 3^n \det(A)$ where n is the size of A. Thus the statement is false except when $n = 1$ or $\det(A) = 0$.

(d) True. If $\det(A) = 0$, the matrix is singular and so the system $A\mathbf{x} = \mathbf{0}$ has infinitely many solutions.

D8. (a) True. If A is invertible, then $\det(A) \neq 0$. Since $\det(ABA) = \det(A)\det(B)\det(A)$ it follows that if A is invertible and $\det(ABA) = 0$, then $\det(B) = 0$.

(b) True. If $A = A^{-1}$, then since $\det(A^{-1}) = \frac{1}{\det(A)}$, it follows that $(\det(A))^2 = 1$ and so $\det(A) = \pm 1$.

(c) True. If the reduced row echelon form of A has a row of zeros, then A is not invertible.

(d) True. Since $\det(A^T) = \det(A)$, it follows that $\det(AA^T) = \det(A)\det(A^T) = (\det(A))^2 \geq 0$.

(e) True. If $\det(A) \neq 0$ then A is invertible, and an invertible matrix can always be written as a product of elementary matrices.

D9. If $A = A^2$, then $\det(A) = \det(A^2) = (\det(A))^2$ and so $\det(A) = 0$ or $\det(A) = 1$. If $A = A^3$, then $\det(A) = \det(A^3) = (\det(A))^3$ and so $\det(A) = 0$ or $\det(A) = \pm 1$.

D10. Each elementary product of this matrix must include a factor that comes from the 3×3 block of zeros on the upper right. Thus all of the elementary products are zero. It follows that $\det(A) = 0$, no matter what values are assigned to the starred quantities.

D11. This permutation of the columns of an $n \times n$ matrix A can be attained via a sequence of $n - 1$ column interchanges which successively move the first column to the right by one position (i.e. interchange columns 1 and 2, then interchange columns 2 and 3, etc.). Thus the determinant of the resulting matrix is equal to $(-1)^{n-1} \det(A)$.

WORKING WITH PROOFS

P1. If $\mathbf{x} = \begin{bmatrix} x_1 \\ x_2 \\ \vdots \\ x_n \end{bmatrix}$ and $\mathbf{y} = \begin{bmatrix} y_1 \\ y_2 \\ \vdots \\ y_n \end{bmatrix}$ then, using cofactor expansions along the jth column, we have

$$\det(C) = (x_1 + y_1)C_{1j} + (x_2 + y_2)C_{2j} + \cdots + (x_n + y_n)C_{nj}$$

$$= (x_1 C_{1j} + x_2 C_{2j} + \cdots + x_n C_{nj}) + (y_1 C_{1j} + y_2 C_{2j} + \cdots + y_n C_{nj}) = \det(A) + \det(B)$$

P2. Suppose A is a square matrix, and B is the matrix that is obtained from A by adding k times the ith row to the jth row. Then, expanding along the jth row of B, we have

$$\det(B) = (a_{j1} + ka_{i1})C_{j1} + (a_{j2} + ka_{i2})C_{j2} + \cdots + (a_{jn} + ka_{in})C_{jn}$$

$$= (a_{j1}C_{j1} + a_{j2}C_{j2} + \cdots + a_{jn}C_{jn}) + k(a_{i1}C_{j1} + a_{i2}C_{j2} + \cdots + a_{in}C_{jn})$$

$$= \det(A) + k\det(C)$$

where C is the matrix obtained from A by replacing the jth row by a copy of the ith row. Since C has two identical rows, it follows that $\det(C) = 0$, and so $\det(B) = \det(A)$.

EXERCISE SET 4.3

1. The matrix of cofactors from A is $C = \begin{bmatrix} -3 & 3 & -2 \\ 5 & -4 & 2 \\ 5 & -5 & 3 \end{bmatrix}$, and $\det(A) = (2)(-3) + (5)(3) + (5)(-2) = -1$.

Thus $\text{adj}(A) = \begin{bmatrix} -3 & 5 & 5 \\ 3 & -4 & -5 \\ -2 & 2 & 3 \end{bmatrix}$ and $A^{-1} = -\text{adj}(A) = \begin{bmatrix} 3 & -5 & -5 \\ -3 & 4 & 5 \\ 2 & -2 & -3 \end{bmatrix}$.

3. The matrix of cofactors from A is $C = \begin{bmatrix} 2 & 0 & 0 \\ 6 & 4 & 0 \\ 4 & 6 & 2 \end{bmatrix}$, and $\det(A) = (2)(2) + (-3)(0) + (5)(0) = 4$. Thus

$$\text{adj}(A) = \begin{bmatrix} 2 & 6 & 4 \\ 0 & 4 & 6 \\ 0 & 0 & 2 \end{bmatrix} \text{ and } A^{-1} = \frac{1}{4}\text{adj}(A) = \frac{1}{4}\begin{bmatrix} 2 & 6 & 4 \\ 0 & 4 & 6 \\ 0 & 0 & 2 \end{bmatrix} = \begin{bmatrix} \frac{1}{2} & \frac{3}{2} & 1 \\ 0 & 1 & \frac{3}{2} \\ 0 & 0 & \frac{1}{2} \end{bmatrix}.$$

5. $x_1 = \dfrac{\begin{vmatrix} 3 & -3 \\ 5 & 1 \end{vmatrix}}{\begin{vmatrix} 7 & -3 \\ 3 & 1 \end{vmatrix}} = \dfrac{18}{16} = \dfrac{9}{8}$ $x_2 = \dfrac{\begin{vmatrix} 7 & 3 \\ 3 & 5 \end{vmatrix}}{\begin{vmatrix} 7 & -3 \\ 3 & 1 \end{vmatrix}} = \dfrac{26}{16} = \dfrac{13}{8}$

7. $x = \dfrac{\begin{vmatrix} 6 & -4 & 1 \\ -1 & -1 & 2 \\ -20 & 2 & -3 \end{vmatrix}}{\begin{vmatrix} 1 & -4 & 1 \\ 4 & -1 & 2 \\ 2 & 2 & -3 \end{vmatrix}} = \dfrac{144}{-55}$ $y = \dfrac{\begin{vmatrix} 1 & 6 & 1 \\ 4 & -1 & 2 \\ 2 & -20 & -3 \end{vmatrix}}{\begin{vmatrix} 1 & -4 & 1 \\ 4 & -1 & 2 \\ 2 & 2 & -3 \end{vmatrix}} = \dfrac{61}{-55}$ $z = \dfrac{\begin{vmatrix} 1 & -4 & 6 \\ 4 & -1 & -1 \\ 2 & 2 & -20 \end{vmatrix}}{\begin{vmatrix} 1 & -4 & 1 \\ 4 & -1 & 2 \\ 2 & 2 & -3 \end{vmatrix}} = \dfrac{-230}{-55} = \dfrac{46}{11}$

9. $x_1 = \dfrac{\begin{vmatrix} -32 & -4 & 2 & 1 \\ 14 & -1 & 7 & 9 \\ 11 & 1 & 3 & 1 \\ -4 & -2 & 1 & -4 \end{vmatrix}}{\begin{vmatrix} -1 & -4 & 2 & 1 \\ 2 & -1 & 7 & 9 \\ -1 & 1 & 3 & 1 \\ 1 & -2 & 1 & -4 \end{vmatrix}} = \dfrac{-2115}{-423} = 5 \qquad x_2 = \dfrac{\begin{vmatrix} -1 & -32 & 2 & 1 \\ 2 & 14 & 7 & 9 \\ -1 & 11 & 3 & 1 \\ 1 & -4 & 1 & -4 \end{vmatrix}}{\begin{vmatrix} -1 & -4 & 2 & 1 \\ 2 & -1 & 7 & 9 \\ -1 & 1 & 3 & 1 \\ 1 & -2 & 1 & -4 \end{vmatrix}} = \dfrac{-3384}{-423} = 8$

$x_3 = \dfrac{\begin{vmatrix} -1 & -4 & -32 & 1 \\ 2 & -1 & 14 & 9 \\ -1 & 1 & 11 & 1 \\ 1 & -2 & -4 & -4 \end{vmatrix}}{-423} = \dfrac{-1269}{-423} = 3 \qquad x_4 = \dfrac{\begin{vmatrix} -1 & -4 & 2 & -32 \\ 2 & -1 & 7 & 14 \\ -1 & 1 & 3 & 11 \\ 1 & -2 & 1 & -4 \end{vmatrix}}{-423} = \dfrac{423}{-423} = -1$

11. $x = \dfrac{\begin{vmatrix} 1 & 3 & 4 \\ 2 & -2 & -1 \\ 4 & 1 & 1 \end{vmatrix}}{\begin{vmatrix} 2 & 3 & 4 \\ 1 & -2 & -1 \\ 3 & 1 & 1 \end{vmatrix}} = \dfrac{21}{14} = \dfrac{3}{2}$

13. The matrix of cofactors is $C = \begin{bmatrix} \cos\theta & \sin\theta & 0 \\ -\sin\theta & \cos\theta & 0 \\ 0 & 0 & 1 \end{bmatrix}$, and $\det(A) = \cos^2\theta + \sin^2\theta = 1$. Thus A is invertible

and $A^{-1} = \text{adj}(A) = \begin{bmatrix} \cos\theta & -\sin\theta & 0 \\ \sin\theta & \cos\theta & 0 \\ 0 & 0 & 1 \end{bmatrix}$.

15. The coefficient matrix is $A = \begin{bmatrix} 3 & 3 & 1 \\ 4 & k & 2 \\ 2k & 2k & k \end{bmatrix}$, and $\det(A) = k(k-4)$. Thus the system has a unique

solution if $k \neq 0$ and $k \neq 4$. In this case the solution is given by:

$$x = \dfrac{\begin{vmatrix} 1 & 3 & 1 \\ 2 & k & 2 \\ 1 & 2k & k \end{vmatrix}}{k(k-4)} = \dfrac{(k-1)(k-6)}{k(k-4)} \qquad y = \dfrac{\begin{vmatrix} 3 & 1 & 1 \\ 4 & 2 & 2 \\ 2k & 1 & k \end{vmatrix}}{k(k-4)} = \dfrac{2(k-1)}{k(k-4)}$$

$$z = \dfrac{\begin{vmatrix} 3 & 3 & 1 \\ 4 & k & 2 \\ 2k & 2k & 1 \end{vmatrix}}{k(k-4)} = -\dfrac{(2k-3)(k-4)}{k(k-4)} = -\dfrac{2k-3}{k}$$

17. We have $\det(A) = y^3 - x^2 y = y(y^2 - x^2)$. Thus A is invertible if and only if $y \neq 0$ and $y \neq \pm x$. The formula for the inverse is $A^{-1} = \dfrac{1}{y(y^2 - x^2)} \begin{bmatrix} -xy & 0 & y^2 \\ y^2 & 0 & -xy \\ -xy & y^2 - x^2 & x^2 \end{bmatrix}$.

19. $|\det(A)| = |1 + 2| = 3$

21. $|\det(A)| = |(0 - 4 + 6) - (0 + 4 + 2)| = |-4| = 4$

23. The parallelogram has the vectors $\overrightarrow{P_1 P_2} = (3, 2)$ and $\overrightarrow{P_1 P_4} = (3, 1)$ as adjacent sides. Let $A = \begin{bmatrix} 3 & 3 \\ 2 & 1 \end{bmatrix}$. Then, from Theorem 4.3.5, the area of the parallelogram is $|\det(A)| = |3 - 6| = 3$.

25. area $\triangle ABC = \dfrac{1}{2} \begin{vmatrix} 2 & 0 & 1 \\ 3 & 4 & 1 \\ -1 & 2 & 1 \end{vmatrix} = 7$

27. $V = |\det(A)|$ where $A = \begin{bmatrix} 2 & 0 & 2 \\ -6 & 4 & 2 \\ 2 & -2 & -4 \end{bmatrix}$; thus $V = |-16| = 16$.

29. The vectors lie in the same plane if and only if the parallelepiped that they determine is degenerate in the sense that its "volume" is zero. In this example, we have

$$V = \begin{vmatrix} -1 & 3 & 5 \\ -2 & 0 & -4 \\ 1 & -2 & 0 \end{vmatrix} = 16$$

and so the vectors do not lie in the same plane.

31. $\mathbf{a} = \pm \dfrac{1}{\sqrt{5}}(0, 2, 1)$

33. $\mathbf{u} \times \mathbf{v} = \begin{vmatrix} \mathbf{i} & \mathbf{j} & \mathbf{k} \\ 2 & 3 & -6 \\ 2 & 3 & 6 \end{vmatrix} = 36\mathbf{i} - 24\mathbf{j}$ $\sin\theta = \dfrac{\|\mathbf{u} \times \mathbf{v}\|}{\|\mathbf{u}\|\|\mathbf{v}\|} = \dfrac{\sqrt{1296 + 576}}{\sqrt{49}\sqrt{49}} = \dfrac{\sqrt{1872}}{49} = \dfrac{12\sqrt{13}}{49}$

35. (a) $\mathbf{v} \times \mathbf{w} = \begin{vmatrix} \mathbf{i} & \mathbf{j} & \mathbf{k} \\ 0 & 2 & -3 \\ 2 & 6 & 7 \end{vmatrix} = (14 + 18)\mathbf{i} - (0 + 6)\mathbf{j} + (0 - 4)\mathbf{k} = 32\mathbf{i} - 6\mathbf{j} - 4\mathbf{k}$

(b) $\mathbf{u} \times (\mathbf{v} \times \mathbf{w}) = \begin{vmatrix} \mathbf{i} & \mathbf{j} & \mathbf{k} \\ 3 & 2 & -1 \\ 32 & -6 & -4 \end{vmatrix} = (-8 - 6)\mathbf{i} - (-12 + 32)\mathbf{j} + (-18 - 64)\mathbf{k} = -14\mathbf{i} - 20\mathbf{j} - 82\mathbf{k}$

(c) $\mathbf{u} \times \mathbf{v} = \begin{vmatrix} \mathbf{i} & \mathbf{j} & \mathbf{k} \\ 3 & 2 & -1 \\ 0 & 2 & -3 \end{vmatrix} = (-6 + 2)\mathbf{i} - (-9 + 0)\mathbf{j} + (6 - 0)\mathbf{k} = -4\mathbf{i} + 9\mathbf{j} + 6\mathbf{k}$

$(\mathbf{u} \times \mathbf{v}) \times \mathbf{w} = \begin{vmatrix} \mathbf{i} & \mathbf{j} & \mathbf{k} \\ -4 & 9 & 6 \\ 2 & 6 & 7 \end{vmatrix} = (63 - 36)\mathbf{i} - (-28 - 12)\mathbf{j} + (-24 - 18)\mathbf{k} = 27\mathbf{i} + 40\mathbf{j} - 42\mathbf{k}$

37. (a) $\mathbf{u} \times \mathbf{v} = \begin{vmatrix} \mathbf{i} & \mathbf{j} & \mathbf{k} \\ -6 & 4 & 2 \\ 3 & 1 & 5 \end{vmatrix} = 18\mathbf{i} + 36\mathbf{j} - 18\mathbf{k} = (18, 36, -18)$ is orthogonal to both $\mathbf{u}$ and $\mathbf{v}$.

(b) $\mathbf{u} \times \mathbf{v} = \begin{vmatrix} \mathbf{i} & \mathbf{j} & \mathbf{k} \\ -2 & 1 & 5 \\ 3 & 0 & -3 \end{vmatrix} = -3\mathbf{i} + 9\mathbf{j} - 3\mathbf{k} = (-3, 9, -3)$ is orthogonal to both $\mathbf{u}$ and $\mathbf{v}$.

39. $\mathbf{u} \times \mathbf{v} = \left(\begin{vmatrix} u_2 & u_3 \\ v_2 & v_3 \end{vmatrix}, -\begin{vmatrix} u_1 & u_3 \\ v_1 & v_3 \end{vmatrix}, \begin{vmatrix} u_1 & u_2 \\ v_1 & v_2 \end{vmatrix} \right) = \left(-\begin{vmatrix} v_2 & v_3 \\ u_2 & u_3 \end{vmatrix}, +\begin{vmatrix} v_1 & v_3 \\ u_1 & u_3 \end{vmatrix}, -\begin{vmatrix} v_1 & v_2 \\ u_1 & u_2 \end{vmatrix} \right) = -\mathbf{v} \times \mathbf{u}$

41. $k(\mathbf{u} \times \mathbf{v}) = k \left(\begin{vmatrix} u_2 & u_3 \\ v_2 & v_3 \end{vmatrix}, -\begin{vmatrix} u_1 & u_3 \\ v_1 & v_3 \end{vmatrix}, \begin{vmatrix} u_1 & u_2 \\ v_1 & v_2 \end{vmatrix} \right) = \left(\begin{vmatrix} ku_2 & ku_3 \\ v_2 & v_3 \end{vmatrix}, -\begin{vmatrix} ku_1 & ku_3 \\ v_1 & v_3 \end{vmatrix}, \begin{vmatrix} ku_1 & ku_2 \\ v_1 & v_2 \end{vmatrix} \right)$
$= (k\mathbf{u}) \times \mathbf{v}$

43. (a) $\mathbf{u} \times \mathbf{v} = \begin{vmatrix} \mathbf{i} & \mathbf{j} & \mathbf{k} \\ 1 & -1 & 2 \\ 0 & 3 & 1 \end{vmatrix} = -7\mathbf{i} - \mathbf{j} + 3\mathbf{k} \qquad A = \|\mathbf{u} \times \mathbf{v}\| = \sqrt{49 + 1 + 9} = \sqrt{59}$

(b) $\mathbf{u} \times \mathbf{v} = \begin{vmatrix} \mathbf{i} & \mathbf{j} & \mathbf{k} \\ 2 & 3 & 0 \\ -1 & 2 & -2 \end{vmatrix} = -6\mathbf{i} + 4\mathbf{j} + 7\mathbf{k} \qquad A = \|\mathbf{u} \times \mathbf{v}\| = \sqrt{36 + 16 + 49} = \sqrt{101}$

45. $\overrightarrow{P_1P_2} \times \overrightarrow{P_1P_3} = \begin{vmatrix} \mathbf{i} & \mathbf{j} & \mathbf{k} \\ -1 & -5 & 2 \\ 2 & 0 & 3 \end{vmatrix} = -15\mathbf{i} + 7\mathbf{j} + 10\mathbf{k}$

$A = \tfrac{1}{2} \left\| \overrightarrow{P_1P_2} \times \overrightarrow{P_1P_3} \right\| = \tfrac{1}{2}\sqrt{225 + 49 + 100} = \frac{\sqrt{374}}{2}$

47. Recall that the dot product distributes across addition, i.e. $(\mathbf{a} + \mathbf{d}) \cdot \mathbf{u} = \mathbf{a} \cdot \mathbf{u} + \mathbf{d} \cdot \mathbf{u}$. Thus, with $\mathbf{u} = \mathbf{b} \times \mathbf{c}$, it follows that $(\mathbf{a} + \mathbf{d}) \cdot (\mathbf{b} \times \mathbf{c}) = \mathbf{a} \cdot (\mathbf{b} \times \mathbf{c}) + \mathbf{d} \cdot (\mathbf{b} \times \mathbf{c})$.

49. The vector $\overrightarrow{AB} \times \overrightarrow{AC} = \begin{vmatrix} \mathbf{i} & \mathbf{j} & \mathbf{k} \\ 1 & 1 & -3 \\ -1 & 3 & -1 \end{vmatrix} = 8\mathbf{i} + 4\mathbf{j} + 4\mathbf{k}$ is perpendicular to $\overrightarrow{AB}$ and $\overrightarrow{AC}$, and thus is perpendicular to the plane determined by the points A, B, and C.

51. (a) We have $\text{adj}(A) = \begin{bmatrix} 34 & -21 & 1 \\ -14 & 7 & 0 \\ 1 & 0 & -1 \end{bmatrix}$, and so $A^{-1} = \frac{1}{\det(A)}\text{adj}(A) = -\tfrac{1}{7}\begin{bmatrix} 34 & -21 & 1 \\ -14 & 7 & 0 \\ 1 & 0 & -1 \end{bmatrix}$.

(b) The reduced row echelon form of $[A \mid I]$ is $\begin{bmatrix} 1 & 0 & 0 & \mid & -\frac{34}{7} & 3 & -\frac{1}{7} \\ 0 & 1 & 0 & \mid & 2 & -1 & 0 \\ 0 & 0 & 1 & \mid & -\frac{1}{7} & 0 & \frac{1}{7} \end{bmatrix}$; thus $A^{-1} = \begin{bmatrix} -\frac{34}{7} & 3 & -\frac{1}{7} \\ 2 & -1 & 0 \\ -\frac{1}{7} & 0 & \frac{1}{7} \end{bmatrix}$.

(c) The method used in (b) requires much less computation.

53. From Theorem 4.3.9, we know that $\mathbf{v} \times \mathbf{w}$ is orthogonal to the plane determined by $\mathbf{v}$ and $\mathbf{w}$. Thus a vector lies in the plane determined by $\mathbf{v}$ and $\mathbf{w}$ if and only if it is orthogonal to $\mathbf{v} \times \mathbf{w}$. Therefore, since $\mathbf{u} \times (\mathbf{v} \times \mathbf{w})$ is orthogonal to $\mathbf{v} \times \mathbf{w}$, it follows that $\mathbf{u} \times (\mathbf{v} \times \mathbf{w})$ lies in the plane determined by $\mathbf{v}$ and $\mathbf{w}$.

55. If A is upper triangular, and if $j > i$, then the submatrix that remains when the ith row and jth column of A are deleted is upper triangular and has a zero on its main diagonal; thus C_{ij} (the ijth cofactor of A) must be zero if $j > i$. It follows that the cofactor matrix C is lower triangular, and so $\mathrm{adj}(A) = C^T$ is upper triangular. Thus, if A is invertible and upper triangular, then $A^{-1} = \frac{1}{\det(A)}\mathrm{adj}(A)$ is also upper triangular.

57. The polynomial $p(x) = ax^3 + bx^2 + cx + d$ passes through the points $(0,1), (1,-1), (2,-1),$ and $(3,7)$ if and only if

$$
\begin{aligned}
d &= 1 \\
a + b + c + d &= -1 \\
8a + 4b + 2c + d &= -1 \\
27a + 9b + 3c + d &= 7
\end{aligned}
$$

Using Cramers Rule, the solution of this system is given by

$$
a = \frac{\begin{vmatrix} 1 & 0 & 0 & 1 \\ -1 & 1 & 1 & 1 \\ -1 & 4 & 2 & 1 \\ 7 & 9 & 3 & 1 \end{vmatrix}}{\begin{vmatrix} 0 & 0 & 0 & 1 \\ 1 & 1 & 1 & 1 \\ 8 & 4 & 2 & 1 \\ 27 & 9 & 3 & 1 \end{vmatrix}} = \frac{12}{12} = 1
\qquad
b = \frac{\begin{vmatrix} 0 & 1 & 0 & 1 \\ 1 & -1 & 1 & 1 \\ 8 & -1 & 2 & 1 \\ 27 & 7 & 3 & 1 \end{vmatrix}}{\begin{vmatrix} 0 & 0 & 0 & 1 \\ 1 & 1 & 1 & 1 \\ 8 & 4 & 2 & 1 \\ 27 & 9 & 3 & 1 \end{vmatrix}} = \frac{-24}{12} = -2
$$

$$
c = \frac{\begin{vmatrix} 0 & 0 & 1 & 1 \\ 1 & 1 & -1 & 1 \\ 8 & 4 & -1 & 1 \\ 27 & 9 & 7 & 1 \end{vmatrix}}{12} = \frac{-12}{12} = -1
\qquad
d = \frac{\begin{vmatrix} 0 & 0 & 0 & 1 \\ 1 & 1 & 1 & -1 \\ 8 & 4 & 2 & -1 \\ 27 & 9 & 3 & 7 \end{vmatrix}}{12} = \frac{12}{12} = 1
$$

Thus the interpolating polynomial is $p(x) = x^3 - 2x^2 - x + 1$.

DISCUSSION AND DISCOVERY

D1. **(a)** The vector $\mathbf{w} = \mathbf{v} \times (\mathbf{u} \times \mathbf{v})$ is orthogonal to both $\mathbf{v}$ and $\mathbf{u} \times \mathbf{v}$; thus $\mathbf{w}$ is orthogonal to $\mathbf{v}$ and lies in the plane determined by $\mathbf{u}$ and $\mathbf{v}$.

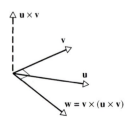

(b) Since **w** is orthogonal to **v**, we have $\mathbf{v} \cdot \mathbf{w} = 0$. On the other hand, $\mathbf{u} \cdot \mathbf{w} = \|\mathbf{u}\|\|\mathbf{w}\| \cos\theta = \|\mathbf{u}\|\|\mathbf{w}\| \sin(\frac{\pi}{2} - \theta)$ where θ is the angle between **u** and **w**. It follows that $|\mathbf{u} \cdot \mathbf{w}|$ is equal to the area of the parallelogram having **u** and **v** as adjacent edges.

D2. No. For example, let $\mathbf{u} = (1,0,0)$, $\mathbf{v} = (0,1,0)$, and $\mathbf{w} = (1,1,0)$. Then $\mathbf{u} \times \mathbf{v} = \mathbf{u} \times \mathbf{w} = (0,0,1)$, but $\mathbf{v} \neq \mathbf{w}$.

D3. $(\mathbf{u} \cdot \mathbf{v}) \times \mathbf{w}$ does not make sense since the first factor is a scalar rather than a vector.

D4. If either **u** or **v** is the zero vector, then $\mathbf{u} \times \mathbf{v} = \mathbf{0}$. If **u** and **v** are nonzero then, from Theorem 4.3.10, we have $\|\mathbf{u} \times \mathbf{v}\| = \|\mathbf{u}\|\|\mathbf{v}\| \sin\theta$ where θ is the angle between **u** and **v**. Thus if $\mathbf{u} \times \mathbf{v} = \mathbf{0}$, with **u** and **v** not zero, then $\sin\theta = 0$ and so **u** and **v** are parallel.

D5. The associative law of multiplication is not valid for the cross product; that is $\mathbf{u} \times (\mathbf{v} \times \mathbf{w})$ is not in general the same as $(\mathbf{u} \times \mathbf{v}) \times \mathbf{w}$.

D6. Let $A = \begin{bmatrix} c & -(1-c) \\ 1-c & c \end{bmatrix}$. Then $\det(A) = c^2 + (1-c)^2 = 2c^2 - 2c + 1 \neq 0$ for all values of c. Thus, for every c, the system has a unique solution given by

$$x_1 = \frac{\begin{vmatrix} 3 & -(1-c) \\ -4 & c \end{vmatrix}}{2c^2 - 2c + 1} = \frac{7c-4}{2c^2 - 2c + 1} \qquad x_2 = \frac{\begin{vmatrix} c & 3 \\ (1-c) & -4 \end{vmatrix}}{2c^2 - 2c + 1} = \frac{-c-3}{2c^2 - 2c + 1}$$

D7. **(c)** The solution by Gauss-Jordan elimination requires much less computation.

D8. **(a)** True. As was shown in the proof of Theorem 4.3.3, we have $A \operatorname{adj}(A) = \det(A)I$.
    (b) False. In addition, the determinant of the coefficient matrix must be nonzero.
    (c) True. In fact we have $\operatorname{adj}(A) = \det(A)A^{-1}$ and so $(\operatorname{adj}(A))^{-1} = \frac{1}{\det(A)}A$.
    (d) False. For example, if $A = \begin{bmatrix} 1 & 1 \\ 0 & 0 \end{bmatrix}$ then $\operatorname{adj}(A) = \begin{bmatrix} 0 & -1 \\ 0 & 1 \end{bmatrix}$.
    (e) True. Both sides are equal to $\begin{vmatrix} u_1 & u_2 & u_3 \\ v_1 & v_2 & v_3 \\ w_1 & w_2 & w_3 \end{vmatrix}$.

WORKING WITH PROOFS

P1. We have $\mathbf{u} \cdot \mathbf{v} = \|\mathbf{u}\|\|\mathbf{v}\| \cos\theta$ and $\|\mathbf{u} \times \mathbf{v}\| = \|\mathbf{u}\|\|\mathbf{v}\| \sin\theta$; thus $\tan\theta = \frac{\|\mathbf{u} \times \mathbf{v}\|}{\mathbf{u} \cdot \mathbf{v}}$.

P2. The angle between the vectors is $\theta = \alpha - \beta$; thus $\mathbf{u} \cdot \mathbf{v} = \|\mathbf{u}\|\|\mathbf{v}\| \cos(\alpha - \beta)$ or $\cos(\alpha - \beta) = \frac{\mathbf{u} \cdot \mathbf{v}}{\|\mathbf{u}\|\|\mathbf{v}\|}$.

P3. **(a)** Using properties of cross products from Theorem 4.3.8, we have

$$(\mathbf{u} + k\mathbf{v}) \times \mathbf{v} = (\mathbf{u} \times \mathbf{v}) + (k\mathbf{v} \times \mathbf{v}) = (\mathbf{u} \times \mathbf{v}) + k(\mathbf{v} \times \mathbf{v}) = (\mathbf{u} \times \mathbf{v}) + k\mathbf{0} = \mathbf{u} \times \mathbf{v}$$

    (b) Using part (a) of Exercise 50, we have

$$\mathbf{u} \cdot (\mathbf{v} \times \mathbf{w}) = \begin{vmatrix} u_1 & u_2 & u_3 \\ v_1 & v_2 & v_3 \\ w_1 & w_2 & w_3 \end{vmatrix} = -\begin{vmatrix} v_1 & v_2 & v_3 \\ u_1 & u_2 & u_3 \\ w_1 & w_2 & w_3 \end{vmatrix} = -\mathbf{v} \cdot (\mathbf{u} \times \mathbf{w}) = -(\mathbf{u} \times \mathbf{w}) \cdot \mathbf{v}$$

P4. If **a**, **b**, **c**, and **d** all lie in the same plane, then $\mathbf{a} \times \mathbf{b}$ and $\mathbf{c} \times \mathbf{d}$ are both perpendicular to that plane, and thus parallel to each other. It follows that $(\mathbf{a} \times \mathbf{b}) \times (\mathbf{c} \times \mathbf{d}) = \mathbf{0}$.

P5. Let $Q_1 = (x_1, y_1, 1)$, $Q_2 = (x_2, y_2, 1)$, $Q_3 = (x_3, y_3, 1)$, and let T denote the tetrahedron in R^3 having the vectors OQ_1, OQ_2, OQ_3, as adjacent edges. The base of this tetrahedron lies in the plane $z = 1$ and is congruent to the triangle $\Delta P_1 P_2 P_3$; thus $\mathrm{vol}(T) = \frac{1}{3}\mathrm{area}(\Delta P_1 P_2 P_3)$. On the other hand, $\mathrm{vol}(T)$ is equal to $\frac{1}{6}$ times the volume of the parallelepiped having OQ_1, OQ_2, OQ_3, as adjacent edges and, from part (b) of Exercise 50, the latter is equal to $OQ_1 \cdot (OQ_2 \times OQ_3)$. Thus

$$\mathrm{area}(\Delta P_1 P_2 P_3) = 3\mathrm{vol}(T) = \frac{1}{2}OQ_1 \cdot (OQ_2 \times OQ_3) = \frac{1}{2}\begin{vmatrix} x_1 & y_1 & 1 \\ x_2 & y_2 & 1 \\ x_3 & y_3 & 1 \end{vmatrix}$$

EXERCISE SET 4.4

1. (a) The matrix $A = \begin{bmatrix} 1 & 0 \\ 1 & 0 \end{bmatrix}$ has nontrivial fixed points since $\det(I - A) = \begin{vmatrix} 0 & 0 \\ 1 & -1 \end{vmatrix} = 0$. The fixed points are the solutions of the system $(I - A)\mathbf{x} = 0$, which can be expressed in vector form as $\mathbf{x} = t\begin{bmatrix} 1 \\ 1 \end{bmatrix}$ where $-\infty < t < \infty$.

(b) The matrix $B = \begin{bmatrix} 0 & 1 \\ 1 & 0 \end{bmatrix}$ has nontrivial fixed points since $\det(I - B) = \begin{vmatrix} 1 & -1 \\ -1 & 1 \end{vmatrix} = 0$. The fixed points are the solutions of the system $(I - B)\mathbf{x} = 0$, which can be expressed in vector form as $\mathbf{x} = t\begin{bmatrix} 1 \\ 1 \end{bmatrix}$ where $-\infty < t < \infty$.

3. We have $A\mathbf{x} = \begin{bmatrix} 4 & 0 & 1 \\ 2 & 3 & 2 \\ 1 & 0 & 4 \end{bmatrix}\begin{bmatrix} 1 \\ 2 \\ 1 \end{bmatrix} = \begin{bmatrix} 5 \\ 10 \\ 5 \end{bmatrix} = 5\mathbf{x}$; thus $\mathbf{x} = \begin{bmatrix} 1 \\ 2 \\ 1 \end{bmatrix}$ is an eigenvector of A corresponding to the eigenvalue $\lambda = 5$.

5. (a) The characteristic equation of $A = \begin{bmatrix} 3 & 0 \\ 8 & -1 \end{bmatrix}$ is $\det(\lambda I - A) = \begin{vmatrix} \lambda - 3 & 0 \\ -8 & \lambda + 1 \end{vmatrix} = (\lambda - 3)(\lambda + 1) = 0$. Thus $\lambda = 3$ and $\lambda = -1$ are eigenvalues of A; each has algebraic multiplicity 1.

(b) The characteristic equation is $\begin{vmatrix} \lambda - 10 & 9 \\ -4 & \lambda + 2 \end{vmatrix} = (\lambda - 10)(\lambda + 2) + 36 = (\lambda - 4)^2 = 0$. Thus $\lambda = 4$ is the only eigenvalue; it has algebraic multiplicity 2.

(c) The characteristic equation is $\begin{vmatrix} \lambda - 2 & 0 \\ -1 & \lambda - 2 \end{vmatrix} = (\lambda - 2)^2 = 0$. Thus $\lambda = 2$ is the only eigenvalue; it has algebraic multiplicity 2.

7. (a) The characteristic equation is $\begin{vmatrix} \lambda - 4 & 0 & -1 \\ 2 & \lambda - 1 & 0 \\ 2 & 0 & \lambda - 1 \end{vmatrix} = \lambda^3 - 6\lambda^2 + 11\lambda - 6 = (\lambda - 1)(\lambda - 2)(\lambda - 3) = 0$. Thus $\lambda = 1$, $\lambda = 2$, and $\lambda = 3$ are eigenvalues; each has algebraic multiplicity 1.

(b) The characteristic equation is $\begin{vmatrix} \lambda - 4 & 5 & 5 \\ -\frac{2}{5} & \lambda - 1 & 1 \\ -\frac{6}{5} & 3 & \lambda + 1 \end{vmatrix} = \lambda^3 - 4\lambda^2 + 4\lambda = \lambda(\lambda - 2)^2 = 0$. Thus $\lambda = 0$ and $\lambda = 2$ are eigenvalues; $\lambda = 0$ has algebraic multiplicity 1, and $\lambda = 2$ has multiplicity 2.

(c) The characteristic equation is $\begin{vmatrix} \lambda-3 & -4 & 1 \\ 1 & \lambda+2 & -1 \\ -3 & -9 & \lambda \end{vmatrix} = \lambda^3 - \lambda^2 - 8\lambda + 12 = (\lambda+3)(\lambda-2)^2 = 0$. Thus $\lambda = -3$ and $\lambda = 2$ are eigenvalues; $\lambda = -3$ has multiplicity 1, and $\lambda = 2$ has multiplicity 2.

9. (a) The eigenspace corresponding to $\lambda = 3$ is found by solving the system $\begin{bmatrix} 0 & 0 \\ -8 & 4 \end{bmatrix} \begin{bmatrix} x \\ y \end{bmatrix} = \begin{bmatrix} 0 \\ 0 \end{bmatrix}$. This yields the general solution $x = t$, $y = 2t$; thus the eigenspace consists of all vectors of the form $\begin{bmatrix} x \\ y \end{bmatrix} = t \begin{bmatrix} 1 \\ 2 \end{bmatrix}$. Geometrically, this is the line $y = 2x$ in the xy-plane.

The eigenspace corresponding to $\lambda = -1$ is found by solving the system $\begin{bmatrix} -4 & 0 \\ -8 & 0 \end{bmatrix} \begin{bmatrix} x \\ y \end{bmatrix} = \begin{bmatrix} 0 \\ 0 \end{bmatrix}$. This yields the general solution $x = 0$, $y = t$; thus the eigenspace consists of all vectors of the form $\begin{bmatrix} x \\ y \end{bmatrix} = t \begin{bmatrix} 0 \\ 1 \end{bmatrix}$. Geometrically, this is the line $x = 0$ (y-axis).

(b) The eigenspace corresponding to $\lambda = 4$ is found by solving the system $\begin{bmatrix} -6 & 9 \\ -4 & 6 \end{bmatrix} \begin{bmatrix} x \\ y \end{bmatrix} = \begin{bmatrix} 0 \\ 0 \end{bmatrix}$. This yields the general solution $x = 3t$, $y = 2t$; thus the eigenspace consist of all vectors of the form $\begin{bmatrix} x \\ y \end{bmatrix} = t \begin{bmatrix} 3 \\ 2 \end{bmatrix}$. Geometrically, this is the line $y = \frac{2}{3}x$.

(c) The eigenspace corresponding to $\lambda = 2$ is found by solving the system $\begin{bmatrix} 0 & 0 \\ -1 & 0 \end{bmatrix} \begin{bmatrix} x \\ y \end{bmatrix} = \begin{bmatrix} 0 \\ 0 \end{bmatrix}$. This yields the general solution $x = 0$, $y = t$; thus the eigenspace consists of all vectors of the form $\begin{bmatrix} x \\ y \end{bmatrix} = t \begin{bmatrix} 0 \\ 1 \end{bmatrix}$. Geometrically, this is the line $x = 0$.

11. (a) The eigenspace corresponding to $\lambda = 1$ is obtained by solving $\begin{bmatrix} -3 & 0 & -1 \\ 2 & 0 & 0 \\ 2 & 0 & 0 \end{bmatrix} \begin{bmatrix} x \\ y \\ z \end{bmatrix} = \begin{bmatrix} 0 \\ 0 \\ 0 \end{bmatrix}$. This yields the general solution $x = 0$, $y = t$, $z = 0$; thus the eigenspace consists of all vectors of the form $\begin{bmatrix} x \\ y \\ z \end{bmatrix} = t \begin{bmatrix} 0 \\ 1 \\ 0 \end{bmatrix}$; this corresponds to a line through the origin (the y-axis) in R^3. Similarly, the eigenspace corresponding to $\lambda = 2$ consists of all vectors of the form $\begin{bmatrix} x \\ y \\ z \end{bmatrix} = t \begin{bmatrix} 1 \\ -2 \\ -2 \end{bmatrix}$, and the eigenspace corresponding to $\lambda = 3$ consists of all vectors of the form $\begin{bmatrix} x \\ y \\ z \end{bmatrix} = t \begin{bmatrix} -1 \\ 1 \\ 1 \end{bmatrix}$.

(b) The eigenspace corresponding to $\lambda = 0$ is found by solving the system $\begin{bmatrix} -4 & 5 & 5 \\ -\frac{2}{5} & -1 & 1 \\ -\frac{6}{5} & 3 & 1 \end{bmatrix} \begin{bmatrix} x \\ y \\ z \end{bmatrix} = \begin{bmatrix} 0 \\ 0 \\ 0 \end{bmatrix}$. This yields the general solution $x = 5t$, $y = t$, $z = 3t$. Thus the eigenspace consists of all vectors of the form $\begin{bmatrix} x \\ y \\ z \end{bmatrix} = t \begin{bmatrix} 5 \\ 1 \\ 3 \end{bmatrix}$; this is the line through the origin and the point $(5, 1, 3)$.

The eigenspace corresponding to $\lambda = 2$ is found by solving the system $\begin{bmatrix} -2 & 5 & 5 \\ -\frac{2}{5} & 1 & 1 \\ -\frac{6}{5} & 3 & 3 \end{bmatrix} \begin{bmatrix} x \\ y \\ z \end{bmatrix} = \begin{bmatrix} 0 \\ 0 \\ 0 \end{bmatrix}$. This yields $\begin{bmatrix} x \\ y \\ z \end{bmatrix} = s \begin{bmatrix} 5 \\ 0 \\ 2 \end{bmatrix} + t \begin{bmatrix} 5 \\ 2 \\ 0 \end{bmatrix}$, which corresponds to a plane through the origin.

(c)   The eigenspace corresponding to $\lambda = -3$ is found by solving $\begin{bmatrix} -6 & -4 & 1 \\ 1 & -1 & -1 \\ -3 & -9 & -3 \end{bmatrix} \begin{bmatrix} x \\ y \\ z \end{bmatrix} = \begin{bmatrix} 0 \\ 0 \\ 0 \end{bmatrix}$. This yields

$\begin{bmatrix} x \\ y \\ z \end{bmatrix} = t \begin{bmatrix} -1 \\ 1 \\ -2 \end{bmatrix}$, which corresponds to a line through the origin. The eigenspace corresponding to

$\lambda = 2$ is found by solving $\begin{bmatrix} -1 & -4 & 1 \\ 1 & 4 & -1 \\ -3 & -9 & 2 \end{bmatrix} \begin{bmatrix} x \\ y \\ z \end{bmatrix} = \begin{bmatrix} 0 \\ 0 \\ 0 \end{bmatrix}$. This yields $\begin{bmatrix} x \\ y \\ z \end{bmatrix} = t \begin{bmatrix} -1 \\ 1 \\ 3 \end{bmatrix}$, which also corre-

sponds to a line through the origin.

13. (a)   The characteristic polynomial is $p(\lambda) = (\lambda + 1)(\lambda - 5)$. The eigenvalues are $\lambda = -1$ and $\lambda = 5$.

(b)   The characteristic polynomial is $p(\lambda) = (\lambda - 3)(\lambda - 7)(\lambda - 1)$. The eigenvalues are $\lambda = 3$, $\lambda = 7$, and $\lambda = 1$.

(c)   The characteristic polynomial is $p(\lambda) = (\lambda + \frac{1}{3})^2(\lambda - 1)(\lambda - \frac{1}{2})$. The eigenvalues are $\lambda = -\frac{1}{3}$ (with multiplicity 2), $\lambda = 1$, and $\lambda = \frac{1}{2}$.

15.   Using the block diagonal structure, the characteristic polynomial of the given matrix is

$$ p(\lambda) = \begin{vmatrix} \lambda - 2 & -3 & 0 & 0 \\ 1 & \lambda - 6 & 0 & 0 \\ 0 & 0 & \lambda + 2 & -5 \\ 0 & 0 & -1 & \lambda - 2 \end{vmatrix} = \begin{vmatrix} \lambda - 2 & -3 \\ 1 & \lambda - 6 \end{vmatrix} \begin{vmatrix} \lambda + 2 & -5 \\ -1 & \lambda - 2 \end{vmatrix} $$

$$ = [(\lambda - 2)(\lambda - 6) + 3][(\lambda + 2)(\lambda - 2) - 5] = (\lambda^2 - 8\lambda + 15)(\lambda^2 - 9) = (\lambda - 5)(\lambda - 3)^2(\lambda + 3) $$

Thus the eigenvalues are $\lambda = 5$, $\lambda = 3$ (with multiplicity 2), and $\lambda = -3$.

17.   The characteristic polynomial of A is

$$ p(\lambda) = \det(\lambda I - A) = \begin{vmatrix} \lambda + 1 & 2 & 2 \\ -1 & \lambda - 2 & -1 \\ 1 & 1 & \lambda \end{vmatrix} = (\lambda + 1)(\lambda - 1)^2 $$

thus the eigenvalues are $\lambda = -1$ and $\lambda = 1$ (with multiplicity 2). The eigenspace corresponding to $\lambda = -1$ is obtained by solving the system

$$ \begin{bmatrix} 0 & 2 & 2 \\ -1 & -3 & -1 \\ 1 & 1 & -1 \end{bmatrix} \begin{bmatrix} x \\ y \\ z \end{bmatrix} = \begin{bmatrix} 0 \\ 0 \\ 0 \end{bmatrix} $$

which yields $\begin{bmatrix} x \\ y \\ z \end{bmatrix} = t \begin{bmatrix} 2 \\ -1 \\ 1 \end{bmatrix}$. Similarly, the eigenspace corresponding to $\lambda = 1$ is obtained by solving

$$ \begin{bmatrix} 2 & 2 & 2 \\ -1 & -1 & -1 \\ 1 & 1 & 1 \end{bmatrix} \begin{bmatrix} x \\ y \\ z \end{bmatrix} = \begin{bmatrix} 0 \\ 0 \\ 0 \end{bmatrix} $$

which has the general solution $x = t$, $y = -t - s$, $z = s$; or (in vector form) $\begin{bmatrix} x \\ y \\ z \end{bmatrix} = s \begin{bmatrix} 0 \\ -1 \\ 1 \end{bmatrix} + t \begin{bmatrix} 1 \\ -1 \\ 0 \end{bmatrix}$.

The eigenvalues of A^{25} are $\lambda = (-1)^{25} = -1$ and $\lambda = (1)^{25} = 1$. Corresponding eigenvectors are the same as above.

19. The characteristic polynomial of A is $p(\lambda) = \lambda^3 - \lambda^2 - 5\lambda - 3 = (\lambda - 3)(\lambda + 1)^2$; thus the eigenvalues are $\lambda_1 = 3$, $\lambda_2 = -1$, $\lambda_3 = -1$. We have $\det(A) = 3$ and $\operatorname{tr}(A) = 1$. Thus $\det(A) = 3 = (3)(-1)(-1) = \lambda_1 \lambda_2 \lambda_3$ and $\operatorname{tr}(A) = 1 = (3) + (-1) + (-1) = \lambda_1 + \lambda_2 + \lambda_3$.

21. The eigenvalues are $\lambda = 0$ and $\lambda = 5$, with associated eigenvectors $\begin{bmatrix} -2 \\ 1 \end{bmatrix}$ and $\begin{bmatrix} 1 \\ 2 \end{bmatrix}$ respectively. Thus the eigenspaces correspond to the perpendicular lines $y = -\frac{1}{2}x$ and $y = 2x$.

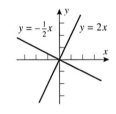

23. The invariant lines, if any, correspond to eigenspaces of the matrix.

 (a) The eigenvalues are $\lambda = 2$ and $\lambda = 3$, with associated eigenvectors $\begin{bmatrix} 1 \\ 2 \end{bmatrix}$ and $\begin{bmatrix} 1 \\ 1 \end{bmatrix}$ respectively. Thus the lines $y = 2x$ and $y = x$ are invariant under the given matrix.

 (b) This matrix has no real eigenvalues, so there are no invariant lines.

 (c) The only eigenvalue is $\lambda = 2$ (multiplicity 2), with associated eigenvector $\begin{bmatrix} 1 \\ 0 \end{bmatrix}$. Thus the line $y = 0$ is invariant under the given matrix.

25. The characteristic polynomial of A is $p(\lambda) = \lambda^2 - (b + 3)\lambda + (3b - 2a)$, so A has the stated eigenvalues if and only if $p(2) = p(5) = 0$. This leads to the equations

$$-2a + b = 2$$
$$a + b = 5$$

from which we conclude that $a = 1$ and $b = 4$.

27. If $A^2 = I$, then $A(\mathbf{x} + A\mathbf{x}) = A\mathbf{x} + A^2\mathbf{x} = A\mathbf{x} + \mathbf{x} = \mathbf{x} + A\mathbf{x}$; thus $\mathbf{y} = \mathbf{x} + A\mathbf{x}$ is an eigenvector of A corresponding to $\lambda = 1$. Similarly, $\mathbf{z} = \mathbf{x} - A\mathbf{x}$ is an eigenvector of A corresponding to $\lambda = -1$.

29. (a) Using Formula (22), the characteristic equation of A is $\lambda^2 + (a + d)\lambda + (ad - bc) = 0$. This is a quadratric equation with discriminant

$$(a + d)^2 - 4(ad - bc) = a^2 + 2ad + d^2 - 4ad + 4bc = (a - d)^2 + 4bc$$

Thus the eigenvalues of A are given by $\lambda = \frac{1}{2}[(a + d) \pm \sqrt{(a - d)^2 + 4bc}]$.

(b) If $(a-d)^2 + 4bc > 0$ then, from (a), the characteristic equation has two distinct real roots.

(c) If $(a-d)^2 + 4bc = 0$ then, from (a), there is one real eigenvalue (of multiplicity 2).

(d) If $(a-d)^2 + 4bc < 0$ then, from (a), there are no real eigenvalues.

31. If the characteristic polynomial of A is $p(\lambda) = \lambda^2 + 3\lambda - 4 = (\lambda - 1)(\lambda + 4)$ then the eigenvalues of A are $\lambda_1 = 1$ and $\lambda_2 = -4$.

(a) From Exercise P3 below, A^{-1} has eigenvalues $\lambda_1 = 1$ and $\lambda_2 = -\frac{1}{4}$.

(b) From (a), together with Theorem 4.4.6, it follows that A^{-3} has eigenvalues $\lambda_1 = (1)^3 = 1$ and $\lambda_2 = (-\frac{1}{4})^3 = -\frac{1}{64}$.

(c) From P4 below, $A - 4I$ has eigenvalues $\lambda_1 = 1 - 4 = -3$ and $\lambda_2 = -4 - 4 = -8$.

(d) From P5 below, $5A$ has eigenvalues $\lambda_1 = 5$ and $\lambda_2 = -20$.

(e) From P2(a) below, the eigenvalues of $4A^T + 2I = (4A + 2I)^T$ are the same as those of $4A + 2I$; namely $\lambda_1 = 4(1) + 2 = 6$ and $\lambda_2 = 4(-4) + 2 = -14$.

33. (a) The characteristic polynomial of the matrix C is

$$p(\lambda) = \det(\lambda I - C) = \begin{vmatrix} \lambda & 0 & 0 & \cdots & 0 & c_0 \\ -1 & \lambda & 0 & \cdots & 0 & c_1 \\ 0 & -1 & \lambda & \cdots & 0 & c_2 \\ \vdots & \vdots & \vdots & & \vdots & \vdots \\ 0 & 0 & 0 & \cdots & -1 & \lambda + c_{n-1} \end{vmatrix}$$

Add λ times the second row to the first row, then expand by cofactors along the first column:

$$p(\lambda) = \begin{vmatrix} 0 & \lambda^2 & 0 & \cdots & 0 & c_0 + c_1\lambda \\ -1 & \lambda & 0 & \cdots & 0 & c_1 \\ 0 & -1 & \lambda & \cdots & 0 & c_2 \\ \vdots & \vdots & \vdots & & \vdots & \vdots \\ 0 & 0 & 0 & \cdots & -1 & \lambda + c_{n-1} \end{vmatrix} = \begin{vmatrix} \lambda^2 & 0 & \cdots & 0 & c_0 + c_1\lambda \\ -1 & \lambda & \cdots & 0 & c_2 \\ \vdots & \vdots & & \vdots & \vdots \\ 0 & 0 & \cdots & -1 & \lambda + c_{n-1} \end{vmatrix}$$

Add λ^2 times the second row to the first row, then expand by cofactors along the first column.

$$p(\lambda) = \begin{vmatrix} 0 & \lambda^2 & \cdots & 0 & c_0 + c_1\lambda + c_2\lambda^2 \\ -1 & \lambda & \cdots & 0 & c_2 \\ \vdots & \vdots & & \vdots & \vdots \\ 0 & 0 & \cdots & -1 & \lambda + c_{n-1} \end{vmatrix} = \begin{vmatrix} \lambda^2 & \cdots & 0 & c_0 + c_1\lambda + c_2\lambda^2 \\ \vdots & & \vdots & \vdots \\ 0 & \cdots & -1 & \lambda + c_{n-1} \end{vmatrix}$$

Continuing in this fashion for $n - 2$ steps, we obtain

$$p(\lambda) = \begin{vmatrix} \lambda^{n-1} & c_0 + c_1\lambda + c_2\lambda^2 + \cdots + c_{n-2}\lambda^{n-2} \\ -1 & \lambda + c_{n-1} \end{vmatrix}$$

$$= c_0 + c_1\lambda + c_2\lambda^2 + \cdots + c_{n-2}\lambda^{n-2} + c_{n-1}\lambda^{n-1} + \lambda^n$$

(b) The matrix $C = \begin{bmatrix} 0 & 0 & 0 & -2 \\ 1 & 0 & 0 & 3 \\ 0 & 1 & 0 & -1 \\ 0 & 0 & 1 & 5 \end{bmatrix}$ has $p(\lambda) = 2 - 3\lambda + \lambda^2 - 5\lambda^3 + \lambda^4$ as its characteristic polynomial.

DISCUSSION AND DISCOVERY

D1. (a) The characteristic polynomial $p(\lambda)$ has degree 6; thus A is a 6×6 matrix.

(b) Yes. From Theorem 4.4.12, we have $\det(A) = (1)(3)^2(4)^3 = 576 \neq 0$; thus A is invertible.

D2. If A is square matrix all of whose entries are the same then $\det(A) = 0$; thus $A\mathbf{x} = \mathbf{0}$ has nontrivial solutions and $\lambda = 0$ is an eigenvalue of A.

D3. Using Formula (22), the characteristic polynomial of A is $p(\lambda) = \lambda^2 - 4\lambda + 4 = (\lambda - 2)^2$. Thus $\lambda = 2$ is the only eigenvalue of A (it has multiplicity 2).

D4. The eigenvalues of A (with multiplicity) are 3, 3, and $-2, -2, -2$. Thus, from Theorem 4.4.12, we have $\det(A) = (3)(3)(-2)(-2)(-2) = -72$ and $\operatorname{tr}(A) = 3 + 3 - 2 - 2 - 2 = 0$.

D5. The matrix $A = \begin{bmatrix} a & b \\ c & d \end{bmatrix}$ satisfies the condition $\operatorname{tr}(A) = \det(A)$ if and only if $a + d = ad - bc$. If $d = 1$ then this is equation is satisfied if and only if $bc = -1$, e.g., $A = \begin{bmatrix} 0 & -1 \\ 1 & 1 \end{bmatrix}$. If $d \neq 1$, then the equation is satisfied if and only if $a = \frac{d+bc}{d-1}$, e.g., $A = \begin{bmatrix} 1 & -1 \\ 1 & 2 \end{bmatrix}$.

D6. The characteristic polynomial of A factors as $p(\lambda) = (\lambda - 1)(\lambda + 2)^3$; thus the eigenvalues of A are $\lambda = 1$ and $\lambda = -2$. It follows from Theorem 4.4.6 that the eigenvalues of A^2 are $\lambda = (1)^2 = 1$ and $\lambda = (-2)^2 = 4$.

D7. (a) False. For example, $\mathbf{x} = \mathbf{0}$ satisfies this condition for any A and λ. The correct statement is that if $A\mathbf{x} = \lambda\mathbf{x}$ for some nonzero vector $\mathbf{x}$, then $\mathbf{x}$ is an eigenvector of A.

(b) True. If λ is an eigenvalue of A, then λ^2 is an eigenvalue of A^2; thus $(\lambda^2 I - A^2)\mathbf{x} = \mathbf{0}$ has nontrivial solutions.

(c) False. If $\lambda = 0$ is an eigenvalue of A, then the system $A\mathbf{x} = \mathbf{0}$ has nontrivial solutions; thus A is not invertible and so the row vectors and column vectors of A are linearly dependent. [The statement becomes true if "independent" is replaced by "dependent".]

(d) False. For example, $A = \begin{bmatrix} 1 & 0 \\ 1 & 2 \end{bmatrix}$ has eigenvalues $\lambda = 1$ and $\lambda = 2$. [But it is true that a symmetric matrix has real eigenvalues.]

D8. (a) False. For example, the reduced row echelon form of $A = \begin{bmatrix} 1 & 0 \\ 1 & 2 \end{bmatrix}$ is $I = \begin{bmatrix} 1 & 0 \\ 0 & 1 \end{bmatrix}$.

(b) True. We have $A(\mathbf{x}_1 + \mathbf{x}_2) = \lambda_1\mathbf{x}_1 + \lambda_2\mathbf{x}_2$ and, if $\lambda_1 \neq \lambda_2$ it can be shown (since $\mathbf{x}_1$ and $\mathbf{x}_2$ must be linearly independent) that $\lambda_1\mathbf{x}_1 + \lambda_2\mathbf{x}_2 \neq \beta(\mathbf{x}_1 + \mathbf{x}_2)$ for any value of β.

(c) True. The characteristic polynomial of A is a cubic polynomial, and every cubic polynomial has at least one real root.

(d) True. If $p(\lambda) = \lambda^n + 1$, then $\det(A) = (-1)^n p(0) = \pm 1 \neq 0$; thus A is invertible.

WORKING WITH PROOFS

P1. If $A = \begin{bmatrix} a & b \\ c & d \end{bmatrix}$, then $A^2 = \begin{bmatrix} a^2 + bc & ab + bd \\ ca + dc & cb + d^2 \end{bmatrix}$ and $\operatorname{tr}(A)A = (a + d)\begin{bmatrix} a & b \\ c & d \end{bmatrix} = \begin{bmatrix} a^2 + da & ab + db \\ ac + dc & ad + d^2 \end{bmatrix}$; thus

$$A^2 - \operatorname{tr}(A)A = \begin{bmatrix} bc - ad & 0 \\ 0 & cb - ad \end{bmatrix} = \begin{bmatrix} -\det(A) & 0 \\ 0 & -\det(A) \end{bmatrix} = -\det(A)I$$

and so $p(A) = A^2 - \operatorname{tr}(A)A + \det(A)I = 0$.

P2. **(a)** Using previously established properties, we have

$$\det(\lambda I - A^T) = \det(\lambda I^T - A^T) = \det((\lambda I - A)^T) = \det(\lambda I - A)$$

Thus A and A^T have the same characteristic polynomial.

(b) The eigenvalues are 2 and 3 in each case. The eigenspace of A corresponding to $\lambda = 2$ is obtained by solving the system $\begin{bmatrix} 0 & 0 \\ -2 & -1 \end{bmatrix} \begin{bmatrix} x \\ y \end{bmatrix} = \begin{bmatrix} 0 \\ 0 \end{bmatrix}$; whereas the eigenspace of A^T corresponding to $\lambda = 2$ is obtained by solving $\begin{bmatrix} 0 & -2 \\ 0 & -1 \end{bmatrix} \begin{bmatrix} x \\ y \end{bmatrix} = \begin{bmatrix} 0 \\ 0 \end{bmatrix}$. Thus the eigenspace of A corresponds to the line $y = -2x$; whereas the eigenspace of A^T corresponds to $y = 0$. Similarly, for $\lambda = 3$, the eigenspace of A corresponds to $x = 0$; whereas the eigenspace of A^T corresponds to $y = \frac{1}{2}x$.

P3. Suppose that $A\mathbf{x} = \lambda\mathbf{x}$ where $\mathbf{x} \neq \mathbf{0}$ and A is invertible. Then $\mathbf{x} = A^{-1}A\mathbf{x} = A^{-1}\lambda\mathbf{x} = \lambda A^{-1}\mathbf{x}$ and, since $\lambda \neq 0$ (because A is invertible), it follows that $A^{-1}\mathbf{x} = \frac{1}{\lambda}\mathbf{x}$. Thus $\frac{1}{\lambda}$ is an eigenvalue of A^{-1} and $\mathbf{x}$ is a corresponding eigenvector.

P4. Suppose that $A\mathbf{x} = \lambda\mathbf{x}$ where $\mathbf{x} \neq \mathbf{0}$. Then $(A - sI)\mathbf{x} = A\mathbf{x} - sI\mathbf{x} = \lambda\mathbf{x} - s\mathbf{x} = (\lambda - s)\mathbf{x}$. Thus $\lambda - s$ is an eigenvalue of $A - sI$ and $\mathbf{x}$ is a corresponding eigenvector.

P5. Suppose that $A\mathbf{x} = \lambda\mathbf{x}$ where $\mathbf{x} \neq \mathbf{0}$. Then $(sA)\mathbf{x} = s(A\mathbf{x}) = s(\lambda\mathbf{x}) = (s\lambda)\mathbf{x}$. Thus $s\lambda$ is an eigenvalue of sA and $\mathbf{x}$ is a corresponding eigenvector.

P6. If the matrix $A = \begin{bmatrix} a & b \\ c & d \end{bmatrix}$ is symmetric, then $c = b$ and so $(a - d)^2 + 4bc = (a - d)^2 + 4b^2$.

In the case that A has a repeated eigenvalue, we must have $(a - d)^2 + 4b^2 = 0$ and so $a = d$ and $b = 0$. Thus the only symmetric 2×2 matrices with repeated eigenvalues are those of the form $A = aI$. Such a matrix has $\lambda = a$ as its only eigenvalue, and the corresponding eigenspace is R^2. This proves part (a) of Theorem 4.4.11.

If $(a - d)^2 + 4b^2 > 0$, then A has two distinct real eigenvalues λ_1 and λ_2, with corresponding eigenvectors $\mathbf{x}_1$ and $\mathbf{x}_2$, given by:

$$\lambda_1 = \tfrac{1}{2}[(a + d) + \sqrt{(a - d)^2 + 4b^2}] \quad \mathbf{x}_1 = \begin{bmatrix} -b \\ a - \lambda_1 \end{bmatrix}$$

$$\lambda_2 = \tfrac{1}{2}[(a + d) - \sqrt{(a - d)^2 + 4b^2}] \quad \mathbf{x}_2 = \begin{bmatrix} -b \\ a - \lambda_1 \end{bmatrix}$$

The eigenspaces correspond to the lines $y = m_1x$ and $y = m_2x$ where $m_j = \frac{a - \lambda_j}{-b}$, $j = 1, 2$. Since

$$(a - \lambda_1)(a - \lambda_2) = (\tfrac{1}{2}[(a - d) + \sqrt{(a - d)^2 + 4b^2}])(\tfrac{1}{2}[(a - d) - \sqrt{(a - d)^2 + 4b^2}])$$
$$= \tfrac{1}{4}[(a - d)^2 - (a - d)^2 - 4b^2] = -b^2$$

we have $m_1m_2 = -1$; thus the eigenspaces correspond to perpendicular lines. This proves part (b) of Theorem 4.4.11.

Note. It is not possible to have $(a - d)^2 + 4b^2 < 0$; thus the eigenvalues of a 2×2 symmetric matrix must necessarily be real.

P7. Suppose that $A\mathbf{x} = \lambda\mathbf{x}$ and $B\mathbf{x} = \mathbf{x}$. Then we have $AB\mathbf{x} = A(B\mathbf{x}) = A(\mathbf{x}) = \lambda\mathbf{x}$ and $BA\mathbf{x} = B(A\mathbf{x}) = B(\lambda\mathbf{x}) = \lambda\mathbf{x}$. Thus λ is an eigenvalue of both AB and BA, and $\mathbf{x}$ is a corresponding eigenvector.

Matrix Models

EXERCISE SET 5.1

1. (a) and (c) are stochastic matrices. (b) and (d) are not stochastic.

3. Using Formula (11), we have $x_1 = Px_0 = \begin{bmatrix} 0.5 & 0.6 \\ 0.5 & 0.4 \end{bmatrix} \begin{bmatrix} 0.5 \\ 0.5 \end{bmatrix} = \begin{bmatrix} 0.55 \\ 0.45 \end{bmatrix}$, $x_2 = \begin{bmatrix} 0.545 \\ 0.455 \end{bmatrix}$, $x_3 = \begin{bmatrix} 0.5455 \\ 0.4545 \end{bmatrix}$, and

$x_4 = \begin{bmatrix} 0.54545 \\ 0.45455 \end{bmatrix}$. Using Formula (12), we have $x_4 = P^4 x_0 = \begin{bmatrix} 0.5455 & 0.5454 \\ 0.4545 & 0.4546 \end{bmatrix} \begin{bmatrix} 0.5 \\ 0.5 \end{bmatrix} = \begin{bmatrix} 0.54545 \\ 0.45455 \end{bmatrix}$.

5. (a) The entries of P are positive; thus P is a regular stochastic matrix.

    (b) All positive powers of P have a zero in the upper right corner; thus P is not regular.

    (c) The entries of $P^2 = \begin{bmatrix} \frac{21}{25} & \frac{1}{5} \\ \frac{4}{25} & \frac{4}{5} \end{bmatrix}$ are all positive; thus P is regular.

7. To find the steady-state vector q, we must solve the system $(I - P)q = 0$ subject to the additional condition that q be a probability vector. The system $(I - P)q = 0$ can be written as

$$\begin{bmatrix} \frac{3}{4} & -\frac{2}{3} \\ \frac{3}{4} & -\frac{2}{3} \end{bmatrix} \begin{bmatrix} q_1 \\ q_2 \end{bmatrix} = \begin{bmatrix} 0 \\ 0 \end{bmatrix}$$

which has general solution $q_1 = \frac{8}{9}t$, $q_2 = t$. In order for q to be a probability vector, we must have

$1 = q_1 + q_2 = \frac{8}{9}t + t = \frac{17}{9}t$, or $t = \frac{9}{17}$. Thus the steady-state vector is $q = \begin{bmatrix} \frac{8}{17} \\ \frac{9}{17} \end{bmatrix}$.

9. The system $(I - P)q = 0$ can be written as $\begin{bmatrix} \frac{1}{2} & -\frac{1}{2} & 0 \\ -\frac{1}{4} & \frac{1}{2} & -\frac{1}{3} \\ -\frac{1}{4} & 0 & \frac{1}{3} \end{bmatrix} \begin{bmatrix} q_1 \\ q_2 \\ q_3 \end{bmatrix} = \begin{bmatrix} 0 \\ 0 \\ 0 \end{bmatrix}$ which has general solution

$q_1 = \frac{4}{3}t$, $q_2 = \frac{4}{3}t$, $q_3 = t$. For q to be a probability vector, we must have $1 = q_1 + q_2 + q_3 = \frac{11}{3}t$, or

$t = \frac{3}{11}$. Thus the steady-state vector is $q = \begin{bmatrix} \frac{4}{11} \\ \frac{4}{11} \\ \frac{3}{11} \end{bmatrix}$.

11. (a) The probability of transition from state 1 to state 1.

    (b) The probability of transition from state 2 to state 1.

    (c) The probability of transition from state 1 to state 2 is $p_{21} = 0.8$.

    (d) The initial state vector is $x(0) = \begin{bmatrix} 0.5 \\ 0.5 \end{bmatrix}$, and so $x(1) = \begin{bmatrix} 0.2 & 0.1 \\ 0.8 & 0.9 \end{bmatrix} \begin{bmatrix} 0.5 \\ 0.5 \end{bmatrix} = \begin{bmatrix} 0.15 \\ 0.85 \end{bmatrix}$. Thus, after the first transition, the probability of the system being in state 2 at the next observation is 0.85.

13. (a) $P = \begin{bmatrix} 0.95 & 0.55 \\ 0.05 & 0.45 \end{bmatrix}$

    (b) $P^2 = \begin{bmatrix} 0.93 & 0.77 \\ 0.07 & 0.23 \end{bmatrix}$; thus if air quality is good today, the probability that it will be good two days from now is 0.93.

(c) $P^3 = \begin{bmatrix} 0.922 & 0.858 \\ 0.078 & 0.142 \end{bmatrix}$; thus if air quality is bad today, the probability that it will be bad three days from now is 0.142.

(d) The initial state vector is $\mathbf{x}(0) = \begin{bmatrix} 0.2 \\ 0.8 \end{bmatrix}$, and so $\mathbf{x}(1) = P\mathbf{x}(0) = \begin{bmatrix} 0.63 \\ 0.37 \end{bmatrix}$. Thus the probablility that the air quality will be good tomorrow is 0.63.

15. This process can be described by a Markov chain with transition matrix $P = \begin{bmatrix} 0.95 & 0.03 \\ 0.05 & 0.97 \end{bmatrix}$, and initial vector $\mathbf{x}(0) = \begin{bmatrix} 0.8 \\ 0.2 \end{bmatrix}$ which represents the percentage of the total population of 125,000 that initially lives in the city (80% of total) and in the suburbs (20% of the total).

(a) The state vector $\mathbf{x}(1)$ is $\mathbf{x}(1) = P\mathbf{x}(0) = \begin{bmatrix} 0.766 \\ 0.234 \end{bmatrix}$ which, upon multiplying by 125,000, corresponds to populations of $\begin{bmatrix} 95750 \\ 29250 \end{bmatrix}$. Similarly, $\mathbf{x}(2) = P\mathbf{x}(1) = \begin{bmatrix} 0.73472 \\ 0.26528 \end{bmatrix}$ which corresponds to $\begin{bmatrix} 91840 \\ 33160 \end{bmatrix}$, etc. Proceeding in this fashion, one constructs the following table showing the populations of the city and its suburbs over a five-year period:

Year	1	2	3	4	5
City	95750	91840	88243	84933	81889
Suburbs	29250	33160	36757	40067	43111

(b) The system $(I - P)\mathbf{q} = \mathbf{0}$ can be written as $\begin{bmatrix} 0.05 & -0.03 \\ -0.05 & 0.03 \end{bmatrix} \begin{bmatrix} q_1 \\ q_2 \end{bmatrix} = \begin{bmatrix} 0 \\ 0 \end{bmatrix}$ which has general solution $q_1 = \frac{3}{5}t$, $q_2 = t$. For $\mathbf{q}$ to be a probability vector, we must have $1 = q_1 + q_2 = \frac{8}{5}t$, or $t = \frac{5}{8}$. Thus the steady-state vector is $\mathbf{q} = \begin{bmatrix} \frac{3}{8} \\ \frac{5}{8} \end{bmatrix} = \begin{bmatrix} 0.375 \\ 0.625 \end{bmatrix}$ which corresponds to populations of $\begin{bmatrix} 46875 \\ 78125 \end{bmatrix}$.

17. This process can be modeled by a Markov chain with transition matrix $P = \begin{bmatrix} \frac{1}{10} & \frac{1}{5} & \frac{3}{5} \\ \frac{4}{5} & \frac{3}{10} & \frac{1}{5} \\ \frac{1}{10} & \frac{1}{2} & \frac{1}{5} \end{bmatrix}$, and initial vector $\mathbf{x}(0) = \begin{bmatrix} 1 \\ 0 \\ 0 \end{bmatrix}$.

(a) The state vector $\mathbf{x}(2)$ is $\mathbf{x}(2) = P^2\mathbf{x}(0) = \begin{bmatrix} \frac{23}{100} & \frac{19}{50} & \frac{11}{50} \\ \frac{17}{50} & \frac{7}{20} & \frac{29}{50} \\ \frac{43}{100} & \frac{27}{100} & \frac{1}{5} \end{bmatrix} \begin{bmatrix} 1 \\ 0 \\ 0 \end{bmatrix} = \begin{bmatrix} \frac{23}{100} \\ \frac{17}{50} \\ \frac{43}{100} \end{bmatrix}$. Thus the probability that the car will be at location 1 after two rentals is $\frac{23}{100} = 0.23$.

(b) The system $(I - P)\mathbf{q} = \mathbf{0}$ can be written as $\begin{bmatrix} \frac{9}{10} & -\frac{1}{5} & -\frac{3}{5} \\ -\frac{4}{5} & \frac{7}{10} & -\frac{1}{5} \\ -\frac{1}{10} & -\frac{1}{2} & \frac{4}{5} \end{bmatrix} \begin{bmatrix} q_1 \\ q_2 \\ q_3 \end{bmatrix} = \begin{bmatrix} 0 \\ 0 \\ 0 \end{bmatrix}$ which has general solution $q_1 = \frac{46}{47}t$, $q_2 = \frac{66}{47}t$, $q_3 = t$. In order for $\mathbf{q}$ to be a probability vector, we must have $1 = q_1 + q_2 + q_3 = \frac{159}{47}t$, or $t = \frac{47}{159}$. Thus the steady-state vector is $\mathbf{q} = \begin{bmatrix} \frac{46}{159} \\ \frac{33}{53} \\ \frac{47}{159} \end{bmatrix}$.

(c)   The number of parking spaces required at each of the locations is found by multiplying $\mathbf{q}$ by

120 and rounding to the nearest integer. This yields $120\mathbf{q} \approx \begin{bmatrix} 35 \\ 50 \\ 35 \end{bmatrix}$.

DISCUSSION AND DISCOVERY

D1. $P = \begin{bmatrix} \frac{7}{10} & \frac{1}{10} & \frac{1}{5} \\ \frac{1}{5} & \frac{3}{10} & \frac{1}{2} \\ \frac{1}{10} & \frac{3}{5} & \frac{3}{10} \end{bmatrix}$, and the steady-state vector is $\mathbf{q} = \begin{bmatrix} \frac{1}{3} \\ \frac{1}{3} \\ \frac{1}{3} \end{bmatrix}$.

D2. $MP = M = \begin{bmatrix} 1 & 1 & \cdots & 1 \end{bmatrix}$

D3.   The vector $\mathbf{q}$ is a fixed point of P, i.e. $P\mathbf{q} = \mathbf{q}$. Thus $P^k\mathbf{q} = \mathbf{q}$ for every positive integer k.

D4.   (a)   From Theorem 5.1.3 it follows that $P^k\mathbf{e}_i \to \mathbf{q}$ as $k \to \infty$, for each $i = 1, 2, \ldots, n$.
　　　　 (b)   The ith column of P^k is $P^k\mathbf{e}_i$; thus each column vector of P^k approaches $\mathbf{q}$ as $k \to \infty$.

WORKING WITH PROOFS

P1.   If P and Q are stochastic matrices, then the columns of P and Q are probability vectors, i.e. $p_{ij} \geq 0$, $q_{ij} \geq 0$, and $\sum_{i=1}^{n} p_{ij} = \sum_{i=1}^{n} q_{ij} = 1$ for each $j = 1, \ldots, n$. It follows that the entries of PQ are nonnegative, and that

$$\sum_{i=1}^{n}(PQ)_{ij} = \sum_{i=1}^{n}\sum_{k=1}^{n} p_{ik}q_{kj} = \sum_{k=1}^{n}\left(\sum_{i=1}^{n} p_{ik}\right) q_{kj} = \sum_{k=1}^{n}(1)q_{kj} = 1$$

for each $j = 1, \ldots, n$. Thus the product PQ is a stochastic matrix.

P2.   (a)   Let $\mathbf{u} = \begin{bmatrix} 1 \\ 1 \\ \vdots \\ 1 \end{bmatrix}$. If the row sums of P are all equal to 1, then $P\mathbf{u} = \mathbf{u}$; thus $\mathbf{q} = \frac{1}{k}\mathbf{u}$ is the

unique probability vector for which $P\mathbf{q} = \mathbf{q}$. Conversely, if $P\mathbf{q} = \mathbf{q}$ where $\mathbf{q} = \frac{1}{k}\mathbf{u}$, then the row sums of P must all be equal to 1.

(b) $\mathbf{q} = \begin{bmatrix} \frac{1}{3} \\ \frac{1}{3} \\ \frac{1}{3} \end{bmatrix}$

(c)   If P is a stochastic matrix that is regular and symmetric, then the row sums of P are all equal to 1. Thus the steady-state vector is $\mathbf{q} = \frac{1}{k}\mathbf{u}$ where $\mathbf{u}$ is as in part (a).

P3.   Suppose $P = [p_{ij}]$ is a stochastic matrix whose entries p_{ij} are all greater than or equal to ρ. Then, for each i and j, we have

$$(P^2)_{ij} = \sum_{k=1}^{n} p_{ik}p_{kj} \geq \sum_{k=1}^{n} \rho p_{kj} = \rho \sum_{k=1}^{n} p_{kj} = (\rho)(1) = \rho$$

Thus the entries of P^2 are all greater than or equal to ρ.

EXERCISE SET 5.2

1. **(a)** A consumption matrix for this economy is $C = \begin{bmatrix} 0.50 & 0.25 \\ 0.25 & 0.10 \end{bmatrix}$.

    (b) We have $(I - C)^{-1} = \begin{bmatrix} 0.50 & -0.25 \\ -0.25 & 0.90 \end{bmatrix}^{-1} = \frac{1}{0.3875}\begin{bmatrix} 0.90 & 0.25 \\ 0.25 & 0.50 \end{bmatrix}$, and so the production needed to provide customers \$7,000 worth of mechanical work and \$14,000 worth of body work is given by

    $$\mathbf{x} = (I - C)^{-1}\mathbf{d} = \frac{1}{0.385}\begin{bmatrix} 0.9 & 0.25 \\ 0.25 & 0.5 \end{bmatrix}\begin{bmatrix} 7000 \\ 14000 \end{bmatrix} \approx \begin{bmatrix} \$25,290 \\ \$22,581 \end{bmatrix}$$

3. **(a)** A consumption matrix for this economy is $C = \begin{bmatrix} 0.1 & 0.6 & 0.4 \\ 0.3 & 0.2 & 0.3 \\ 0.4 & 0.1 & 0.2 \end{bmatrix}$.

    (b) The Leontief equation $(I - C)\mathbf{x} = \mathbf{d}$ is represented by the augmented matrix

    $$\begin{bmatrix} 0.9 & -0.6 & -0.4 & \vdots & 1930 \\ -0.3 & 0.8 & -0.3 & \vdots & 3860 \\ -0.4 & -0.1 & 0.8 & \vdots & 5790 \end{bmatrix}$$

    The reduced row echelon form of this matrix is

    $$\begin{bmatrix} 1 & 0 & 0 & \vdots & 31500 \\ 0 & 1 & 0 & \vdots & 26500 \\ 0 & 0 & 1 & \vdots & 26300 \end{bmatrix}$$

    Thus the required production vector is $\mathbf{x} = \begin{bmatrix} \$31,500 \\ \$26,500 \\ \$26,300 \end{bmatrix}$.

5. The Leontief equation $(I - C)\mathbf{x} = \mathbf{d}$ for this economy is $\begin{bmatrix} 0.9 & -0.3 \\ -0.5 & 0.6 \end{bmatrix}\begin{bmatrix} x_1 \\ x_2 \end{bmatrix} = \begin{bmatrix} 50 \\ 60 \end{bmatrix}$; thus the required production vector is

    $$\mathbf{x} = (I - C)^{-1}\mathbf{d} = \frac{1}{0.39}\begin{bmatrix} 0.6 & 0.3 \\ 0.5 & 0.9 \end{bmatrix}\begin{bmatrix} 50 \\ 60 \end{bmatrix} = \frac{1}{0.39}\begin{bmatrix} 48 \\ 79 \end{bmatrix} \approx \begin{bmatrix} 123.08 \\ 202.56 \end{bmatrix}$$

7. **(a)** The column sums of C are all less than 1; thus the economy is productive.
    (b) The row sums of C are all less than 1; thus the economy is productive.

9. If $C = \begin{bmatrix} 0.50 & 0 & 0.25 \\ 0.20 & 0.80 & 0.10 \\ 1 & 0.40 & 0 \end{bmatrix}$, then $(I - C)^{-1} = \begin{bmatrix} 0.50 & 0 & -0.25 \\ -0.20 & 0.20 & -0.10 \\ -1 & -0.40 & 1 \end{bmatrix}^{-1} = \begin{bmatrix} 16 & 10 & 5 \\ 30 & 25 & 10 \\ 28 & 20 & 10 \end{bmatrix}$ has nonnegative entries; thus the economy is productive. This does not violate Theorem 5.2.1.

DISCUSSION AND DISCOVERY

D1. **(a)** The Leontief equation for this economy is $\begin{bmatrix} \frac{1}{2} & 0 \\ 0 & 0 \end{bmatrix}\begin{bmatrix} x_1 \\ x_2 \end{bmatrix} = \begin{bmatrix} d_1 \\ d_2 \end{bmatrix}$. If $d_1 = 2$ and $d_2 = 0$, the system is consistent with general solution $x_1 = 4$, $x_2 = t$ $(0 \le t < \infty)$. If $d_1 = 2$ and $d_2 = 1$ the system is inconsistent.

(b) The consumption matrix $C = \begin{bmatrix} \frac{1}{2} & 0 \\ 0 & 1 \end{bmatrix}$ indicates that the entire output of the second sector is consumed in producing that output; thus there is nothing left to satisfy any outside demand. Mathematically, the Leontief matrix $I - C = \begin{bmatrix} \frac{1}{2} & 0 \\ 0 & 0 \end{bmatrix}$ is not invertible.

D2. The first sector must produce the largest dollar amount.

D3. We have $\det(I - C) = (1 - c_{11}) - c_{21}c_{12}$. Thus, if $c_{21}c_{12} < 1 - c_{11}$, it follows that $\det(I - C) > 0$. From this we conclude that the matrix $I - C$ is invertible, and $(I - C)^{-1} = \frac{1}{\det(I-C)} \begin{bmatrix} 1 & c_{12} \\ c_{21} & 1 - c_{11} \end{bmatrix}$ has nonnegative entries. This shows that the economy is productive. Thus the Leontief equation $(I - C)\mathbf{x} = \mathbf{d}$ has a unique solution for every demand vector $\mathbf{d}$.

WORKING WITH PROOFS

P1. (a) The equation $(I - C)^{-1}\mathbf{d} = \mathbf{x}$ can be written as

$$d_1\mathbf{c}_1 + d_2\mathbf{c}_2 + \cdots + d_j\mathbf{c}_j + \cdots + d_n\mathbf{c}_n = \mathbf{x}$$

where $\mathbf{c}_1, \mathbf{c}_2, \ldots, \mathbf{c}_j, \ldots, \mathbf{c}_n$ are the column vectors of $(I - C)^{-1}$. Thus replacing d_j by $d_j + 1$ corresponds to replacing $\mathbf{x}$ by $\mathbf{x} + \mathbf{c}_j$.

(b) The jth column of the matrix $(I - C)^{-1}$ represents the additional production that is required in order to satisfy one additional unit of demand for the jth sector.

P2. If C has row sums less than 1, then C^T has column sums less than 1. Thus, from the discussion preceding Theorem 5.2.1, it follows that $I - C^T$ is invertible and that $(I - C^T)^{-1}$ has nonnegative entries. Since $I - C = (I^T - C^T)^T = (I - C^T)^T$, we can conclude that $I - C$ is invertible and that

$$(I - C)^{-1} = ((I - C^T)^T)^{-1} = ((I - C^T)^{-1})^T$$

has nonnegative entries.

P3. If k is the nilpotency index of the consumption matrix C then, from Theorem 3.6.6, $I - C$ is invertible with $(I - C)^{-1} = I + C + C^2 + \cdots + C^{k-1}$. Thus the matrix $(I - C)^{-1}$ has nonnegative entries and so the economy is productive.

EXERCISE SET 5.3

1. Using Jacobi iteration: We begin by solving the first equation for x_1 and the second equation for x_2. This yields

$$x_1 = \tfrac{7}{2} - \tfrac{1}{2}x_2 \qquad x_1 = 3.5 - 0.5x_2$$
$$\text{or}$$
$$x_2 = -\tfrac{1}{2} + \tfrac{1}{2}x_1 \qquad x_2 = -0.5 + 0.5x_1$$

which can be written in matrix form as

$$\begin{bmatrix} x_1 \\ x_2 \end{bmatrix} = \begin{bmatrix} 0 & -0.5 \\ 0.5 & 0 \end{bmatrix} \begin{bmatrix} x_1 \\ x_2 \end{bmatrix} + \begin{bmatrix} 3.5 \\ -0.5 \end{bmatrix}$$

Starting with $\mathbf{x}_0 = \begin{bmatrix} 0 \\ 0 \end{bmatrix}$, the first iterate is $\mathbf{x}_1 = \begin{bmatrix} 0 & -0.5 \\ 0.5 & 0 \end{bmatrix} \begin{bmatrix} 0 \\ 0 \end{bmatrix} + \begin{bmatrix} 3.5 \\ -0.5 \end{bmatrix} = \begin{bmatrix} 3.5 \\ -0.5 \end{bmatrix}$, the second iterate

is $\mathbf{x}_2 = \begin{bmatrix} 0 & -0.5 \\ 0.5 & 0 \end{bmatrix} \begin{bmatrix} 3.5 \\ -0.5 \end{bmatrix} + \begin{bmatrix} 3.5 \\ -0.5 \end{bmatrix} = \begin{bmatrix} 3.75 \\ 1.25 \end{bmatrix}$, and $\mathbf{x}_3 = \begin{bmatrix} 0 & -0.5 \\ 0.5 & 0 \end{bmatrix} \begin{bmatrix} 3.75 \\ 1.25 \end{bmatrix} + \begin{bmatrix} 3.5 \\ -0.5 \end{bmatrix} = \begin{bmatrix} 2.875 \\ 1.375 \end{bmatrix}$. The

exact solution is $x_1 = 3$, $x_2 = 1$.

Using Gauss-Seidel iteration: Starting with $x_1 = x_2 = 0$, we use the equations above to compute successive values, using the latest updated values at each step. The first Gauss-Seidel iterate is

$$x_1 = 3.5 - (0.5)(0) = 3.5$$
$$x_2 = -0.5 + (0.5)(3.5) = 1.25$$

the second iterate is

$$x_1 = 3.5 - (0.5)(1.25) = 2.875$$
$$x_2 = -0.5 + (0.5)(2.875) = 0.9375$$

and the third iterate is

$$x_1 = 3.5 - (0.5)(0.9375) = 3.03125$$
$$x_2 = 0.5 + (0.5)(3.03125) = 1.015625$$

3. Using Jacobi iteration: The equations for Jacobi iteration are

$$x_1 = \tfrac{1}{10}(3 - x_2 - 2x_3) = 0.3 - 0.1x_2 - 0.2x_3$$
$$x_2 = \tfrac{1}{10}\left(\tfrac{3}{2} - x_1 + x_3\right) = 0.15 - 0.1x_1 + 0.1x_3$$
$$x_3 = \tfrac{1}{10}(-9 - 2x_1 - x_2) = -0.9 - 0.2x_1 - 0.1x_2$$

which can be written in matrix form as

$$\begin{bmatrix} x_1 \\ x_2 \\ x_3 \end{bmatrix} = \begin{bmatrix} 0 & -0.1 & -0.2 \\ -0.1 & 0 & 0.1 \\ -0.2 & -0.1 & 0 \end{bmatrix} \begin{bmatrix} x_1 \\ x_2 \\ x_3 \end{bmatrix} + \begin{bmatrix} 0.3 \\ 0.15 \\ -0.9 \end{bmatrix}$$

Starting with $\mathbf{x}_0 = \begin{bmatrix} 0 \\ 0 \\ 0 \end{bmatrix}$, the first iterates is $\mathbf{x}_1 = \begin{bmatrix} 0 & -0.1 & -0.2 \\ -0.1 & 0 & 0.1 \\ -0.2 & -0.1 & 0 \end{bmatrix} \begin{bmatrix} 0 \\ 0 \\ 0 \end{bmatrix} + \begin{bmatrix} 0.3 \\ 0.15 \\ -0.9 \end{bmatrix} = \begin{bmatrix} 0.3 \\ 0.15 \\ -0.9 \end{bmatrix}$, the

second iterate is $\mathbf{x}_2 = \begin{bmatrix} 0 & -0.1 & -0.2 \\ -0.1 & 0 & 0.1 \\ -0.2 & -0.1 & 0 \end{bmatrix} \begin{bmatrix} 0.3 \\ 0.15 \\ -0.9 \end{bmatrix} + \begin{bmatrix} 0.3 \\ 0.15 \\ -0.9 \end{bmatrix} = \begin{bmatrix} 0.465 \\ 0.03 \\ -0.975 \end{bmatrix}$, and the third iterate is

$\mathbf{x}_3 = \begin{bmatrix} 0 & -0.1 & -0.2 \\ -0.1 & 0 & 0.1 \\ -0.2 & -0.1 & 0 \end{bmatrix} \begin{bmatrix} 0.465 \\ 0.03 \\ -0.975 \end{bmatrix} + \begin{bmatrix} 0.3 \\ 0.15 \\ -0.9 \end{bmatrix} = \begin{bmatrix} 0.4920 \\ 0.0060 \\ -0.9960 \end{bmatrix}$. The exact solution is $x_1 = \tfrac{1}{2}$, $x_2 = 0$,

$x_3 = -1$.

Using Gauss-Seidel iteration: Starting with $x_1 = x_2 = x_3 = 0$, we use the equations above to compute successive values, using the latest updated values at each step. The first Gauss-Seidel iterate is

$$x_1 = 0.3 - 0.1(0) - 0.2(0) = 0.3$$
$$x_2 = 0.15 - 0.1(0.3) + 0.1(0) = 0.12$$
$$x_3 = -0.9 - 0.2(0.3) - 0.1(0.12) = -0.972$$

the second iterate is

$$x_1 = 0.3 - 0.1(0.12) - 0.2(-0.972) = 0.4824$$
$$x_2 = 0.15 - 0.1(0.4824) + 0.1(-0.972) = 0.00456$$
$$x_3 = -0.9 - 0.2(0.4824) - 0.1(0.00456) = -0.996936$$

and the third iterate is

$$x_1 = 0.3 - 0.1(0.00456) - 0.2(-0.996936) = 0.4989312$$
$$x_2 = 0.15 - 0.1(0.4989312) + 0.1(-0.996936) = 0.00041328$$
$$x_3 = -0.9 - 0.2(0.4989312) - 0.1(0.00041328) = -0.999827568$$

5. (a) In the second row, we have $|-2| = 2$. This matrix is not strictly diagonally dominant. (b) and (c) are strictly diagonally dominant.

7. This matrix is not strictly diagonally dominant, but can be made so by permuting the rows as follows:
$$\begin{bmatrix} 5 & 2 & 2 \\ 3 & -7 & -3 \\ -1 & 4 & -6 \end{bmatrix}$$

9. This matrix is not strictly diagonally dominant, and cannot be made so by any permutation of the rows.

WORKING WITH PROOFS

P1. (a) $\displaystyle \|aA\|_\infty = \max_{1 \le i \le m} \left[\sum_{j=1}^{n} |aa_{ij}| \right] = \max_{1 \le i \le m} \left[\sum_{j=1}^{n} |a| \, |a_{ij}| \right] = \max_{1 \le i \le m} \left[|a| \sum_{j=1}^{n} |a_{ij}| \right]$

$\displaystyle = |a| \max_{1 \le i \le m} \left[\sum_{j=1}^{n} |a_{ij}| \right] = |a| \, \|A\|_\infty$

(b) $\displaystyle \|A + B\|_\infty = \max_{1 \le i \le m} \left[\sum_{j=1}^{n} |a_{ij} + b_{ij}| \right] \le \max_{1 \le i \le m} \left[\sum_{j=1}^{n} (|a_{ij}| + |b_{ij}|) \right]$

$\displaystyle \le \max_{1 \le i \le m} \left[\sum_{j=1}^{n} |a_{ij}| \right] + \max_{1 \le i \le m} \left[\sum_{j=1}^{n} |b_{ij}| \right] = \|A\|_\infty + \|B\|_\infty$

(c) If $\mathbf{x}$ is a $1 \times n$ row vector and B is an $n \times r$ matrix, then $\mathbf{x}B = \begin{bmatrix} \mathbf{x} \cdot \mathbf{c}_1(B) & \mathbf{x} \cdot \mathbf{c}_2(B) & \cdots & \mathbf{x} \cdot \mathbf{c}_r(B) \end{bmatrix}$ is a $1 \times r$ row vector, and

$$\|\mathbf{x}B\|_\infty = \sum_{j=1}^{r} |\mathbf{x} \cdot \mathbf{c}_j(B)| = \sum_{j=1}^{r} \left| \sum_{k=1}^{n} x_k b_{kj} \right| \le \sum_{j=1}^{r} \sum_{k=1}^{n} |x_k| \, |b_{kj}| = \sum_{k=1}^{n} |x_k| \sum_{j=1}^{r} |b_{kj}|$$

$$\le \|B\|_\infty \sum_{k=1}^{n} |x_k| = \|B\|_\infty \, \|\mathbf{x}\|_\infty$$

(d) If A is an $m \times n$ matrix and B is an $n \times r$ matrix, then

$$\|AB\|_\infty = \max_{1 \le i \le m} \left[\sum_{j=1}^{r} \left| \sum_{k=1}^{n} a_{ik} b_{kj} \right| \right] \le \max_{1 \le i \le m} \left[\sum_{j=1}^{r} \sum_{k=1}^{n} |a_{ik}| \, |b_{kj}| \right]$$

$$= \max_{1 \le i \le m} \left[\sum_{k=1}^{n} |a_{ik}| \sum_{j=1}^{r} |b_{kj}| \right] \le \max_{1 \le i \le m} \left[\sum_{k=1}^{n} |a_{ik}| \right] \|B\|_\infty = \|A\|_\infty \, \|B\|_\infty$$

P2. (a) The sequence $\{\mathbf{x}_k\}$ is defined recursively by $\mathbf{x}_k = M\mathbf{x}_{k-1} + \mathbf{y}$, where $\mathbf{x}_0$ is arbitrary. We will prove that the formula

$$\mathbf{x}_k = M^k\mathbf{x}_0 + (I + M + M^2 + \cdots + M^{k-1})\mathbf{y}$$

is valid for every positive integer $k \geq 1$. Our proof is by induction on k.

Step 1. We have $\mathbf{x}_1 = M\mathbf{x}_0 + \mathbf{y} = M\mathbf{x}_0 + I\mathbf{y}$; thus the formula is valid for $k = 1$.

Step 2 (induction step). Suppose the formula is valid for a fixed integer $j \geq 1$. Then

$$\mathbf{x}_j = M^j\mathbf{x}_0 + (I + M + M^2 + \cdots + M^{j-1})\mathbf{y}$$

and so

$$\begin{aligned}
\mathbf{x}_{j+1} = M\mathbf{x}_j + \mathbf{y} &= M\left[M^j\mathbf{x}_0 + (I + M + M^2 + \cdots + M^{j-1})\mathbf{y}\right] + \mathbf{y} \\
&= M^{j+1}\mathbf{x}_0 + (M + M^2 + M^3 + \cdots + M^j)\mathbf{y} + \mathbf{y} \\
&= M^{j+1}\mathbf{x}_0 + (I + M + M^2 + \cdots + M^j)\mathbf{y}
\end{aligned}$$

This shows that if the formula is valid for $k = j$, then it is also valid for $k = j + 1$.

(b) If $\|M\|_\infty < 1$, then $\left\|M^k\mathbf{x}_0\right\|_\infty \leq \|M\|_\infty^k \|\mathbf{x}_0\|_\infty \to 0$ as $k \to \infty$. Thus, from part (a), together with Theorem 3.6.7, we have

$$\mathbf{x}_k \to (I + M + M^2 + M^3 + \cdots)\mathbf{y} = (I - M)^{-1}\mathbf{y}$$

as $k \to \infty$.

P3. (a) If $A = [a_{ij}]$, then the product $D^{-1}(D - A)$ is

$$\begin{bmatrix} \frac{1}{a_{11}} & 0 & 0 & 0 \\ 0 & \frac{1}{a_{22}} & 0 & 0 \\ 0 & 0 & \ddots & 0 \\ 0 & 0 & 0 & \frac{1}{a_{nn}} \end{bmatrix} \begin{bmatrix} 0 & -a_{12} & \cdots & -a_{1n} \\ -a_{21} & 0 & \cdots & -a_{2n} \\ \vdots & \vdots & \ddots & \vdots \\ -a_{n1} & -a_{n2} & \cdots & 0 \end{bmatrix} = \begin{bmatrix} 0 & -\frac{a_{12}}{a_{11}} & \cdots & -\frac{a_{1n}}{a_{11}} \\ -\frac{a_{21}}{a_{22}} & 0 & \cdots & -\frac{a_{2n}}{a_{22}} \\ \vdots & \vdots & \ddots & \vdots \\ -\frac{a_{n1}}{a_{nn}} & -\frac{a_{n2}}{a_{nn}} & \cdots & 0 \end{bmatrix}$$

and, since A is strictly diagonally dominant, we have

$$\sum_{j \neq k} \left| -\frac{a_{kj}}{a_{kk}} \right| = \frac{1}{|a_{kk}|} \sum_{j \neq k} |a_{kj}| < 1$$

for each $k = 1, 2, \ldots, n$. Thus $\left\|D^{-1}(D - A)\right\|_\infty < 1$.

(b) Using Formula (4), the Jacobi iterates for the system $A\mathbf{x} = \mathbf{b}$ are given by

$$\mathbf{x}_{k+1} = D^{-1}(D - A)\mathbf{x}_k + D^{-1}\mathbf{b}$$

and, since $\left\|D^{-1}(D - A)\right\|_\infty < 1$, it follows from Exercise P2 that

$$\mathbf{x}_k \to (I - D^{-1}(D - A))^{-1}D^{-1}\mathbf{b} = A^{-1}\mathbf{b}$$

as $k \to \infty$.

P4. (a) If A has the property that the absolute value of each of its diagonal entries is greater then the sum of the absolute values of the remaining entries in the same column, then

$$\sum_{i \neq j} \left| -\frac{a_{ij}}{a_{jj}} \right| = \frac{1}{|a_{jj}|} \sum_{i \neq j} |a_{ij}| < 1$$

for each $i = 1, 2, \ldots, n$. Thus, as in Exercise P3(a), we have $\left\|D^{-1}(D - A)\right\|_1 < 1$.

(b) The Jacobi iterates for the system $A\mathbf{x} = \mathbf{b}$ are given by

$$\mathbf{x}_{k+1} = D^{-1}(D - A)\mathbf{x}_k + D^{-1}\mathbf{b}$$

and, since $\left\|D^{-1}(D - A)\right\|_1 < 1$, it follows that

$$\mathbf{x}_k \to (I - D^{-1}(D - A))^{-1}D^{-1}\mathbf{b} = A^{-1}\mathbf{b}$$

as $k \to \infty$.

EXERCISE SET 5.4

1. (a) $\lambda_3 = -8$ is a dominant eigenvalue   (b) There is no dominant eigenvalue.

3. The characteristic polynomial of A is $\begin{vmatrix} \lambda + 7 & 12 \\ -8 & \lambda - 13 \end{vmatrix} = (\lambda + 7)(\lambda - 13) + 96 = (\lambda - 1)\ (\lambda - 5);$
thus $\lambda = 5$ is a dominant eigenvalue. The corresponding dominant eigenvectors are found by
solving the system $\begin{bmatrix} 12 & 12 \\ -8 & -8 \end{bmatrix}\begin{bmatrix} x_1 \\ x_2 \end{bmatrix} = \begin{bmatrix} 0 \\ 0 \end{bmatrix}$, which leads to $\mathbf{x} = t\begin{bmatrix} -1 \\ 1 \end{bmatrix}$. Thus the unit eigenvector to
which the sequence $\mathbf{x}_1, \mathbf{x}_2, \ldots, \mathbf{x}_k, \ldots$ converges is

$$\mathbf{u} = \begin{bmatrix} -\frac{1}{\sqrt{2}} \\ \frac{1}{\sqrt{2}} \end{bmatrix} \approx \begin{bmatrix} -0.7071 \\ 0.7071 \end{bmatrix}$$

We have $A\mathbf{x}_1 \approx \begin{bmatrix} -7 & -12 \\ 8 & 13 \end{bmatrix}\begin{bmatrix} -0.6585 \\ 0.7526 \end{bmatrix} = \begin{bmatrix} -4.4214 \\ 4.5155 \end{bmatrix}$, and so

$$A\mathbf{x}_1 \cdot \mathbf{x}_1 \approx (-4.4214)(-0.6585) + (4.5155)(0.7526) \approx 6.3097$$

Similarly $A\mathbf{x}_2 \cdot \mathbf{x}_2 \approx 5.2101$, $A\mathbf{x}_3 \cdot \mathbf{x}_3 \approx 5.0404$, $A\mathbf{x}_4 \cdot \mathbf{x}_4 \approx 5.0080$, $\ldots$, etc. This sequence converges to the dominant eigenvalue $\lambda = 5$.

5. The characteristic polynomial of A is $p(\lambda) = \begin{vmatrix} \lambda - 5 & 1 \\ 1 & \lambda + 1 \end{vmatrix} = (\lambda - 5)(\lambda + 1) - 1 = \lambda^2 - 4\lambda - 6;$ thus
the eigenvalues are $2 \pm \sqrt{10}$ and $\lambda = 2 + \sqrt{10} \approx 5.16228$ is a dominant eigenvalue. A corresponding
dominant eigenvector is found by solving the system $\begin{bmatrix} -3 + \sqrt{10} & 1 \\ 1 & 3 + \sqrt{10} \end{bmatrix}\begin{bmatrix} x_1 \\ x_2 \end{bmatrix} = \begin{bmatrix} 0 \\ 0 \end{bmatrix}$, which yields
the dominant (scaled) eigenvector $\mathbf{x} = \begin{bmatrix} 1 \\ 3 - \sqrt{10} \end{bmatrix} \approx \begin{bmatrix} 1 \\ -0.16227 \end{bmatrix}$.

The power sequence with maximum entry scaling can be computed as follows:

$$A\mathbf{x}_0 = \begin{bmatrix} 5 & -1 \\ -1 & -1 \end{bmatrix}\begin{bmatrix} 0 \\ 1 \end{bmatrix} = \begin{bmatrix} -1 \\ -1 \end{bmatrix} \qquad \mathbf{x}_1 = \frac{A\mathbf{x}_0}{\max(A\mathbf{x}_0)} = \frac{1}{(-1)}\begin{bmatrix} -1 \\ -1 \end{bmatrix} = \begin{bmatrix} 1 \\ 1 \end{bmatrix}$$

$$A\mathbf{x}_1 = \begin{bmatrix} 5 & -1 \\ -1 & -1 \end{bmatrix}\begin{bmatrix} 1 \\ 1 \end{bmatrix} = \begin{bmatrix} 4 \\ -2 \end{bmatrix} \qquad \mathbf{x}_2 = \frac{A\mathbf{x}_1}{\max(A\mathbf{x}_1)} = \frac{1}{4}\begin{bmatrix} 4 \\ -2 \end{bmatrix} = \begin{bmatrix} 1 \\ -\frac{1}{2} \end{bmatrix} = \begin{bmatrix} 1 \\ -0.5 \end{bmatrix}$$

$$A\mathbf{x}_2 = \begin{bmatrix} 5 & -1 \\ -1 & -1 \end{bmatrix}\begin{bmatrix} 1 \\ -\frac{1}{2} \end{bmatrix} = \begin{bmatrix} \frac{11}{2} \\ -\frac{1}{2} \end{bmatrix} \qquad \mathbf{x}_3 = \frac{A\mathbf{x}_2}{\max(A\mathbf{x}_2)} = \frac{2}{11}\begin{bmatrix} \frac{11}{2} \\ -\frac{1}{2} \end{bmatrix} = \begin{bmatrix} 1 \\ -\frac{1}{11} \end{bmatrix} \approx \begin{bmatrix} 1 \\ -0.091 \end{bmatrix}$$

$$A\mathbf{x}_3 = \begin{bmatrix} 5 & -1 \\ -1 & -1 \end{bmatrix}\begin{bmatrix} 1 \\ -\frac{1}{11} \end{bmatrix} = \begin{bmatrix} \frac{56}{11} \\ -\frac{10}{11} \end{bmatrix} \qquad \mathbf{x}_4 = \frac{A\mathbf{x}_3}{\max(A\mathbf{x}_3)} = \frac{11}{56}\begin{bmatrix} \frac{56}{11} \\ -\frac{10}{11} \end{bmatrix} = \begin{bmatrix} 1 \\ -\frac{5}{28} \end{bmatrix} \approx \begin{bmatrix} 1 \\ -0.179 \end{bmatrix}$$

$$A\mathbf{x}_4 = \begin{bmatrix} 5 & -1 \\ -1 & -1 \end{bmatrix}\begin{bmatrix} 1 \\ -\frac{5}{28} \end{bmatrix} = \begin{bmatrix} \frac{145}{28} \\ -\frac{23}{28} \end{bmatrix} \qquad \mathbf{x}_5 = \frac{A\mathbf{x}_4}{\max(A\mathbf{x}_4)} = \frac{28}{145}\begin{bmatrix} \frac{145}{28} \\ -\frac{23}{28} \end{bmatrix} = \begin{bmatrix} 1 \\ -\frac{23}{145} \end{bmatrix} \approx \begin{bmatrix} 1 \\ -0.159 \end{bmatrix}$$

The corresponding Rayleigh quotients are:

$$\lambda^{(1)} = \frac{A\mathbf{x}_1 \cdot \mathbf{x}_1}{\mathbf{x}_1 \cdot \mathbf{x}_1} = \frac{(A\mathbf{x}_1)^T \mathbf{x}_1}{\mathbf{x}_1^T \mathbf{x}_1} = \frac{2}{2} = 1$$

$$\lambda^{(2)} = \frac{A\mathbf{x}_2 \cdot \mathbf{x}_2}{\mathbf{x}_2 \cdot \mathbf{x}_2} = \frac{(A\mathbf{x}_2)^T \mathbf{x}_2}{\mathbf{x}_2^T \mathbf{x}_2} = \frac{\frac{23}{4}}{\frac{5}{4}} = \frac{23}{5} = 4.6$$

$$\lambda^{(3)} = \frac{A\mathbf{x}_3 \cdot \mathbf{x}_3}{\mathbf{x}_3 \cdot \mathbf{x}_3} = \frac{(A\mathbf{x}_3)^T \mathbf{x}_3}{\mathbf{x}_3^T \mathbf{x}_3} = \frac{\frac{626}{121}}{\frac{122}{121}} = \frac{313}{61} \approx 5.1311$$

$$\lambda^{(4)} = \frac{A\mathbf{x}_4 \cdot \mathbf{x}_4}{\mathbf{x}_4 \cdot \mathbf{x}_4} = \frac{(A\mathbf{x}_4)^T \mathbf{x}_4}{\mathbf{x}_4^T \mathbf{x}_4} = \frac{\frac{4175}{784}}{\frac{809}{784}} = \frac{4175}{809} \approx 5.1607$$

The sequence $\lambda^{(1)}, \lambda^{(2)}, \ldots, \lambda^{(k)}, \ldots$ approaches the dominant eigenvalue $\lambda = 2 + \sqrt{10} \approx 5.16228$, and $\mathbf{x}_1, \mathbf{x}_2, \ldots, \mathbf{x}_k, \ldots$ approaches the dominant eigenvector $\mathbf{x} = \begin{bmatrix} 1 \\ 3 - \sqrt{10} \end{bmatrix} \approx \begin{bmatrix} 1 \\ -0.16227 \end{bmatrix}$.

7. The characteristic polynomial of A is $p(\lambda) = \begin{vmatrix} \lambda - 1 & -2 \\ -2 & \lambda - 1 \end{vmatrix} = (\lambda - 1)^2 - 4 = (\lambda - 3)(\lambda + 1)$; thus $\lambda = 3$ is a dominant eigenvalue. The corresponding dominant eigenvectors are found by solving the system $\begin{bmatrix} 2 & -2 \\ -2 & 2 \end{bmatrix} \begin{bmatrix} x_1 \\ x_2 \end{bmatrix} = \begin{bmatrix} 0 \\ 0 \end{bmatrix}$, which yields $\mathbf{x} = t \begin{bmatrix} 1 \\ 1 \end{bmatrix}$.

Using the given power sequence, together with (10), we have

$$\mathbf{x}_1 = \frac{A\mathbf{x}_0}{\max(A\mathbf{x}_0)} = \frac{1}{2} \begin{bmatrix} 1 \\ 2 \end{bmatrix} = \begin{bmatrix} 0.5 \\ 1 \end{bmatrix}$$

$$\mathbf{x}_2 = \frac{A^2\mathbf{x}_0}{\max(A^2\mathbf{x}_0)} = \frac{1}{5} \begin{bmatrix} 5 \\ 4 \end{bmatrix} = \begin{bmatrix} 1 \\ 0.8 \end{bmatrix}$$

$$\mathbf{x}_3 = \frac{A^3\mathbf{x}_0}{\max(A^3\mathbf{x}_0)} = \frac{1}{14} \begin{bmatrix} 13 \\ 14 \end{bmatrix} \approx \begin{bmatrix} 0.92857 \\ 1 \end{bmatrix}$$

$$\mathbf{x}_4 = \frac{A^4\mathbf{x}_0}{\max(A^4\mathbf{x}_0)} = \frac{1}{41} \begin{bmatrix} 41 \\ 40 \end{bmatrix} \approx \begin{bmatrix} 1 \\ 0.97561 \end{bmatrix}$$

$$\mathbf{x}_5 = \frac{A^5\mathbf{x}_0}{\max(A^5\mathbf{x}_0)} = \frac{1}{122} \begin{bmatrix} 121 \\ 122 \end{bmatrix} \approx \begin{bmatrix} 0.99180 \\ 1 \end{bmatrix}$$

This sequence of vectors approaches the dominant eigenvector $\mathbf{x} = \begin{bmatrix} 1 \\ 1 \end{bmatrix}$.

From (10), it follows that the Rayleigh quotients can be also computed directly from the power sequence as follows:

$$\frac{A\mathbf{x}_k \cdot \mathbf{x}_k}{\mathbf{x}_k \cdot \mathbf{x}_k} = \frac{A(\frac{A^k\mathbf{x}_0}{\max(A^k\mathbf{x}_0)}) \cdot (\frac{A^k\mathbf{x}_0}{\max(A^k\mathbf{x}_0)})}{(\frac{A^k\mathbf{x}_0}{\max(A^k\mathbf{x}_0)}) \cdot (\frac{A^k\mathbf{x}_0}{\max(A^k\mathbf{x}_0)})} = \frac{A^{k+1}\mathbf{x}_0 \cdot A^k\mathbf{x}_0}{A^k\mathbf{x}_0 \cdot A^k\mathbf{x}_0}$$

Thus the first four Rayleigh quotients are

$$\lambda^{(1)} = \frac{A^2\mathbf{x}_0 \cdot A\mathbf{x}_0}{A\mathbf{x}_0 \cdot A\mathbf{x}_0} = \frac{13}{5} = 2.6, \qquad \lambda^{(2)} = \frac{A^3\mathbf{x}_0 \cdot A^2\mathbf{x}_0}{A^2\mathbf{x}_0 \cdot A^2\mathbf{x}_0} = \frac{121}{41} \approx 2.95122$$

$$\lambda^{(3)} = \frac{A^4\mathbf{x}_0 \cdot A^3\mathbf{x}_0}{A^3\mathbf{x}_0 \cdot A^3\mathbf{x}_0} = \frac{1093}{365} \approx 2.99452, \quad \lambda^{(4)} = \frac{A^5\mathbf{x}_0 \cdot A^4\mathbf{x}_0}{A^4\mathbf{x}_0 \cdot A^4\mathbf{x}_0} = \frac{9841}{3281} \approx 2.99939$$

This sequence approaches the dominant eigenvalue $\lambda = 3$.

9. The matrix $A = \begin{bmatrix} -1 & 0 \\ 0 & 0 \end{bmatrix}$ has $\lambda = -1$ and $\lambda = 0$ as its eigenvalues; thus the dominant eigenvalue, $\lambda = -1$, is negative. Let $\mathbf{x}_0 = \begin{bmatrix} a \\ b \end{bmatrix}$, with $a \neq 0$ and $a^2 + b^2 = 1$ (to make $\mathbf{x}_0$ a unit vector). Then $A\mathbf{x}_0 = \begin{bmatrix} -a \\ 0 \end{bmatrix}$; thus $\mathbf{x}_1 = \frac{A\mathbf{x}_0}{\|A\mathbf{x}_0\|} = \frac{1}{|a|}\begin{bmatrix} -a \\ 0 \end{bmatrix} = \begin{bmatrix} -1 \\ 0 \end{bmatrix}$ if $a > 0$, and $\mathbf{x}_1 = \begin{bmatrix} 1 \\ 0 \end{bmatrix}$ if $a < 0$. We then have $A\mathbf{x}_1 = \begin{bmatrix} 1 \\ 0 \end{bmatrix}$ if $a > 0$, and $A\mathbf{x}_1 = \begin{bmatrix} -1 \\ 0 \end{bmatrix}$ if $a < 0$; thus $\mathbf{x}_2 = \begin{bmatrix} 1 \\ 0 \end{bmatrix}$ if $a > 0$, and $\mathbf{x}_2 = \begin{bmatrix} -1 \\ 0 \end{bmatrix}$ if $a < 0$. This pattern continues. Thus in either case, $a > 0$ or $a < 0$, the normalized power sequence $\{\mathbf{x}_k\}$ alternates between $\begin{bmatrix} -1 \\ 0 \end{bmatrix}$ and $\begin{bmatrix} 1 \\ 0 \end{bmatrix}$, and because of this does not converge to a specific dominant eigenvector.

DISCUSSION AND DISCOVERY

D1. If $A = \begin{bmatrix} 0 & 1 \\ 1 & 0 \end{bmatrix}$, then the eigenvalues are $\lambda = \pm 1$, so there is no dominant eigenvalue. Starting with $\mathbf{x}_0 = \begin{bmatrix} a \\ b \end{bmatrix}$, we have

$$A\mathbf{x}_0 = \begin{bmatrix} b \\ a \end{bmatrix} \qquad \mathbf{x}_1 = \frac{A\mathbf{x}_0}{\|A\mathbf{x}_0\|} = \frac{1}{\sqrt{a^2 + b^2}}\begin{bmatrix} b \\ a \end{bmatrix}$$

$$A^2\mathbf{x}_0 = \begin{bmatrix} a \\ b \end{bmatrix} \qquad \mathbf{x}_2 = \frac{A^2\mathbf{x}_0}{\|A^2\mathbf{x}_0\|} = \frac{1}{\sqrt{a^2 + b^2}}\begin{bmatrix} a \\ b \end{bmatrix}$$

$$A^3\mathbf{x}_0 = \begin{bmatrix} b \\ a \end{bmatrix} \qquad \mathbf{x}_3 = \frac{A^3\mathbf{x}_0}{\|A^3\mathbf{x}_0\|} = \frac{1}{\sqrt{a^2 + b^2}}\begin{bmatrix} b \\ a \end{bmatrix}$$

$$A^4\mathbf{x}_0 = \begin{bmatrix} a \\ b \end{bmatrix} \qquad \mathbf{x}_4 = \frac{A^4\mathbf{x}_0}{\|A^4\mathbf{x}_0\|} = \frac{1}{\sqrt{a^2 + b^2}}\begin{bmatrix} a \\ b \end{bmatrix}$$

$$\vdots \qquad\qquad \vdots$$

Thus, unless $a = b$, the normalized power sequence $\mathbf{x}_1, \mathbf{x}_2, \mathbf{x}_3, \mathbf{x}_4, \ldots$ alternates between $\frac{1}{\sqrt{a^2+b^2}}\begin{bmatrix} b \\ a \end{bmatrix}$ and $\frac{1}{\sqrt{a^2+b^2}}\begin{bmatrix} a \\ b \end{bmatrix}$, and does not appoach an eigenvector of A. If $a = b$, then $\mathbf{x}_k = \frac{1}{\sqrt{2}}\begin{bmatrix} 1 \\ 1 \end{bmatrix}$ for all k; this is the unit eigenvector corresponding to $\lambda = 1$.

D2. The Rayleigh quotients will converge to the dominant eigenvalue $\lambda_4 = -8.1$. However, since the ratio $\frac{|\lambda_4|}{|\lambda_1|} = \frac{8.1}{8} = 1.0125$ is very close to 1, the rate of convergence can be expected to be quite slow.

WORKING WITH PROOFS

P1. First we note that, for every vector $\mathbf{x}$ in R^n, we have

$$A^T A \mathbf{x} \cdot \mathbf{x} = (A^T A \mathbf{x})^T \mathbf{x} = (\mathbf{x}^T A^T A) \mathbf{x} = (\mathbf{x}^T A^T)(A \mathbf{x}) = (A \mathbf{x})^T (A \mathbf{x}) = (A \mathbf{x}) \cdot (A \mathbf{x}) = \| A \mathbf{x} \|^2$$

It follows that if $A^T A \mathbf{x} = \lambda \mathbf{x}$ with $\mathbf{x} \neq \mathbf{0}$, then $\| A \mathbf{x} \|^2 = A^T A \mathbf{x} \cdot \mathbf{x} = \lambda \mathbf{x} \cdot \mathbf{x} = \lambda \| \mathbf{x} \|^2$ and so $\lambda \geq 0$. Thus the eigenvalues of $A^T A$ must be nonnegative. Similarly, one shows that the eigenvalues of $A A^T$ are nonnegative. Next we note that the matrices $A^T A$ and $A A^T$ are both symmetric. Thus, from the comment following Example 1, each of these matrices has n linearly independent eigenvectors. Finally, since $A \neq 0$, the corresponding eigenvalues cannot all be zero. Thus each of these matrices has a largest positive eigenvalue, and this is its dominant eigenvalue.

P2. We must show that $\mathbf{x}_k = \frac{A^k \mathbf{x}_0}{\| A^k \mathbf{x}_0 \|}$ for each positive integer k. The proof is by induction on k.

Step 1. The statement is clearly true for $k = 1$, since $\mathbf{x}_1 = \frac{A \mathbf{x}_0}{\| A \mathbf{x}_0 \|} = \frac{A^1 \mathbf{x}_0}{\| A^1 \mathbf{x}_0 \|}$.

Step 2 (induction step). Suppose the statement is true for a fixed integer $j \geq 1$. Then

$$\mathbf{x}_{j+1} = \frac{A \mathbf{x}_j}{\| A \mathbf{x}_j \|} = \frac{A \left(\dfrac{A^j \mathbf{x}_0}{\| A^j \mathbf{x}_0 \|} \right)}{\left\| A \left(\dfrac{A^j \mathbf{x}_0}{\| A^j \mathbf{x}_0 \|} \right) \right\|} = \frac{\dfrac{1}{\| A^j \mathbf{x}_0 \|} A(A^j \mathbf{x}_0)}{\dfrac{1}{\| A^j \mathbf{x}_0 \|} \| A(A^j \mathbf{x}_0) \|} = \frac{A^{j+1} \mathbf{x}_0}{\| A^{j+1} \mathbf{x}_0 \|}$$

and so the statement is also true for $j + 1$. This shows that if the statement is true for a fixed integer $k = j$, it is also true for $k = j + 1$. These two steps complete the proof by induction.

P3. We must show that $\mathbf{x}_k = \frac{A^k \mathbf{x}_0}{\max(A^k \mathbf{x}_0)}$ for each positive integer k. The proof is by induction on k.

Step 1. The statement is clearly true for $k = 1$, since $\mathbf{x}_1 = \frac{A \mathbf{x}_0}{\max(A \mathbf{x}_0)} = \frac{A^1 \mathbf{x}_0}{\max(A^1 \mathbf{x}_0)}$.

Step 2 (induction step). Suppose the statement is true for a fixed integer $k = j$, where $j \geq 1$. Then

$$\mathbf{x}_{j+1} = \frac{A \mathbf{x}_j}{\max(A \mathbf{x}_j)} = \frac{A \left(\dfrac{A^j \mathbf{x}_0}{\max(A^j \mathbf{x}_0)} \right)}{\max \left(A \left(\dfrac{A^j \mathbf{x}_0}{\max(A^j \mathbf{x}_0)} \right) \right)} = \frac{\dfrac{1}{\max(A^j \mathbf{x}_0)} A(A^j \mathbf{x}_0)}{\dfrac{1}{\max(A^j \mathbf{x}_0)} \max(A(A^j \mathbf{x}_0))} = \frac{A^{j+1} \mathbf{x}_0}{\max(A^{j+1} \mathbf{x}_0)}$$

and so the statement is also true for $k = j + 1$. This shows that if the statement is true for an integer $k = j$, it is also true for $k = j + 1$. These two steps complete the proof by induction.

CHAPTER 6
Linear Transformations

EXERCISE SET 6.1

1. (a) $T_A : R^2 \to R^3$; domain $= R^2$, codomain $= R^3$

(b) $T_A : R^3 \to R^2$; domain $= R^3$, codomain $= R^2$

(c) $T_A : R^3 \to R^3$; domain $= R^3$, codomain $= R^3$

3. The domain of T is R^2, the codomain of T is R^3, and $T(1, -2) = (-1, 2, 3)$.

5. (a) $T(\mathbf{x}) = \begin{bmatrix} 1 & 2 \\ 3 & 4 \end{bmatrix} \begin{bmatrix} 3 \\ -2 \end{bmatrix} = \begin{bmatrix} -1 \\ 1 \end{bmatrix}$
(b) $T(\mathbf{x}) = \begin{bmatrix} -1 & 2 & 0 \\ 3 & 1 & 5 \end{bmatrix} \begin{bmatrix} -1 \\ 1 \\ 3 \end{bmatrix} = \begin{bmatrix} 3 \\ 13 \end{bmatrix}$

7. (a) We have $T_A(\mathbf{x}) = \mathbf{b}$ if and only if $\mathbf{x}$ is a solution of the linear system

$$\begin{bmatrix} 1 & 2 & 0 \\ 0 & -1 & 3 \\ 2 & 5 & -3 \end{bmatrix} \begin{bmatrix} x_1 \\ x_2 \\ x_3 \end{bmatrix} = \begin{bmatrix} 1 \\ -1 \\ 3 \end{bmatrix}$$

The reduced row echelon form of the augmented matrix of the above system is

$$\begin{bmatrix} 1 & 0 & 6 & -1 \\ 0 & 1 & -3 & 1 \\ 0 & 0 & 0 & 0 \end{bmatrix}$$

and it follows that the system has the general solution $x_1 = -1 - 6t$, $x_2 = 1 + 3t$, $x_3 = t$.

Thus any vector of the form $\mathbf{x} = \begin{bmatrix} -1 \\ 1 \\ 0 \end{bmatrix} + t \begin{bmatrix} -6 \\ 3 \\ 1 \end{bmatrix}$ will have the property that $T_A(\mathbf{x}) = \begin{bmatrix} 1 \\ -1 \\ 3 \end{bmatrix}$.

(b) We have $T_A(\mathbf{x}) = \mathbf{b}$ if and only if $\mathbf{x}$ is a solution of the linear system

$$\begin{bmatrix} 1 & 2 & 0 \\ 0 & -1 & 3 \\ 2 & 5 & -3 \end{bmatrix} \begin{bmatrix} x_1 \\ x_2 \\ x_3 \end{bmatrix} = \begin{bmatrix} 2 \\ 1 \\ 1 \end{bmatrix}$$

The reduced row echelon form of the augmented matrix of the above system is

$$\begin{bmatrix} 1 & 0 & 6 & -1 \\ 0 & 1 & -3 & 1 \\ 0 & 0 & 0 & 1 \end{bmatrix}$$

and from the last row we see that the system is inconsistent. Thus there is no vector $\mathbf{x}$ in R^3 for which $T_A(\mathbf{x}) = \begin{bmatrix} 2 \\ 1 \\ 1 \end{bmatrix}$.

9. (a), (c), and (d) are linear transformations. (b) is not linear; neither homogeneous nor additive.

11. (a) and (c) are linear transformations. (b) is not linear; neither homogeneous nor additive.

13. This transformation can be written in matrix form as $\begin{bmatrix} w_1 \\ w_2 \end{bmatrix} = \begin{bmatrix} 3 & -2 & 4 \\ 5 & -8 & 1 \end{bmatrix} \begin{bmatrix} x_1 \\ x_2 \\ x_3 \end{bmatrix}$; it is a linear transformation with domain R^3 and codomain R^2.

15. This transformation can be written in matrix form as $\begin{bmatrix} w_1 \\ w_2 \\ w_3 \end{bmatrix} = \begin{bmatrix} 5 & -1 & 1 \\ -1 & 1 & 7 \\ 2 & -4 & -1 \end{bmatrix} \begin{bmatrix} x_1 \\ x_2 \\ x_3 \end{bmatrix}$; it is a linear transformation with domain R^3 and codomain R^3.

17. $[T] = [T(\mathbf{e}_1) \ \ T(\mathbf{e}_2)] = \begin{bmatrix} 3 & -1 \\ 2 & 3 \\ 4 & 0 \end{bmatrix}$

19. (a) We have $T(1,0) = (-1,0)$ and $T(0,1) = (1,1)$; thus the standard matrix is $[T] = \begin{bmatrix} -1 & 1 \\ 0 & 1 \end{bmatrix}$. Using the matrix, we have $T\left(\begin{bmatrix} -1 \\ 4 \end{bmatrix}\right) = \begin{bmatrix} -1 & 1 \\ 0 & 1 \end{bmatrix} \begin{bmatrix} -1 \\ 4 \end{bmatrix} = \begin{bmatrix} 5 \\ 4 \end{bmatrix}$. This agrees with direct calculation using the given formula: $T(-1,4) = (-(-1)+4, 4) = (5,4)$.

(b) We have $T(1,0,0) = (2,0,0)$, $T(0,1,0) = (-1,1,0)$, and $T(0,0,1) = (1,1,0)$; thus the standard matrix is $[T] = \begin{bmatrix} 2 & -1 & 1 \\ 0 & 1 & 1 \\ 0 & 0 & 0 \end{bmatrix}$. Using the matrix, we have $T\left(\begin{bmatrix} 2 \\ 1 \\ -3 \end{bmatrix}\right) = \begin{bmatrix} 2 & -1 & 1 \\ 0 & 1 & 1 \\ 0 & 0 & 0 \end{bmatrix} \begin{bmatrix} 2 \\ 1 \\ -3 \end{bmatrix} = \begin{bmatrix} 0 \\ -2 \\ 0 \end{bmatrix}$. This agrees with direct calculation using the given formula:

$$T(2,1,-3) = (2(2) - (1) + (-3), 1 + (-3), 0) = (0, -2, 0)$$

21. (a) The standard matrix of the transformation is $[T] = \begin{bmatrix} 3 & 5 & -1 \\ 4 & -1 & 1 \\ 3 & 2 & -1 \end{bmatrix}$.

(b) If $\mathbf{x} = (-1, 2, 4)$ then, using the equations, we have

$$T(\mathbf{x}) = (3(-1) + 5(2) - (4), 4(-1) - (2) + (4), 3(-1) + 2(2) - (4)) = (3, -2, -3)$$

On the other hand, using the matrix, we have

$$T(\mathbf{x}) = T\left(\begin{bmatrix} -1 \\ 2 \\ 4 \end{bmatrix}\right) = \begin{bmatrix} 3 & 5 & -1 \\ 4 & -1 & 1 \\ 3 & 2 & -1 \end{bmatrix} \begin{bmatrix} -1 \\ 2 \\ 4 \end{bmatrix} = \begin{bmatrix} 3 \\ -2 \\ -3 \end{bmatrix}$$

23. (a) $\begin{bmatrix} 1 & 0 \\ 0 & -1 \end{bmatrix} \begin{bmatrix} -2 \\ 1 \end{bmatrix} = \begin{bmatrix} -2 \\ -1 \end{bmatrix}$

(b) $\begin{bmatrix} 0 & -1 \\ -1 & 0 \end{bmatrix} \begin{bmatrix} -2 \\ 1 \end{bmatrix} = \begin{bmatrix} -1 \\ 2 \end{bmatrix}$

(c) $\begin{bmatrix} 1 & 0 \\ 0 & 0 \end{bmatrix} \begin{bmatrix} -2 \\ 1 \end{bmatrix} = \begin{bmatrix} -2 \\ 0 \end{bmatrix}$

(d) $\begin{bmatrix} 0 & 0 \\ 0 & 1 \end{bmatrix} \begin{bmatrix} -2 \\ 1 \end{bmatrix} = \begin{bmatrix} 0 \\ 1 \end{bmatrix}$

25. (a) $\begin{bmatrix} \frac{1}{\sqrt{2}} & -\frac{1}{\sqrt{2}} \\ \frac{1}{\sqrt{2}} & \frac{1}{\sqrt{2}} \end{bmatrix} \begin{bmatrix} 3 \\ 4 \end{bmatrix} = \begin{bmatrix} -\frac{1}{\sqrt{2}} \\ \frac{7}{\sqrt{2}} \end{bmatrix} \approx \begin{bmatrix} -0.707 \\ 4.950 \end{bmatrix}$

(b) $\begin{bmatrix} 0 & -1 \\ 1 & 0 \end{bmatrix} \begin{bmatrix} 3 \\ 4 \end{bmatrix} = \begin{bmatrix} -4 \\ 3 \end{bmatrix}$

(c) $\begin{bmatrix} -1 & 0 \\ 0 & -1 \end{bmatrix} \begin{bmatrix} 3 \\ 4 \end{bmatrix} = \begin{bmatrix} -3 \\ -4 \end{bmatrix}$

(d) $\begin{bmatrix} \frac{\sqrt{3}}{2} & \frac{1}{2} \\ -\frac{1}{2} & \frac{\sqrt{3}}{2} \end{bmatrix} \begin{bmatrix} 3 \\ 4 \end{bmatrix} = \begin{bmatrix} \frac{3\sqrt{3}}{2} + 2 \\ -\frac{3}{2} + 2\sqrt{3} \end{bmatrix} \approx \begin{bmatrix} 4.598 \\ 1.964 \end{bmatrix}$

27. (a) $\begin{bmatrix} -\frac{1}{2} & \frac{\sqrt{3}}{2} \\ \frac{\sqrt{3}}{2} & \frac{1}{2} \end{bmatrix} \begin{bmatrix} 3 \\ 4 \end{bmatrix} = \begin{bmatrix} -\frac{3}{2} + 2\sqrt{3} \\ \frac{3\sqrt{3}}{2} + 2 \end{bmatrix} \approx \begin{bmatrix} 1.964 \\ 4.598 \end{bmatrix}$ (b) $\begin{bmatrix} \frac{1}{4} & \frac{\sqrt{3}}{4} \\ \frac{\sqrt{3}}{4} & \frac{3}{4} \end{bmatrix} \begin{bmatrix} 3 \\ 4 \end{bmatrix} = \begin{bmatrix} \frac{3}{4} + \sqrt{3} \\ \frac{3\sqrt{3}}{4} + 3 \end{bmatrix} \approx \begin{bmatrix} 2.482 \\ 4.299 \end{bmatrix}$

29. The matrix $A = \begin{bmatrix} -\frac{1}{\sqrt{2}} & -\frac{1}{\sqrt{2}} \\ \frac{1}{\sqrt{2}} & -\frac{1}{\sqrt{2}} \end{bmatrix}$ corresponds to $R_\theta = \begin{bmatrix} \cos\theta & -\sin\theta \\ \sin\theta & \cos\theta \end{bmatrix}$ where $\theta = \frac{3\pi}{4}(135°)$.

31. (a) $H_L = \begin{bmatrix} \cos 2\theta & \sin 2\theta \\ \sin 2\theta & -\cos 2\theta \end{bmatrix} = \begin{bmatrix} \cos^2\theta - \sin^2\theta & 2\sin\theta\cos\theta \\ 2\sin\theta\cos\theta & \sin^2\theta - \cos^2\theta \end{bmatrix}$ where $\cos\theta = \frac{1}{\sqrt{1+m^2}}$ and $\sin\theta = \frac{m}{\sqrt{1+m^2}}$.

Thus $H_L = \frac{1}{1+m^2}\begin{bmatrix} 1-m^2 & 2m \\ 2m & m^2-1 \end{bmatrix}$.

(b) $P_L = \begin{bmatrix} \cos^2\theta & \sin\theta\cos\theta \\ \sin\theta\cos\theta & \sin^2\theta \end{bmatrix} = \frac{1}{1+m^2}\begin{bmatrix} 1 & m \\ m & m^2 \end{bmatrix}$

33. (a) We have $m = 3$; thus $H = H_L = \frac{1}{10}\begin{bmatrix} -8 & 6 \\ 6 & 8 \end{bmatrix} = \frac{1}{5}\begin{bmatrix} -4 & 3 \\ 3 & 4 \end{bmatrix}$ and the reflection of $\mathbf{x} = \begin{bmatrix} 4 \\ 3 \end{bmatrix}$ about

the line $y = 3x$ is given by $H\left(\begin{bmatrix} 4 \\ 3 \end{bmatrix}\right) = \frac{1}{5}\begin{bmatrix} -4 & 3 \\ 3 & 4 \end{bmatrix}\begin{bmatrix} 4 \\ 3 \end{bmatrix} = \frac{1}{5}\begin{bmatrix} -7 \\ 24 \end{bmatrix} = \begin{bmatrix} -1.4 \\ 4.8 \end{bmatrix}$.

(b) We have $m = 3$; thus $P = P_L = \frac{1}{10}\begin{bmatrix} 1 & 3 \\ 3 & 9 \end{bmatrix}$ and $P\left(\begin{bmatrix} 4 \\ 3 \end{bmatrix}\right) = \frac{1}{10}\begin{bmatrix} 1 & 3 \\ 3 & 9 \end{bmatrix}\begin{bmatrix} 4 \\ 3 \end{bmatrix} = \frac{1}{10}\begin{bmatrix} 13 \\ 39 \end{bmatrix} = \begin{bmatrix} 1.3 \\ 3.9 \end{bmatrix}$.

35. (a) $T_A(\mathbf{e}_1) = \mathbf{c}_1(A) = \begin{bmatrix} -1 \\ 2 \\ 4 \end{bmatrix}$ $\quad T_A(\mathbf{e}_2) = \mathbf{c}_2(A) = \begin{bmatrix} 3 \\ 1 \\ 5 \end{bmatrix}$ $\quad T_A(\mathbf{e}_3) = \mathbf{c}_3(A) = \begin{bmatrix} 0 \\ 2 \\ -3 \end{bmatrix}$

(b) $T_A(\mathbf{e}_1 + \mathbf{e}_2 + \mathbf{e}_3) = T_A(\mathbf{e}_1) + T_A(\mathbf{e}_2) + T_A(\mathbf{e}_3) = \begin{bmatrix} -1 \\ 2 \\ 4 \end{bmatrix} + \begin{bmatrix} 3 \\ 1 \\ 5 \end{bmatrix} + \begin{bmatrix} 0 \\ 2 \\ -3 \end{bmatrix} = \begin{bmatrix} 2 \\ 5 \\ 6 \end{bmatrix}$

(c) $T_A(7\mathbf{e}_3) = 7T_A(\mathbf{e}_3) = 7\begin{bmatrix} 0 \\ 2 \\ -3 \end{bmatrix} = \begin{bmatrix} 0 \\ 14 \\ -21 \end{bmatrix}$

37. (a) $[T] = \begin{bmatrix} 1 & 0 & 0 \\ 0 & 0 & 1 \\ 0 & 1 & 0 \end{bmatrix}$ (b) $[T] = \begin{bmatrix} 0 & 1 & 0 \\ 1 & 0 & 0 \\ 0 & 0 & 1 \end{bmatrix}$ (c) $[T] = \begin{bmatrix} 0 & 0 & 1 \\ 0 & 1 & 0 \\ 1 & 0 & 0 \end{bmatrix}$

39. $T(x, y) = (-x, 0)$; thus $[T] = \begin{bmatrix} -1 & 0 \\ 0 & 0 \end{bmatrix}$.

DISCUSSION AND DISCOVERY

D1. (a) False. For example, $T(x, y) = (x^2, y^2)$ satisfies $T(\mathbf{0}) = \mathbf{0}$ but is not linear.

(b) True. Such a transformation is both homogeneous and additive.

(c) True. This is the transformation with standard matrix $[T] = -I$.

(d) True. The zero transformation $T(x, y) = (0, 0)$ is the only linear transformation with this property.

(e) False. Such a transformation cannot be linear since $T(\mathbf{0}) = \mathbf{v}_0 \neq \mathbf{0}$.

D2. The eigenvalues of A are $\lambda = 1$ and $\lambda = -1$ with corresponding eigenspaces given (respectively)

by $t\begin{bmatrix} 1 \\ 1 \end{bmatrix}$ and $t\begin{bmatrix} -1 \\ 1 \end{bmatrix}$, where $-\infty < t < \infty$.

D3. From familiar trigonometric identities, we have $A = \begin{bmatrix} \cos 2\theta & -\sin 2\theta \\ \sin 2\theta & \cos 2\theta \end{bmatrix} = R_{2\theta}$. Thus multiplication by A corresponds to rotation about the origin through the angle 2θ.

D4. If $A = R_\theta = \begin{bmatrix} \cos \theta & -\sin \theta \\ \sin \theta & \cos \theta \end{bmatrix}$, then $A^T = \begin{bmatrix} \cos \theta & \sin \theta \\ -\sin \theta & \cos \theta \end{bmatrix} = \begin{bmatrix} \cos(-\theta) & -\sin(-\theta) \\ \sin(-\theta) & \cos(-\theta) \end{bmatrix} = R_{-\theta}$. Thus multiplication by A^T corresponds to rotation through the angle $-\theta$.

D5. Since $T(0) = \mathbf{x}_0 \neq \mathbf{0}$, this transformation is not linear. Geometrically, it corresponds to a rotation followed by a translation.

D6. If $b = 0$, then f is both additive and homogeneous. If $b \neq 0$, then f is neither additive nor homogeneous.

D7. Since T is linear, we have $T(\mathbf{x}_0 + t\mathbf{v}) = T(\mathbf{x}_0) + tT(\mathbf{v})$. Thus, if $T(\mathbf{v}) \neq \mathbf{0}$, the image of the line $\mathbf{x} = \mathbf{x}_0 + t\mathbf{v}$ is the line $\mathbf{y} = \mathbf{y}_0 + t\mathbf{w}$ where $\mathbf{y}_0 = T(\mathbf{x}_0)$ and $\mathbf{w} = T(\mathbf{v})$. If $T(\mathbf{v}) = \mathbf{0}$, then the image of $\mathbf{x} = \mathbf{x}_0 + t\mathbf{v}$ is the point $\mathbf{y}_0 = T(\mathbf{x}_0)$.

EXERCISE SET 6.2

1. $A^T A = \begin{bmatrix} \frac{3}{5} & \frac{4}{5} \\ -\frac{4}{5} & \frac{3}{5} \end{bmatrix} \begin{bmatrix} \frac{3}{5} & -\frac{4}{5} \\ \frac{4}{5} & \frac{3}{5} \end{bmatrix} = \begin{bmatrix} 1 & 0 \\ 0 & 1 \end{bmatrix}$; thus A is orthogonal and $A^{-1} = A^T = \begin{bmatrix} \frac{3}{5} & \frac{4}{5} \\ -\frac{4}{5} & \frac{3}{5} \end{bmatrix}$.

3. $A^T A = \begin{bmatrix} \frac{4}{5} & -\frac{9}{25} & \frac{12}{25} \\ 0 & \frac{4}{5} & \frac{3}{5} \\ -\frac{3}{5} & -\frac{12}{25} & \frac{16}{25} \end{bmatrix} \begin{bmatrix} \frac{4}{5} & 0 & -\frac{3}{5} \\ -\frac{9}{25} & \frac{4}{5} & -\frac{12}{25} \\ \frac{12}{25} & \frac{3}{5} & \frac{16}{25} \end{bmatrix} = \begin{bmatrix} 1 & 0 & 0 \\ 0 & 1 & 0 \\ 0 & 0 & 1 \end{bmatrix}$; thus A is orthogonal and $A^{-1} = A^T =$

$\begin{bmatrix} \frac{4}{5} & -\frac{9}{25} & \frac{12}{25} \\ 0 & \frac{4}{5} & \frac{3}{5} \\ -\frac{3}{5} & -\frac{12}{25} & \frac{16}{25} \end{bmatrix}$.

5. (a) $A^T A = \begin{bmatrix} -\frac{1}{\sqrt{2}} & -\frac{1}{\sqrt{2}} \\ \frac{1}{\sqrt{2}} & -\frac{1}{\sqrt{2}} \end{bmatrix} \begin{bmatrix} -\frac{1}{\sqrt{2}} & \frac{1}{\sqrt{2}} \\ -\frac{1}{\sqrt{2}} & -\frac{1}{\sqrt{2}} \end{bmatrix} = \begin{bmatrix} 1 & 0 \\ 0 & 1 \end{bmatrix}$; thus A is orthogonal. We have $\det(A) = 1$, and

$A = \begin{bmatrix} -\frac{1}{\sqrt{2}} & \frac{1}{\sqrt{2}} \\ -\frac{1}{\sqrt{2}} & -\frac{1}{\sqrt{2}} \end{bmatrix} = R_\theta$ where $\theta = \frac{3\pi}{4}$. Thus multiplication by A corresponds to counterclockwise rotation about the origin through the angle $\frac{3\pi}{4}$.

(b) $A^T A = \begin{bmatrix} -\frac{1}{2} & \frac{\sqrt{3}}{2} \\ \frac{\sqrt{3}}{2} & \frac{1}{2} \end{bmatrix} \begin{bmatrix} -\frac{1}{2} & \frac{\sqrt{3}}{2} \\ \frac{\sqrt{3}}{2} & \frac{1}{2} \end{bmatrix} = \begin{bmatrix} 1 & 0 \\ 0 & 1 \end{bmatrix}$; thus A is orthogonal. We have $\det(A) = -1$, and so

$A = \begin{bmatrix} -\frac{1}{2} & \frac{\sqrt{3}}{2} \\ \frac{\sqrt{3}}{2} & \frac{1}{2} \end{bmatrix} = H_\theta$ where $\theta = \frac{\pi}{3}$. Thus multiplication by A corresponds to reflection about the line origin through the origin making an angle $\frac{\pi}{3}$ with the positive x-axis.

7. (a) $A = \begin{bmatrix} \frac{1}{5} & 0 \\ 0 & \frac{1}{5} \end{bmatrix}$ **(b)** $A = \begin{bmatrix} \frac{1}{3} & 0 \\ 0 & 1 \end{bmatrix}$ **(c)** $A = \begin{bmatrix} 1 & 0 \\ 0 & 6 \end{bmatrix}$ **(d)** $A = \begin{bmatrix} 1 & 3 \\ 0 & 1 \end{bmatrix}$

9. (a) Expansion in the x-direction with factor 3. **(b)** Contraction with factor $\frac{1}{4}$.
(c) Shear in the x-direction with factor 4. **(d)** Shear in the y-direction with factor -4.

11. The action of T on the standard unit vectors is $e_1 = \begin{bmatrix} 1 \\ 0 \end{bmatrix} \rightarrow \begin{bmatrix} 3 \\ 0 \end{bmatrix} \rightarrow \begin{bmatrix} 0 \\ 3 \end{bmatrix} \rightarrow \begin{bmatrix} 0 \\ 3 \end{bmatrix} = Te_1$ and $e_2 = $

$\begin{bmatrix} 0 \\ 1 \end{bmatrix} \rightarrow \begin{bmatrix} 0 \\ 3 \end{bmatrix} \rightarrow \begin{bmatrix} 3 \\ 0 \end{bmatrix} \rightarrow \begin{bmatrix} 0 \\ 0 \end{bmatrix} = Te_2$; thus $[T] = [Te_1 \ Te_2] = \begin{bmatrix} 0 & 0 \\ 3 & 0 \end{bmatrix}$.

13. The action of T on the standard unit vectors is $e_1 = \begin{bmatrix} 1 \\ 0 \\ 0 \end{bmatrix} \rightarrow \begin{bmatrix} 1 \\ 0 \\ 0 \end{bmatrix} \rightarrow \begin{bmatrix} 1 \\ 0 \\ 0 \end{bmatrix} = Te_1$, $e_2 = \begin{bmatrix} 0 \\ 1 \\ 0 \end{bmatrix} \rightarrow \begin{bmatrix} 0 \\ 0 \\ 0 \end{bmatrix} \rightarrow$

$\begin{bmatrix} 0 \\ 0 \\ 0 \end{bmatrix} = Te_2$, and $e_3 = \begin{bmatrix} 0 \\ 0 \\ 1 \end{bmatrix} \rightarrow \begin{bmatrix} 0 \\ 0 \\ 1 \end{bmatrix} \rightarrow \begin{bmatrix} 0 \\ 0 \\ 0 \end{bmatrix} = Te_3$; thus $[T] = \begin{bmatrix} 1 & 0 & 0 \\ 0 & 0 & 0 \\ 0 & 0 & 0 \end{bmatrix}$.

15. (a) (b) (c)

17. (a) (b) (c)

19. (a) $\begin{bmatrix} 1 & 0 & 0 \\ 0 & 1 & 0 \\ 0 & 0 & -1 \end{bmatrix} \begin{bmatrix} 2 \\ 5 \\ 3 \end{bmatrix} = \begin{bmatrix} 2 \\ 5 \\ -3 \end{bmatrix}$ (b) $\begin{bmatrix} 1 & 0 & 0 \\ 0 & -1 & 0 \\ 0 & 0 & 1 \end{bmatrix} \begin{bmatrix} 2 \\ 5 \\ 3 \end{bmatrix} = \begin{bmatrix} 2 \\ -5 \\ 3 \end{bmatrix}$ (c) $\begin{bmatrix} -1 & 0 & 0 \\ 0 & 1 & 0 \\ 0 & 0 & 1 \end{bmatrix} \begin{bmatrix} 2 \\ 5 \\ 3 \end{bmatrix} = \begin{bmatrix} -2 \\ 5 \\ 3 \end{bmatrix}$

21. (a) $[T] = \begin{bmatrix} 1 & 0 & 0 \\ 0 & 0 & -1 \\ 0 & 1 & 0 \end{bmatrix}$ (b) $[T] = \begin{bmatrix} 0 & 0 & 1 \\ 0 & 1 & 0 \\ -1 & 0 & 0 \end{bmatrix}$ (c) $[T] = \begin{bmatrix} 0 & 1 & 0 \\ -1 & 0 & 0 \\ 0 & 0 & 1 \end{bmatrix}$

23. (a) $[R] = \begin{bmatrix} 1 & 0 & 0 \\ 0 & \frac{\sqrt{3}}{2} & -\frac{1}{2} \\ 0 & \frac{1}{2} & \frac{\sqrt{3}}{2} \end{bmatrix}$; thus $R\left(\begin{bmatrix} -2 \\ 1 \\ 2 \end{bmatrix} \right) = \begin{bmatrix} 1 & 0 & 0 \\ 0 & \frac{\sqrt{3}}{2} & -\frac{1}{2} \\ 0 & \frac{1}{2} & \frac{\sqrt{3}}{2} \end{bmatrix} \begin{bmatrix} -2 \\ 1 \\ 2 \end{bmatrix} = \begin{bmatrix} -2 \\ \frac{\sqrt{3}}{2} - 1 \\ \frac{1}{2} + \sqrt{3} \end{bmatrix}$.

(b) $[R] = \begin{bmatrix} \frac{1}{\sqrt{2}} & 0 & \frac{1}{\sqrt{2}} \\ 0 & 1 & 0 \\ -\frac{1}{\sqrt{2}} & 0 & \frac{1}{\sqrt{2}} \end{bmatrix}$; thus $R\left(\begin{bmatrix} -2 \\ 1 \\ 2 \end{bmatrix} \right) = \begin{bmatrix} \frac{1}{\sqrt{2}} & 0 & \frac{1}{\sqrt{2}} \\ 0 & 1 & 0 \\ -\frac{1}{\sqrt{2}} & 0 & \frac{1}{\sqrt{2}} \end{bmatrix} \begin{bmatrix} -2 \\ 1 \\ 2 \end{bmatrix} = \begin{bmatrix} 0 \\ 1 \\ 2\sqrt{2} \end{bmatrix}$.

(c) $[R] = \begin{bmatrix} \frac{1}{2} & \frac{\sqrt{3}}{2} & 0 \\ -\frac{\sqrt{3}}{2} & \frac{1}{2} & 0 \\ 0 & 0 & 1 \end{bmatrix}$; thus $R\left(\begin{bmatrix} -2 \\ 1 \\ 2 \end{bmatrix} \right) = \begin{bmatrix} \frac{1}{2} & \frac{\sqrt{3}}{2} & 0 \\ -\frac{\sqrt{3}}{2} & \frac{1}{2} & 0 \\ 0 & 0 & 1 \end{bmatrix} \begin{bmatrix} -2 \\ 1 \\ 2 \end{bmatrix} = \begin{bmatrix} -1 + \frac{\sqrt{3}}{2} \\ \sqrt{3} + \frac{1}{2} \\ 2 \end{bmatrix}$.

25. The matrix A is a rotation matrix since it is orthogonal and $\det(A) = 1$. The axis of rotation is found by solving the system $(I - A)\mathbf{x} = \mathbf{0}$, which is $\begin{bmatrix} 0 & 0 & 0 \\ 0 & 1 & 1 \\ 0 & -1 & 1 \end{bmatrix} \begin{bmatrix} x \\ y \\ z \end{bmatrix} = \begin{bmatrix} 0 \\ 0 \\ 0 \end{bmatrix}$. A general solution of this system is $\mathbf{x} = \begin{bmatrix} x \\ y \\ z \end{bmatrix} = t \begin{bmatrix} 1 \\ 0 \\ 0 \end{bmatrix}$; thus the matrix A corresponds to a rotation about the x-axis. Choosing the positive orientation and comparing with Table 6.2.6, we see that the angle of rotation is $\theta = \frac{\pi}{2}$.

27. We have $\text{tr}(A) = 1$, and so Formula (17) reduces to $\mathbf{v} = A\mathbf{x} + A^T\mathbf{x}$. Taking $\mathbf{x} = \mathbf{e}_1$, this results in

$$\mathbf{v} = A\mathbf{x} + A^T\mathbf{x} = (A + A^T)\mathbf{x} = \begin{bmatrix} 2 & 0 & 0 \\ 0 & 0 & 0 \\ 0 & 0 & 0 \end{bmatrix}\begin{bmatrix} 1 \\ 0 \\ 0 \end{bmatrix} = \begin{bmatrix} 2 \\ 0 \\ 0 \end{bmatrix}$$

From this we conclude that the x-axis is the axis of rotation. Finally, using Formula (16), the rotation angle is determined by $\cos\theta = \frac{\text{tr}(A)-1}{2} = 0$, and so $\theta = \frac{\pi}{2}$.

29. **(a)** We have $T_1\left(\begin{bmatrix} x \\ y \\ z \end{bmatrix}\right) = \begin{bmatrix} x \\ 0 \\ 0 \end{bmatrix} = \begin{bmatrix} 1 & 0 & 0 \\ 0 & 0 & 0 \\ 0 & 0 & 0 \end{bmatrix}\begin{bmatrix} x \\ y \\ z \end{bmatrix}$; thus T_1 is a linear operator and $M_1 = \begin{bmatrix} 1 & 0 & 0 \\ 0 & 0 & 0 \\ 0 & 0 & 0 \end{bmatrix}$.

Similarly, $M_2 = \begin{bmatrix} 0 & 0 & 0 \\ 0 & 1 & 0 \\ 0 & 0 & 0 \end{bmatrix}$ and $M_3 = \begin{bmatrix} 0 & 0 & 0 \\ 0 & 0 & 0 \\ 0 & 0 & 1 \end{bmatrix}$.

(b) If $\mathbf{x} = (x, y, z)$, then $T_1\mathbf{x} \cdot (\mathbf{x} - T_1\mathbf{x}) = (x, 0, 0) \cdot (0, y, z) = 0$; thus $T_1\mathbf{x}$ and $\mathbf{x} - T_1\mathbf{x}$ are orthogonal for every vector $\mathbf{x}$ in R^3. Similarly for T_2 and T_3.

31. If $\mathbf{u} = (1, 0, 0)$ then, on substituting $a = 1$ and $b = c = 0$ into Formula (13), we have

$$R_{\mathbf{u},\theta} = \begin{bmatrix} 1 & 0 & 0 \\ 0 & \cos\theta & -\sin\theta \\ 0 & \sin\theta & \cos\theta \end{bmatrix}$$

The other entries of Table 6.2.6 are obtained similarly.

DISCUSSION AND DISCOVERY

D1. **(a)** The unit square is mapped onto the segment $0 \le x \le 2$ along the x-axis ($y = 0$).

(b) The unit square is mapped onto the segment $0 \le y \le 3$ along the y-axis ($x = 0$).

(c) The unit square is mapped onto the rectangle $0 \le x \le 2$, $0 \le y \le 3$.

(d) The unit square is rotated about the origin through an angle $\theta = -\frac{\pi}{6}$ (30 degrees clockwise).

D2. In order for the matrix to be orthogonal we must have

$$a^2 + b^2 + c^2 = 1$$
$$\tfrac{1}{\sqrt{2}}a + \tfrac{1}{\sqrt{6}}b + \tfrac{1}{\sqrt{3}}c = 0$$
$$-\tfrac{1}{\sqrt{2}}a + \tfrac{1}{\sqrt{6}}b + \tfrac{1}{\sqrt{3}}c = 0$$

and from this it follows that $a = 0$, $b = -\frac{2}{\sqrt{6}}$, $c = \frac{1}{\sqrt{3}}$ or $a = 0$, $b = \frac{2}{\sqrt{6}}$, $c = -\frac{1}{\sqrt{3}}$. These are the only possibilities.

D3. The two columns of this matrix are orthogonal for any values of a and b. Thus, for the matrix to be orthogonal, all that is required is that the column vectors be of length 1. Thus a and b must satisfy $(a + b)^2 + (a - b)^2 = 1$, or (equivalently) $2a^2 + 2b^2 = 1$.

D4. If A is an orthogonal matrix and $A\mathbf{x} = \lambda\mathbf{x}$, then $\|\mathbf{x}\| = \|A\mathbf{x}\| = \|\lambda\mathbf{x}\| = |\lambda|\,\|\mathbf{x}\|$. Thus the eigenvalues of A (if any) must be of absolute value 1.

D5. **(a)** Vectors parallel to the line $y = x$ will be eigenvectors corresponding to the eigenvalue $\lambda = 1$. Vectors perpendicular to $y = x$ will be eigenvectors corresponding to the eigenvalue $\lambda = -1$.

(b) Every nonzero vector is an eigenvector corresponding to the eigenvalue $\lambda = \frac{1}{2}$.

D6. The shear in the x-direction with factor -2; thus $T(x,y) = (x - 2y, y)$ and $[T] = \begin{bmatrix} 1 & -2 \\ 0 & 1 \end{bmatrix}$.

D7. From the polarization identity, we have $\mathbf{x} \cdot \mathbf{y} = \frac{1}{4}(\|\mathbf{x} + \mathbf{y}\|^2 - \|\mathbf{x} - \mathbf{y}\|^2) = \frac{1}{4}(16 - 4) = 3$.

D8. If $\|\mathbf{x} + \mathbf{y}\| = \|\mathbf{x} - \mathbf{y}\|$ then the parallelogram having $\mathbf{x}$ and $\mathbf{y}$ as adjacent edges has diagonals of equal length and must therefore be a rectangle.

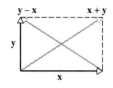

EXERCISE SET 6.3

1. **(a)** $\ker(T) = \{(0, y)| - \infty < y < \infty\}$ (y-axis), $\operatorname{ran}(T) = \{(x, 0)| - \infty < x < \infty\}$ (x-axis). The transformation T is neither 1-1 nor onto.

   (b) $\ker(T) = \{(x, 0, 0)| - \infty < x < \infty\}$ (x-axis), $\operatorname{ran}(T) = \{(0, y, z)| - \infty < y, z < \infty\}$ (yz-plane). The transformation T is neither 1-1 nor onto.

   (c) $\ker(T) = \left\{ \begin{bmatrix} 0 \\ 0 \end{bmatrix} \right\}$, $\operatorname{ran}(T) = R^2$. The transformation T is both 1-1 and onto.

   (d) $\ker(T) = \left\{ \begin{bmatrix} 0 \\ 0 \\ 0 \end{bmatrix} \right\}$, $\operatorname{ran}(T) = R^3$. The transformation T is both 1-1 and onto.

3. The kernel of the transformation is the solution set of the system $A\mathbf{x} = \mathbf{0}$. The augmented matrix of this system can be reduced to $\begin{bmatrix} 1 & 1 & 0 \\ 0 & 0 & 0 \end{bmatrix}$; thus the solution set consists of all vectors of the form $\mathbf{x} = t \begin{bmatrix} -1 \\ 1 \end{bmatrix}$ where $-\infty < t < \infty$.

5. The kernel of the transformation is the solution set of $\begin{bmatrix} 1 & 2 & -3 \\ -1 & -2 & 3 \\ 4 & 4 & 4 \end{bmatrix} \begin{bmatrix} x \\ y \\ z \end{bmatrix} = \begin{bmatrix} 0 \\ 0 \\ 0 \end{bmatrix}$. The augmented matrix of this system can be reduced to $\begin{bmatrix} 1 & 0 & 5 & | & 0 \\ 0 & 1 & -4 & | & 0 \\ 0 & 0 & 0 & | & 0 \end{bmatrix}$; thus the solution set consists of all vectors of the form $\mathbf{x} = t \begin{bmatrix} -5 \\ 4 \\ 1 \end{bmatrix}$ where $-\infty < t < \infty$.

7. The kernel of T is equal to the solution space of $\begin{bmatrix} 1 & 0 & -1 \\ 0 & 1 & -1 \\ 1 & -1 & 0 \\ 1 & 1 & 1 \end{bmatrix} \begin{bmatrix} x \\ y \\ z \end{bmatrix} = \begin{bmatrix} 0 \\ 0 \\ 0 \end{bmatrix}$. The augmented matrix of this system can be reduced to $\begin{bmatrix} 1 & 0 & 0 & | & 0 \\ 0 & 1 & 0 & | & 0 \\ 0 & 0 & 1 & | & 0 \\ 0 & 0 & 0 & | & 0 \end{bmatrix}$; thus the system has only the trivial solution and so $\ker(T) = \left\{ \begin{bmatrix} 0 \\ 0 \\ 0 \end{bmatrix} \right\}$.

9. (a) The vector $\mathbf{b}$ is in the column space of A if and only if the linear system $A\mathbf{x} = \mathbf{b}$ is consistent. The augmented matrix of $A\mathbf{x} = \mathbf{b}$ is

$$\left[\begin{array}{ccc|c} 1 & 2 & 0 & 1 \\ -2 & 2 & 1 & 1 \\ 1 & 8 & 1 & 1 \end{array}\right]$$

and the reduced row echelon form of this matrix is

$$\left[\begin{array}{ccc|c} 1 & 0 & -\frac{1}{3} & 0 \\ 0 & 1 & \frac{1}{6} & 0 \\ 0 & 0 & 0 & 1 \end{array}\right]$$

From this we conclude that the system is inconsistent; thus $\mathbf{b}$ is not in the column space of A.

(b) The augmented matrix of the system $A\mathbf{x} = \mathbf{b}$ is

$$\left[\begin{array}{ccc|c} 1 & 2 & 0 & 3 \\ -2 & 2 & 1 & -1 \\ 1 & 8 & 1 & 8 \end{array}\right]$$

and the reduced row echelon form of this matrix is

$$\left[\begin{array}{ccc|c} 1 & 0 & -\frac{1}{3} & \frac{4}{3} \\ 0 & 1 & \frac{1}{6} & \frac{5}{6} \\ 0 & 0 & 0 & 0 \end{array}\right]$$

From this we conclude that the system is consistent, with general solution

$$\mathbf{x} = \begin{bmatrix} \frac{4}{3} + \frac{1}{3}t \\ \frac{5}{6} - \frac{1}{6}t \\ t \end{bmatrix} = \begin{bmatrix} \frac{4}{3} \\ \frac{5}{6} \\ 0 \end{bmatrix} + t \begin{bmatrix} \frac{1}{3} \\ -\frac{1}{6} \\ 1 \end{bmatrix}$$

Taking $t = 0$, the vector $\mathbf{b}$ can expressed as a linear combination of the column vectors of A as follows:

$$\begin{bmatrix} 3 \\ -1 \\ 8 \end{bmatrix} = \mathbf{b} = \frac{4}{3}\mathbf{c}_1(A) + \frac{5}{6}\mathbf{c}_2(A) + 0\mathbf{c}_3(A) = \frac{4}{3}\begin{bmatrix} 1 \\ -2 \\ 1 \end{bmatrix} + \frac{5}{6}\begin{bmatrix} 2 \\ 2 \\ 8 \end{bmatrix}$$

11. The vector $\mathbf{w}$ is in the range of the linear operator T if and only if the linear system

$$\begin{aligned} 2x - y &= 3 \\ x + z &= 3 \\ y - z &= 0 \end{aligned}$$

is consistent. The augmented matrix of this system can be reduced to

$$\left[\begin{array}{ccc|c} 1 & 0 & 0 & 2 \\ 0 & 1 & 0 & 1 \\ 0 & 0 & 1 & 1 \end{array}\right]$$

Thus the system has a unique solution $\mathbf{x} = (2, 1, 1)$, and we have $T\mathbf{x} = T(2, 1, 1) = (3, 3, 0) = \mathbf{w}$.

13. The operator can be written as $\begin{bmatrix} w_1 \\ w_2 \end{bmatrix} = \begin{bmatrix} 2 & -3 \\ 5 & 1 \end{bmatrix} \begin{bmatrix} x_1 \\ x_2 \end{bmatrix}$; thus the standard matrix is $A = \begin{bmatrix} 2 & -3 \\ 5 & 1 \end{bmatrix}$.

Since $\det(A) = 17 \neq 0$, the operator is both 1-1 and onto.

15. The operator can be written as $\begin{bmatrix} w_1 \\ w_2 \\ w_3 \end{bmatrix} = \begin{bmatrix} -1 & 3 & 2 \\ 2 & 0 & 4 \\ 1 & 3 & 6 \end{bmatrix} \begin{bmatrix} x_1 \\ x_2 \\ x_3 \end{bmatrix}$; thus the standard matrix is $A = \begin{bmatrix} -1 & 3 & 2 \\ 2 & 0 & 4 \\ 1 & 3 & 6 \end{bmatrix}$. Since $\det(A) = 0$, the operator is neither 1-1 nor onto.

17. The operator can be written as $\mathbf{w} = T_A \mathbf{x}$, where $A = \begin{bmatrix} 4 & -2 \\ 2 & -1 \end{bmatrix}$. Since $\det(A) = 0$, T_A is not onto.

The range of T_A consists of all vectors of the form $\mathbf{w} = t \begin{bmatrix} 2 \\ 1 \end{bmatrix}$ where $-\infty < t < \infty$. In particular, $\mathbf{w} = \begin{bmatrix} 1 \\ 1 \end{bmatrix}$ is not in the range of T_A.

19. (a) The linear transformation $T_A : R^2 \to R^3$ is one-to-one if and only if the linear system $A\mathbf{x} = \mathbf{0}$ has only the trivial solution. The augmented matrix of $A\mathbf{x} = \mathbf{0}$ is

$$\begin{bmatrix} 1 & -1 & | & 0 \\ 2 & 0 & | & 0 \\ 3 & -4 & | & 0 \end{bmatrix}$$

and the reduced row echelon form of this matrix is

$$\begin{bmatrix} 1 & 0 & | & 0 \\ 0 & 1 & | & 0 \\ 0 & 0 & | & 0 \end{bmatrix}$$

From this we conclude that $A\mathbf{x} = \mathbf{0}$ has only the trivial solution, and so T_A is one-to-one.

(b) The augmented matrix of the system $A\mathbf{x} = \mathbf{0}$ is

$$\begin{bmatrix} 1 & 2 & 3 & | & 0 \\ -1 & 0 & -4 & | & 0 \end{bmatrix}$$

and the reduced row echelon form of this matrix is

$$\begin{bmatrix} 1 & 0 & 4 & | & 0 \\ 0 & 1 & -\frac{1}{2} & | & 0 \end{bmatrix}$$

From this we conclude that $A\mathbf{x} = \mathbf{0}$ has the general solution $\mathbf{x} = t \begin{bmatrix} -4 \\ \frac{1}{2} \\ 1 \end{bmatrix}$. In particular, the system $A\mathbf{x} = \mathbf{0}$ has nontrivial solutions and so the transformation $T_A : R^3 \to R^2$ is not one-to-one.

21. (a) The augmented matrix of the system $A\mathbf{x} = \mathbf{b}$ can be row reduced as follows:

$$\begin{bmatrix} 1 & -2 & -1 & 3 & | & b_1 \\ 2 & 4 & 6 & -2 & | & b_2 \\ 3 & 0 & 3 & 3 & | & b_3 \end{bmatrix} \to \begin{bmatrix} 1 & -2 & -1 & 3 & | & b_1 \\ 0 & 8 & 8 & -8 & | & b_2 - 2b_1 \\ 0 & 6 & 6 & -6 & | & b_3 - 3b_1 \end{bmatrix}$$

$$\rightarrow \begin{bmatrix} 1 & -2 & -1 & 3 & | & b_1 \\ 0 & 1 & 1 & -1 & | & \frac{1}{8}b_2 - \frac{1}{4}b_1 \\ 0 & 0 & 0 & 0 & | & \frac{1}{6}b_3 - \frac{1}{8}b_2 - \frac{1}{4}b_1 \end{bmatrix}$$

It follows that $A\mathbf{x} = \mathbf{b}$ is consistent if and only if $\frac{1}{6}b_3 - \frac{1}{8}b_2 - \frac{1}{4}b_1 = 0$, or $6b_1 + 3b_2 - 4b_3 = 0$.

(b) The range of the transformation T_A consists of all vectors $\mathbf{b}$ of the form

$$\mathbf{b} = \begin{bmatrix} -\frac{1}{2}s + \frac{2}{3}t \\ s \\ t \end{bmatrix} = s\begin{bmatrix} -\frac{1}{2} \\ 1 \\ 0 \end{bmatrix} + t\begin{bmatrix} \frac{2}{3} \\ 0 \\ 1 \end{bmatrix}$$

Note. This is just one possibility; it was obtained by solving for b_1 in terms of b_2 and b_3 and then making b_2 and b_3 into parameters.

(c) The augmented matrix of the system $A\mathbf{x} = \mathbf{0}$ can be row reduced to

$$\begin{bmatrix} 1 & 0 & 1 & 1 & | & 0 \\ 0 & 1 & 1 & -1 & | & 0 \\ 0 & 0 & 0 & 0 & | & 0 \end{bmatrix}$$

Thus the kernel of T_A (i.e. the solution space of $A\mathbf{x} = \mathbf{0}$) consists of all vectors of the form

$$\mathbf{x} = \begin{bmatrix} -s-t \\ -s+t \\ s \\ t \end{bmatrix} = s\begin{bmatrix} -1 \\ -1 \\ 1 \\ 0 \end{bmatrix} + t\begin{bmatrix} -1 \\ 1 \\ 0 \\ 1 \end{bmatrix}$$

DISCUSSION AND DISCOVERY

D1. **(a)** True. If T is one-to-one, then $T\mathbf{x} = \mathbf{0}$ if and only if $\mathbf{x} = \mathbf{0}$; thus $T(\mathbf{u} - \mathbf{v}) = \mathbf{0}$ implies $\mathbf{u} - \mathbf{v} = \mathbf{0}$ and $\mathbf{u} = \mathbf{v}$.

(b) True. If $T : R^n \to R^n$ is onto, then (from Theorem 6.3.14) it is one-to-one and so the argument given in part (a) applies.

(c) True. See Theorem 6.3.15.

(d) True. If T_A is not one-to-one, then the homogeneous linear system $A\mathbf{x} = \mathbf{0}$ has infinitely many nontrivial solutions.

(e) True. The standard matrix of a shear operator T is of the form $A = \begin{bmatrix} 1 & k \\ 0 & 1 \end{bmatrix}$ or $A = \begin{bmatrix} 1 & 0 \\ k & 1 \end{bmatrix}$. In either case, we have $\det(A) = 1 \neq 0$ and so $T = T_A$ is one-to-one.

D2. No. The transformation is not one-to-one since $T(\mathbf{v}) = \mathbf{a} \times \mathbf{v} = \mathbf{0}$ for all vectors $\mathbf{v}$ that are parallel to $\mathbf{a}$.

D3. The transformation $T_A : R^n \to R^m$ is onto.

D4. No (assuming $\mathbf{v}_0$ is not a scalar multiple of $\mathbf{v}$). The line $\mathbf{x} = \mathbf{v}_0 + t\mathbf{v}$ does not pass through the origin and thus is not a subspace of R^n. It follows from Theorem 6.3.7 that this line cannot be equal to range of a linear operator.

WORKING WITH PROOFS

P1. If $B\mathbf{x} = \mathbf{0}$, then $(AB)\mathbf{x} = A(B\mathbf{x}) = A\mathbf{0} = \mathbf{0}$; thus $\mathbf{x}$ is in the nullspace of AB.

EXERCISE SET 6.4

1. $[T_B \circ T_A] = BA = \begin{bmatrix} 2 & -3 & 3 \\ 5 & 0 & 1 \\ 6 & 1 & 7 \end{bmatrix} \begin{bmatrix} 1 & -2 & 0 \\ 4 & 1 & -3 \\ 5 & 2 & 4 \end{bmatrix} = \begin{bmatrix} 5 & -1 & 21 \\ 10 & -8 & 4 \\ 45 & 3 & 25 \end{bmatrix}$

$[T_A \circ T_B] = AB = \begin{bmatrix} 1 & -2 & 0 \\ 4 & 1 & -3 \\ 5 & 2 & 4 \end{bmatrix} \begin{bmatrix} 2 & -3 & 3 \\ 5 & 0 & 1 \\ 6 & 1 & 7 \end{bmatrix} = \begin{bmatrix} -8 & -3 & 1 \\ -5 & -15 & -8 \\ 44 & -11 & 45 \end{bmatrix}$

3. (a) $[T_1] = \begin{bmatrix} 1 & 1 \\ 1 & -1 \end{bmatrix}$ $[T_2] = \begin{bmatrix} 3 & 0 \\ 2 & 4 \end{bmatrix}$

   (b) $[T_2 \circ T_1] = \begin{bmatrix} 3 & 0 \\ 2 & 4 \end{bmatrix} \begin{bmatrix} 1 & 1 \\ 1 & -1 \end{bmatrix} = \begin{bmatrix} 3 & 3 \\ 6 & -2 \end{bmatrix}$ $[T_1 \circ T_2] = \begin{bmatrix} 1 & 1 \\ 1 & -1 \end{bmatrix} \begin{bmatrix} 3 & 0 \\ 2 & 4 \end{bmatrix} = \begin{bmatrix} 5 & 4 \\ 1 & -4 \end{bmatrix}$

   (c) $T_2(T_1(x_1, x_2)) = (3x_1 + 3x_2, 6x_1 - 2x_2)$ $T_1(T_2(x_1, x_2)) = (5x_1 + 4x_2, x_1 - 4x_2)$

5. (a) The standard matrix for the rotation is $A_1 = \begin{bmatrix} 0 & -1 \\ 1 & 0 \end{bmatrix}$ and the standard matrix for the reflection is $A_2 = \begin{bmatrix} 0 & 1 \\ 1 & 0 \end{bmatrix}$. Thus the standard matrix for the rotation followed by the reflection is

   $$A_2 A_1 = \begin{bmatrix} 0 & 1 \\ 1 & 0 \end{bmatrix} \begin{bmatrix} 0 & -1 \\ 1 & 0 \end{bmatrix} = \begin{bmatrix} 1 & 0 \\ 0 & -1 \end{bmatrix}$$

   (b) The standard matrix for the projection followed by the contraction is

   $$A_2 A_1 = \begin{bmatrix} \frac{1}{2} & 0 \\ 0 & \frac{1}{2} \end{bmatrix} \begin{bmatrix} 0 & 0 \\ 0 & 1 \end{bmatrix} = \begin{bmatrix} 0 & 0 \\ 0 & \frac{1}{2} \end{bmatrix}$$

   (c) The standard matrix for the reflection followed by the dilation is

   $$A_2 A_1 = \begin{bmatrix} 3 & 0 \\ 0 & 3 \end{bmatrix} \begin{bmatrix} 1 & 0 \\ 0 & -1 \end{bmatrix} = \begin{bmatrix} 3 & 0 \\ 0 & -3 \end{bmatrix}$$

7. (a) The standard matrix for the reflection followed by the projection is

   $$A_2 A_1 = \begin{bmatrix} 1 & 0 & 0 \\ 0 & 0 & 0 \\ 0 & 0 & 1 \end{bmatrix} \begin{bmatrix} -1 & 0 & 0 \\ 0 & 1 & 0 \\ 0 & 0 & 1 \end{bmatrix} = \begin{bmatrix} -1 & 0 & 0 \\ 0 & 0 & 0 \\ 0 & 0 & 1 \end{bmatrix}$$

   (b) The standard matrix for the rotation followed by the dilation is

   $$A_2 A_1 = \begin{bmatrix} \sqrt{2} & 0 & 0 \\ 0 & \sqrt{2} & 0 \\ 0 & 0 & \sqrt{2} \end{bmatrix} \begin{bmatrix} \frac{1}{\sqrt{2}} & 0 & \frac{1}{\sqrt{2}} \\ 0 & 1 & 0 \\ -\frac{1}{\sqrt{2}} & 0 & \frac{1}{\sqrt{2}} \end{bmatrix} = \begin{bmatrix} 1 & 0 & 1 \\ 0 & \sqrt{2} & 0 \\ -1 & 0 & 1 \end{bmatrix}$$

   (c) The standard matrix for the projection followed by the reflection is

   $$A_2 A_1 = \begin{bmatrix} -1 & 0 & 0 \\ 0 & 1 & 0 \\ 0 & 0 & 1 \end{bmatrix} \begin{bmatrix} 1 & 0 & 0 \\ 0 & 1 & 0 \\ 0 & 0 & 0 \end{bmatrix} = \begin{bmatrix} -1 & 0 & 0 \\ 0 & 1 & 0 \\ 0 & 0 & 0 \end{bmatrix}$$

9. **(a)** The standard matrix for the composition is

$$A_3A_2A_1 = \begin{bmatrix} \frac{1}{4} & 0 & 0 \\ 0 & \frac{1}{4} & 0 \\ 0 & 0 & \frac{1}{4} \end{bmatrix} \begin{bmatrix} \frac{\sqrt{3}}{2} & -\frac{1}{2} & 0 \\ \frac{1}{2} & \frac{\sqrt{3}}{2} & 0 \\ 0 & 0 & 1 \end{bmatrix} \begin{bmatrix} 1 & 0 & 0 \\ 0 & \frac{\sqrt{3}}{2} & -\frac{1}{2} \\ 0 & \frac{1}{2} & \frac{\sqrt{3}}{2} \end{bmatrix} = \begin{bmatrix} \frac{\sqrt{3}}{8} & -\frac{\sqrt{3}}{16} & \frac{1}{16} \\ \frac{1}{8} & \frac{3}{16} & -\frac{\sqrt{3}}{16} \\ 0 & \frac{1}{8} & \frac{\sqrt{3}}{8} \end{bmatrix}$$

(b) The standard matrix for the composition is

$$A_3A_2A_1 = \begin{bmatrix} 0 & 0 & 0 \\ 0 & 1 & 0 \\ 0 & 0 & 1 \end{bmatrix} \begin{bmatrix} 1 & 0 & 0 \\ 0 & -1 & 0 \\ 0 & 0 & 1 \end{bmatrix} \begin{bmatrix} 1 & 0 & 0 \\ 0 & 1 & 0 \\ 0 & 0 & -1 \end{bmatrix} = \begin{bmatrix} 0 & 0 & 0 \\ 0 & -1 & 0 \\ 0 & 0 & -1 \end{bmatrix}$$

11. **(a)** We have $A = \begin{bmatrix} 2 & 0 \\ 0 & 3 \end{bmatrix} = \begin{bmatrix} 2 & 0 \\ 0 & 1 \end{bmatrix}\begin{bmatrix} 1 & 0 \\ 0 & 3 \end{bmatrix}$; thus multiplication by A corresponds to expansion by a factor of 3 in the y-direction and expansion by a factor of 2 in the x-direction.

(b) We have $A = \begin{bmatrix} 2 & 1 \\ 1 & 0 \end{bmatrix} = \begin{bmatrix} 1 & 2 \\ 0 & 1 \end{bmatrix}\begin{bmatrix} 0 & 1 \\ 1 & 0 \end{bmatrix}$; thus multiplication by A corresponds to reflection about

the line $y = x$ followed by a shear in the x-direction with factor 2.
Note. The factorization of A as a product of elementary matrices is not unique. This is just one possibility.

(c) We have $A = \begin{bmatrix} 0 & -2 \\ 4 & 0 \end{bmatrix} = \begin{bmatrix} 0 & 1 \\ 1 & 0 \end{bmatrix}\begin{bmatrix} 4 & 0 \\ 0 & 1 \end{bmatrix}\begin{bmatrix} 1 & 0 \\ 0 & 2 \end{bmatrix}\begin{bmatrix} 1 & 0 \\ 0 & -1 \end{bmatrix}$; thus multiplication by A corresponds to

reflection about the x-axis, followed by expansion in the y-direction by a factor of 2, expansion in the x-direction by a factor of 4, and reflection about the line $y = x$.

(d) We have $A = \begin{bmatrix} 1 & -3 \\ 4 & 6 \end{bmatrix} = \begin{bmatrix} 1 & 0 \\ 4 & 1 \end{bmatrix}\begin{bmatrix} 1 & 0 \\ 0 & 18 \end{bmatrix}\begin{bmatrix} 1 & -3 \\ 0 & 1 \end{bmatrix}$; thus multiplication by A corresponds to a shear

in the x-direction with factor -3, followed by expansion in the y-direction with factor 18, and a shear in the y-direction with factor 4.

13. **(a)** Reflection of R^2 about the x-axis.
(b) Rotation of R^2 about the origin through an angle of $-\frac{\pi}{4}$.
(c) Contraction of R^2 by a factor of $\frac{1}{3}$.
(d) Expansion of R^2 in the y-direction with factor 2.

15. The standard matrix for the operator T is $A = \begin{bmatrix} 1 & 2 \\ 1 & 1 \end{bmatrix}$. Since A is invertible, T is one-to-one and

the standard matrix for T^{-1} is $A^{-1} = \begin{bmatrix} -1 & 2 \\ 1 & -1 \end{bmatrix}$; thus $T^{-1}(w_1, w_2) = (-w_1 + 2w_2, w_1 - w_2)$.

17. The standard matrix for the operator T is $A = \begin{bmatrix} 0 & -1 \\ 1 & 0 \end{bmatrix}$. Since A is invertible, T is one-to-one and

the standard matrix for T^{-1} is $A^{-1} = \begin{bmatrix} 0 & 1 \\ -1 & 0 \end{bmatrix}$; thus $T^{-1}(w_1, w_2) = (w_2, -w_1)$.

19. The standard matrix for the operator T is $A = \begin{bmatrix} 1 & -2 & 2 \\ 2 & 1 & 1 \\ 1 & 1 & 0 \end{bmatrix}$. Since A is invertible, T is one-to-one

and the standard matrix for T^{-1} is $A^{-1} = \begin{bmatrix} 1 & -2 & 4 \\ -1 & 2 & -3 \\ -1 & 3 & -5 \end{bmatrix}$; thus the formula for T^{-1} is

$$T^{-1}(w_1, w_2, w_3) = (w_1 - 2w_2 + 4w_3, -w_1 + 2w_2 - 3w_3, -w_1 + 3w_2 - 5w_3)$$

21. The standard matrix for the operator T is $A = \begin{bmatrix} 1 & 4 & -1 \\ 2 & 7 & 1 \\ 1 & 3 & 0 \end{bmatrix}$. Since A is invertible, T is one-to-one

and the standard matrix for T^{-1} is $A^{-1} = \begin{bmatrix} -\frac{3}{2} & -\frac{3}{2} & \frac{11}{2} \\ \frac{1}{2} & \frac{1}{2} & -\frac{3}{2} \\ -\frac{1}{2} & \frac{1}{2} & -\frac{1}{2} \end{bmatrix}$; thus the formula for T^{-1} is

$$T^{-1}(w_1, w_2, w_3) = (-\tfrac{3}{2}w_1 - \tfrac{3}{2}w_2 + \tfrac{11}{2}w_3, \tfrac{1}{2}w_1 + \tfrac{1}{2}w_2 - \tfrac{3}{2}w_3, -\tfrac{1}{2}w_1 + \tfrac{1}{2}w_2 - \tfrac{1}{2}w_3)$$

23. **(a)** It is easy to see directly (from the geometric definitions) that $T_1 \circ T_2 = 0 = T_2 \circ T_1$. This

also follows from $[T_1][T_2] = \begin{bmatrix} 1 & 0 \\ 0 & 0 \end{bmatrix}\begin{bmatrix} 0 & 0 \\ 0 & 1 \end{bmatrix} = \begin{bmatrix} 0 & 0 \\ 0 & 0 \end{bmatrix}$ and $[T_2][T_1] = \begin{bmatrix} 0 & 0 \\ 0 & 1 \end{bmatrix}\begin{bmatrix} 1 & 0 \\ 0 & 0 \end{bmatrix} = \begin{bmatrix} 0 & 0 \\ 0 & 0 \end{bmatrix}$.

(b) It is easy to see directly that the composition of T_1 and T_2 (in either order) corresponds to rotation about the origin through the angle $\theta_1 + \theta_2$; thus $T_1 \circ T_2 = T_2 \circ T_1$. This also follows from the computation carried out in Example 1.

(c) We have $[T_1] = \begin{bmatrix} 1 & 0 & 0 \\ 0 & \cos\theta_1 & -\sin\theta_1 \\ 0 & \sin\theta_1 & \cos\theta_1 \end{bmatrix}$ and $[T_2] = \begin{bmatrix} \cos\theta_2 & -\sin\theta_2 & 0 \\ \sin\theta_2 & \cos\theta_2 & 0 \\ 0 & 0 & 1 \end{bmatrix}$; thus

$$[T_1 \circ T_2] = [T_1][T_2] = \begin{bmatrix} \cos\theta_2 & -\sin\theta_2 & 0 \\ \cos\theta_1\sin\theta_2 & \cos\theta_1\cos\theta_2 & -\sin\theta_1 \\ \sin\theta_1\sin\theta_2 & \sin\theta_1\cos\theta_2 & \cos\theta_1 \end{bmatrix}$$

$$[T_2 \circ T_1] = [T_2][T_1] = \begin{bmatrix} \cos\theta_2 & -\cos\theta_1\sin\theta_2 & \sin\theta_1\sin\theta_2 \\ \sin\theta_2 & \cos\theta_1\cos\theta_2 & -\sin\theta_1\cos\theta_2 \\ 0 & \sin\theta_1 & \cos\theta_1 \end{bmatrix} \neq [T_1 \circ T_2]$$

and it follows that $T_1 \circ T_2 \neq T_2 \circ T_1$.

25. We have $H_{\pi/3} = \begin{bmatrix} -\frac{1}{2} & \frac{\sqrt{3}}{2} \\ \frac{\sqrt{3}}{2} & \frac{1}{2} \end{bmatrix}$ and $H_{\pi/6} = \begin{bmatrix} \frac{1}{2} & \frac{\sqrt{3}}{2} \\ \frac{\sqrt{3}}{2} & -\frac{1}{2} \end{bmatrix}$. Thus the standard matrix for the composition is

$$H_{\pi/6}H_{\pi/3} = \begin{bmatrix} \frac{1}{2} & \frac{\sqrt{3}}{2} \\ \frac{\sqrt{3}}{2} & -\frac{1}{2} \end{bmatrix}\begin{bmatrix} -\frac{1}{2} & \frac{\sqrt{3}}{2} \\ \frac{\sqrt{3}}{2} & \frac{1}{2} \end{bmatrix} = \begin{bmatrix} \frac{1}{2} & \frac{\sqrt{3}}{2} \\ -\frac{\sqrt{3}}{2} & \frac{1}{2} \end{bmatrix} = R_{-\pi/3}$$

27. The image of the unit square is the parallelogram having the

vectors $T(\mathbf{e}_1) = \begin{bmatrix} 1 \\ 1 \end{bmatrix}$ and $T(\mathbf{e}_2) = \begin{bmatrix} -1 \\ 2 \end{bmatrix}$ as adjacent sides. The area

of this parallelogram is $|\det(A)| = \left|\det\begin{bmatrix} 1 & -1 \\ 1 & 2 \end{bmatrix}\right| = |2 + 1| = 3$.

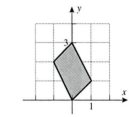

DISCUSSION AND DISCOVERY

D1. **(a)** True. If $T_1(\mathbf{x}) = \mathbf{0}$ with $\mathbf{x} \neq \mathbf{0}$, then $T_2(T_1(\mathbf{x})) = T_2(\mathbf{0}) = \mathbf{0}$; thus $T_2 \circ T_1$ is not one-to-one.

(b) False. If $T_2(\text{ran}(T_1)) = R^k$ then $T_2 \circ T_1$ is onto.

(c) False. If $\mathbf{x}$ is in R^n, then $T_2(T_1(\mathbf{x})) = \mathbf{0}$ if and only if $T_1(\mathbf{x})$ belongs to the kernel of T_2; thus if T_1 is one-to-one and $\text{ran}(T_1) \cap \text{ker}(T_2) = \{\mathbf{0}\}$, then the transformation $T_2 \circ T_1$ will be one-to-one.

(d) True. If $\text{ran}(T_2) \neq R^k$ then $\text{ran}(T_2 \circ T_1) \subseteq \text{ran}(T_2) \neq R^k$.

D2. We have $R_\beta = \begin{bmatrix} \cos\beta & -\sin\beta \\ \sin\beta & \cos\beta \end{bmatrix}$, $H_0 = \begin{bmatrix} 1 & 0 \\ 0 & -1 \end{bmatrix}$, and $R_\beta^{-1} = R_{-\beta} = \begin{bmatrix} \cos\beta & \sin\beta \\ -\sin\beta & \cos\beta \end{bmatrix}$. Thus

$$R_\beta H_0 R_\beta^{-1} = \begin{bmatrix} \cos^2\beta - \sin^2\beta & 2\sin\beta\cos\beta \\ 2\sin\beta\cos\beta & \sin^2\beta - \cos^2\beta \end{bmatrix} = \begin{bmatrix} \cos 2\beta & \sin 2\beta \\ \sin 2\beta & -\cos 2\beta \end{bmatrix} = H_\beta$$

and so multiplication by the matrix $R_\beta H_0 R_\beta^{-1}$ corresponds to reflection about the line L.

D3. From Example 2, we have $H_{\theta_1} H_{\theta_2} = R_{2(\theta_2 - \theta_1)}$. Since $\theta = 2(\theta - \tfrac{1}{2}\theta)$, it follows that $R_\theta = H_\theta H_{\theta/2}$. Thus every rotation can be expressed as a composition of two reflections.

EXERCISE SET 6.5

1.

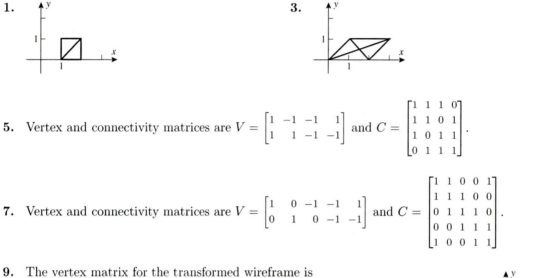

3.

5. Vertex and connectivity matrices are $V = \begin{bmatrix} 1 & -1 & -1 & 1 \\ 1 & 1 & -1 & -1 \end{bmatrix}$ and $C = \begin{bmatrix} 1 & 1 & 1 & 0 \\ 1 & 1 & 0 & 1 \\ 1 & 0 & 1 & 1 \\ 0 & 1 & 1 & 1 \end{bmatrix}$.

7. Vertex and connectivity matrices are $V = \begin{bmatrix} 1 & 0 & -1 & -1 & 1 \\ 0 & 1 & 0 & -1 & -1 \end{bmatrix}$ and $C = \begin{bmatrix} 1 & 1 & 0 & 0 & 1 \\ 1 & 1 & 1 & 0 & 0 \\ 0 & 1 & 1 & 1 & 0 \\ 0 & 0 & 1 & 1 & 1 \\ 1 & 0 & 0 & 1 & 1 \end{bmatrix}$.

9. The vertex matrix for the transformed wireframe is

$$V_1 = AV = \begin{bmatrix} 2 & 0 \\ 0 & 3 \end{bmatrix} \begin{bmatrix} 1 & 0 & -1 & -1 & 1 \\ 0 & 1 & 0 & -1 & -1 \end{bmatrix} = \begin{bmatrix} 2 & 0 & -2 & -2 & 2 \\ 0 & 3 & 0 & -3 & -3 \end{bmatrix}$$

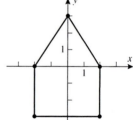

and the connectivity matrix is unchanged. The transformed image is obtained by expansion in the x-direction with factor 2 and expansion in the y-direction with factor 3.

11. The vertex matrix for the transformed wireframe is

$$V_1 = AV = \begin{bmatrix} 1 & 2 \\ 0 & 1 \end{bmatrix} \begin{bmatrix} 1 & 0 & -1 & -1 & 1 \\ 0 & 1 & 0 & -1 & -1 \end{bmatrix} = \begin{bmatrix} 1 & 2 & -1 & -3 & -1 \\ 0 & 1 & 0 & -1 & -1 \end{bmatrix}$$

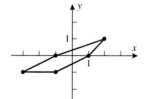

and the connectivity matrix is unchanged. The transformed image is obtained by a shear in the x-direction with factor 2.

13. The vertex matrix of the original wireframe is $V_1 = \begin{bmatrix} 1 & 0 & -1 & -1 & 1 \\ 0 & 1 & 0 & -1 & -1 \end{bmatrix}$. Thus the vertex matrix for the translated wireframe can be obtained in homogeneous coordinates from the product

$$\begin{bmatrix} 1 & 0 & 1 \\ 0 & 1 & 1 \\ \hline 0 & 0 & 1 \end{bmatrix} \begin{bmatrix} 1 & 0 & -1 & -1 & 1 \\ 0 & 1 & 0 & -1 & -1 \\ \hline 1 & 1 & 1 & 1 & 1 \end{bmatrix} = \begin{bmatrix} 2 & 1 & 0 & 0 & 2 \\ 1 & 2 & 1 & 0 & 0 \\ \hline 1 & 1 & 1 & 1 & 1 \end{bmatrix}$$

where, upon dropping the 1's in the third row, we have $V_2 = \begin{bmatrix} 2 & 1 & 0 & 0 & 2 \\ 1 & 2 & 1 & 0 & 0 \end{bmatrix}$. The translated image is obtained by moving points 1 unit in the x-direction and 1 unit in the y-direction.

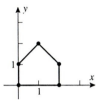

15. The vertex matrix of the original wireframe is $V_1 = \begin{bmatrix} 1 & 0 & -1 & -1 & 1 \\ 0 & 1 & 0 & -1 & -1 \end{bmatrix}$. Thus the vertex matrix for the translated wireframe can be obtained in homogeneous coordinates from the matrix product

$$\begin{bmatrix} 1 & 0 & 2 \\ 0 & 1 & -1 \\ 0 & 0 & 1 \end{bmatrix} \begin{bmatrix} 1 & 0 & -1 & -1 & 1 \\ 0 & 1 & 0 & -1 & -1 \\ 1 & 1 & 1 & 1 & 1 \end{bmatrix} = \begin{bmatrix} 3 & 2 & 1 & 1 & 3 \\ -1 & 0 & -1 & -2 & -2 \\ 1 & 1 & 1 & 1 & 1 \end{bmatrix}$$

where, upon dropping the 1's in the third row, we have $V_2 = \begin{bmatrix} 3 & 2 & 1 & 1 & 3 \\ -1 & 0 & -1 & -2 & -2 \end{bmatrix}$. The translated image is obtained by moving points 2 units in the x-direction and 1 unit in the negative y-direction.

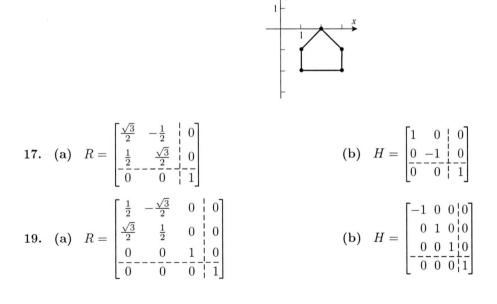

17. (a) $R = \begin{bmatrix} \frac{\sqrt{3}}{2} & -\frac{1}{2} & 0 \\ \frac{1}{2} & \frac{\sqrt{3}}{2} & 0 \\ 0 & 0 & 1 \end{bmatrix}$

(b) $H = \begin{bmatrix} 1 & 0 & 0 \\ 0 & -1 & 0 \\ 0 & 0 & 1 \end{bmatrix}$

19. (a) $R = \begin{bmatrix} \frac{1}{2} & -\frac{\sqrt{3}}{2} & 0 & 0 \\ \frac{\sqrt{3}}{2} & \frac{1}{2} & 0 & 0 \\ 0 & 0 & 1 & 0 \\ 0 & 0 & 0 & 1 \end{bmatrix}$

(b) $H = \begin{bmatrix} -1 & 0 & 0 & 0 \\ 0 & 1 & 0 & 0 \\ 0 & 0 & 1 & 0 \\ 0 & 0 & 0 & 1 \end{bmatrix}$

21. (a) $\begin{bmatrix} 1 & 0 & 1 \\ 0 & 1 & 2 \\ 0 & 0 & 1 \end{bmatrix}$

(b) $\begin{bmatrix} 1 & 0 & 1 \\ 0 & 1 & 2 \\ 0 & 0 & 1 \end{bmatrix} \begin{bmatrix} 3 \\ 4 \\ 1 \end{bmatrix} = \begin{bmatrix} 4 \\ 6 \\ 1 \end{bmatrix}$; thus the image of the point $\mathbf{x} = (3,4)$ is $(4,6)$.

23. (a)

$$\begin{bmatrix} 1 & 0 & 0 & | & 5 \\ 0 & 1 & 0 & | & 3 \\ 0 & 0 & 1 & | & -1 \\ \hline 0 & 0 & 0 & | & 1 \end{bmatrix}$$

(b)

$$\begin{bmatrix} 1 & 0 & 0 & | & 5 \\ 0 & 1 & 0 & | & 3 \\ 0 & 0 & 1 & | & -1 \\ \hline 0 & 0 & 0 & | & 1 \end{bmatrix} \begin{bmatrix} 5 \\ -4 \\ 1 \\ 1 \end{bmatrix} = \begin{bmatrix} 10 \\ -1 \\ 0 \\ 1 \end{bmatrix}; \text{ thus the image of } \mathbf{x} = (5, -4, 1) \text{ is } (10, -1, 0).$$

25. $A = TR = \begin{bmatrix} 1 & 0 & | & 3 \\ 0 & 1 & | & -1 \\ \hline 0 & 0 & | & 1 \end{bmatrix} \begin{bmatrix} \frac{1}{2} & -\frac{\sqrt{3}}{2} & | & 0 \\ \frac{\sqrt{3}}{2} & \frac{1}{2} & | & 0 \\ \hline 0 & 0 & | & 1 \end{bmatrix} = \begin{bmatrix} \frac{1}{2} & -\frac{\sqrt{3}}{2} & | & 3 \\ \frac{\sqrt{3}}{2} & \frac{1}{2} & | & -1 \\ \hline 0 & 0 & | & 1 \end{bmatrix}$

27. $A = TB = \begin{bmatrix} 1 & 0 & 0 & | & 2 \\ 0 & 1 & 0 & | & -3 \\ 0 & 0 & 1 & | & 5 \\ \hline 0 & 0 & 0 & | & 1 \end{bmatrix} \begin{bmatrix} -1 & 0 & 0 & | & 0 \\ 0 & 2 & 1 & | & 0 \\ 1 & 1 & 0 & | & 0 \\ \hline 0 & 0 & 0 & | & 1 \end{bmatrix} = \begin{bmatrix} -1 & 0 & 0 & | & 2 \\ 0 & 2 & 1 & | & -3 \\ 1 & 1 & 0 & | & 5 \\ \hline 0 & 0 & 0 & | & 1 \end{bmatrix}$

DISCUSSION AND DISCOVERY

D1. The matrix that performs the specified rotation is

$$R = \begin{bmatrix} 1 & 0 & | & 2 \\ 0 & 1 & | & -1 \\ \hline 0 & 0 & | & 1 \end{bmatrix} \begin{bmatrix} \frac{\sqrt{3}}{2} & -\frac{1}{2} & | & 0 \\ \frac{1}{2} & \frac{\sqrt{3}}{2} & | & 0 \\ \hline 0 & 0 & | & 1 \end{bmatrix} \begin{bmatrix} 1 & 0 & | & -2 \\ 0 & 1 & | & 1 \\ \hline 0 & 0 & | & 1 \end{bmatrix} = \begin{bmatrix} \frac{\sqrt{3}}{2} & -\frac{1}{2} & | & -\sqrt{3}+\frac{3}{2} \\ \frac{1}{2} & \frac{\sqrt{3}}{2} & | & -2+\frac{\sqrt{3}}{2} \\ \hline 0 & 0 & | & 1 \end{bmatrix}$$

and we have $\begin{bmatrix} \frac{\sqrt{3}}{2} & -\frac{1}{2} & | & -\sqrt{3}+\frac{3}{2} \\ \frac{1}{2} & \frac{\sqrt{3}}{2} & | & -2+\frac{\sqrt{3}}{2} \\ \hline 0 & 0 & | & 1 \end{bmatrix} \begin{bmatrix} x \\ y \\ 1 \end{bmatrix} = \begin{bmatrix} \frac{\sqrt{3}}{2}x - \frac{1}{2}y - \sqrt{3} + \frac{3}{2} \\ \frac{1}{2}x + \frac{\sqrt{3}}{2}y - 2 + \frac{\sqrt{3}}{2} \\ \hline 1 \end{bmatrix}$. Thus the image of the point (x, y)

under the rotation is $(\frac{\sqrt{3}}{2}x - \frac{1}{2}y - \sqrt{3} + \frac{3}{2}, \frac{1}{2}x + \frac{\sqrt{3}}{2}y - 2 + \frac{\sqrt{3}}{2})$.

D2. (a) $H = \begin{bmatrix} 1 & 0 & | & x_0 \\ 0 & 1 & | & y_0 \\ 0 & 0 & | & 1 \end{bmatrix} \begin{bmatrix} \cos 2\theta & \sin 2\theta & | & 0 \\ \sin 2\theta & -\cos 2\theta & | & 0 \\ \hline 0 & 0 & | & 1 \end{bmatrix} \begin{bmatrix} 1 & 0 & | & -x_0 \\ 0 & 1 & | & -y_0 \\ \hline 0 & 0 & | & 1 \end{bmatrix}$

(b) The matrix that performs the specified reflection is

$$H = \begin{bmatrix} 1 & 0 & | & 2 \\ 0 & 1 & | & -1 \\ \hline 0 & 0 & | & 1 \end{bmatrix} \begin{bmatrix} \frac{1}{2} & \frac{\sqrt{3}}{2} & | & 0 \\ \frac{\sqrt{3}}{2} & -\frac{1}{2} & | & 0 \\ \hline 0 & 0 & | & 1 \end{bmatrix} \begin{bmatrix} 1 & 0 & | & -2 \\ 0 & 1 & | & 1 \\ \hline 0 & 0 & | & 1 \end{bmatrix} = \begin{bmatrix} \frac{1}{2} & \frac{\sqrt{3}}{2} & | & 1+\frac{\sqrt{3}}{2} \\ \frac{\sqrt{3}}{2} & -\frac{1}{2} & | & -\sqrt{3}-\frac{3}{2} \\ \hline 0 & 0 & | & 1 \end{bmatrix}$$

and we have $\begin{bmatrix} \frac{1}{2} & \frac{\sqrt{3}}{2} & | & 1+\frac{\sqrt{3}}{2} \\ \frac{\sqrt{3}}{2} & -\frac{1}{2} & | & -\sqrt{3}-\frac{3}{2} \\ \hline 0 & 0 & | & 1 \end{bmatrix} \begin{bmatrix} x \\ y \\ 1 \end{bmatrix} = \begin{bmatrix} \frac{1}{2}x + \frac{\sqrt{3}}{2}y + 1 + \frac{\sqrt{3}}{2} \\ \frac{\sqrt{3}}{2}x - \frac{1}{2}y - \sqrt{3} - \frac{3}{2} \\ \hline 1 \end{bmatrix}$. Thus the image of the point

(x, y) under the reflection is $(\frac{1}{2}x + \frac{\sqrt{3}}{2}y + 1 + \frac{\sqrt{3}}{2}, \frac{\sqrt{3}}{2}x - \frac{1}{2}y - \sqrt{3} - \frac{3}{2})$.

D3. Following the given suggestions, we conclude that the matrix H in homogeneous coordinates that reflects R^2 about the line with slope m that passes through the point (x_0, y_0) is

$$H = \begin{bmatrix} 1 & 0 & x_0 \\ 0 & 1 & y_0 \\ \hline 0 & 0 & 1 \end{bmatrix} \begin{bmatrix} \frac{1-m^2}{1+m^2} & \frac{2m}{1+m^2} & 0 \\ \frac{2m}{1+m^2} & \frac{m^2-1}{1+m^2} & 0 \\ \hline 0 & 0 & 1 \end{bmatrix} \begin{bmatrix} 1 & 0 & -x_0 \\ 0 & 1 & -y_0 \\ \hline 0 & 0 & 1 \end{bmatrix}$$

Thus the matrix that performs the specified reflection is

$$H = \begin{bmatrix} 1 & 0 & 1 \\ 0 & 1 & 2 \\ \hline 0 & 0 & 1 \end{bmatrix} \begin{bmatrix} \frac{3}{5} & \frac{4}{5} & 0 \\ \frac{4}{5} & -\frac{3}{5} & 0 \\ \hline 0 & 0 & 1 \end{bmatrix} \begin{bmatrix} 1 & 0 & -1 \\ 0 & 1 & -2 \\ \hline 0 & 0 & 1 \end{bmatrix} = \begin{bmatrix} \frac{3}{5} & \frac{4}{5} & -\frac{6}{5} \\ \frac{4}{5} & -\frac{3}{5} & \frac{12}{5} \\ \hline 0 & 0 & 1 \end{bmatrix}$$

We have $\begin{bmatrix} \frac{3}{5} & \frac{4}{5} & -\frac{6}{5} \\ \frac{4}{5} & -\frac{3}{5} & \frac{12}{5} \\ \hline 0 & 0 & 1 \end{bmatrix} \begin{bmatrix} x \\ y \\ 1 \end{bmatrix} = \begin{bmatrix} \frac{3}{5}x + \frac{4}{5}y - \frac{6}{5} \\ \frac{4}{5}x - \frac{3}{5}y + \frac{12}{5} \\ \hline 1 \end{bmatrix}$; thus the image of the point $(3, 4)$ under the reflection

is $\left(\frac{9}{5} + \frac{16}{5} - \frac{6}{5}, \frac{12}{5} - \frac{12}{5} + \frac{12}{5}\right) = \left(\frac{19}{5}, \frac{12}{5}\right)$.

D4. The matrix that performs the specified transformation is

$$S = \begin{bmatrix} 1 & 0 & x_0 \\ 0 & 1 & y_0 \\ \hline 0 & 0 & 1 \end{bmatrix} \begin{bmatrix} \lambda_1 & 0 & 0 \\ 0 & \lambda_2 & 0 \\ \hline 0 & 0 & 1 \end{bmatrix} \begin{bmatrix} 1 & 0 & -x_0 \\ 0 & 1 & -y_0 \\ \hline 0 & 0 & 1 \end{bmatrix} = \begin{bmatrix} \lambda_1 & 0 & -\lambda_1 x_0 + x_0 \\ 0 & \lambda_2 & -\lambda_2 y_0 + y_0 \\ \hline 0 & 0 & 1 \end{bmatrix}$$

We have $\begin{bmatrix} \lambda_1 & 0 & -\lambda_1 x_0 + x_0 \\ 0 & \lambda_2 & -\lambda_2 y_0 + y_0 \\ \hline 0 & 0 & 1 \end{bmatrix} \begin{bmatrix} x \\ y \\ 1 \end{bmatrix} = \begin{bmatrix} \lambda_1 x - \lambda_1 x_0 + x_0 \\ \lambda_2 y - \lambda_2 y_0 + y_0 \\ \hline 1 \end{bmatrix} = \begin{bmatrix} \lambda_1(x - x_0) + x_0 \\ \lambda_2(y - y_0) + y_0 \\ \hline 1 \end{bmatrix}$; thus the image of the

point (x, y) under this transformation is $(x_0 + \lambda_1(x - x_0), y_0 + \lambda_2(y - y_0))$.

D5. **(a)** The matrix that performs the operations in the specified order is

$$A = \begin{bmatrix} \cos\theta & -\sin\theta & 0 \\ \sin\theta & \cos\theta & 0 \\ \hline 0 & 0 & 1 \end{bmatrix} \begin{bmatrix} 1 & 0 & x_0 \\ 0 & 1 & y_0 \\ \hline 0 & 0 & 1 \end{bmatrix} = \begin{bmatrix} \cos\theta & -\sin\theta & x_0\cos\theta - y_0\sin\theta \\ \sin\theta & \cos\theta & x_0\sin\theta + y_0\cos\theta \\ \hline 0 & 0 & 1 \end{bmatrix}$$

(b) The matrix that performs the operations in the opposite order is

$$B = \begin{bmatrix} 1 & 0 & x_0 \\ 0 & 1 & y_0 \\ \hline 0 & 0 & 1 \end{bmatrix} \begin{bmatrix} \cos\theta & -\sin\theta & 0 \\ \sin\theta & \cos\theta & 0 \\ \hline 0 & 0 & 1 \end{bmatrix} = \begin{bmatrix} \cos\theta & -\sin\theta & x_0 \\ \sin\theta & \cos\theta & y_0 \\ \hline 0 & 0 & 1 \end{bmatrix}$$

(c) Since $A \neq B$, we conclude that the operations (rotation and translation) do not commute.

CHAPTER 7
Dimension and Structure

EXERCISE SET 7.1

1. (a) $\mathbf{v}_2 = 2\mathbf{v}_1$ (b) $\mathbf{v}_3 = \mathbf{v}_1 - \mathbf{v}_2$ (c) $\mathbf{v}_4 = -\mathbf{v}_1 + 2\mathbf{v}_2 + 6\mathbf{v}_3$

3. $\mathbf{v}_3 = -\frac{7}{3}\mathbf{v}_1 + \frac{2}{3}\mathbf{v}_2$

5. (a) Any one of the vectors $(1,2)$, $(\frac{1}{2}, 1)$, $(-1, -2)$ forms a basis for the line $y = 2x$.
   (b) Any two of the vectors $(1, -1, 0)$, $(2, 0, -1)$, $(0, 2, -1)$ form a basis for $x + y + 2z = 0$.

7. The augmented matrix of the system is $\begin{bmatrix} 3 & 1 & 1 & 1 & \vdots & 0 \\ 5 & -1 & 1 & -1 & \vdots & 0 \end{bmatrix}$ and the reduced row echelon form of this matrix is $\begin{bmatrix} 1 & 0 & \frac{1}{4} & 0 & \vdots & 0 \\ 0 & 1 & \frac{1}{4} & 1 & \vdots & 0 \end{bmatrix}$. Thus a general solution of the system is

$$\mathbf{x} = (-\tfrac{1}{4}s, -\tfrac{1}{4}s - t, s, t) = s(-\tfrac{1}{4}, -\tfrac{1}{4}, 1, 0) + t(0, -1, 0, 1)$$

where $-\infty < s, t < \infty$. The solution space is 2-dimensional with canonical basis $\{\mathbf{v}_1, \mathbf{v}_2\}$ where $\mathbf{v}_1 = (-\frac{1}{4}, -\frac{1}{4}, 1, 0)$ and $\mathbf{v}_2 = (0, -1, 0, 1)$.

9. The augmented matrix of the system is $\begin{bmatrix} 2 & 2 & -1 & 0 & 1 & \vdots & 0 \\ -1 & -1 & 2 & -3 & 1 & \vdots & 0 \\ 1 & 1 & -2 & 0 & -1 & \vdots & 0 \\ 0 & 0 & 1 & 1 & 1 & \vdots & 0 \end{bmatrix}$ and the reduced row echelon form of this matrix is $\begin{bmatrix} 1 & 1 & 0 & 0 & 1 & \vdots & 0 \\ 0 & 0 & 1 & 0 & 1 & \vdots & 0 \\ 0 & 0 & 0 & 1 & 0 & \vdots & 0 \\ 0 & 0 & 0 & 0 & 0 & \vdots & 0 \end{bmatrix}$. Thus the general solution is

$$\mathbf{x} = (-s - t, s, -t, 0, t) = s(-1, 1, 0, 0, 0) + t(-1, 0, -1, 0, 1)$$

where $-\infty < s, t < \infty$. The solution space is 2-dimensional with canonical basis $\{\mathbf{v}_1, \mathbf{v}_2\}$ where $\mathbf{v}_1 = (-1, 1, 0, 0, 0)$ and $\mathbf{v}_2 = (-1, 0, -1, 0, 1)$.

11. (a) The hyperplane $(1, 2, -3)^{\perp}$ consists of all vectors $\mathbf{x} = (x, y, z)$ in R^3 satisfying the equation $x + 2y - 3z = 0$. Using $y = s$ and $z = t$ as free variables, the solutions of this equation can be written in the form

$$\mathbf{x} = (-2s + 3t, s, t) = s(-2, 1, 0) + t(3, 0, 1)$$

where $-\infty < s, t < \infty$. Thus the vectors $\mathbf{v}_1 = (-2, 1, 0)$ and $\mathbf{v}_2 = (3, 0, 1)$ form a basis for the hyperplane.

   (b) The hyperplane $(2, -1, 4, 1)^{\perp}$ consists of all vectors $\mathbf{x} = (x_1, x_2, x_3, x_4)$ in R^4 satisfying the equation $2x_1 - x_2 + 4x_3 + x_4 = 0$. Using $x_1 = r$, $x_3 = s$ and $x_4 = t$ as free variables, the solutions of this equation can be written in the form

$$\mathbf{x} = (r, 2r + 4s + t, s, t) = r(1, 2, 0, 0) + s(0, 4, 1, 0) + t(0, 1, 0, 1)$$

where $-\infty < r, s, t < \infty$. Thus the vectors $\mathbf{v}_1 = (1, 2, 0, 0)$, $\mathbf{v}_2 = (0, 4, 1, 0)$, and $\mathbf{v}_3 = (0, 1, 0, 1)$ form a basis for the hyperplane.

DISCUSSION AND DISCOVERY

D1. (a) Assuming $A \neq 0$, the dimension of the solution space of $A\mathbf{x} = \mathbf{0}$ is at most $n - 1$.

(b) A hyperplane in R^6 has dimension 5.

(c) The subspaces of R^5 have dimensions $0, 1, 2, 3, 4,$ or 5.

(d) The vectors $\mathbf{v}_1 = (1, 0, 1, 0)$, $\mathbf{v}_2 = (1, 1, 0, 0)$, and $\mathbf{v}_3 = (1, 1, 1, 0)$ are linearly independent, thus they span a 3-dimensional subspace of R^4.

D2. Yes, they are linearly independent. If we write the vectors in the order

$$\mathbf{v}_4 = (0, 0, 0, 0, 1, {}^*), \mathbf{v}_3 = (0, 0, 0, 1, {}^*, {}^*), \mathbf{v}_2 = (0, 0, 1, {}^*, {}^*, {}^*), \mathbf{v}_1 = (1, {}^*, {}^*, {}^*, {}^*, {}^*)$$

then, because of the positions of the leading 1s, it is clear that none of these vectors can be written as a linear combination of the preceding ones in the list.

D3. False. Such a set is linearly dependent since, when the set is written in reverse order, some vector is a linear combination of its predecessors.

D4. The solution space of $A\mathbf{x} = \mathbf{0}$ has positive dimension if and only if the system has nontrivial solutions, and this occurs if and only if $\det(A) = 0$. Since $\det(A) = t^2 + 7t + 12 = (t + 3)(t + 4)$, it follows that the solution space has positive dimension if and only if $t = -3$ or $t = -4$. The solution space has dimension 1 in each case.

WORKING WITH PROOFS

P1. (a) If $S = \{\mathbf{v}_1, \mathbf{v}_2, \ldots, \mathbf{v}_k\}$ is a linearly dependent set in R^n then there are scalars $c_1, c_2, \ldots, c_k$ (not all zero) such that $c_1\mathbf{v}_1 + c_2\mathbf{v}_2 + \cdots + c_k\mathbf{v}_k = \mathbf{0}$. It follows that the set $S' = \{\mathbf{v}_1, \mathbf{v}_2, \ldots, \mathbf{v}_k, \mathbf{w}_1, \ldots, \mathbf{w}_r\}$ is linearly dependent since $c_1\mathbf{v}_1 + c_2\mathbf{v}_2 + \cdots + c_k\mathbf{v}_k + 0\mathbf{w}_1 + \cdots + 0\mathbf{w}_r = \mathbf{0}$ is a nontrivial dependency relation among its elements.

(b) If $S = \{\mathbf{v}_1, \mathbf{v}_2, \ldots, \mathbf{v}_k\}$ is a linearly independent set in R^n then there is no nontrivial dependency relation among its elements, i.e., if $c_1\mathbf{v}_1 + c_2\mathbf{v}_2 + \cdots + c_k\mathbf{v}_k = \mathbf{0}$ then $c_1 = c_2 = \cdots = c_k = 0$. It follows that if S' is any nonempty subset of S then S' must also be linearly independent since a nontrivial dependency among its elements would also be a nontrivial dependency among the elements of S.

P2. If $k \neq 0$, then $(k\mathbf{a}) \cdot \mathbf{x} = k(\mathbf{a} \cdot \mathbf{x}) = 0$ if and only if $\mathbf{a} \cdot \mathbf{x} = 0$; thus $(k\mathbf{a})^\perp = \mathbf{a}^\perp$.

EXERCISE SET 7.2

1. (a) A basis for R^2 must contain exactly two vectors (three are too many); any set of three vectors in R^2 is linearly dependent.

(b) A basis for R^3 must contain exactly three vectors (two are not enough); a set of two vectors cannot span R^3.

(c) The vectors $\mathbf{v}_1$ and $\mathbf{v}_2$ are linearly dependent ($\mathbf{v}_2 = 2\mathbf{v}_1$).

3. (a) The vectors $\mathbf{v}_1 = (2, 1)$ and $\mathbf{v}_2 = (0, 3)$ are linearly independent since neither is a scalar multiple of the other; thus these two vectors form a basis for R^2.

(b) The vector $\mathbf{v}_2 = (-7, 8, 0)$ is not a scalar multiple of $\mathbf{v}_1 = (4, 1, 0)$, and $\mathbf{v}_3 = (1, 1, 1)$ is not a linear combination of $\mathbf{v}_1$ and $\mathbf{v}_2$ since any such linear combination would have 0 in the third component. Thus $\mathbf{v}_1, \mathbf{v}_2,$ and $\mathbf{v}_3$ are linearly independent, and it follows that these vectors from a basis for R^3.

5. (a) The matrix having $\mathbf{v}_1$, $\mathbf{v}_2$, $\mathbf{v}_3$ as its column vectors is $A = \begin{bmatrix} 3 & 4 & 5 \\ 2 & 1 & 2 \\ -4 & -2 & -3 \end{bmatrix}$. Since $\det(A) = -5 \neq 0$,

the column vectors are linearly independent and hence form a basis for R^3.

(b) The matrix having $\mathbf{v}_1$, $\mathbf{v}_2$, and $\mathbf{v}_3$ as its column vectors is $A = \begin{bmatrix} 3 & 8 & 2 \\ 4 & 7 & -1 \\ -5 & -2 & 8 \end{bmatrix}$. Since $\det(A) = 0$,

the column vectors are linearly dependent and hence do not form a basis for R^3.

7. (a) An arbitrary vector $\mathbf{x} = (x, y, z)$ in R^3 can be written as

$$\mathbf{x} = x\mathbf{v}_1 + \tfrac{1}{2}y\mathbf{v}_2 + \tfrac{1}{3}z\mathbf{v}_3 + 0\mathbf{v}_4$$

thus the vectors $\mathbf{v}_1$, $\mathbf{v}_2$, $\mathbf{v}_3$, and $\mathbf{v}_4$ span R^3. On the other hand, since

$$\mathbf{v}_4 = \mathbf{v}_1 + \tfrac{1}{2}\mathbf{v}_2 + \tfrac{1}{3}\mathbf{v}_3$$

the vectors $\mathbf{v}_1$, $\mathbf{v}_2$, $\mathbf{v}_3$, and $\mathbf{v}_4$ are linearly dependent and do not form a basis for R^3.

(b) The vector equation $\mathbf{v} = c_1\mathbf{v}_1 + c_2\mathbf{v}_2 + c_3\mathbf{v}_3 + c_4\mathbf{v}_4$ is equivalent to the linear system

$$\begin{bmatrix} 1 & 0 & 0 & 1 \\ 0 & 2 & 0 & 1 \\ 0 & 0 & 3 & 1 \end{bmatrix} \begin{bmatrix} c_1 \\ c_2 \\ c_3 \\ c_4 \end{bmatrix} = \begin{bmatrix} 1 \\ 2 \\ 3 \end{bmatrix}$$

and, setting $c_4 = t$, a general solution of this system is given by

$$c_1 = 1 - t, c_2 = 1 - \tfrac{1}{2}t, c_3 = 1 - \tfrac{1}{3}t, c_4 = t$$

where $-\infty < t < \infty$. Thus

$$\mathbf{v} = (1 - t)\mathbf{v}_1 + (1 - \tfrac{1}{2}t)\mathbf{v}_2 + (1 - \tfrac{1}{3}t)\mathbf{v}_3 + t\mathbf{v}_4$$

for any value of t. In particular, corresponding to $t = 0$, $t = 1$, and $t = -6$, we have $\mathbf{v} = \mathbf{v}_1 + \mathbf{v}_2 + \mathbf{v}_3$, $\mathbf{v} = \tfrac{1}{2}\mathbf{v}_2 + \tfrac{2}{3}\mathbf{v}_3 + \mathbf{v}_4$, and $\mathbf{v} = 7\mathbf{v}_1 + 4\mathbf{v}_2 + 3\mathbf{v}_3 - 6\mathbf{v}_4$.

9. The vector $\mathbf{v}_2 = (1, -2, -2)$ is not a scalar multiple of $\mathbf{v}_1 = (-1, 2, 3)$; thus $S = \{\mathbf{v}_1, \mathbf{v}_2\}$ is a linearly independent set. Let A be the matrix having $\mathbf{v}_1$, $\mathbf{v}_2$, and $\mathbf{e}_1$ as its columns, i.e.,

$$A = \begin{bmatrix} -1 & 1 & 1 \\ 2 & -2 & 0 \\ 3 & -2 & 0 \end{bmatrix}$$

Then $\det(A) = 2 \neq 0$, and it follows from this that $S' = \{\mathbf{v}_1, \mathbf{v}_2, \mathbf{e}_1\}$ is a linearly independent set and hence a basis for R^3. [Similarly, it can be shown that $\{\mathbf{v}_1, \mathbf{v}_2, \mathbf{e}_2\}$ is a basis for R^3. On the other hand, the set $\{\mathbf{v}_1, \mathbf{v}_2, \mathbf{e}_3\}$ is linearly dependent and thus not a basis for R^3.]

11. We have $\mathbf{v}_3 = \mathbf{v}_1 + \mathbf{v}_2$, and if we delete the vector $\mathbf{v}_3$ from the set S then the remaining vectors are

linearly independent since $\det \begin{bmatrix} 1 & 3 & 1 \\ 2 & -1 & 3 \\ -2 & 1 & 6 \end{bmatrix} = -63 \neq 0$. Thus $S' = \{\mathbf{v}_1, \mathbf{v}_2, \mathbf{v}_4\}$ is a basis for R^3.

13. Since $\det \begin{bmatrix} 1 & 1 & 1 \\ 0 & 1 & 1 \\ 0 & 0 & 1 \end{bmatrix} = 1 \neq 0$, the vectors v_1, v_2, v_3 (column vectors of the matrix) form a basis for R^3.

The vector equation $v = c_1 v_1 + c_2 v_2 + c_3 v_3$ is equivalent to the linear system

$$\begin{bmatrix} 1 & 1 & 1 \\ 0 & 1 & 1 \\ 0 & 0 & 1 \end{bmatrix} \begin{bmatrix} c_1 \\ c_2 \\ c_3 \end{bmatrix} = \begin{bmatrix} 2 \\ 5 \\ 1 \end{bmatrix}$$

which, from back substitution, has the solution $c_3 = 1, c_2 = 5 - c_3 = 4, c_1 = 2 - c_2 - c_3 = -3$. Thus $v = -3v_1 + 4v_2 + v_3$.

15. (a) Since $u \cdot n = (1)(2) + (2)(0) + (-1)(1) = 1 \neq 0$, the vector u is not orthogonal to n. Thus V is not contained in W.

(b) The line V is parallel to $u = (2, -1, 1)$, and the plane W has normal vector $n = (1, 3, 1)$. Since $u \cdot n = (2)(1) + (-1)(3) + (1)(1) = 0$, the vector u is orthogonal to n. Thus V is contained in W.

17. (a) The vector equation $c_1(1, 1, 0) + c_2(0, 1, 2) + c_3(2, 1, 3) = (3, 2, -1)$ is equivalent to the linear system with augmented matrix

$$\begin{bmatrix} 1 & 0 & 2 & | & 3 \\ 1 & 1 & 1 & | & 2 \\ 0 & 2 & 3 & | & -1 \end{bmatrix}$$

The reduced row echelon form of this matrix is

$$\begin{bmatrix} 1 & 0 & 0 & | & \frac{13}{5} \\ 0 & 1 & 0 & | & -\frac{4}{5} \\ 0 & 0 & 1 & | & \frac{1}{5} \end{bmatrix}$$

Thus $(3, 2, -1) = \frac{13}{5}(1, 1, 0) - \frac{4}{5}(0, 1, 2) + \frac{1}{5}(2, 1, 3)$ and, by linearity, it follows that

$$T(3, 2, -1) = \frac{13}{5}(2, 1, -1) - \frac{4}{5}(1, 0, 2) + \frac{1}{5}(4, 1, 0) = \frac{1}{5}(26, 13, -20)$$

(b) The vector equation $c_1(1, 1, 0) + c_2(0, 1, 2) + c_3(2, 1, 3) = (a, b, c)$ is equivalent to the linear system with augmented matrix

$$\begin{bmatrix} 1 & 0 & 2 & | & a \\ 1 & 1 & 1 & | & b \\ 0 & 2 & 3 & | & c \end{bmatrix}$$

The reduced row echelon form of this matrix is

$$\begin{bmatrix} 1 & 0 & 0 & | & \frac{1}{5}a + \frac{4}{5}b - \frac{2}{5}c \\ 0 & 1 & 0 & | & -\frac{3}{5}a + \frac{3}{5}b + \frac{1}{5}c \\ 0 & 0 & 1 & | & \frac{2}{5}a - \frac{2}{5}b + \frac{1}{5}c \end{bmatrix}$$

Thus $(a, b, c) = (\frac{1}{5}a + \frac{4}{5}b - \frac{2}{5}c)(1, 1, 0) + (-\frac{3}{5}a + \frac{3}{5}b + \frac{1}{5}c)(0, 1, 2) + (\frac{2}{5}a - \frac{2}{5}b + \frac{1}{5}c)(2, 1, 3)$ and

$$T(a, b, c) = (\frac{1}{5}a + \frac{4}{5}b - \frac{2}{5}c)(2, 1, -1) + (-\frac{3}{5}a + \frac{3}{5}b + \frac{1}{5}c)(1, 0, 2) + (\frac{2}{5}a - \frac{2}{5}b + \frac{1}{5}c)(4, 1, 0)$$

$$= (\frac{7}{5}a + \frac{3}{5}b + \frac{1}{5}c, \frac{1}{5}a + \frac{4}{5}b - \frac{2}{5}c, -a + c)$$

(c) From the formula above, we have $[T] = \begin{bmatrix} \frac{7}{5} & \frac{3}{5} & \frac{1}{5} \\ \frac{1}{5} & \frac{4}{5} & -\frac{2}{5} \\ -1 & 0 & 1 \end{bmatrix}$.

19. Since det $\begin{bmatrix} 1 & 3 & 4 \\ -7 & -2 & -3 \\ -5 & 8 & 5 \end{bmatrix} = -100 \neq 0$, the vectors $\mathbf{v}_1 = (1, -7, -5)$, $\mathbf{v}_2 = (3, -2, 8)$, $\mathbf{v}_3 = (4, -3, 5)$ form

a basis for R^3. The vector equation $c_1\mathbf{v}_1 + c_2\mathbf{v}_2 + c_3\mathbf{v}_3 = \mathbf{x}$ is equivalent to the linear system with augmented matrix

$$\begin{bmatrix} 1 & 3 & 4 & \vdots & x \\ -7 & -2 & -3 & \vdots & y \\ -5 & 8 & 5 & \vdots & z \end{bmatrix}$$

and the reduced row echelon form of this matrix is

$$\begin{bmatrix} 1 & 0 & 0 & \vdots & -\frac{7}{50}x - \frac{17}{100}y + \frac{1}{100}z \\ 0 & 1 & 0 & \vdots & -\frac{1}{2}x - \frac{1}{4}y + \frac{1}{4}z \\ 0 & 0 & 1 & \vdots & \frac{33}{50}x + \frac{23}{100}y - \frac{19}{100}z \end{bmatrix}$$

Thus a general vector $\mathbf{x} = (x, y, z)$ can be expressed in terms of the basis vectors $\mathbf{v}_1$, $\mathbf{v}_2$, $\mathbf{v}_3$ as follows:

$$\mathbf{x} = \left(-\tfrac{7}{50}x - \tfrac{17}{100}y + \tfrac{1}{100}z\right)\mathbf{v}_1 + \left(-\tfrac{1}{2}x - \tfrac{1}{4}y + \tfrac{1}{4}z\right)\mathbf{v}_2 + \left(\tfrac{33}{50}x + \tfrac{23}{100}y - \tfrac{19}{100}z\right)\mathbf{v}_3$$

DISCUSSION AND DISCOVERY

D1. **(a)** True. Any set of more than n vectors in R^n is linearly dependent.

(b) True. Any set of less than n vectors cannot be spanning set for R^n.

(c) True. If every vector in R^n can be expressed in exactly one way as a linear combination of the vectors in S, then S is a basis for R^n and thus must contain exactly n vectors.

(d) True. If $A\mathbf{x} = \mathbf{0}$ has infinitely many solutions, then $\det(A) = 0$ and so the row vectors of A are linearly dependent.

(e) True. If $V \subseteq W$ (or $W \subseteq V$) and if $\dim(V) = \dim(W)$, then $V = W$.

D2. No. If $S = \{\mathbf{v}_1, \mathbf{v}_2, \ldots, \mathbf{v}_n\}$ is a linearly dependent set in R^n, then S is not a spanning set for R^n; thus it is not possible to create a basis by forming linear combinations of the vectors in S.

D3. Each such operator corresponds to (and is determined by) a permutation of the vectors in the basis B. Thus there are a total of $n!$ such operators.

D4. Let A be the matrix having the vectors $\mathbf{v}_1$ and $\mathbf{v}_2$ as its columns. Then

$$\det(A) = (\sin^2\alpha - \sin^2\beta) - (\cos^2\alpha - \cos^2\beta) = -\cos 2\alpha + \cos 2\beta$$

and $\det(A) \neq 0$ if and only if $\cos 2\alpha \neq \cos 2\beta$, i.e., if and only if $\alpha \neq \pm\beta + k\pi$ where $k = 0, \pm1, \pm2, \ldots$. For these values of α and β, the vectors $\mathbf{v}_1$ and $\mathbf{v}_2$ form a basis for R^2.

D5. Suppose W is a subspace of R^n and $\dim(W) = k$. If $S = \{w_1, w_2, \ldots, w_j\}$ is a spanning set for W, then either S is a basis for W (in which case S contains exactly k vectors) or, from Theorem 7.2.2, a basis for W can be obtained by removing appropriate vectors from S. Thus the number of elements in a spanning set must be at least k, and the smallest possible number is k.

WORKING WITH PROOFS

P1. Let V be a nonzero subspace of R^n, and let $\mathbf{v}_1$ be a nonzero vector in V. If $\dim(V) = 1$, then $S = \{\mathbf{v}_1\}$ is a basis for V. Otherwise, from Theorem 7.2.2(b), a basis for V can be obtained by adding appropriate vectors from V to the set S.

P2. Let $\{\mathbf{v}_1, \mathbf{v}_2, \ldots, \mathbf{v}_n\}$ be a basis for R^n and, for k any integer between 1 and n, let $V = \text{span}\{\mathbf{v}_1, \mathbf{v}_2, \ldots, \mathbf{v}_k\}$. Then $S = \{\mathbf{v}_1, \mathbf{v}_2, \ldots, \mathbf{v}_k\}$ is a basis for V and so $\dim(V) = k$. The subspace $V = \{\mathbf{0}\}$ has dimension 0.

P3. Let $S = \{\mathbf{v}_1, \mathbf{v}_2, \ldots, \mathbf{v}_n\}$. Since every vector in R^n can be written as a linear combination of vectors in S, we have $\text{span}(S) = R^n$. Moreover, from the uniqueness, if $c_1\mathbf{v}_1 + c_2\mathbf{v}_2 + \cdots + c_n\mathbf{v}_n = \mathbf{0}$ then $c_1 = c_2 = \cdots = c_n = 0$. Thus the vectors $\mathbf{v}_1, \mathbf{v}_2, \ldots, \mathbf{v}_n$ span R^n and are linearly independent, i.e., $S = \{\mathbf{v}_1, \mathbf{v}_2, \ldots, \mathbf{v}_n\}$ is a basis for R^n.

P4. Since we know that $\dim(R^n) = n$, it suffices to show that the vectors $T(\mathbf{v}_1), T(\mathbf{v}_2), \ldots, T(\mathbf{v}_n)$ are linearly independent. This follows from the fact that if $c_1 T(\mathbf{v}_1) + c_2 T(\mathbf{v}_2) + \cdots + c_n T(\mathbf{v}_n) = \mathbf{0}$ then

$$T(c_1\mathbf{v}_1 + c_2\mathbf{v}_2 + \cdots + c_n\mathbf{v}_n) = c_1 T(\mathbf{v}_1) + c_2 T(\mathbf{v}_2) + \cdots + c_n T(\mathbf{v}_n) = \mathbf{0}$$

and so, since T is one-to-one, we must have $c_1\mathbf{v}_1 + c_2\mathbf{v}_2 + \cdots + c_n\mathbf{v}_n = \mathbf{0}$. Since $\mathbf{v}_1, \mathbf{v}_2, \ldots, \mathbf{v}_n$ are linearly independent it follows from this that $c_1 = c_2 = \cdots = c_n = 0$. Thus $T(\mathbf{v}_1), T(\mathbf{v}_2), \ldots, T(\mathbf{v}_n)$ are linearly independent.

P5. Since $B = \{\mathbf{v}_1, \mathbf{v}_2, \ldots, \mathbf{v}_n\}$ is a basis for R^n, every vector $\mathbf{x}$ in R^n can be expressed as a linear combination $\mathbf{x} = c_1\mathbf{v}_1 + c_2\mathbf{v}_2 + \cdots + c_n\mathbf{v}_n$ for exactly one choice of scalars $c_1, c_2, \ldots, c_n$. Thus it makes sense to *define* a transformation $T : R^n \to R^n$ by setting

$$T(\mathbf{x}) = T(c_1\mathbf{v}_1 + c_2\mathbf{v}_2 + \cdots + c_n\mathbf{v}_n) = c_1\mathbf{w}_1 + c_2\mathbf{w}_2 + \cdots + c_n\mathbf{w}_n$$

It is easy to check that T is linear. For example, if $\mathbf{x} = \sum_{j=1}^{n} c_j\mathbf{v}_j$ and $\mathbf{y} = \sum_{j=1}^{n} d_j\mathbf{v}_j$, then

$$T(\mathbf{x} + \mathbf{y}) = T\left(\sum_{j=1}^{n} (c_j + d_j)\mathbf{v}_j\right) = \sum_{j=1}^{n} (c_j + d_j)\mathbf{w}_j = \sum_{j=1}^{n} c_j\mathbf{w}_j + \sum_{j=1}^{n} d_j\mathbf{w}_j = T\mathbf{x} + T\mathbf{y}$$

and so T is additive. Finally, the transformation T has the property that $T\mathbf{v}_j = \mathbf{w}_j$ for each $j = 1, \ldots, n$, and it is clear from the defining formula that T is uniquely determined by this property.

P6. (a) Since $\{\mathbf{u}_1, \mathbf{u}_2, \mathbf{u}_3\}$ has the correct number of elements, we need only show that the vectors are linearly independent. Suppose c_1, c_2, c_3 are scalars such that $c_1\mathbf{u}_1 + c_2\mathbf{u}_2 + c_3\mathbf{u}_3 = \mathbf{0}$. Then $(c_1 + c_2 + c_3)\mathbf{v}_1 + (c_2 + c_3)\mathbf{v}_2 + c_3\mathbf{v}_3 = \mathbf{0}$ and, since $\{\mathbf{v}_1, \mathbf{v}_2, \mathbf{v}_3\}$ is a linearly independent set, we must have $c_1 + c_2 + c_3 = c_2 + c_3 = c_3 = 0$. It follows that $c_1 = c_2 = c_3 = 0$, and this shows that $\mathbf{u}_1, \mathbf{u}_2, \mathbf{u}_3$ are linearly independent.

(b) If $\{\mathbf{v}_1, \mathbf{v}_2, \ldots, \mathbf{v}_n\}$ is a basis for R^n, then so is $\{\mathbf{u}_1, \mathbf{u}_2, \ldots, \mathbf{u}_n\}$ where

$$\mathbf{u}_1 = \mathbf{v}_1, \quad \mathbf{u}_2 = \mathbf{v}_1 + \mathbf{v}_2, \quad \mathbf{u}_3 = \mathbf{v}_1 + \mathbf{v}_2 + \mathbf{v}_3, \quad \ldots, \quad \mathbf{u}_n = \mathbf{v}_1 + \mathbf{v}_2 + \mathbf{v}_3 + \cdots + \mathbf{v}_n$$

P7. Suppose $\mathbf{x}$ is an eigenvector of A. Then $\mathbf{x} \neq \mathbf{0}$, and $A\mathbf{x} = \lambda\mathbf{x}$ for some scalar λ. It follows that $\text{span}\{\mathbf{x}, A\mathbf{x}\} = \text{span}\{\mathbf{x}, \lambda\mathbf{x}\} = \text{span}\{\mathbf{x}\}$ and so $\text{span}\{\mathbf{x}, A\mathbf{x}\}$ has dimension 1.

Conversely, suppose that $\text{span}\{\mathbf{x}, A\mathbf{x}\}$ has dimension 1. Then the vectors $\mathbf{x} \neq \mathbf{0}$ and $A\mathbf{x}$ are linearly dependent; thus there exist scalars c_1 and c_2, not both zero, such that $c_1\mathbf{x} + c_2 A\mathbf{x} = \mathbf{0}$. We note further that $c_2 \neq 0$, for if $c_2 = 0$ then since $\mathbf{x} \neq \mathbf{0}$ we would have $c_1 = 0$ also. Thus $A x = \lambda x$ where $\lambda = -c_1/c_2$.

P8. Suppose $S = \{\mathbf{v}_1, \mathbf{v}_2, \ldots, \mathbf{v}_k\}$ is a basis for V, where $V \subset W$ and $\dim(V) = \dim(W)$. Then S is a linearly independent set in W, and it follows that S must be basis for W. Otherwise, from Theorem 7.2.2, a basis for W could be obtained by adding additional vectors from W to S and this would violate the assumption that $\dim(V) = \dim(W)$. Finally, since S is a basis for W and $S \subset V$, we must have $W = \mathrm{span}(S) \subset V$ and so $W = V$.

EXERCISE SET 7.3

1. The orthogonal complement of $S = \{\mathbf{v}_1, \mathbf{v}_2\}$ is the solution set of the system

$$\begin{bmatrix} 1 & 1 & 3 \\ 0 & 2 & -1 \end{bmatrix} \begin{bmatrix} x \\ y \\ z \end{bmatrix} = \begin{bmatrix} 0 \\ 0 \end{bmatrix}$$

A general solution of this system is given by $x = -\frac{7}{2}t$, $y = \frac{1}{2}t$, $z = t$ or $(x, y, z) = t(-\frac{7}{2}, \frac{1}{2}, 1)$. Thus $S^{\perp}$ is the line through the origin that is parallel to the vector $(-\frac{7}{2}, \frac{1}{2}, 1)$.

A vector that is orthogonal to both $\mathbf{v}_1$ and $\mathbf{v}_2$ is $\mathbf{w} = \mathbf{v}_1 \times \mathbf{v}_2 = \begin{vmatrix} \mathbf{i} & \mathbf{j} & \mathbf{k} \\ 1 & 1 & 3 \\ 0 & 2 & -1 \end{vmatrix} = -7\mathbf{i} + \mathbf{j} + 2\mathbf{k} = (-7, 1, 2)$.

Note the vector $\mathbf{w}$ is parallel to the one obtained in our first solution. Thus $S^{\perp}$ is the line through the origin that is parallel to $(-7, 1, 2)$ or, equivalently, parallel to $(-\frac{7}{2}, \frac{1}{2}, 1)$.

3. We have $\mathbf{u} \cdot \mathbf{v}_1 = (-1)(6) + (1)(2) + (0)(7) + (2)(2) = 0$. Similarly, $\mathbf{u} \cdot \mathbf{v}_2 = 0$ and $\mathbf{u} \cdot \mathbf{v}_3 = 0$. Thus $\mathbf{u}$ is orthogonal to any linear combination of the vectors $\mathbf{v}_1$, $\mathbf{v}_2$, and $\mathbf{v}_3$, i.e., $\mathbf{u}$ belongs to the orthogonal complement of $W = \mathrm{span}\{\mathbf{v}_1, \mathbf{v}_2, \mathbf{v}_3\}$.

5. The line $y = 2x$ corresponds to vectors of the form $\mathbf{u} = t(1, 2)$, i.e., $W = \mathrm{span}\{(1, 2)\}$. Thus $W^{\perp}$ corresponds to the line $y = -\frac{1}{2}x$ or, equivalently, to vectors of the form $\mathbf{w} = s(2, -1)$.

7. The line W corresponds to scalar multiples of the vector $\mathbf{u} = (2, -5, 4)$; thus a vector $\mathbf{w} = (x, y, z)$ is in $W^{\perp}$ if and only if $\mathbf{u} \cdot \mathbf{w} = 2x - 5y + 4z = 0$. Parametric equations for this plane are:

$$x = \tfrac{5}{2}s - 2t, \quad y = s, \quad z = t$$

9. Let A be the matrix having the given vectors as its rows. The reduced row echelon form of A is

$$U = \begin{bmatrix} 1 & 0 & -16 \\ 0 & 1 & -19 \\ 0 & 0 & 0 \end{bmatrix}$$

Thus the vectors $\mathbf{w}_1 = (1, 0, -16)$ and $\mathbf{w}_2 = (0, 1, -19)$ form a basis for the row space of A or, equivalently, for the subspace W spanned by the given vectors.
We have $W^{\perp} = \mathrm{row}(A)^{\perp} = \mathrm{null}(A) = \mathrm{null}(U)$. Thus, from the above, we conclude that $W^{\perp}$ consists of all vectors of the form $\mathbf{x} = (16t, 19t, t)$, i.e., the vector $\mathbf{u} = (16, 19, 1)$ forms a basis for $W^{\perp}$.

11. Let A be the matrix having the given vectors as its rows. The reduced row echelon form of A is

$$R = \begin{bmatrix} 1 & 0 & 0 & 0 \\ 0 & 1 & 0 & 0 \\ 0 & 0 & 1 & 1 \\ 0 & 0 & 0 & 0 \end{bmatrix}$$

Thus the vectors $\mathbf{w}_1 = (1,0,0,0)$, $\mathbf{w}_2 = (0,1,0,0)$, and $\mathbf{w}_3 = (0,0,1,1)$ form a basis for the row space of A or, equivalently, for the space W spanned by the given vectors.
We have $W^\perp = \mathrm{row}(A)^\perp = \mathrm{null}(A) = \mathrm{null}(R)$. Thus, from the above, we conclude that $W^\perp$ consists of all vectors of the form $\mathbf{x} = (0,0,-t,t)$, i.e., the vector $\mathbf{u} = (0,0,-1,1)$ forms a basis for $W^\perp$.

13. Let A be the matrix having the given vectors as its rows. The reduced row echelon form of A is

$$R = \begin{bmatrix} 1 & 0 & 0 & 2 & 0 \\ 0 & 1 & 0 & 3 & 0 \\ 0 & 0 & 1 & 4 & 0 \\ 0 & 0 & 0 & 0 & 1 \end{bmatrix}$$

Thus the vectors $\mathbf{w}_1 = (1,0,0,2,0)$, $\mathbf{w}_2 = (0,1,0,3,0)$, $\mathbf{w}_3 = (0,0,1,4,0)$ and $\mathbf{w}_4 = (0,0,0,0,1)$ form a basis for the row space of A or, equivalently, for the space W spanned by the given vectors.
We have $W^\perp = \mathrm{row}(A)^\perp = \mathrm{null}(A) = \mathrm{null}(R)$. Thus, from the above, we conclude that $W^\perp$ consists of all vectors of the form $\mathbf{x} = (-2t,-3t,-4t,t,0)$, i.e., the vector $\mathbf{u} = (-2,-3,-4,1,0)$ forms a basis for $W^\perp$.

15. In Exercise 9 we found that the vector $\mathbf{u} = (16,19,1)$ forms a basis for $W^\perp$; thus $W = W^{\perp\perp}$ consists of the set of all vectors which are orthogonal to $\mathbf{u}$, i.e., vectors $\mathbf{x} = (x,y,z)$ satisfying $16x + 19y + z = 0$.

17. In Exercise 11 we found that the vector $\mathbf{u} = (0,0,-1,1)$ forms a basis for $W^\perp$; thus $W = W^{\perp\perp}$ consists of the set of all vectors which are orthogonal to $\mathbf{u}$, i.e., vectors $\mathbf{x} = (x_1,x_2,x_3,x_4)$ satisfying $-x_3 + x_4 = 0$.

19. *Solution* 1. Let A be the matrix having the given vectors as its columns. Then a (column) vector $\mathbf{b}$ is a linear combination of $\mathbf{v}_1$, $\mathbf{v}_2$, $\mathbf{v}_3$ if and only if the linear system $A\mathbf{x} = \mathbf{b}$ is consistent. The augmented matrix of this system is

$$\begin{bmatrix} 1 & 5 & 7 & | & b_1 \\ -1 & -4 & -6 & | & b_2 \\ 3 & -4 & 2 & | & b_3 \end{bmatrix}$$

and a row echelon form for this matrix is

$$\begin{bmatrix} 1 & 5 & 7 & | & b_1 \\ 0 & 1 & 1 & | & b_2 + b_1 \\ 0 & 0 & 0 & | & b_3 + 19b_2 + 16b_1 \end{bmatrix}$$

From this we conclude that $\mathbf{b} = (b_1,b_2,b_3)$ lies in the space spanned by the given vectors if and only if $16b_1 + 19b_2 + b_3 = 0$.

Solution 2. The matrix $\begin{bmatrix} \mathbf{v}_1 \\ \mathbf{v}_2 \\ \mathbf{v}_3 \\ \hline \mathbf{b} \end{bmatrix} = \begin{bmatrix} 1 & -1 & 3 \\ 5 & -4 & -4 \\ 7 & -6 & 2 \\ \hline b_1 & b_2 & b_3 \end{bmatrix}$ can be row reduced to $\begin{bmatrix} 1 & 0 & -16 \\ 0 & 1 & -19 \\ 0 & 0 & 0 \\ \hline b_1 & b_2 & b_3 \end{bmatrix}$, and then further reduced to

$$\begin{bmatrix} 1 & 0 & -16 \\ 0 & 1 & -19 \\ 0 & 0 & 0 \\ \hline 0 & 0 & 16b_1 + 19b_2 + b_3 \end{bmatrix}$$

From this we conclude that $W = \text{span}\{\mathbf{v}_1, \mathbf{v}_2, \mathbf{v}_3\}$ has dimension 2, and that $\mathbf{b} = (b_1, b_2, b_3)$ is in W if and only if $16b_1 + 19b_2 + b_3 = 0$.

Solution 3. In Exercise 9 we found that the vector $\mathbf{u} = (16, 19, 1)$ forms a basis for $W^\perp$. Thus $\mathbf{b} = (b_1, b_2, b_3)$ is in $W = W^{\perp\perp}$ if and only if $\mathbf{u} \cdot \mathbf{b} = 0$, i.e., if and only if $16b_1 + 19b_2 + b_3 = 0$.

21. *Solution* 1. Let A be the matrix having the given vectors as its columns. Then a (column) vector $\mathbf{b}$ is a linear combination of $\mathbf{v}_1$, $\mathbf{v}_2$, $\mathbf{v}_3$, $\mathbf{v}_4$ if and only if the linear system $A\mathbf{x} = \mathbf{b}$ is consistent. The augmented matrix of this system is

$$\begin{bmatrix} 1 & 0 & -2 & 4 & \vdots & b_1 \\ 1 & 0 & 0 & 2 & \vdots & b_2 \\ 0 & 1 & 2 & -1 & \vdots & b_3 \\ 0 & 1 & 2 & -1 & \vdots & b_4 \end{bmatrix}$$

and a row reduced form for this matrix is

$$\begin{bmatrix} 1 & 0 & -2 & 4 & \vdots & b_1 \\ 0 & 1 & 2 & -1 & \vdots & b_3 \\ 0 & 0 & 2 & -2 & \vdots & -b_1 + b_2 \\ 0 & 0 & 0 & 0 & \vdots & -b_3 + b_4 \end{bmatrix}$$

Thus a vector $\mathbf{b} = (b_1, b_2, b_3)$ lies in the space spanned by the given vectors if and only if $-b_3 + b_4 = 0$.

Solution 2. The matrix $\begin{bmatrix} \mathbf{v}_1 \\ \mathbf{v}_2 \\ \mathbf{v}_3 \\ \mathbf{v}_4 \\ \hline \mathbf{b} \end{bmatrix} = \begin{bmatrix} 1 & 1 & 0 & 0 \\ 0 & 0 & 1 & 1 \\ -2 & 0 & 2 & 2 \\ 4 & 2 & -1 & -1 \\ \hline b_1 & b_2 & b_3 & b_4 \end{bmatrix}$ can be row reduced to $\begin{bmatrix} 1 & 0 & 0 & 0 \\ 0 & 1 & 0 & 0 \\ 0 & 0 & 1 & 1 \\ 0 & 0 & 0 & 0 \\ \hline b_1 & b_2 & b_3 & b_4 \end{bmatrix}$, and then

further reduced to

$$\begin{bmatrix} 1 & 0 & 0 & 0 \\ 0 & 1 & 0 & 0 \\ 0 & 0 & 1 & 1 \\ 0 & 0 & 0 & 0 \\ \hline 0 & 0 & 0 & -b_3 + b_4 \end{bmatrix}$$

From this we conclude that $W = \text{span}\{\mathbf{v}_1, \mathbf{v}_2, \mathbf{v}_3, \mathbf{v}_4\}$ has dimension 3, and that $\mathbf{b} = (b_1, b_2, b_3)$ is in W if and only if $-b_3 + b_4 = 0$.

Solution 3. In Exercise 11 we found that the vector $\mathbf{u} = (0, 0, -1, 1)$ forms a basis for $W^\perp$. Thus $\mathbf{b} = (b_1, b_2, b_3, b_4)$ is in $W = W^{\perp\perp}$ if and only if $\mathbf{u} \cdot \mathbf{b} = 0$, i.e., if and only if $-b_3 + b_4 = 0$.

23. The augmented matrix $[\mathbf{v}_1 \quad \mathbf{v}_2 \quad \mathbf{v}_3 \quad \mathbf{v}_4 \,|\, \mathbf{b}_1 \,|\, \mathbf{b}_2 \,|\, \mathbf{b}_3]$ is

$$\begin{bmatrix} 1 & -2 & -1 & 0 & \vdots & -2 & \vdots & 0 & \vdots & -2 \\ 1 & 0 & 1 & 2 & \vdots & 4 & \vdots & -2 & \vdots & 2 \\ 0 & 1 & 2 & -1 & \vdots & 2 & \vdots & -3 & \vdots & -1 \\ -1 & 1 & 1 & 1 & \vdots & 2 & \vdots & -1 & \vdots & 1 \\ 2 & 3 & -1 & 1 & \vdots & 5 & \vdots & 5 & \vdots & 0 \end{bmatrix}$$

and the reduced row echelon form of this matrix is

$$\begin{bmatrix} 1 & 0 & 0 & 0 & 1 & 0 & 0 \\ 0 & 1 & 0 & 0 & 1 & 1 & 0 \\ 0 & 0 & 1 & 0 & 1 & -2 & 0 \\ 0 & 0 & 0 & 1 & 1 & 0 & 0 \\ 0 & 0 & 0 & 0 & 0 & 0 & 1 \end{bmatrix}$$

From this we conclude that the vectors $\mathbf{b}_1$ and $\mathbf{b}_2$ lie in span$\{\mathbf{v}_1, \mathbf{v}_2, \mathbf{v}_3, \mathbf{v}_4\}$, but $\mathbf{b}_3$ does not.

25. The reduced row echelon form of the matrix A is

$$R = \begin{bmatrix} 1 & 3 & 0 & 4 & 0 & 0 \\ 0 & 0 & 1 & 2 & 0 & 0 \\ 0 & 0 & 0 & 0 & 1 & 0 \\ 0 & 0 & 0 & 0 & 0 & 1 \end{bmatrix}$$

Thus the vectors $\mathbf{r}_1 = (1, 3, 0, 4, 0, 0)$, $\mathbf{r}_2 = (0, 0, 1, 2, 0, 0)$, $\mathbf{r}_3 = (0, 0, 0, 0, 1, 0)$, $\mathbf{r}_4 = (0, 0, 0, 0, 0, 1)$, form a basis for the row space of A. We also conclude from an inspection of R that the null space of A (solutions of $A\mathbf{x} = \mathbf{0}$) consists of vectors of the form

$$\mathbf{x} = s(-3, 1, 0, 0, 0, 0) + t(-4, 0, -2, 1, 0, 0)$$

Thus the vectors $\mathbf{n}_1 = (-3, 1, 0, 0, 0, 0)$ and $\mathbf{n}_2 = (-4, 0, -2, 1, 0, 0)$ form a basis for the null space of A. It is easy to check that $\mathbf{r}_i \cdot \mathbf{n}_j = 0$ for all i, j. Thus row(A) and null(A) are orthogonal subspaces of R^6.

27. The vectors $\mathbf{r}_1 = (1, 3, 0, 4, 0, 0)$, $\mathbf{r}_2 = (0, 0, 1, 2, 0, 0)$, $\mathbf{r}_3 = (0, 0, 0, 0, 1, 0)$, $\mathbf{r}_4 = (0, 0, 0, 0, 0, 1)$ form a basis for the row space of A (see Exercise 25).

29. The reduced row echelon forms of A and B are $\begin{bmatrix} 1 & 0 & 3 & 0 \\ 0 & 1 & 2 & 0 \\ 0 & 0 & 0 & 1 \end{bmatrix}$ and $\begin{bmatrix} 1 & 0 & 3 & 0 \\ 0 & 1 & 2 & 0 \\ 0 & 0 & 0 & 1 \\ 0 & 0 & 0 & 0 \end{bmatrix}$ respectively. Thus the

vectors $\mathbf{r}_1 = (1, 0, 3, 0)$, $\mathbf{r}_2 = (0, 1, 2, 0)$, $\mathbf{r}_3 = (0, 0, 0, 1)$ form a basis for both of the row spaces. It follows that row$(A) = $ row(B).

31. Let B be the matrix having the given vectors as its rows. The reduced row echelon form of B is

$$R = \begin{bmatrix} 1 & 0 & -1 & 2 \\ 0 & 1 & -4 & 0 \end{bmatrix}$$

and from this it follows that a general solution of $B\mathbf{x} = \mathbf{0}$ is given by $\mathbf{x} = s(1, 4, 1, 0) + t(-2, 0, 0, 1)$, where $-\infty < s, t < \infty$. Thus the vectors $\mathbf{w}_1 = (1, 4, 1, 0)$ and $\mathbf{w}_2 = (-2, 0, 0, 1)$ form a basis for the null space of B, and so the matrix

$$A = \begin{bmatrix} 1 & 4 & 1 & 0 \\ -2 & 0 & 0 & 1 \end{bmatrix}$$

has the property that null$(A) = $ row$(A)^{\perp} = $ null$(B)^{\perp} = $ row$(B)^{\perp\perp} = $ row(B).

DISCUSSION AND DISCOVERY

D1. (a) False. This statement is true if and only if the nonzero rows of A are linearly independent, i.e., if there are no additional zero rows in an echelon form.

(b) True. If E is an elementary matrix, then E is invertible and so $EA\mathbf{x} = \mathbf{0}$ if and only if $A\mathbf{x} = \mathbf{0}$.

(c) True. If A has rank n, then there can be no zero rows in an echelon form for A; thus the reduced row echelon form is the identity matrix.

(d) False. For example, if $m = n$ and A is invertible, then $\text{row}(A) = R^n$ and $\text{null}(A) = \{\mathbf{0}\}$.

(e) True. This follows from the fact (Theorem 7.3.3) that $S^\perp$ is a subspace of R^n.

D2. (a) True. If A is invertible, then the rows of A and the columns of A each form a basis for R^n; thus $\text{row}(A) = \text{col}(A) = R^n$.

(b) False. In fact, the opposite is true: If W is a subspace of V, then $V^\perp$ is a subspace of $W^\perp$ since every vector that is orthogonal to V will also be orthogonal to W.

(c) False. The specified condition implies that $\text{row}(A) \subseteq \text{row}(B)$ from which it follows that

$$\text{null}(A) = \text{row}(A)^\perp \supseteq \text{row}(B)^\perp = \text{null}(B)$$

but in general the inclusion can be proper.

(d) False. For example $A = \begin{bmatrix} 0 & 1 \\ 0 & 0 \end{bmatrix}$ and $B = \begin{bmatrix} 0 & 0 \\ 0 & 1 \end{bmatrix}$ have the same row space but different column spaces.

(e) True. This is in fact true for any invertible matrix E. The rows of EA are linear combinations of the rows of A; thus it is always true that $\text{row}(EA) \subseteq \text{row}(A)$. If E is invertible, then $A = E^{-1}(EA)$ and so we also have $\text{row}(A) \subseteq \text{row}(EA)$.

D3. If $\text{null}(A)$ is the line $3x - 5y = 0$, then $\text{row}(A) = \text{null}(A)^\perp$ is the line $5x + 3y = 0$. Thus each row of A must be a scalar multiple of the vector $(3, -5)$, i.e., A is of the form $A = \begin{bmatrix} 3s & -5s \\ 3t & -5t \end{bmatrix}$.

D4. The null space of A corresponds to the kernel of T_A, and the column space of A corresponds to the range of T_A.

D5. If $W = \mathbf{a}^\perp$, then $W^\perp = \mathbf{a}^{\perp\perp} = \text{span}\{\mathbf{a}\}$ is the 1-dimensional subspace spanned by the vector $\mathbf{a}$.

D6. (a) If $\text{null}(A)$ is a line through the origin, then $\text{row}(A) = \text{null}(A)^\perp$ is the plane through the origin that is perpendicular to that line.

(b) If $\text{col}(A)$ is a line through the origin, then $\text{null}(A^T) = \text{col}(A)^\perp$ is the plane through the origin that is perpendicular to that line.

D7. The first two matrices are invertible; thus in each case the null space is $\{\mathbf{0}\}$. The null space of the matrix $A = \begin{bmatrix} 6 & 2 \\ 3 & 1 \end{bmatrix}$ is the line $3x + y = 0$, and the null space of $A = \begin{bmatrix} 0 & 0 \\ 0 & 0 \end{bmatrix}$ is all of R^2.

D8. (a) Since S has equation $y = 3x$, $S^\perp$ has equation $y = -\frac{1}{3}x$, and $S^{\perp\perp} = S$ has equation $y = 3x$.

(b) If $S = \{(1, 2)\}$, then $\text{span}(S)$ has equation $y = 2x$; thus $S^\perp$ has equation $y = -\frac{1}{2}x$ and $S^{\perp\perp} = \text{span}(S)$ has equation $y = 2x$.

D9. No, this is not possible. The row space of an invertible $n \times n$ matrix is all of R^n since its rows form a basis for R^n. On the other hand, the row space of a singular matrix is a proper subspace of R^n since its rows are linearly dependent and do not span all of R^n.

WORKING WITH PROOFS

P1. It is clear, from the definition of row space, that (a) implies (c). Conversely, suppose that (c) holds. Then, since each row of A is a linear combination of the rows of B, it follows that any linear combination of the rows of A can be expressed as a linear combination of the rows of B. This shows that $\text{row}(A) \subseteq \text{row}(B)$, and a similar argument shows that $\text{row}(B) \subseteq \text{row}(A)$. Thus (a) holds.

P2. The row vectors of an invertible matrix are linearly independent and so, since there are exactly n of them, they form a basis for R^n.

P3. If P is invertible, then $(PA)\mathbf{x} = P(A\mathbf{x}) = \mathbf{0}$ if and only if $A\mathbf{x} = \mathbf{0}$; thus the matrices PA and A have the same null space, and so $\text{nullity}(PA) = \text{nullity}(A)$. From Theorem 7.3.8, it follows that PA and A have also have the same row space. Thus $\text{rank}(PA) = \dim(\text{row}(PA)) = \dim(\text{row}(A)) = \text{rank}(A)$.

P4. From Theorem 7.3.4 we have $S^{\perp} = \text{span}(S)^{\perp}$, and $(\text{span}(S)^{\perp})^{\perp} = \text{span}(S)$ since $\text{span}(S)$ is a subspace. Thus $(S^{\perp})^{\perp} = (\text{span}(S)^{\perp})^{\perp} = \text{span}(S)$

P5. We have $AA^{-1} = \begin{bmatrix} \mathbf{r}_1(A)\cdot\mathbf{c}_1(A^{-1}) & \mathbf{r}_1(A)\cdot\mathbf{c}_2(A^{-1}) & \cdots & \mathbf{r}_1(A)\cdot\mathbf{c}_n(A^{-1}) \\ \mathbf{r}_2(A)\cdot\mathbf{c}_1(A^{-1}) & \mathbf{r}_2(A)\cdot\mathbf{c}_2(A^{-1}) & \cdots & \mathbf{r}_2(A)\cdot\mathbf{c}_n(A^{-1}) \\ \vdots & \vdots & \ddots & \vdots \\ \mathbf{r}_n(A)\cdot\mathbf{c}_1(A^{-1}) & \mathbf{r}_n(A)\cdot\mathbf{c}_2(A^{-1}) & \cdots & \mathbf{r}_n(A)\cdot\mathbf{c}_n(A^{-1}) \end{bmatrix} = \begin{bmatrix} 1 & 0 & \cdots & 0 \\ 0 & 1 & \cdots & 0 \\ \vdots & \vdots & \ddots & \vdots \\ 0 & 0 & \cdots & 1 \end{bmatrix} = [\delta_{ij}]$. In par-

ticular, if $i \in \{1, 2, \ldots, k\}$ and $j \in \{k+1, k+2, \ldots, n\}$, then $i < j$ and so $\mathbf{r}_i(A)\cdot\mathbf{c}_j(A^{-1}) = 0$. This shows that the first k rows of A and the last $n-k$ columns of A^{-1} are orthogonal.

EXERCISE SET 7.4

1. The reduced row echelon form for A is

$$\begin{bmatrix} 1 & 0 & -16 \\ 0 & 1 & -19 \\ 0 & 0 & 0 \end{bmatrix}$$

Thus $\text{rank}(A) = 2$, and a general solution of $A\mathbf{x} = \mathbf{0}$ is given by $\mathbf{x} = (16t, 19t, t) = t(16, 19, 1)$. It follows that $\text{nullity}(A) = 1$. Thus $\text{rank}(A) + \text{nullity}(A) = 2 + 1 = 3$ the number of columns of A.

3. The reduced row echelon form for A is

$$\begin{bmatrix} 1 & 0 & 1 & -\frac{2}{7} \\ 0 & 1 & 1 & \frac{4}{7} \\ 0 & 0 & 0 & 0 \end{bmatrix}$$

Thus $\text{rank}(A) = 2$, and a general solution of $A\mathbf{x} = \mathbf{0}$ is given by $\mathbf{x} = s(-1, -1, 1, 0) + t(\frac{2}{7}, -\frac{4}{7}, 0, 1)$. It follows that $\text{nullity}(A) = 2$. Thus $\text{rank}(A) + \text{nullity}(A) = 2 + 2 = 4$ the number of columns of A.

5. The reduced row echelon form for A is

$$\begin{bmatrix} 1 & 0 & 0 & 2 & \frac{4}{3} \\ 0 & 1 & 0 & 0 & -\frac{1}{6} \\ 0 & 0 & 1 & 0 & -\frac{5}{12} \\ 0 & 0 & 0 & 0 & 0 \\ 0 & 0 & 0 & 0 & 0 \end{bmatrix}$$

Thus rank(A) = 3, and a general solution of $A\mathbf{x} = \mathbf{0}$ is given by

$$\mathbf{x} = s(-2, 0, 0, 1, 0) + t(-\tfrac{4}{3}, \tfrac{1}{6}, \tfrac{5}{12}, 0, 1)$$

It follows that nullity(A) = 2. Thus rank(A) + nullity(A) = 3 + 2 = 5.

7. **(a)** If A is a 5×8 matrix having rank 3, then its nullity must be $8 - 3 = 5$. Thus there are 3 pivot variables and 5 free parameters in a general solution of $A\mathbf{x} = \mathbf{0}$.

   (b) If A is a 7×4 matrix having nullity 2, then its rank must be $4 - 2 = 2$. Thus there are 2 pivot variables and 2 free parameters in a general solution of $A\mathbf{x} = \mathbf{0}$.

   (c) If A is a 6×6 matrix whose row echelon forms have 2 nonzero rows, then A has rank 2 and nullity $6 - 2 = 4$. Thus there are 2 pivot variables and 4 free parameters in a general solution of $A\mathbf{x} = \mathbf{0}$.

9. **(a)** If A is a 5×3 matrix, then the largest possible value for rank(A) is 3 and the smallest possible value for nullity(A) is 0.

   (b) If A is a 3×5 matrix, then the largest possible value for rank(A) is 3 and the smallest possible value for nullity(A) is 2.

   (c) If A is a 4×4 matrix, then the largest possible value for rank(A) is 4 and the smallest possible value for nullity(A) is 0.

11. Let A be the 2×4 matrix having $\mathbf{v}_1$ and $\mathbf{v}_2$ as its rows. Then the reduced row echelon form of A is

$$R = \begin{bmatrix} 1 & 0 & -\tfrac{4}{3} & -\tfrac{5}{3} \\ 0 & 1 & \tfrac{4}{3} & \tfrac{5}{3} \end{bmatrix}$$

and so a general solution of $A\mathbf{x} = \mathbf{0}$ is given by $\mathbf{x} = s(\tfrac{4}{3}, -\tfrac{4}{3}, 1, 0) + t(\tfrac{5}{3}, -\tfrac{5}{3}, 0, 1)$. Thus the vectors

$$\mathbf{v}_1 = (1, 1, 0, 0), \quad \mathbf{v}_2 = (0, 3, 4, 5), \quad \mathbf{w}_3 = (\tfrac{4}{3}, -\tfrac{4}{3}, 1, 0), \quad \mathbf{w}_4 = (\tfrac{5}{3}, -\tfrac{5}{3}, 0, 1)$$

form a basis for R^4.

13. **(a)** This matrix is of rank 1 with $A = \begin{bmatrix} 1 & -7 \\ -2 & 14 \end{bmatrix} = \begin{bmatrix} 1 \\ -2 \end{bmatrix} \begin{bmatrix} 1 & -7 \end{bmatrix} = \mathbf{u}\mathbf{v}^T$.

   (b) This matrix is of rank 2.

   (c) This matrix is of rank 1 with $A = \begin{bmatrix} 1 & 1 & 3 & 3 & -9 \\ -2 & -2 & -6 & -6 & 18 \\ 3 & 3 & 9 & 9 & -27 \end{bmatrix} = \begin{bmatrix} 1 \\ -2 \\ 3 \end{bmatrix} \begin{bmatrix} 1 & 1 & 3 & 3 & -9 \end{bmatrix} = \mathbf{u}\mathbf{v}^T$.

15. $\mathbf{u}\mathbf{u}^T = \begin{bmatrix} 2 \\ 3 \\ 1 \\ 1 \end{bmatrix} \begin{bmatrix} 2 & 3 & 1 & 1 \end{bmatrix} = \begin{bmatrix} 4 & 6 & 2 & 2 \\ 6 & 9 & 3 & 3 \\ 2 & 3 & 1 & 1 \\ 2 & 3 & 1 & 1 \end{bmatrix}$ is a symmetric matrix.

17. The matrix A can be row reduced to upper triangular form as follows:

$$A = \begin{bmatrix} 1 & 1 & t \\ 1 & t & 1 \\ t & 1 & 1 \end{bmatrix} \rightarrow \begin{bmatrix} 1 & 1 & t \\ 0 & t-1 & 1-t \\ 0 & 1-t & 1-t^2 \end{bmatrix} \rightarrow \begin{bmatrix} 1 & 1 & t \\ 0 & t-1 & 1-t \\ 0 & 0 & (2+t)(1-t) \end{bmatrix}$$

If $t = 1$, then the latter has only one nonzero row and so rank(A) = 1. If $t = -2$, then there are two nonzero rows and rank(A) = 2. If $t \neq 1$ or -2, then there are three nonzero rows and rank(A) = 3.

19. If the matrix $\begin{bmatrix} x & y & z \\ 1 & x & y \end{bmatrix}$ has rank 1 then the first row must be a scalar multiple of the second row. Thus $(x, y, z) = t(1, x, y)$, and so $x = t$, $y = tx = t^2$, $z = ty = t^3$.

21. The subspace W, consisting of all vectors of the form $\mathbf{x} = t(2, -1, -3)$, has dimension 1. The subspace $W^\perp$ is the hyperplane consisting of all vectors $\mathbf{x} = (x, y, z)$ which are orthogonal to $(2, -1, 3)$, i.e., which satisfy the equation

$$2x - y - 3z = 0$$

A general solution of the latter is given by

$$\mathbf{x} = (s, 2s - 3t, t) = s(1, 2, 0) + t(0, -3, 1)$$

where $-\infty < t < \infty$. Thus the vectors $(1, 2, 0)$ and $(0, -3, 1)$ form a basis for $W^\perp$, and we have

$$\dim(W) + \dim(W^\perp) = 1 + 2 = 3$$

23. (a) If B is obtained from A by changing only one entry, then $A - B$ has only one nonzero entry (hence only one nonzero row), and so $\text{rank}(A - B) = 1$.

(b) If B is obtained from A by changing only one column (or one row), then $A - B$ has only one nonzero column (or row), and so $\text{rank}(A - B) = 1$.

(c) If B is obtained from A in the specified manner. Then $B - A$ has rank 1. Thus $B - A$ is of the form $\mathbf{u}\mathbf{v}^T$ and $B = A + \mathbf{u}\mathbf{v}^T$.

DISCUSSION AND DISCOVERY

D1. (a) True. For example, if A is $m \times n$ where $m > n$ (more rows than columns) then the rows of A form a set of m vectors in R^n and must therefore be linearly dependent. On the other hand, if $m < n$ then the columns of A must be linearly dependent.

(b) False. If the additional row is a linear combination of the existing rows, then the rank will not be increased.

(c) False. For example, if $m = 1$ then $\text{rank}(A) = 1$ and $\text{nullity}(A) = n - 1$.

(d) True. Such a matrix must have rank less than n; thus $\text{nullity}(A) = n - \text{rank}(A) = 1$.

(e) False. If $A\mathbf{x} = \mathbf{b}$ is inconsistent for some $\mathbf{b}$ then A is not invertible and so $A\mathbf{x} = \mathbf{0}$ has nontrivial solutions; thus $\text{nullity}(A) \geq 1$.

(f) True. We must have $\text{rank}(A) + \text{nullity}(A) = 3$; thus it is not possible to have $\text{rank}(A)$ and $\text{nullity}(A)$ both equal to 1.

D2. If A is $m \times n$, then A^T is $n \times m$ and so, by Theorem 7.4.1, we have $\text{rank}(A^T) + \text{nullity}(A^T) = m$.

D3. If A is a 3×5 matrix, then the number of leading 1's in the reduced row echelon form is at most 3, and (assuming $A \neq 0$) the number of parameters in a general solution of $A\mathbf{x} = \mathbf{0}$ is at most 4.

D4. If A is a 5×3 matrix, then $\text{rank}(A) \leq 3$ and so the number of leading 1's in the reduced row echelon form of A is at most 3. Assuming $A \neq 0$, we have $\text{rank}(A) \geq 1$ and so the number of free parameters in a general solution of $A\mathbf{x} = \mathbf{0}$ is at most 2.

D5. If A is a 3×5 matrix, then the possible values for $\text{rank}(A)$ are 0 (if $A = 0$), 1, 2, or 3, and the corresponding values for $\text{nullity}(A)$ are 5, 4, 3, or 2. If A is a 5×3 matrix, then the possible values for $\text{rank}(A)$ are 0, 1, 2, or 3, and the corresponding values for $\text{nullity}(A)$ are 3, 2, 1, or 0. If A is a 5×5 matrix, then the possible values for $\text{rank}(A)$ are 0, 1, 2, 3, 4, or 5, and the corresponding values for $\text{nullity}(A)$ are 5, 4, 3, 2, 1, or 0.

D6. Assuming $\mathbf{u}$ and $\mathbf{v}$ are nonzero vectors, the rank of $A = \mathbf{u}\mathbf{v}^T$ is 1; thus the nullity is $n - 1$.

D7. Let A be the standard matrix of T. If $\ker(T)$ is a line through the origin, then $\text{nullity}(A) = 1$ and so $\text{rank}(A) = n - 1$. It follows that $\text{ran}(T) = \text{col}(A)$ has dimension $n - 1$ and thus is a hyperplane in R^n.

D8. If $r = 2$ and $s = 1$, then $\text{rank}(A) = 2$. Otherwise, either $r - 2$ or $s - 1$ (or both) is $\neq 0$, and $\text{rank}(A) = 3$. Note that, since the first and fourth rows are linearly independent, $\text{rank}(A)$ can never be 1 (or 0).

D9. If $\lambda \neq 0$, then the reduced row echelon from of A is

$$\begin{bmatrix} 1 & 0 & 0 & 0 \\ 0 & 1 & 0 & \frac{13}{2} \\ 0 & 0 & 1 & -\frac{5}{2} \\ 0 & 0 & 0 & 0 \end{bmatrix}$$

and so $\text{rank}(A) = 3$. On the other hand, if $\lambda = 0$, then the reduced row echelon form is

$$\begin{bmatrix} 1 & 0 & -\frac{1}{2} & \frac{5}{4} \\ 0 & 1 & \frac{5}{2} & \frac{1}{4} \\ 0 & 0 & 0 & 0 \\ 0 & 0 & 0 & 0 \end{bmatrix}$$

and so $\text{rank}(A) = 2$. Thus $\lambda = 0$ is the value for which the matrix A has lowest rank.

D10. Let $A = \begin{bmatrix} 1 & 0 \\ 0 & 0 \end{bmatrix}$ and $B = \begin{bmatrix} 0 & 1 \\ 0 & 0 \end{bmatrix}$. Then $\text{rank}(A) = \text{rank}(B) = 1$, whereas $A^2 = \begin{bmatrix} 1 & 0 \\ 0 & 0 \end{bmatrix}$ has rank 1 and $B^2 = \begin{bmatrix} 0 & 0 \\ 0 & 0 \end{bmatrix}$ has rank 0.

D11. If $AB = 0$ then $\text{rank}(A) + \text{rank}(B) - n = \text{rank}(AB) = 0$, and so $\text{rank}(A) + \text{rank}(B) \leq n$. Since $\text{rank}(B) = n - \text{nullity}(B)$, it follows that $\text{rank}(A) + n - \text{nullity}(B) \leq n$; thus $\text{rank}(A) \leq \text{nullity}(B)$. Similarly, $\text{rank}(B) \leq \text{nullity}(A)$.

WORKING WITH PROOFS

P1. First we note that the matrix $A = \begin{bmatrix} a_{11} & a_{12} & a_{13} \\ a_{21} & a_{22} & a_{23} \end{bmatrix}$ fails to be of rank 2 if and only if one of the rows is a scalar multiple of the other (which includes the case where one or both is a row of zeros). Thus it is sufficient to prove that the latter is equivalent to the condition that

$$(\#) \quad \begin{vmatrix} a_{11} & a_{12} \\ a_{21} & a_{22} \end{vmatrix} = \begin{vmatrix} a_{11} & a_{13} \\ a_{21} & a_{23} \end{vmatrix} = \begin{vmatrix} a_{12} & a_{13} \\ a_{22} & a_{23} \end{vmatrix} = 0$$

Suppose that one of the rows of A is a scalar multiple of the other. Then the same is true of each of the 2×2 matrices that appear in $(\#)$ and so each of these determinants is equal to zero. Suppose, conversely, that the condition $(\#)$ holds. Then we have

$$a_{11}a_{22} - a_{12}a_{21} = a_{11}a_{23} - a_{13}a_{21} = a_{12}a_{23} - a_{13}a_{22} = 0$$

Without loss of generality we may assume that the first row of A is not a row of zeros, and further (interchange columns if necessary) that $a_{11} \neq 0$. We then have $a_{22} = (\frac{a_{21}}{a_{11}})a_{12}$ and $a_{23} = (\frac{a_{21}}{a_{11}})a_{13}$. Thus $(a_{21}, a_{22}, a_{23}) = (\frac{a_{21}}{a_{11}})(a_{11}, a_{12}, a_{13})$, and so the second row is a scalar multiple of the first row.

P2. If A is of rank 1, then $A = \mathbf{x}\mathbf{y}^T$ for some (nonzero) column vectors $\mathbf{x}$ and $\mathbf{y}$. If, in addition, A is symmetric, then we also have $A = A^T = (\mathbf{x}\mathbf{y}^T)^T = \mathbf{y}\mathbf{x}^T$. From this it follows that

$$\mathbf{x}(\mathbf{y}^T\mathbf{x}) = (\mathbf{x}\mathbf{y}^T)\mathbf{x} = (\mathbf{y}\mathbf{x}^T)\mathbf{x} = \mathbf{y}(\mathbf{x}^T\mathbf{x}) = \mathbf{y}\|\mathbf{x}\|^2$$

Since $\mathbf{x}$ and $\mathbf{y}$ are nonzero we have $\mathbf{x}^T\mathbf{y} = \mathbf{y}^T\mathbf{x} = \mathbf{y} \cdot \mathbf{x} \neq 0$ and $\mathbf{x} = \frac{\mathbf{x}^T\mathbf{x}}{\mathbf{x}^T\mathbf{y}}\mathbf{y}$. If $\mathbf{x}^T\mathbf{y} > 0$, it follows that $A = \mathbf{y}\mathbf{x}^T = \mathbf{y}(\frac{\mathbf{x}^T\mathbf{x}}{\mathbf{x}^T\mathbf{y}}\mathbf{y})^T = (\frac{\mathbf{x}^T\mathbf{x}}{\mathbf{x}^T\mathbf{y}})\mathbf{y}\mathbf{y}^T = (\sqrt{\frac{\mathbf{x}^T\mathbf{x}}{\mathbf{x}^T\mathbf{y}}}\mathbf{y})(\sqrt{\frac{\mathbf{x}^T\mathbf{x}}{\mathbf{x}^T\mathbf{y}}}\mathbf{y})^T = \mathbf{u}\mathbf{u}^T$. If $\mathbf{x}^T\mathbf{y} < 0$, then the corresponding formula is $A = -\mathbf{u}\mathbf{u}^T$ where $\mathbf{u} = \sqrt{-\frac{\mathbf{x}^T\mathbf{x}}{\mathbf{x}^T\mathbf{y}}}\mathbf{y}$.

P3. $AB = [\mathbf{c}_1(A) \quad \mathbf{c}_2(A) \quad \cdots \quad \mathbf{c}_k(A)]\begin{bmatrix} \mathbf{r}_1(B) \\ \mathbf{r}_2(B) \\ \vdots \\ \mathbf{r}_k(B) \end{bmatrix} = \mathbf{c}_1(A)\mathbf{r}_1(B) + \mathbf{c}_2(A)\mathbf{r}_2(B) + \cdots + \mathbf{c}_k(A)\mathbf{r}_k(B)$, and

each of the products $\mathbf{c}_j(A)\mathbf{r}_j(B)$ is a rank 1 matrix.

P4. Since the set $V \cup W$ contains n vectors, it suffices to show that $V \cup W$ is a linearly independent set. Suppose then that $c_1, c_2, \ldots, c_k$ and $d_1, d_2, \ldots, d_{n-k}$ are scalars with the property that

$$c_1\mathbf{v}_1 + c_2\mathbf{v}_2 + \cdots + c_k\mathbf{v}_k + d_1\mathbf{w}_1 + d_2\mathbf{w}_2 + \cdots + d_{n-k}\mathbf{w}_{n-k} = \mathbf{0}$$

Then the vector $\mathbf{n} = c_1\mathbf{v}_1 + c_2\mathbf{v}_2 + \cdots + c_k\mathbf{v}_k = -(d_1\mathbf{w}_1 + d_2\mathbf{w}_2 + \cdots + d_{n-k}\mathbf{w}_{n-k})$ belongs both to $V = \text{row}(A)$ and to $W = \text{null}(A) = \text{row}(A)^\perp$. It follows that $\mathbf{n} \cdot \mathbf{n} = \|\mathbf{n}\|^2 = 0$ and so $\mathbf{n} = \mathbf{0}$. Thus we simultaneously have $c_1\mathbf{v}_1 + c_2\mathbf{v}_2 + \cdots + c_k\mathbf{v}_k = \mathbf{0}$ and $d_1\mathbf{w}_1 + d_2\mathbf{w}_2 + \cdots + d_{n-k}\mathbf{w}_{n-k} = \mathbf{0}$. Since the vectors $\mathbf{v}_1, \mathbf{v}_2, \ldots, \mathbf{v}_k$ are linearly independent it follows that $c_1 = c_2 = \cdots = c_k = 0$, and since $\mathbf{w}_1, \mathbf{w}_2, \ldots, \mathbf{w}_{n-k}$ are linearly independent it follows that $d_1 = d_2 = \cdots = d_{n-k} = 0$. This shows that $V \cup W$ is a linearly independent set.

P5. From the inequality $\text{rank}(A) + \text{rank}(B) - n \leq \text{rank}(AB) \leq \text{rank}(A)$ it follows that

$$n - \text{rank}(A) \leq n - \text{rank}(AB) \leq 2n - \text{rank}(A) - \text{rank}(B)$$

which, using Theorem 7.4.1, is the same as

$$\text{nullity}(A) \leq \text{nullity}(AB) \leq \text{nullity}(A) + \text{nullity}(B)$$

Similarly, $\text{nullity}(B) \leq \text{nullity}(AB) \leq \text{nullity}(A) + \text{nullity}(B)$.

P6. Suppose $\lambda = 1 + \mathbf{v}^T A^{-1}\mathbf{u} \neq 0$, and let $X = A^{-1} - \frac{A^{-1}\mathbf{u}\mathbf{v}^T A^{-1}}{\lambda}$. Then

$$BX = (A + \mathbf{u}\mathbf{v}^T)\left(A^{-1} - \frac{A^{-1}\mathbf{u}\mathbf{v}^T A^{-1}}{\lambda}\right) = I + \mathbf{u}\mathbf{v}^T A^{-1} - \frac{\mathbf{u}\mathbf{v}^T A^{-1} + \mathbf{u}\mathbf{v}^T A^{-1}\mathbf{u}\mathbf{v}^T A^{-1}}{\lambda}$$

$$= I + \mathbf{u}\mathbf{v}^T A^{-1} - \frac{\mathbf{u}\mathbf{v}^T A^{-1} + \mathbf{u}(\mathbf{v}^T A^{-1}\mathbf{u})\mathbf{v}^T A^{-1}}{\lambda} = I + \mathbf{u}\mathbf{v}^T A^{-1} - \frac{1 + \mathbf{v}^T A^{-1}\mathbf{u}}{\lambda}\mathbf{u}\mathbf{v}^T A^{-1} = I$$

and so B is invertible with $B^{-1} = X = A^{-1} - \frac{A^{-1}\mathbf{u}\mathbf{v}^T A^{-1}}{\lambda} = A^{-1} - \frac{A^{-1}\mathbf{u}\mathbf{v}^T A^{-1}}{1 + \mathbf{v}^T A^{-1}\mathbf{u}}$.

Note. If $\mathbf{v}^T A^{-1}\mathbf{u} = -1$, then $BA^{-1}\mathbf{u} = (A + \mathbf{u}\mathbf{v}^T)A^{-1}\mathbf{u} = \mathbf{u} + \mathbf{u}(\mathbf{v}^T A^{-1}\mathbf{u}) = \mathbf{u} - \mathbf{u} = \mathbf{0}$; thus BA^{-1} is singular and so B is singular.

EXERCISE SET 7.5

1. The reduced row echelon form of A is $\begin{bmatrix} 1 & 0 & -7 & 4 & -5 \\ 0 & 1 & -9 & 5 & -6 \\ 0 & 0 & 0 & 0 & 0 \\ 0 & 0 & 0 & 0 & 0 \end{bmatrix}$ and the reduced row echelon form of

A^T is $\begin{bmatrix} 1 & 0 & -1 & 2 \\ 0 & 1 & 1 & 1 \\ 0 & 0 & 0 & 0 \\ 0 & 0 & 0 & 0 \\ 0 & 0 & 0 & 0 \end{bmatrix}$. Thus $\dim(\text{row}(A)) = 2$ and $\dim(\text{col}(A) = \dim(\text{row}(A^T)) = 2$. It follows that

$\dim(\text{null}(A)) = 5 - 2 = 3$, and $\dim(\text{null}(A^T)) = 4 - 2 = 2$. Since $\dim(\text{null}(A)) = 3$, there are 3 free parameters in a general solution of $A\mathbf{x} = \mathbf{0}$.

3. The reduced row echelon form of A is $\begin{bmatrix} 1 & 0 & 0 \\ 0 & 1 & 0 \\ 0 & 0 & 1 \\ 0 & 0 & 0 \end{bmatrix}$ and the reduced row echelon form of A^T is $\begin{bmatrix} 1 & 0 & 2 & 0 \\ 0 & 1 & -1 & 0 \\ 0 & 0 & 0 & 1 \end{bmatrix}$.

Thus $\text{rank}(A) = 3$ and $\text{rank}(A^T) = 3$.

5. (a) $\dim(\text{row}(A)) = \dim(\text{col}(A)) = \text{rank}(A) = 3$,
$\dim(\text{null}(A)) = 3 - 3 = 0$, $\dim(\text{null}(A^T)) = 3 - 3 = 0$.

(b) $\dim(\text{row}(A)) = \dim(\text{col}(A)) = \text{rank}(A) = 2$,
$\dim(\text{null}(A)) = 3 - 2 = 1$, $\dim(\text{null}(A^T)) = 3 - 2 = 1$.

(c) $\dim(\text{row}(A)) = \dim(\text{col}(A)) = \text{rank}(A) = 1$,
$\dim(\text{null}(A)) = 3 - 1 = 2$, $\dim(\text{null}(A^T)) = 3 - 1 = 2$.

(d) $\dim(\text{row}(A)) = \dim(\text{col}(A)) = \text{rank}(A) = 2$,
$\dim(\text{null}(A)) = 9 - 2 = 7$, $\dim(\text{null}(A^T)) = 5 - 2 = 3$.

(e) $\dim(\text{row}(A)) = \dim(\text{col}(A)) = \text{rank}(A) = 2$,
$\dim(\text{null}(A)) = 5 - 2 = 3$, $\dim(\text{null}(A^T)) = 9 - 2 = 7$.

7. (a) Since $\text{rank}(A) = \text{rank}[A\,|\,\mathbf{b}] = 3$ the system is consistent. The number of parameters in a general solution is $n - r = 3 - 3 = 0$; i.e., the system has a unique solution.

(b) Since $\text{rank}(A) \neq \text{rank}[A\,|\,\mathbf{b}]$ the system is inconsistent.

(c) Since $\text{rank}(A) = \text{rank}[A\,|\,\mathbf{b}] = 1$ the system is consistent. The number of parameters in a general solution is $n - r = 3 - 1 = 2$.

(d) Since $\text{rank}(A) = \text{rank}[A\,|\,\mathbf{b}] = 2$ the system is consistent. The number of parameters in a general solution is $n - r = 9 - 2 = 7$.

9. (a) This matrix has full column rank because the two column vectors are not scalar multiples of each other; it does not have full row rank because any three vectors in R^2 are linearly dependent.

(b) This matrix does not have full row rank because the 2nd row is a scalar multiple of the 1st row; it does not have full column rank since any three vectors in R^2 are linearly dependent.

(c) This matrix has full row rank because the two row vectors are not scalar multiples of each other; it does not have full column rank because any three vectors in R^2 are linearly dependent.

(d) This (square) matrix is invertible, it has full row rank and full column rank.

11. (a) $\det(A^T A) = \det \begin{bmatrix} 6 & -1 \\ -1 & 25 \end{bmatrix} = 149 \neq 0$ and $\det(AA^T) = \det \begin{bmatrix} 10 & 2 & 11 \\ 2 & 4 & -2 \\ 11 & -2 & 17 \end{bmatrix} = 0$; thus $A^T A$ is invert-

ible and AA^T is not invertible. This corresponds to the fact (see Exercise 9a) that A has full column rank but not full row rank.

(b) $\det(A^T A) = \det \begin{bmatrix} 5 & 20 & -10 \\ 20 & 80 & -40 \\ -10 & -40 & 20 \end{bmatrix} = 0$ and $\det(AA^T) = \det \begin{bmatrix} 21 & 42 \\ 42 & 84 \end{bmatrix} = 0$; thus neither $A^T A$ nor AA^T
is invertible. This corresponds to the fact that A does not have full column rank nor full row rank.

(c) $\det(A^T A) = \det \begin{bmatrix} 10 & 3 & 14 \\ 3 & 1 & 5 \\ 14 & 5 & 26 \end{bmatrix} = 0$ and $\det(AA^T) = \det \begin{bmatrix} 35 & 2 \\ 2 & 2 \end{bmatrix} = 66 \neq 0$; thus $A^T A$ is not invertible
but AA^T is invertible. This corresponds to the fact that A has does not have full column rank but does have full row rank.

(d) $\det(A^T A) = \det \begin{bmatrix} 5 & 4 \\ 4 & 16 \end{bmatrix} = 64 \neq 0$ and $\det(AA^T) = \det \begin{bmatrix} 4 & 2 \\ 2 & 17 \end{bmatrix} = 64 \neq 0$; thus $A^T A$ and AA^T are
both invertible. This corresponds to the fact that A has full column rank and full row rank.

13. The augmented matrix of $A\mathbf{x} = \mathbf{b}$ can be row reduced to $\begin{bmatrix} 1 & 0 & \vdots & \frac{1}{3}b_3 \\ 0 & 1 & \vdots & \frac{1}{2}b_1 - \frac{1}{6}b_3 \\ 0 & 0 & \vdots & b_1 + b_2 \end{bmatrix}$. Thus the system $A\mathbf{x} = \mathbf{b}$

is either inconsistent (if $b_1 + b_2 \neq 0$), or has exactly one solution (if $b_1 + b_2 = 0$). Note that the latter includes the case $b_1 = b_2 = b_3 = 0$; thus the system $A\mathbf{x} = \mathbf{0}$ has only the trivial solution.

15. If $A = \begin{bmatrix} 1 & 2 \\ 2 & 4 \\ -1 & -2 \end{bmatrix}$, then $A^T A = \begin{bmatrix} 6 & 12 \\ 12 & 24 \end{bmatrix}$. It is clear from inspection that the rows of A and $A^T A$

are multiples of the single vector $\mathbf{u} = (1, 2)$. Thus $\text{row}(A) = \text{row}(A^T A)$ is the 1-dimensional space consisting of all scalar multiples of $\mathbf{u}$. Similarly, $\text{null}(A) = \text{null}(A^T A)$ is the 1-dimensional space consisting of all vectors $\mathbf{v}$ in R^2 which are orthogonal to $\mathbf{u}$, i.e., all vectors of the form $\mathbf{v} = s(-2, 1)$.

17. The augmented matrix of the system $A\mathbf{x} = \mathbf{b}$ can be reduced to

$$\begin{bmatrix} 1 & -3 & \vdots & b_1 \\ 0 & 1 & \vdots & b_2 - b_1 \\ 0 & 0 & \vdots & b_3 - 4b_2 + 3b_1 \\ 0 & 0 & \vdots & b_4 + b_2 - 2b_1 \\ 0 & 0 & \vdots & b_5 - 8b_2 + 7b_1 \end{bmatrix}$$

thus the system will be inconsistent unless $(b_1, b_2, b_3, b_4, b_5)$ satisfies the equations $b_3 = -3b_1 + 4b_2$, $b_4 = 2b_1 - b_2$, and $b_5 = -7b_1 + 8b_2$, where b_1 and b_2 can assume any values.

DISCUSSION AND DISCOVERY

D1. If A is a 7×5 matrix with rank 3, then A^T also has rank 3; thus $\dim(\text{row}(A^T)) = \dim(\text{col}(A^T)) = 3$ and $\dim(\text{null}(A^T)) = 7 - 3 = 4$.

D2. If A has rank k then, from Theorems 7.5.2 and 7.5.9, we have $\dim(\text{row}(A^T A)) = \text{rank}(A^T A) = \text{rank}(A^T) = \text{rank}(A) = k$ and $\dim(\text{row}(AA^T)) = \text{rank}(AA^T) = \text{rank}(A) = k$.

D3. If $A^T \mathbf{x} = \mathbf{0}$ has only the trivial solution then, from Theorem 7.5.11, A has full row rank. Thus, if A is $m \times n$, we must have $n \geq m$ and $\dim(\text{row}(A)) = \dim(\text{col}(A)) = m$.

D4. (a) False. The row space and column space always have the same dimension.
    (b) False. It is always true that $\text{rank}(A) = \text{rank}(A^T)$, whether A is square or not.

(c) True. Under these assumptions, the system $A\mathbf{x} = \mathbf{b}$ is consistent (for any $\mathbf{b}$) and so the matrices A and $[A \mid \mathbf{b}]$ have the same rank.

(d) True. If an $m \times n$ matrix A has full row rank and full column rank, then $m = \dim(\text{row}(A)) = \text{rank}(A) = \dim(\text{col}(A)) = n$.

(e) True. If $A^T A$ and AA^T are both invertible then, from Theorem 7.5.10, A has full column rank and full row rank; thus A is square.

(f) True. The rank of a 3×3 matrix is 0, 1, 2, or 3 and the corresponding nullity is 3, 2, 1, or 0.

D5. (a) The solutions of the system are given by $\mathbf{x} = (b - s - t, s, t)$ where $-\infty < s, t < \infty$. This does not violate Theorem 7.5.7(b).

(b) The solutions can be expressed as $(b, 0, 0) + s(-1, 1, 0) + t(-1, 0, 1)$, where $(b, 0, 0)$ is a particular solution and $s(-1, 1, 0) + t(-1, 0, 1)$ is a general solution of the corresponding homogeneous system.

D6. (a) If A is 3×5, then the columns of A are a set of five vectors in R^3 and thus are linearly dependent.

(b) If A is 5×3, then the rows of A are a set of 5 vectors in R^3 and thus are linearly dependent.

(c) If A is $m \times n$, with $m \neq n$, then either the columns of A are linearly dependent or the rows of A are linearly dependent (or both).

WORKING WITH PROOFS

P1. From Theorem 7.5.8(a) we have $\text{null}(A^T A) = \text{null}(A)$. Thus if A is $m \times n$, then $A^T A$ is $n \times n$ and so $\text{rank}(A^T A) = n - \text{nullity}(A^T A) = n - \text{nullity}(A) = \text{rank}(A)$. Similarly, $\text{null}(AA^T) = \text{null}(A^T)$ and so $\text{rank}(AA^T) = m - \text{nullity}(AA^T) = m - \text{nullity}(A^T) = \text{rank}(A^T) = \text{rank}(A)$.

P2. As above, we have $\text{rank}(A^T A) = n - \text{nullity}(A^T A) = n - \text{nullity}(A) = \text{rank}(A)$.

P3. (a) Since $\text{null}(A^T A) = \text{null}(A)$, we have $\text{row}(A) = \text{null}(A)^{\perp} = \text{null}(A^T A)^{\perp} = \text{row}(A^T A)$.

(b) Since $A^T A$ is symmetric, we have $\text{col}(A^T A) = \text{row}(A^T A) = \text{row}(A) = \text{col}(A^T)$.

P4. If A is $m \times n$ where $m < n$, then the columns of A form a set of n vectors in R^m and thus are linearly dependent. Similarly, if $m > n$, then the rows of A form a set of m vectors in R^n and thus are linearly dependent.

P5. If $\text{rank}(A^2) = \text{rank}(A)$ then $\dim(\text{null}(A^2)) = n - \text{rank}(A^2) = n - \text{rank}(A) = \dim(\text{null}(A))$ and, since $\text{null}(A) \subseteq \text{null}(A^2)$, it follows that $\text{null}(A) = \text{null}(A^2)$.
Suppose now that $\mathbf{y}$ belongs to $\text{null}(A) \cap \text{col}(A)$. Then $\mathbf{y} = A\mathbf{x}$ for some $\mathbf{x}$ in R^n and $A\mathbf{y} = \mathbf{0}$. Since $A^2\mathbf{x} = A\mathbf{y} = \mathbf{0}$, it follows that the vector $\mathbf{x}$ belongs to $\text{null}(A^2) = \text{null}(A)$, and so $\mathbf{y} = A\mathbf{x} = \mathbf{0}$. This shows that $\text{null}(A) \cap \text{col}(A) = \{\mathbf{0}\}$.

P6. First we prove that if A is a nonzero matrix with rank k, then A has at least one invertible $k \times k$ submatrix, and all submatrices of larger size are singular. The proof is organized as suggested:

Step 1. If A is an $m \times n$ matrix with rank k, then $\dim(\text{col}(A)) = k$ and so A has k linearly independent columns. Let B be the $m \times k$ submatrix of A having these vectors as its columns. This matrix also has rank k and thus has k linearly independent rows. Let C be the $k \times k$ submatrix of B having these vectors as its rows. Then C is an invertible $k \times k$ submatrix of A.

Step 2. Suppose D is an $r \times r$ submatrix of A with $r > k$. Then, since $\dim(\text{col}(A)) = k < r$, the columns of A which contain those of D must be linearly dependent. It follows that the columns of D are linearly dependent since a nontrivial linear dependence among the containing columns results in a nontrivial linear dependence among the columns of D. Thus D is singular.

Conversely, we prove that if the largest invertible submatrix of A is $k \times k$, then A has rank k.

Step 1. Let C be an invertible $k \times k$ submatrix of A. Then the columns of C are linearly independent and so the columns of A that contain the columns of C are also linearly independent. This shows that $\mathrm{rank}(A) = \dim(\mathrm{col}(A)) \geq k$.

Step 2. Suppose $\mathrm{rank}(A) = r > k$. Then $\dim(\mathrm{col}(A)) = r$, and so A has r linearly independent columns. Let B be the $m \times r$ submatrix of A having these vectors as its columns. Then B also has rank r and thus has r linearly independent rows. Let C be the submatrix of B having these vectors as its rows. Then C is a nonsingular $r \times r$ submatrix of A. Thus the assumption that $\mathrm{rank}(A) > k$ has led to a contradiction. This, together with Step 1, shows that $\mathrm{rank}(A) = k$.

P7. If A is invertible then so is A^T. Thus, using the cited exercise and Theorem 7.5.2, we have

$$\mathrm{rank}(CP) = \mathrm{rank}((CP)^T) = \mathrm{rank}(P^T C^T) = \mathrm{rank}(C^T) = \mathrm{rank}(C)$$

and from this it also follows that $\mathrm{nullity}(CP) = n - \mathrm{rank}(CP) = n - \mathrm{rank}(C) = \mathrm{nullity}(C)$.

EXERCISE SET 7.6

1. A row echelon from for A is $B = \begin{bmatrix} 1 & -1 & 3 \\ 0 & 1 & -19 \\ 0 & 0 & 0 \end{bmatrix}$; thus the first two columns of A are the pivot columns and these column vectors form a basis for $\mathrm{col}(A)$. A row echelon from for A^T is $C = \begin{bmatrix} 1 & 5 & 7 \\ 0 & 1 & 1 \\ 0 & 0 & 0 \end{bmatrix}$; thus $\dim(\mathrm{row}(A)) = 2$, and the first two rows of A form a basis for $\mathrm{row}(A)$.

3. The matrix A can be row reduced to $B = \begin{bmatrix} 1 & 4 & 5 & 2 \\ 0 & 7 & 7 & 4 \\ 0 & 0 & 0 & 0 \end{bmatrix}$; thus the first two columns of A are the pivot columns and these column vectors form a basis for $\mathrm{col}(A)$. The matrix A^T can be row reduced to $C = \begin{bmatrix} 1 & 2 & 1 \\ 0 & 1 & -1 \\ 0 & 0 & 0 \\ 0 & 0 & 0 \end{bmatrix}$; thus $\dim(\mathrm{row}(A)) = 2$, and the first two rows of A form a basis for $\mathrm{row}(A)$.

5. The reduced row echelon form of A is $B = \begin{bmatrix} 1 & 0 & 0 & 2 & \frac{4}{3} \\ 0 & 1 & 0 & 0 & -\frac{1}{6} \\ 0 & 0 & 1 & 0 & -\frac{5}{12} \\ 0 & 0 & 0 & 0 & 0 \\ 0 & 0 & 0 & 0 & 0 \end{bmatrix}$ and the reduced row echelon form of A^T is $C = \begin{bmatrix} 1 & 0 & 0 & 1 & -2 \\ 0 & 1 & 0 & 0 & 1 \\ 0 & 0 & 1 & 1 & 0 \\ 0 & 0 & 0 & 0 & 0 \\ 0 & 0 & 0 & 0 & 0 \end{bmatrix}$. Thus the first three columns of A form a basis for $\mathrm{col}(A)$, and the first three rows of A form a basis for $\mathrm{row}(A)$.

7. Proceeding as in Example 2, we first form the matrix having the given vectors as its columns:

$$A = \begin{bmatrix} 1 & 4 & 1 \\ 2 & 0 & 10 \\ -1 & -6 & -11 \\ 1 & 2 & 3 \end{bmatrix}$$

The reduced row echelon form of A is:

$$R = \begin{bmatrix} 1 & 0 & 0 \\ 0 & 1 & 0 \\ 0 & 0 & 1 \\ 0 & 0 & 0 \end{bmatrix}$$

From this we conclude that all three columns of A are pivot columns; thus the vectors $\mathbf{v}_1$, $\mathbf{v}_2$, $\mathbf{v}_3$ are linearly independent and form a basis for the space which they span (the column space of A).

9. Proceeding as in Example 2, we first form the matrix having the given vectors as its columns:

$$A = \begin{bmatrix} 1 & 2 & -1 & 0 \\ -2 & -4 & 1 & -1 \\ 0 & 0 & 2 & 2 \\ 3 & 6 & 0 & 3 \end{bmatrix}$$

The reduced row echelon form of A is:

$$R = \begin{bmatrix} 1 & 2 & 0 & 1 \\ 0 & 0 & 1 & 1 \\ 0 & 0 & 0 & 0 \\ 0 & 0 & 0 & 0 \end{bmatrix}$$

From this we conclude that the 1st and 3rd columns of A are the pivot columns; thus $\{\mathbf{v}_1, \mathbf{v}_3\}$ is a basis for $W = \text{col}(A)$. Since $\mathbf{c}_2(R) = 2\mathbf{c}_1(R)$ and $\mathbf{c}_4(R) = \mathbf{c}_1(R) + \mathbf{c}_3(R)$, we also conclude that $\mathbf{v}_2 = 2\mathbf{v}_1$ and $\mathbf{v}_4 = \mathbf{v}_1 + \mathbf{v}_3$.

11. The matrix $[A \mid I_3] = \begin{bmatrix} 2 & 0 & -1 & \vdots & 1 & 0 & 0 \\ 4 & 0 & -2 & \vdots & 0 & 1 & 0 \\ 0 & 0 & 0 & \vdots & 0 & 0 & 1 \end{bmatrix}$ can be row reduced (and further partitioned) to

$$\begin{bmatrix} V & \vdots & E_1 \\ \hline O & \vdots & E_2 \end{bmatrix} = \begin{bmatrix} 1 & 0 & -\frac{1}{2} & \vdots & \frac{1}{2} & 0 & 0 \\ \hline 0 & 0 & 0 & \vdots & 1 & -\frac{1}{2} & 0 \\ 0 & 0 & 0 & \vdots & 0 & 0 & 1 \end{bmatrix}$$

Thus the vectors $\mathbf{v}_1 = (1, -\frac{1}{2}, 0)$ and $\mathbf{v}_2 = (0, 0, 1)$ form a basis for $\text{null}(A^T)$.

13. The reduced row echelon form of A is $R = \begin{bmatrix} 1 & 0 & -1 & -4 \\ 0 & 1 & -3 & -7 \\ 0 & 0 & 0 & 0 \\ 0 & 0 & 0 & 0 \end{bmatrix}$. From this we conclude that the first two

columns of A are the pivot columns, and so these two columns form a basis for $\text{col}(A)$. The first two rows of R form a basis for $\text{row}(A) = \text{row}(R)$. We have $A\mathbf{x} = \mathbf{0}$ if and only if $R\mathbf{x} = \mathbf{0}$, and the general solution of the latter is $\mathbf{x} = (s + 4t, 3s + 7t, s, t) = s(1, 3, 1, 0) + t(4, 7, 0, 1)$. Thus $(1, 3, 1, 0)$ and $(4, 7, 0, 1)$ form a basis for $\text{null}(A)$.

The reduced row echelon form of the partitioned matrix $[A \mid I_4]$ is

$$\left[\begin{array}{cccc|cccc}
1 & 0 & -1 & -4 & 0 & 0 & \frac{1}{4} & -\frac{3}{4} \\
0 & 1 & -3 & -7 & 0 & 0 & 0 & -1 \\
0 & 0 & 0 & 0 & 1 & 0 & -\frac{1}{4} & -\frac{1}{4} \\
0 & 0 & 0 & 0 & 0 & 1 & \frac{1}{2} & -\frac{1}{2}
\end{array}\right]$$

Thus the vectors $(1, 0, -\frac{1}{4}, -\frac{1}{4})$ and $(0, 1, \frac{1}{2}, -\frac{1}{2})$ form a basis for $\text{null}(A^T)$.

15. (a) The 1st, 3rd, and 4th columns of A are the pivot columns; thus these vectors form a basis for $\text{col}(A)$.

(b) The first three rows of A form a basis for $\text{row}(A)$.

(c) We have $A\mathbf{x} = \mathbf{0}$ if and only if $R\mathbf{x} = \mathbf{0}$, i.e., if and only if $\mathbf{x}$ is of the form

$$\mathbf{x} = (-4t, t, 0, 0) = t(-4, 1, 0, 0)$$

Thus the vector $\mathbf{u} = (-4, 1, 0, 0)$ forms a basis for $\text{null}(A)$.

(d) We have $A^T\mathbf{x} = \mathbf{0}$ if and only if $C\mathbf{x} = \mathbf{0}$, i.e., if and only if $\mathbf{x}$ is of the form

$$\mathbf{x} = (-\tfrac{1}{4}t, -\tfrac{3}{5}s, 0, s, t) = t(-\tfrac{1}{4}, 0, 0, 0, 1) + s(0, -\tfrac{3}{5}, 0, 1, 0)$$

Thus the vectors $\mathbf{v}_1 = (-\frac{1}{4}, 0, 0, 0, 1)$ and $\mathbf{v}_2 = (0, -\frac{3}{5}, 0, 1, 0)$ form a basis for $\text{null}(A^T)$.

17. The reduced row echelon form of $A = \begin{bmatrix} 2 & -4 & 2 & 6 \\ 1 & 1 & -4 & -2 \\ 3 & -9 & 8 & 14 \\ 0 & 6 & -10 & -10 \end{bmatrix}$ is $R_0 = \begin{bmatrix} 1 & 0 & -\frac{7}{3} & -\frac{1}{3} \\ 0 & 1 & -\frac{5}{3} & -\frac{5}{3} \\ 0 & 0 & 0 & 0 \\ 0 & 0 & 0 & 0 \end{bmatrix}$. Thus the column-

row factorization is

$$A = CR = \begin{bmatrix} 2 & -4 \\ 1 & 1 \\ 3 & -9 \\ 0 & 6 \end{bmatrix} \begin{bmatrix} 1 & 0 & -\frac{7}{3} & -\frac{1}{3} \\ 0 & 1 & -\frac{5}{3} & -\frac{5}{3} \end{bmatrix}$$

and the corresponding column-row expansion is

$$A = \begin{bmatrix} 2 \\ 1 \\ 3 \\ 0 \end{bmatrix} \begin{bmatrix} 1 & 0 & -\frac{7}{3} & -\frac{1}{3} \end{bmatrix} + \begin{bmatrix} -4 \\ 1 \\ -9 \\ 6 \end{bmatrix} \begin{bmatrix} 0 & 1 & -\frac{5}{3} & -\frac{5}{3} \end{bmatrix}$$

DISCUSSION AND DISCOVERY

D1. (a) If A is a 3×5 matrix, then the number of leading 1's in a row echelon form of A is at most 3, the number of parameters in a general solution of $A\mathbf{x} = \mathbf{0}$ is at most 4 (assuming $A \neq O$), the rank of A is at most 3, the rank of A^T is at most 3, and the nullity of A^T is at most 2 (assuming $A \neq O$).

(b) If A is a 5×3 matrix, then the number of leading 1's in a row echelon form of A is at most 3, the number of parameters in a general solution of $A\mathbf{x} = \mathbf{0}$ is at most 2 (assuming $A \neq O$), the rank of A is at most 3, the rank of A^T is at most 3, and the nullity of A^T is at most 4 (assuming $A \neq O$).

(c)   If A is a 4×4 matrix, then the number of leading 1's in a row echelon form of A is at most 4, the number of parameters in a general solution of $A\mathbf{x} = \mathbf{0}$ is at most 3 (assuming $A \neq O$), the rank of A is at most 4, the rank of A^T is at most 4, and the nullity of A^T is at most 3 (assuming $A \neq O$).

(d)   If A is an $m \times n$ matrix, then the number of leading 1's in a row echelon form of A is at most m, the number of parameters in a general solution of $A\mathbf{x} = \mathbf{0}$ is at most $n - 1$ (assuming $A \neq O$), the rank of A is at most $\min\{m, n\}$, the rank of A^T is at most $\min\{m, n\}$, and the nullity of A^T is at most $m - 1$ (assuming $A \neq O$).

D2. The pivot columns of a matrix A are those columns that correspond to the columns of a row echelon form R of A which contain a leading 1. For example,

$$A = \begin{bmatrix} 0 & 0 & -2 & 0 & 7 & 12 \\ 2 & 4 & -10 & 6 & 12 & 28 \\ 2 & 4 & -5 & 6 & -5 & -1 \end{bmatrix} \rightarrow R = \begin{bmatrix} 1 & 2 & 0 & 3 & 0 & 7 \\ 0 & 0 & 1 & 0 & 0 & 1 \\ 0 & 0 & 0 & 0 & 1 & 2 \end{bmatrix}$$

and so the 1st, 3rd, and 5th columns of A are the pivot columns.

D3. The vectors $\mathbf{v}_1 = (4, 0, 1, -4, 5)$ and $\mathbf{v}_2 = (0, 1, 0, 0, 1)$ form a basis for $\text{null}(A^T)$.

D4. (a) $\{\mathbf{a}_1, \mathbf{a}_2, \mathbf{a}_4\}$ is a basis for $\text{col}(A)$.

(b) $\mathbf{a}_3 = 4\mathbf{a}_1 - 3\mathbf{a}_2$ and $\mathbf{a}_5 = 6\mathbf{a}_1 + 7\mathbf{a}_2 + 2\mathbf{a}_4$.

EXERCISE SET 7.7

1.   Using Formula (5) we have $\text{proj}_\mathbf{a}\mathbf{x} = \frac{\mathbf{a} \cdot \mathbf{x}}{\|\mathbf{a}\|^2}\mathbf{a} = (\frac{11}{5})(1, 2) = (\frac{11}{5}, \frac{22}{5})$.
On the other hand, since $\sin\theta = \frac{2}{\sqrt{5}}$ and $\cos\theta = \frac{1}{\sqrt{5}}$, the standard matrix for the projection is $P_\theta = \begin{bmatrix} \frac{1}{5} & \frac{2}{5} \\ \frac{2}{5} & \frac{4}{5} \end{bmatrix}$ and we have $P_\theta\mathbf{x} = \begin{bmatrix} \frac{1}{5} & \frac{2}{5} \\ \frac{2}{5} & \frac{4}{5} \end{bmatrix}\begin{bmatrix} -1 \\ 6 \end{bmatrix} = \begin{bmatrix} \frac{11}{5} \\ \frac{22}{5} \end{bmatrix}$.

3.   A vector parallel to l is $\mathbf{a} = (1, 2)$. Thus, using Formula 5, the projection of $\mathbf{x}$ on l is given by

$$\text{proj}_\mathbf{a}\mathbf{x} = \frac{\mathbf{a} \cdot \mathbf{x}}{\|\mathbf{a}\|^2}\mathbf{a} = (\frac{3}{5})(1, 2) = (\frac{3}{5}, \frac{6}{5})$$

On the other hand, since $\sin\theta = \frac{2}{\sqrt{5}}$ and $\cos\theta = \frac{1}{\sqrt{5}}$, the standard matrix for the projection is $P_\theta = \begin{bmatrix} \frac{1}{5} & \frac{2}{5} \\ \frac{2}{5} & \frac{4}{5} \end{bmatrix}$ and we have $P_\theta\mathbf{x} = \begin{bmatrix} \frac{1}{5} & \frac{2}{5} \\ \frac{2}{5} & \frac{4}{5} \end{bmatrix}\begin{bmatrix} 1 \\ 1 \end{bmatrix} = \begin{bmatrix} \frac{3}{5} \\ \frac{6}{5} \end{bmatrix}$.

5.   The vector component of $\mathbf{x}$ along $\mathbf{a}$ is $\text{proj}_\mathbf{a}\mathbf{x} = \frac{\mathbf{x} \cdot \mathbf{a}}{\|\mathbf{a}\|^2}\mathbf{a} = \frac{1}{5}(0, 2, -1) = (0, \frac{2}{5}, -\frac{1}{5})$, and the component orthogonal to $\mathbf{a}$ is $\mathbf{x} - \text{proj}_\mathbf{a}\mathbf{x} = (1, 1, 1) - (0, \frac{2}{5}, -\frac{1}{5}) = (1, \frac{3}{5}, \frac{6}{5})$.

7.   The vector component of $\mathbf{x}$ along $\mathbf{a}$ is $\text{proj}_\mathbf{a}\mathbf{x} = \frac{\mathbf{x} \cdot \mathbf{a}}{\|\mathbf{a}\|^2}\mathbf{a} = \frac{2}{40}(4, -4, 2, -2) = (\frac{1}{5}, -\frac{1}{5}, \frac{1}{10} - \frac{1}{10})$, and the component orthogonal to $\mathbf{a}$ is $\mathbf{x} - \text{proj}_\mathbf{a}\mathbf{x} = (2, 1, 1, 2) - (\frac{1}{5}, -\frac{1}{5}, \frac{1}{10}, -\frac{1}{10}) = (\frac{9}{5}, \frac{6}{5}, \frac{9}{10}, \frac{21}{10})$.

9. $\|\text{proj}_\mathbf{a}\mathbf{x}\| = \frac{|\mathbf{a} \cdot \mathbf{x}|}{\|\mathbf{a}\|} = \frac{|2 - 6 + 24|}{\sqrt{4 + 9 + 36}} = \frac{20}{\sqrt{49}} = \frac{20}{7}$

11. $\|\text{proj}_\mathbf{a}\mathbf{x}\| = \frac{|\mathbf{a} \cdot \mathbf{x}|}{\|\mathbf{a}\|} = \frac{|-8 + 6 - 2 + 15|}{\sqrt{16 + 4 + 4 + 9}} = \frac{11}{\sqrt{33}} = \frac{\sqrt{33}}{3}$

13. If $\mathbf{a} = (-1, 5, 2)$ then, from Theorem 7.7.3, the standard matrix for the orthogonal projection of R^3 onto span$\{\mathbf{a}\}$ is given by

$$P = \frac{1}{\mathbf{a}^T \mathbf{a}} \mathbf{a}\mathbf{a}^T = \frac{1}{30} \begin{bmatrix} -1 \\ 5 \\ 2 \end{bmatrix} \begin{bmatrix} -1 & 5 & 2 \end{bmatrix} = \frac{1}{30} \begin{bmatrix} 1 & -5 & -2 \\ -5 & 25 & 10 \\ -2 & 10 & 4 \end{bmatrix}$$

We note from inspection that P is symmetric. It is also apparent that P has rank 1 since each of its rows is a scalar multiple of $\mathbf{a}$. Finally, it is easy to check that

$$P^2 = \frac{1}{30} \begin{bmatrix} 1 & -5 & -2 \\ -5 & 25 & 10 \\ -2 & 10 & 4 \end{bmatrix} \frac{1}{30} \begin{bmatrix} 1 & -5 & -2 \\ -5 & 25 & 10 \\ -2 & 10 & 4 \end{bmatrix} = \frac{1}{30} \begin{bmatrix} 1 & -5 & -2 \\ -5 & 25 & 10 \\ -2 & 10 & 4 \end{bmatrix} = P$$

and so P is idempotent.

15. Let $M = \begin{bmatrix} 3 & 2 \\ -4 & 0 \\ 1 & 3 \end{bmatrix}$. Then $M^T M = \begin{bmatrix} 26 & 9 \\ 9 & 13 \end{bmatrix}$ and, from Theorem 7.7.5, the standard matrix for the orthogonal projection of R^3 onto $W = $ span$\{\mathbf{a}_1, \mathbf{a}_2\}$ is given by

$$P = M(M^T M)^{-1} M^T = \begin{bmatrix} 3 & 2 \\ -4 & 0 \\ 1 & 3 \end{bmatrix} \frac{1}{257} \begin{bmatrix} 13 & -9 \\ -9 & 26 \end{bmatrix} \begin{bmatrix} 3 & -4 & 1 \\ 2 & 0 & 3 \end{bmatrix} = \frac{1}{257} \begin{bmatrix} 113 & -84 & 96 \\ -84 & 208 & 56 \\ 96 & 56 & 193 \end{bmatrix}$$

We note from inspection that the matrix P is symmetric. The reduced row echelon form of P is

$$\begin{bmatrix} 1 & 0 & \frac{3}{2} \\ 0 & 1 & \frac{7}{8} \\ 0 & 0 & 0 \end{bmatrix}$$

and from this we conclude that P has rank 2. Finally, it is easy to check that

$$P^2 = \frac{1}{257} \begin{bmatrix} 113 & -84 & 96 \\ -84 & 208 & 56 \\ 96 & 56 & 193 \end{bmatrix} \frac{1}{257} \begin{bmatrix} 113 & -84 & 96 \\ -84 & 208 & 56 \\ 96 & 56 & 193 \end{bmatrix} = \frac{1}{257} \begin{bmatrix} 113 & -84 & 96 \\ -84 & 208 & 56 \\ 96 & 56 & 193 \end{bmatrix} = P$$

and so P is idempotent.

17. The standard matrix for the orthogonal projection of R^3 onto the xz-plane is $P = \begin{bmatrix} 1 & 0 & 0 \\ 0 & 0 & 0 \\ 0 & 0 & 1 \end{bmatrix}$. This agrees with the following computation using Formula (27): Let $M = \begin{bmatrix} 1 & 0 \\ 0 & 0 \\ 0 & 1 \end{bmatrix}$. Then $M^T M = \begin{bmatrix} 1 & 0 \\ 0 & 1 \end{bmatrix}$, and $M(M^T M)^{-1} M^T = \begin{bmatrix} 1 & 0 \\ 0 & 0 \\ 0 & 1 \end{bmatrix} \begin{bmatrix} 1 & 0 \\ 0 & 1 \end{bmatrix} \begin{bmatrix} 1 & 0 & 0 \\ 0 & 0 & 1 \end{bmatrix} = \begin{bmatrix} 1 & 0 & 0 \\ 0 & 0 & 0 \\ 0 & 0 & 1 \end{bmatrix}$.

19. We proceed as in Example 6. The general solution of the equation $x + y + z = 0$ can be written as

$$\begin{bmatrix} x \\ y \\ z \end{bmatrix} = \begin{bmatrix} -s - t \\ s \\ t \end{bmatrix} = s \begin{bmatrix} -1 \\ 1 \\ 0 \end{bmatrix} + t \begin{bmatrix} -1 \\ 0 \\ 1 \end{bmatrix}$$

and so the two column vectors on the right form a basis for the plane. If M is the 3×2 matrix having these vectors as its columns, then $M^T M = \begin{bmatrix} 2 & 1 \\ 1 & 2 \end{bmatrix}$ and the standard matrix of the orthogonal projection onto the plane is

$$P = M(M^T M)^{-1} M^T = \begin{bmatrix} -1 & -1 \\ 1 & 0 \\ 0 & 1 \end{bmatrix} \frac{1}{2} \begin{bmatrix} 2 & -1 \\ -1 & 2 \end{bmatrix} \begin{bmatrix} -1 & 1 & 0 \\ -1 & 0 & 1 \end{bmatrix} = \frac{1}{3} \begin{bmatrix} 2 & -1 & -1 \\ -1 & 2 & -1 \\ -1 & -1 & 2 \end{bmatrix}$$

The orthogonal projection of the vector $\mathbf{v}$ on the plane is $P\mathbf{v} = \frac{1}{3} \begin{bmatrix} 2 & -1 & -1 \\ -1 & 2 & -1 \\ -1 & -1 & 2 \end{bmatrix} \begin{bmatrix} 2 \\ 4 \\ -1 \end{bmatrix} = \frac{1}{3} \begin{bmatrix} 1 \\ 7 \\ -8 \end{bmatrix}$.

21. Let A be the matrix having the given vectors as its columns. The reduced row echelon form of A is

$$\begin{bmatrix} 1 & \frac{1}{2} & 0 & 1 \\ 0 & 0 & 1 & 1 \\ 0 & 0 & 0 & 0 \\ 0 & 0 & 0 & 0 \end{bmatrix}$$

and from this we conclude that the vectors $\mathbf{a}_1$ and $\mathbf{a}_3$ form a basis for the subspace W spanned by the given vectors (column space of A). Let M be the 4×2 matrix having $\mathbf{a}_1$ and $\mathbf{a}_3$ as its columns. Then

$$M^T M = \begin{bmatrix} 4 & -6 & 2 & -4 \\ 1 & 0 & -2 & 5 \end{bmatrix} \begin{bmatrix} 4 & 1 \\ -6 & 0 \\ 2 & -2 \\ -4 & 5 \end{bmatrix} = \begin{bmatrix} 72 & -20 \\ -20 & 30 \end{bmatrix}$$

and the standard matrix for the orthogonal projection of R^4 onto W is given by

$$M(M^T M)^{-1} M^T = \begin{bmatrix} 4 & 1 \\ -6 & 0 \\ 2 & -2 \\ -4 & 5 \end{bmatrix} \frac{1}{1760} \begin{bmatrix} 30 & -20 \\ -20 & 72 \end{bmatrix} \begin{bmatrix} 4 & -6 & 2 & -4 \\ 1 & 0 & -2 & 5 \end{bmatrix}$$

$$= \frac{1}{220} \begin{bmatrix} 89 & -105 & -3 & 25 \\ -105 & 135 & -15 & 15 \\ -3 & -15 & 31 & -75 \\ 25 & 15 & -75 & 185 \end{bmatrix}$$

23. The reduced row echelon from of the matrix A is $R = \begin{bmatrix} 1 & 0 & \frac{1}{2} & -\frac{1}{2} \\ 0 & 1 & \frac{1}{2} & \frac{1}{2} \end{bmatrix}$; thus a general solution of the system $A\mathbf{x} = \mathbf{0}$ can be expressed as

$$\mathbf{x} = \begin{bmatrix} -\frac{1}{2}s + \frac{1}{2}t \\ -\frac{1}{2}s - \frac{1}{2}t \\ s \\ t \end{bmatrix} = s \begin{bmatrix} -\frac{1}{2} \\ -\frac{1}{2} \\ 1 \\ 0 \end{bmatrix} + t \begin{bmatrix} \frac{1}{2} \\ -\frac{1}{2} \\ 0 \\ 1 \end{bmatrix}$$

where the two column vectors on the right hand side form a basis for the solution space. Let B be the matrix having these two vectors as its columns. Then

$$B^T B = \begin{bmatrix} -\frac{1}{2} & -\frac{1}{2} & 1 & 0 \\ \frac{1}{2} & -\frac{1}{2} & 0 & 1 \end{bmatrix} \begin{bmatrix} -\frac{1}{2} & \frac{1}{2} \\ -\frac{1}{2} & -\frac{1}{2} \\ 1 & 0 \\ 0 & 1 \end{bmatrix} = \begin{bmatrix} \frac{3}{2} & 0 \\ 0 & \frac{3}{2} \end{bmatrix} = \frac{3}{2} \begin{bmatrix} 1 & 0 \\ 0 & 1 \end{bmatrix}$$

and the standard matrix for the orthogonal projection from R^4 onto the solution space is

$$P = B(B^T B)^{-1} B^T = \begin{bmatrix} -\frac{1}{2} & \frac{1}{2} \\ -\frac{1}{2} & -\frac{1}{2} \\ 1 & 0 \\ 0 & 1 \end{bmatrix} \frac{2}{3} \begin{bmatrix} 1 & 0 \\ 0 & 1 \end{bmatrix} \begin{bmatrix} -\frac{1}{2} & -\frac{1}{2} & 1 & 0 \\ \frac{1}{2} & -\frac{1}{2} & 0 & 1 \end{bmatrix} = \frac{1}{3} \begin{bmatrix} 1 & 0 & -1 & 1 \\ 0 & 1 & -1 & -1 \\ -1 & -1 & 2 & 0 \\ 1 & -1 & 0 & 2 \end{bmatrix}$$

Thus the orthogonal projection of $\mathbf{v} = (5, 6, 7, 2)$ on the solution space is given (in column form) by

$$P\mathbf{v} = \frac{1}{3} \begin{bmatrix} 1 & 0 & -1 & 1 \\ 0 & 1 & -1 & -1 \\ -1 & -1 & 2 & 0 \\ 1 & -1 & 0 & 2 \end{bmatrix} \begin{bmatrix} 5 \\ 6 \\ 7 \\ 2 \end{bmatrix} = \frac{1}{3} \begin{bmatrix} 0 \\ -3 \\ 3 \\ 3 \end{bmatrix} = \begin{bmatrix} 0 \\ -1 \\ 1 \\ 1 \end{bmatrix}$$

25. The reduced row echelon form of the matrix $A = \begin{bmatrix} 1 & 1 & -1 & -1 \\ 1 & 0 & 1 & 3 \\ 3 & 2 & -1 & 1 \end{bmatrix}$ is $R = \begin{bmatrix} 1 & 0 & 1 & 3 \\ 0 & 1 & -2 & -4 \\ 0 & 0 & 0 & 0 \end{bmatrix}$. From this we conclude that the first two rows of R form a basis for $\text{row}(A)$, and the first two columns of A form a basis for $\text{col}(A)$.

Orthogonal projection of R^4 onto $\text{row}(A)$: Let B be the 4×2 matrix having the first two rows of R as its columns. Then

$$B^T B = \begin{bmatrix} 1 & 0 & 1 & 3 \\ 0 & 1 & -2 & -4 \end{bmatrix} \begin{bmatrix} 1 & 0 \\ 0 & 1 \\ 1 & -2 \\ 3 & -4 \end{bmatrix} = \begin{bmatrix} 11 & -14 \\ -14 & 21 \end{bmatrix}$$

and the standard matrix for the orthogonal projection of R^4 onto $\text{row}(A)$ is given by

$$P_r = B(B^T B)^{-1} B^T = \begin{bmatrix} 1 & 0 \\ 0 & 1 \\ 1 & -2 \\ 3 & -4 \end{bmatrix} \frac{1}{35} \begin{bmatrix} 21 & 14 \\ 14 & 11 \end{bmatrix} \begin{bmatrix} 1 & 0 & 1 & 3 \\ 0 & 1 & -2 & -4 \end{bmatrix} = \frac{1}{35} \begin{bmatrix} 21 & 14 & -7 & 7 \\ 14 & 11 & -8 & -2 \\ -7 & -8 & 9 & 11 \\ 7 & -2 & 11 & 29 \end{bmatrix}$$

Orthogonal projection of R^3 onto $\text{col}(A)$: Let C be the 3×2 matrix having the first two columns of A as its columns. Then

$$C^T C = \begin{bmatrix} 1 & 1 & 3 \\ 1 & 0 & 2 \end{bmatrix} \begin{bmatrix} 1 & 1 \\ 1 & 0 \\ 3 & 2 \end{bmatrix} = \begin{bmatrix} 11 & 7 \\ 7 & 5 \end{bmatrix}$$

and the standard matrix for the orthogonal projection of R^3 onto $\text{col}(A)$ is given by

$$P_c = C(C^T C)^{-1} C^T = \begin{bmatrix} 1 & 1 \\ 1 & 0 \\ 3 & 2 \end{bmatrix} \frac{1}{6} \begin{bmatrix} 5 & -7 \\ -7 & 11 \end{bmatrix} \begin{bmatrix} 1 & 1 & 3 \\ 1 & 0 & 2 \end{bmatrix} = \frac{1}{6} \begin{bmatrix} 2 & -2 & 2 \\ -2 & 5 & 1 \\ 2 & 1 & 5 \end{bmatrix}$$

27. The reduced row echelon form of the matrix $[A \,|\, \mathbf{b}]$ is $[R \,|\, \mathbf{c}] = \begin{bmatrix} 1 & 0 & \frac{1}{5} & \frac{8}{5} & | & -\frac{2}{5} \\ 0 & 1 & -\frac{3}{5} & \frac{1}{5} & | & \frac{1}{5} \\ 0 & 0 & 0 & 0 & | & 0 \end{bmatrix}$. From this we

conclude that the system $A\mathbf{x} = \mathbf{b}$ is consistent, with general solution $\mathbf{x} = \begin{bmatrix} -\frac{2}{5} - \frac{1}{5}s - \frac{8}{5}t \\ \frac{1}{5} + \frac{3}{5}s + \frac{1}{5}t \\ s \\ t \end{bmatrix}$. In particu-

lar, $\mathbf{x}_0 = \begin{bmatrix} -\frac{2}{5} \\ \frac{1}{5} \\ 0 \\ 0 \end{bmatrix}$ is a solution of $A\mathbf{x} = \mathbf{b}$. Let B be the 4×2 matrix having the first two rows of R as

its columns, and let $C = B^T B = \begin{bmatrix} \frac{18}{5} & \frac{1}{5} \\ \frac{1}{5} & \frac{7}{5} \end{bmatrix}$. Then the standard matrix for the orthogonal projection

of R^4 onto $\text{row}(R) = \text{row}(A)$ is $P = BC^{-1}B^T$, and the solution of $A\mathbf{x} = \mathbf{b}$ which lies in $\text{row}(A)$ is
given by

$$\mathbf{x}_{\text{row}(A)} = P\mathbf{x}_0 = \begin{bmatrix} \frac{7}{25} & -\frac{1}{25} & \frac{2}{25} & \frac{11}{25} \\ -\frac{1}{25} & \frac{18}{25} & -\frac{11}{25} & \frac{2}{25} \\ \frac{2}{25} & -\frac{11}{25} & \frac{7}{25} & \frac{1}{25} \\ \frac{11}{25} & \frac{1}{25} & \frac{2}{25} & \frac{18}{25} \end{bmatrix} \begin{bmatrix} -\frac{2}{5} \\ \frac{1}{5} \\ 0 \\ 0 \end{bmatrix} = \begin{bmatrix} -\frac{3}{25} \\ \frac{4}{25} \\ -\frac{3}{25} \\ -\frac{4}{25} \end{bmatrix}$$

29. In Exercise 19 we found that the standard matrix for the orthogonal projection of R^3 onto the plane
W with equation $x + y + z = 0$ is

$$P = \frac{1}{3}\begin{bmatrix} 2 & -1 & -1 \\ -1 & 2 & -1 \\ -1 & -1 & 2 \end{bmatrix}$$

Thus the standard matrix for the orthogonal projection onto $W^\perp$ (the line perpendicular to W) is

$$I - P = \begin{bmatrix} 1 & 0 & 0 \\ 0 & 1 & 0 \\ 0 & 0 & 1 \end{bmatrix} - \frac{1}{3}\begin{bmatrix} 2 & -1 & -1 \\ -1 & 2 & -1 \\ -1 & -1 & 2 \end{bmatrix} = \frac{1}{3}\begin{bmatrix} 1 & 1 & 1 \\ 1 & 1 & 1 \\ 1 & 1 & 1 \end{bmatrix}$$

Note that $W^\perp$ is the 1-dimensional space (line) spanned by the vector $\mathbf{a} = (1, 1, 1)$ and so the computation above is consistent with the formula given in Theorem 7.7.3.

31. Let A be the 4×2 matrix having the vectors $\mathbf{v}_1$ and $\mathbf{v}_2$ as its columns. Then

$$A^T A = \begin{bmatrix} 1 & -2 & 3 & 0 \\ 3 & 4 & -1 & 2 \end{bmatrix}\begin{bmatrix} 1 & 3 \\ -2 & 4 \\ 3 & -1 \\ 0 & 2 \end{bmatrix} = \begin{bmatrix} 14 & -8 \\ -8 & 30 \end{bmatrix}$$

and the standard matrix for the orthogonal projection of R^4 onto the subspace $W = \text{span}\{\mathbf{v}_1, \mathbf{v}_2\}$ is

$$P = A(A^T A)^{-1}A^T = \begin{bmatrix} 1 & 3 \\ -2 & 4 \\ 3 & -1 \\ 0 & 2 \end{bmatrix}\frac{1}{178}\begin{bmatrix} 15 & 4 \\ 4 & 7 \end{bmatrix}\begin{bmatrix} 1 & -2 & 3 & 0 \\ 3 & 4 & -1 & 2 \end{bmatrix} = \frac{1}{89}\begin{bmatrix} 51 & 23 & 28 & 25 \\ 23 & 54 & -31 & 20 \\ 28 & -31 & 59 & 5 \\ 25 & 20 & 5 & 14 \end{bmatrix}$$

Thus the orthogonal projection of the vector $\mathbf{v} = (1,1,1,1)$ onto $W^\perp$ is given (in column form) by

$$\mathbf{v} - P\mathbf{v} = \begin{bmatrix} 1 \\ 1 \\ 1 \\ 1 \end{bmatrix} - \frac{1}{89} \begin{bmatrix} 51 & 23 & 28 & 25 \\ 23 & 54 & -31 & 20 \\ 28 & -31 & 59 & 5 \\ 25 & 20 & 5 & 14 \end{bmatrix} \begin{bmatrix} 1 \\ 1 \\ 1 \\ 1 \end{bmatrix} = \begin{bmatrix} 1 \\ 1 \\ 1 \\ 1 \end{bmatrix} - \frac{1}{89} \begin{bmatrix} 127 \\ 66 \\ 61 \\ 64 \end{bmatrix} = \frac{1}{89} \begin{bmatrix} -38 \\ 23 \\ 28 \\ 25 \end{bmatrix}$$

DISCUSSION AND DISCOVERY

D1. (a) The rank of the standard matrix for the orthogonal projection of R^n onto a line through the origin is 1, and onto its orthogonal complement is $n - 1$.

  (b) If $n \geq 2$, then the rank of the standard matrix for the orthogonal projection of R^n onto a plane through the origin is 2, and onto its orthogonal complement is $n - 2$.

D2. A 5×5 matrix P is the standard matrix for an orthogonal projection of R^5 onto some 3-dimensional subspace if and only if it is symmetric, idempotent, and has rank 3.

D3. If $\mathbf{x}_1 = \text{proj}_{\mathbf{a}} \mathbf{x}$ and $\mathbf{x}_2 = \mathbf{x} - \mathbf{x}_1$, then $\|\mathbf{x}\|^2 = \|\mathbf{x}_1\|^2 + \|\mathbf{x}_2\|^2$ and so

$$\|\mathbf{x}_2\|^2 = \|\mathbf{x}\|^2 - \|\mathbf{x}_1\|^2 = \|\mathbf{x}\|^2 - \left(\frac{|\mathbf{a} \cdot \mathbf{x}|}{\|\mathbf{a}\|} \right)^2 = \|\mathbf{x}\|^2 - \frac{|\mathbf{a} \cdot \mathbf{x}|^2}{\|\mathbf{a}\|^2}$$

Thus $q = \sqrt{\|\mathbf{x}\|^2 - \frac{|\mathbf{a} \cdot \mathbf{x}|^2}{\|\mathbf{a}\|^2}} = \|\mathbf{x}_2\|$ where $\mathbf{x}_2$ is the vector component of $\mathbf{x}$ orthogonal to $\mathbf{a}$.

D4. If P is the standard matrix for the orthogonal projection of R^n on a subspace W, then $P^2 = P$ (P is idempotent) and so $P^k = P$ for all $k \geq 2$. In particular, we have $P^n = P$.

D5. (a) True. Since $\text{proj}_W \mathbf{u}$ belongs to W and $\text{proj}_{W^\perp} \mathbf{u}$ belongs to $W^\perp$, the two vectors are orthogonal.

  (b) False. For example, the matrix $P = \begin{bmatrix} 0 & 1 \\ 0 & 1 \end{bmatrix}$ satisfies $P^2 = P$ but is not symmetric and therefore does not correspond to an orthogonal projection.

  (c) True. See the proof of Theorem 7.7.7.

  (d) True. Since $P^2 = P$, we also have $(I - P)^2 = I - 2P + P^2 = I - P$; thus $I - P$ is idempotent.

  (e) False. In fact, since $\text{proj}_{\text{col}(A)} \mathbf{b}$ belongs to $\text{col}(A)$, the system $A\mathbf{x} = \text{proj}_{\text{col}(A)} \mathbf{b}$ is always consistent.

D6. Since $(W^\perp)^\perp = W$ (Theorem 7.7.8), it follows that $((W^\perp)^\perp)^\perp = W^\perp$.

D7. The matrix $A = \begin{bmatrix} 1 & 1 & 1 \\ 1 & 1 & 1 \\ 1 & 1 & 1 \end{bmatrix}$ is symmetric and has rank 1, but is not the standard matrix of an orthogonal projection. Note that $A^2 = 3A$, so A is not idempotent.

D8. In this case the row space of A is equal to all of R^n. Thus the orthogonal projection of R^n onto $\text{row}(A)$ is the identity transformation, and its matrix is the identity matrix.

D9. Suppose that A is an $n \times n$ idempotent matrix, and that λ is an eigenvalue of A with corresponding eigenvector $\mathbf{x}(\mathbf{x} \neq \mathbf{0})$. Then $A^2\mathbf{x} = A(A\mathbf{x}) = A(\lambda\mathbf{x}) = \lambda^2\mathbf{x}$. On the other hand, since $A^2 = A$, we have $A^2\mathbf{x} = A\mathbf{x} = \lambda\mathbf{x}$. Since $\mathbf{x} \neq \mathbf{0}$, it follows that $\lambda^2 = \lambda$ and so $\lambda = 0$ or 1.

D10. *Using calculus:* The reduced row echelon form of $[A \,|\, \mathbf{b}]$ is $\begin{bmatrix} 1 & 0 & 3 & | & 7 \\ 0 & 1 & 1 & | & 3 \\ 0 & 0 & 0 & | & 0 \end{bmatrix}$; thus the general solution of

$A\mathbf{x} = \mathbf{b}$ is $\mathbf{x} = (7 - 3t, 3 - t, t)$ where $-\infty < t < \infty$. We have

$$\|\mathbf{x}\|^2 = (7 - 3t)^2 + (3 - t)^2 + t^2 = 58 - 48t + 11t^2$$

and so the solution vector of smallest length corresponds to $\frac{d}{dt}[\|\mathbf{x}\|^2] = -48 + 22t = 0$, i.e., to $t = \frac{24}{11}$. We conclude that $\mathbf{x}_{\text{row}} = (7 - \frac{72}{11}, 3 - \frac{24}{11}, \frac{24}{11}) = (\frac{5}{11}, \frac{9}{11}, \frac{24}{11})$.

Using an orthogonal projection: The solution $\mathbf{x}_{\text{row}}$ is equal to the orthogonal projection of any solution of $A\mathbf{x} = \mathbf{b}$, e.g., $\mathbf{x} = (7, 3, 0)$, onto the row space of A. From the row reduction alluded to above, we see that the vectors $\mathbf{v}_1 = (1, 0, 3)$ and $\mathbf{v}_2 = (0, 1, 1)$ form a basis for the row space of A. Let B be the 3×2 matrix having these vector as its columns. Then $B^T B = \begin{bmatrix} 10 & 3 \\ 3 & 2 \end{bmatrix}$, and the standard matrix for the orthogonal projection of R^3 onto $W = \text{row}(A)$ is given by

$$P = B(B^T B)^{-1} B^T = \begin{bmatrix} 1 & 0 \\ 0 & 1 \\ 3 & 1 \end{bmatrix} \frac{1}{11} \begin{bmatrix} 2 & -3 \\ -3 & 10 \end{bmatrix} \begin{bmatrix} 1 & 0 & 3 \\ 0 & 1 & 1 \end{bmatrix} = \frac{1}{11} \begin{bmatrix} 2 & -3 & 3 \\ -3 & 10 & 1 \\ 3 & 1 & 10 \end{bmatrix}$$

Finally, in agreement with the calculus solution, we have

$$\mathbf{x}_{\text{row}} = P\mathbf{x} = \frac{1}{11} \begin{bmatrix} 2 & -3 & 3 \\ -3 & 10 & 1 \\ 3 & 1 & 10 \end{bmatrix} \begin{bmatrix} 7 \\ 3 \\ 0 \end{bmatrix} = \frac{1}{11} \begin{bmatrix} 5 \\ 9 \\ 24 \end{bmatrix}$$

D11. The rows of R form a basis for the row space of A, and $G = R^T$ has these vectors as its columns. Thus, from Theorem 7.7.5, $G(G^T G)^{-1} G^T$ is the standard matrix for the orthogonal projection of R^n onto $W = \text{row}(A)$.

WORKING WITH PROOFS

P1. If $\mathbf{x}$ and $\mathbf{y}$ are vectors in R^n and if α and β are scalars, then $\mathbf{a} \cdot (\alpha\mathbf{x} + \beta\mathbf{y}) = \alpha(\mathbf{a} \cdot \mathbf{x}) + \beta(\mathbf{a} \cdot \mathbf{y})$. Thus

$$T(\alpha\mathbf{x} + \beta\mathbf{y}) = \frac{\mathbf{a} \cdot (\alpha\mathbf{x} + \beta\mathbf{y})}{\|\mathbf{a}\|^2}\mathbf{a} = \alpha\frac{(\mathbf{a} \cdot \mathbf{x})}{\|\mathbf{a}\|^2}\mathbf{a} + \beta\frac{(\mathbf{a} \cdot \mathbf{y})}{\|\mathbf{a}\|^2}\mathbf{a} = \alpha T(\mathbf{x}) + \beta T(\mathbf{y})$$

which shows that T is linear.

P2. If $\mathbf{b} = t\mathbf{a}$, then $\mathbf{b}^T \mathbf{b} = \mathbf{b} \cdot \mathbf{b} = (t\mathbf{a}) \cdot (t\mathbf{a}) = t^2 \mathbf{a} \cdot \mathbf{a} = t^2 \mathbf{a}^T \mathbf{a}$ and (similarly) $\mathbf{b}\mathbf{b}^T = t^2 \mathbf{a}\mathbf{a}^T$; thus

$$\frac{1}{\mathbf{b}^T \mathbf{b}}\mathbf{b}\mathbf{b}^T = \frac{1}{t^2 \mathbf{a}^T \mathbf{a}}t^2 \mathbf{a}\mathbf{a}^T = \frac{1}{\mathbf{a}^T \mathbf{a}}\mathbf{a}\mathbf{a}^T$$

P3. Let P be a symmetric $n \times n$ matrix that is idempotent and has rank k. Then $W = \text{col}(P)$ is a k-dimensional subspace of R^n. We will show that P is the standard matrix for the orthogonal projection of R^n onto W, i.e., that $P\mathbf{x} = \text{proj}_W \mathbf{x}$ for all $\mathbf{x}$ in R^n. To this end, we first note that $P\mathbf{x}$ belongs to W and that

$$\mathbf{x} = P\mathbf{x} + (\mathbf{x} - P\mathbf{x}) = P\mathbf{x} + (I - P)\mathbf{x}$$

To show that $P\mathbf{x} = \text{proj}_W \mathbf{x}$ it suffices (from Theorem 7.7.4) to show that $(I - P)\mathbf{x}$ belongs to $W^\perp$, and since $W = \text{col}(A) = \text{ran}(P)$, this is equivalent to showing that $P\mathbf{y} \cdot (I - P)\mathbf{x} = 0$ for all

$\mathbf{y}$ in R^n. Finally, since $P^T = P = P^2$ (P is symmetric and idempotent), we have $P(I - P) = P - P^2 = P - P = O$ and so

$$P\mathbf{y} \cdot (I - P)\mathbf{x} = (P\mathbf{y})^T(I - P)\mathbf{x} = \mathbf{y}^T P^T(I - P)\mathbf{x} = \mathbf{y}^T P(I - P)\mathbf{x} = \mathbf{y}^T O\mathbf{x} = 0$$

for every $\mathbf{x}$ and $\mathbf{y}$ in R^n. This completes the proof.

EXERCISE SET 7.8

1. First we note that the columns of A are linearly independent since they are not scalar multiples of each other; thus A has full column rank. It follows from Theorem 7.8.3(b) that the system $A\mathbf{x} = \mathbf{b}$ has a unique least squares solution given by

$$\mathbf{x} = (A^T A)^{-1} A^T \mathbf{b} = \begin{bmatrix} 21 & 25 \\ 25 & 35 \end{bmatrix}^{-1} \begin{bmatrix} 1 & 2 & 4 \\ -1 & 3 & 5 \end{bmatrix} \begin{bmatrix} 2 \\ -1 \\ 5 \end{bmatrix} = \frac{1}{11} \begin{bmatrix} 20 \\ -8 \end{bmatrix}$$

The least squares error vector is

$$\mathbf{b} - A\mathbf{x} = \begin{bmatrix} 2 \\ -1 \\ 5 \end{bmatrix} - \begin{bmatrix} 1 & -1 \\ 2 & 3 \\ 4 & 5 \end{bmatrix} \frac{1}{11} \begin{bmatrix} 20 \\ -8 \end{bmatrix} = \frac{1}{11} \begin{bmatrix} -6 \\ -27 \\ 15 \end{bmatrix}$$

and it is easy to check that this vector is in fact orthogonal to each of the columns of A. For example, $(\mathbf{b} - A\mathbf{x}) \cdot \mathbf{c}_1(A) = \frac{1}{11}[(-6)(1) + (-27)(2) + (15)(4)] = 0$.

3. From Exercise 1, the least squares solution of $A\mathbf{x} = \mathbf{b}$ is $\mathbf{x} = \frac{1}{11}\begin{bmatrix} 20 \\ -8 \end{bmatrix}$; thus

$$A\mathbf{x} = \begin{bmatrix} 1 & -1 \\ 2 & 3 \\ 4 & 5 \end{bmatrix} \frac{1}{11} \begin{bmatrix} 20 \\ -8 \end{bmatrix} = \frac{1}{11} \begin{bmatrix} 28 \\ 16 \\ 40 \end{bmatrix}$$

On the other hand, the standard matrix for the orthogonal projection of R^3 onto col(A) is

$$P = A(A^T A)^{-1} A^T = \begin{bmatrix} 1 & -1 \\ 2 & 3 \\ 4 & 5 \end{bmatrix} \begin{bmatrix} 21 & 25 \\ 25 & 35 \end{bmatrix}^{-1} \begin{bmatrix} 1 & 2 & 4 \\ -1 & 3 & 5 \end{bmatrix} = \frac{1}{220} \begin{bmatrix} 212 & -36 & 20 \\ -36 & 58 & 90 \\ 20 & 90 & 170 \end{bmatrix}$$

and so we have

$$\text{proj}_{\text{col}(A)}\mathbf{b} = P\mathbf{b} = \frac{1}{220} \begin{bmatrix} 212 & -36 & 20 \\ -36 & 58 & 90 \\ 20 & 90 & 170 \end{bmatrix} \begin{bmatrix} 2 \\ -1 \\ 5 \end{bmatrix} = \frac{1}{11} \begin{bmatrix} 28 \\ 16 \\ 40 \end{bmatrix} = A\mathbf{x}$$

5. The least squares solutions of $A\mathbf{x} = \mathbf{b}$ are obtained by solving the associated normal system $A^T A\mathbf{x} = A^T \mathbf{b}$ which is

$$\begin{bmatrix} 24 & 8 \\ 8 & 6 \end{bmatrix} \begin{bmatrix} x_1 \\ x_2 \end{bmatrix} = \begin{bmatrix} 12 \\ 8 \end{bmatrix}$$

Since the matrix on the left is nonsingular, this system has the unique solution

$$\mathbf{x} = \begin{bmatrix} x_1 \\ x_2 \end{bmatrix} = \begin{bmatrix} 24 & 8 \\ 8 & 6 \end{bmatrix}^{-1} \begin{bmatrix} 12 \\ 8 \end{bmatrix} = \frac{1}{80} \begin{bmatrix} 6 & -8 \\ -8 & 24 \end{bmatrix} \begin{bmatrix} 12 \\ 8 \end{bmatrix} = \begin{bmatrix} \frac{1}{10} \\ \frac{6}{5} \end{bmatrix}$$

The error vector is

$$\mathbf{b} - A\mathbf{x} = \begin{bmatrix} 3 \\ 2 \\ 1 \end{bmatrix} - \begin{bmatrix} 2 & 1 \\ 4 & 2 \\ -2 & 1 \end{bmatrix} \begin{bmatrix} \frac{1}{10} \\ \frac{6}{5} \end{bmatrix} = \begin{bmatrix} 3 \\ 2 \\ 1 \end{bmatrix} - \begin{bmatrix} \frac{7}{5} \\ \frac{14}{5} \\ 1 \end{bmatrix} = \begin{bmatrix} \frac{8}{5} \\ -\frac{4}{5} \\ 0 \end{bmatrix}$$

and the least squares error is $\|\mathbf{b} - A\mathbf{x}\| = \sqrt{(\frac{8}{5})^2 + (-\frac{4}{5})^2 + (0)^2} = \sqrt{\frac{80}{25}} = \frac{4\sqrt{5}}{5}$.

7. The least squares solutions of $A\mathbf{x} = \mathbf{b}$ are obtained by solving the normal system $A^T A\mathbf{x} = A^T \mathbf{b}$ which is

$$\begin{bmatrix} 5 & -1 & 4 \\ -1 & 11 & 10 \\ 4 & 10 & 14 \end{bmatrix} \begin{bmatrix} x_1 \\ x_2 \\ x_3 \end{bmatrix} = \begin{bmatrix} -7 \\ 14 \\ 7 \end{bmatrix}$$

The augmented matrix of this system reduces to $\begin{bmatrix} 1 & 0 & 1 & | & -\frac{7}{6} \\ 0 & 1 & 1 & | & \frac{7}{6} \\ 0 & 0 & 0 & | & 0 \end{bmatrix}$; thus there are infinitely many

solutions given by

$$\mathbf{x} = \begin{bmatrix} x_1 \\ x_2 \\ x_3 \end{bmatrix} = \begin{bmatrix} -\frac{7}{6} - t \\ \frac{7}{6} - t \\ t \end{bmatrix} = \begin{bmatrix} -\frac{7}{6} \\ \frac{7}{6} \\ 0 \end{bmatrix} + t \begin{bmatrix} -1 \\ -1 \\ 1 \end{bmatrix}$$

The error vector is

$$\mathbf{b} - A\mathbf{x} = \begin{bmatrix} 7 \\ 0 \\ -7 \end{bmatrix} - \begin{bmatrix} -1 & 3 & 2 \\ 2 & 1 & 3 \\ 0 & 1 & 1 \end{bmatrix} \left(\begin{bmatrix} -\frac{7}{6} \\ \frac{7}{6} \\ 0 \end{bmatrix} + t \begin{bmatrix} -1 \\ -1 \\ 1 \end{bmatrix} \right) = \begin{bmatrix} 7 \\ 0 \\ -7 \end{bmatrix} - \left(\begin{bmatrix} \frac{14}{3} \\ -\frac{7}{6} \\ \frac{7}{6} \end{bmatrix} + \begin{bmatrix} 0 \\ 0 \\ 0 \end{bmatrix} \right) = \begin{bmatrix} \frac{7}{3} \\ \frac{7}{6} \\ -\frac{49}{6} \end{bmatrix}$$

and the least squares error is $\|\mathbf{b} - A\mathbf{x}\| = \sqrt{(\frac{7}{3})^2 + (\frac{7}{6})^2 + (-\frac{49}{6})^2} = \sqrt{\frac{2646}{36}} = \frac{\sqrt{294}}{2}$.

9. The linear model for the given data is $M\mathbf{v} = \mathbf{y}$ where $M = \begin{bmatrix} 1 & 2 \\ 1 & 3 \\ 1 & 5 \\ 1 & 6 \end{bmatrix}$ and $\mathbf{y} = \begin{bmatrix} 1 \\ 2 \\ 3 \\ 4 \end{bmatrix}$. The least squares

solution is obtained by solving the normal system $M^T M\mathbf{v} = M^T \mathbf{y}$ which is

$$\begin{bmatrix} 4 & 16 \\ 16 & 74 \end{bmatrix} \begin{bmatrix} v_1 \\ v_2 \end{bmatrix} = \begin{bmatrix} 10 \\ 47 \end{bmatrix}$$

Since the matrix on the left is nonsingular, this system has a unique solution given by

$$\begin{bmatrix} v_1 \\ v_2 \end{bmatrix} = \begin{bmatrix} 4 & 16 \\ 16 & 74 \end{bmatrix}^{-1} \begin{bmatrix} 10 \\ 47 \end{bmatrix} = \frac{1}{20} \begin{bmatrix} 37 & -8 \\ -8 & 2 \end{bmatrix} \begin{bmatrix} 10 \\ 47 \end{bmatrix} = \frac{1}{10} \begin{bmatrix} -3 \\ 7 \end{bmatrix} = \begin{bmatrix} -\frac{3}{10} \\ \frac{7}{10} \end{bmatrix}$$

Thus the least squares straight line fit to the given data is $y = -\frac{3}{10} + \frac{7}{10}x$.

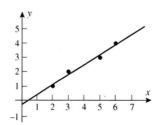

11. The quadratic least squares model for the given data is $M\mathbf{v} = \mathbf{y}$ where $M = \begin{bmatrix} 1 & 0 & 0 \\ 1 & 2 & 4 \\ 1 & 3 & 9 \\ 1 & 3 & 9 \end{bmatrix}$ and $\mathbf{y} = \begin{bmatrix} 1 \\ 0 \\ 1 \\ 2 \end{bmatrix}$.

The least squares solution is obtained by solving the normal system $M^T M \mathbf{v} = M^T \mathbf{y}$ which is

$$\begin{bmatrix} 4 & 8 & 22 \\ 8 & 22 & 62 \\ 22 & 62 & 178 \end{bmatrix} \begin{bmatrix} v_1 \\ v_2 \\ v_3 \end{bmatrix} = \begin{bmatrix} 4 \\ 9 \\ 27 \end{bmatrix}$$

Since the matrix on the left is nonsingular, this system has a unique solution given by

$$\begin{bmatrix} v_1 \\ v_2 \\ v_3 \end{bmatrix} = \begin{bmatrix} 4 & 8 & 22 \\ 8 & 22 & 62 \\ 22 & 62 & 178 \end{bmatrix}^{-1} \begin{bmatrix} 4 \\ 9 \\ 27 \end{bmatrix} = \begin{bmatrix} 1 & -\frac{5}{6} & \frac{1}{6} \\ -\frac{5}{6} & \frac{19}{6} & -1 \\ \frac{1}{6} & -1 & \frac{1}{3} \end{bmatrix} \begin{bmatrix} 4 \\ 9 \\ 27 \end{bmatrix} = \begin{bmatrix} 1 \\ -\frac{11}{6} \\ \frac{2}{3} \end{bmatrix}$$

Thus the least squares quadratic fit to the given data is $y = 1 - \frac{11}{6}x + \frac{2}{3}x^2$.

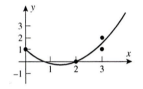

13. The model for the least squares cubic fit to the given data is $M\mathbf{v} = \mathbf{y}$ where

$$M = \begin{bmatrix} 1 & 1 & 1 & 1 \\ 1 & 2 & 4 & 8 \\ 1 & 3 & 9 & 27 \\ 1 & 4 & 16 & 64 \\ 1 & 5 & 25 & 125 \end{bmatrix} \qquad \mathbf{y} = \begin{bmatrix} 4.9 \\ 10.8 \\ 27.9 \\ 60.2 \\ 113.0 \end{bmatrix}$$

The associated normal system $M^T M \mathbf{v} = M^T \mathbf{y}$ is

$$\begin{bmatrix} 5 & 15 & 55 & 225 \\ 15 & 55 & 225 & 979 \\ 55 & 225 & 979 & 4425 \\ 225 & 979 & 4425 & 20515 \end{bmatrix} \begin{bmatrix} a_0 \\ a_1 \\ a_2 \\ a_3 \end{bmatrix} = \begin{bmatrix} 216.8 \\ 916.0 \\ 4087.4 \\ 18822.4 \end{bmatrix}$$

and the solution, written in comma delimited form, is $(a_0, a_1, a_2, a_3) \approx (5.160, -1.864, 0.811, 0.775)$.

15. If $M = \begin{bmatrix} 1 & x_1 \\ 1 & x_2 \\ \vdots & \vdots \\ 1 & x_n \end{bmatrix}$ and $\mathbf{y} = \begin{bmatrix} y_1 \\ y_2 \\ \vdots \\ y_n \end{bmatrix}$, then $M^T M = \begin{bmatrix} 1 & 1 & \cdots & 1 \\ x_1 & x_2 & \cdots & x_n \end{bmatrix} \begin{bmatrix} 1 & x_1 \\ 1 & x_2 \\ \vdots & \vdots \\ 1 & x_n \end{bmatrix} = \begin{bmatrix} n & \sum x_i \\ \sum x_i & \sum x_i^2 \end{bmatrix}$ and

$M^T \mathbf{y} = \begin{bmatrix} 1 & 1 & \cdots & 1 \\ x_1 & x_2 & \cdots & x_n \end{bmatrix} \begin{bmatrix} y_1 \\ y_2 \\ \vdots \\ y_n \end{bmatrix} = \begin{bmatrix} \sum y_i \\ \sum x_i y_i \end{bmatrix}$. Thus the normal system can be written as

$$\begin{bmatrix} n & \sum x_i \\ \sum x_i & \sum x_i^2 \end{bmatrix} \begin{bmatrix} a \\ b \end{bmatrix} = \begin{bmatrix} \sum y_i \\ \sum x_i y_i \end{bmatrix}$$

DISCUSSION AND DISCOVERY

D1. **(a)** The distance from the point $P_0 = (1, -2, 1)$ to the plane W with equation $x + y - z = 0$ is

$$d = \frac{|(1)(1) + (1)(-2) + (-1)(1)|}{\sqrt{(1)^2 + (1)^2 + (-1)^2}} = \frac{2}{\sqrt{3}} = \frac{2\sqrt{3}}{3}$$

and the point in the plane that is closest to P_0 is $Q = (\frac{5}{3}, -\frac{4}{3}, \frac{1}{3})$. The latter is found by computing the orthogonal projection of the vector $\mathbf{b} = \overrightarrow{OP_0}$ onto the plane: The column vectors of the matrix $A = \begin{bmatrix} -1 & 1 \\ 1 & 0 \\ 0 & 1 \end{bmatrix}$ form a basis for W and so the orthogonal projection of $\mathbf{b}$ onto W is given by

$$\text{proj}_W \mathbf{b} = A(A^T A)^{-1} A^T \mathbf{b} = \begin{bmatrix} -1 & 1 \\ 1 & 0 \\ 0 & 1 \end{bmatrix} \frac{1}{3} \begin{bmatrix} 2 & 1 \\ 1 & 2 \end{bmatrix} \begin{bmatrix} -1 & 1 & 0 \\ 1 & 0 & 1 \end{bmatrix} \begin{bmatrix} 1 \\ -2 \\ 1 \end{bmatrix} = \frac{1}{3} \begin{bmatrix} 5 \\ -4 \\ 1 \end{bmatrix}$$

(b) The distance from the point $P_0 = (1, 2, 0, -1)$ to the hyperplane $x_1 - x_2 + 2x_3 - 2x_4 = 0$ is

$$d = \frac{|(1)(1) + (-1)(2) + (2)(0) + (-2)(-1)|}{\sqrt{(1)^2 + (-1)^2 + (2)^2 + (-2)^2}} = \frac{1}{\sqrt{10}} = \frac{\sqrt{10}}{10}$$

and the point in the plane that is closest to P_0 is $Q = (\frac{9}{10}, \frac{21}{10}, -\frac{2}{10}, -\frac{8}{10})$. The latter is found by computing the orthogonal projection of the vector $\mathbf{b} = \overrightarrow{OP_0}$ onto the hyperplane:

$$\text{proj}_W \mathbf{b} = \begin{bmatrix} 1 & -2 & 2 \\ 1 & 0 & 0 \\ 0 & 1 & 0 \\ 0 & 0 & 1 \end{bmatrix} \frac{1}{10} \begin{bmatrix} 9 & 2 & -2 \\ 2 & 6 & 4 \\ -2 & 4 & 4 \end{bmatrix} \begin{bmatrix} 1 & 1 & 0 & 0 \\ -2 & 0 & 1 & 0 \\ 2 & 0 & 0 & 1 \end{bmatrix} \begin{bmatrix} 1 \\ 2 \\ 0 \\ -1 \end{bmatrix} = \frac{1}{10} \begin{bmatrix} 9 \\ 21 \\ -2 \\ -8 \end{bmatrix}$$

D2. **(a)** The vector in $\text{col}(A)$ that is closest to $\mathbf{b}$ is $\text{proj}_{\text{col}(A)} \mathbf{b} = A(A^T A)^{-1} A^T \mathbf{b}$.

(b) The least squares solution of $A\mathbf{x} = \mathbf{b}$ is $\mathbf{x} = (A^T A)^{-1} A^T \mathbf{b}$.

(c) The least squares error vector is $\mathbf{b} - A(A^T A)^{-1} A^T \mathbf{b}$.

(d) The least squares error is $\|\mathbf{b} - A(A^T A)^{-1} A^T \mathbf{b}\|$.

(e) The standard matrix for the orthogonal projection onto $\text{col}(A)$ is $P = A(A^T A)^{-1} A^T$.

D3. From Theorem 7.8.4, a vector $\mathbf{x}$ is a least squares solution of $A\mathbf{x} = \mathbf{b}$ if and only if $\mathbf{b} - A\mathbf{x}$ belongs to $\text{col}(A)^\perp$. We have $A = \begin{bmatrix} 1 & -1 \\ 2 & 3 \\ 4 & 5 \end{bmatrix}$ and $\mathbf{b} - A\mathbf{x} = \begin{bmatrix} 1 \\ 1 \\ s \end{bmatrix} - \begin{bmatrix} 1 & -1 \\ 2 & 3 \\ 4 & 5 \end{bmatrix} \begin{bmatrix} 1 \\ 2 \end{bmatrix} = \begin{bmatrix} 1 \\ 1 \\ s \end{bmatrix} - \begin{bmatrix} -1 \\ 8 \\ 14 \end{bmatrix} = \begin{bmatrix} 2 \\ -7 \\ s - 14 \end{bmatrix}$; thus $\mathbf{b} - A\mathbf{x}$ is orthogonal to $\text{col}(A)$ if and only if

$$(1)(2) + (2)(-7) + (4)(s - 14) = 0 = (-1)(2) + (3)(-7) + (5)(s - 14)$$

$$4s - 68 = 0 = 5s - 93$$

These equations are clearly incompatible and so we conclude that, for any value of s, the vector $\mathbf{x}$ is not a least squares solution of $A\mathbf{x} = \mathbf{b}$.

D4. The given data points nearly fall on a straight line; thus it would be reasonable to perform a linear least squares fit and then use the resulting linear formula $y = a + bx$ to extrapolate to $x = 45$.

D5. The model for this least squares fit is $\begin{bmatrix} 1 & 1 \\ 1 & \frac{1}{3} \\ 1 & \frac{1}{6} \end{bmatrix} \begin{bmatrix} a \\ b \end{bmatrix} = \begin{bmatrix} 7 \\ 3 \\ 1 \end{bmatrix}$, and the corresponding normal system is

$\begin{bmatrix} 3 & \frac{3}{2} \\ \frac{3}{2} & \frac{41}{36} \end{bmatrix} \begin{bmatrix} a \\ b \end{bmatrix} = \begin{bmatrix} 11 \\ \frac{49}{6} \end{bmatrix}$. Thus the least squares solution is

$$\begin{bmatrix} a \\ b \end{bmatrix} = \begin{bmatrix} 3 & \frac{3}{2} \\ \frac{3}{2} & \frac{41}{36} \end{bmatrix}^{-1} \begin{bmatrix} 11 \\ \frac{49}{6} \end{bmatrix} = \begin{bmatrix} \frac{41}{42} & -\frac{9}{7} \\ -\frac{9}{7} & \frac{18}{7} \end{bmatrix} \begin{bmatrix} 11 \\ \frac{49}{6} \end{bmatrix} = \begin{bmatrix} \frac{5}{21} \\ \frac{48}{7} \end{bmatrix}$$

resulting in $y = \frac{5}{21} + \frac{48}{7} \frac{1}{x}$ as the best least squares fit by a curve of this type.

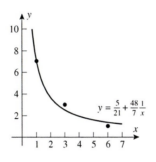

D6. We have $\begin{bmatrix} A & I_m \\ O & A^T \end{bmatrix} \begin{bmatrix} \mathbf{x} \\ \mathbf{r} \end{bmatrix} = \begin{bmatrix} \mathbf{b} \\ \mathbf{0} \end{bmatrix}$ if and only if $A\mathbf{x} + \mathbf{r} = \mathbf{b}$ and $A^T\mathbf{r} = \mathbf{0}$. Note that $A^T\mathbf{r} = \mathbf{0}$ if and only if $\mathbf{r}$ is orthogonal to $\mathrm{col}(A)$. It follows that $\mathbf{b} - A\mathbf{x}$ belongs to $\mathrm{col}(A)^\perp$ and so, from Theorem 7.8.4, $\mathbf{x}$ is a least squares solution of $A\mathbf{x} = \mathbf{b}$ and $\mathbf{r} = \mathbf{b} - A\mathbf{x}$ is the least squares error vector.

WORKING WITH PROOFS

P1. If $A\mathbf{x} = \mathbf{b}$ is consistent, then $\mathbf{b}$ is in the column space of A and any solution of $A\mathbf{x} = \mathbf{b}$ is also a least squares solution (since $\|\mathbf{b} - A\mathbf{x}\| = 0$). If, in addition, the columns of A are linearly independent then there is only one solution of $A\mathbf{x} = \mathbf{b}$ and, from Theorem 7.8.3, the least squares solution is also unique. Thus, in this case, the least squares solution is the same as the exact solution of $A\mathbf{x} = \mathbf{b}$.

P2. If $\mathbf{b}$ is orthogonal to the column space of A, then $\mathrm{proj}_{\mathrm{col}(A)}\mathbf{b} = \mathbf{0}$. Thus, since the columns of A are linearly independent, we have $A\mathbf{x} = \mathrm{proj}_{\mathrm{col}(A)}\mathbf{b} = \mathbf{0}$ if and only if $\mathbf{x} = \mathbf{0}$.

P3. The least squares solutions of $A\mathbf{x} = \mathbf{b}$ are the solutions of the normal system $A^TA\mathbf{x} = A^T\mathbf{b}$. From Theorem 3.5.1, the solution space of the latter is the translated subspace $\hat{\mathbf{x}} + W$ where $\hat{\mathbf{x}}$ is any least squares solution and $W = \mathrm{null}(A^TA) = \mathrm{null}(A)$.

P4. If $\mathbf{w}$ is in W and $\mathbf{w} \neq \mathrm{proj}_W\mathbf{b}$ then, as in the proof of Theorem 7.8.1, we have $\|\mathbf{b} - \mathbf{w}\| > \|\mathbf{b} - \mathrm{proj}_W\mathbf{b}\|$; thus $\mathrm{proj}_W\mathbf{b}$ is the only best approximation to $\mathbf{b}$ from W.

P5. If $a_0, a_1, a_2, \ldots, a_m$ are scalars such that $a_0\mathbf{c}_1(M) + a_1\mathbf{c}_2(M) + a_2\mathbf{c}_3(M) + \cdots + a_m\mathbf{c}_{m+1}(M) = \mathbf{0}$, then

$$a_0 + a_1x_i + a_2x_i^2 + \cdots + a_mx_i^m = 0$$

for each $i = 1, 2, \ldots, n$. Thus each x_i is a root of the polynomial $P(x) = a_0 + a_1x + \cdots + a_mx^m$. But such a polynomial (if not identically zero) can have at most m distinct roots. Thus, if $n > m$ and if at least $m + 1$ of the numbers $x_1, x_2, \ldots, x_n$ are distinct, then $a_0 = a_1 = a_2 = \cdots = a_m = 0$. This shows that the column vectors of M are linearly independent.

P6. If at least $m+1$ of the numbers $x_1, x_2, \ldots, x_n$ are distinct then, from Exercise P5, the column vectors of M are linearly independent; thus M has full column rank and $M^T M$ is invertible.

EXERCISE SET 7.9

1. (a) $\mathbf{v}_1 \cdot \mathbf{v}_2 = (2)(3) + (3)(2) = 12 \neq 0$; thus the vectors $\mathbf{v}_1$, $\mathbf{v}_2$ do not form an orthogonal set.
    (b) $\mathbf{v}_1 \cdot \mathbf{v}_2 = (-1)(1) + (1)(1) = 0$; thus the vectors $\mathbf{v}_1$, $\mathbf{v}_2$ form an orthogonal set. The corresponding orthonormal set is $\mathbf{q}_1 = \frac{\mathbf{v}_1}{\|\mathbf{v}_1\|} = (-\frac{1}{\sqrt{2}}, \frac{1}{\sqrt{2}})$, $\mathbf{q}_2 = \frac{\mathbf{v}_2}{\|\mathbf{v}_2\|} = (\frac{1}{\sqrt{2}}, \frac{1}{\sqrt{2}})$.
    (c) We have $\mathbf{v}_1 \cdot \mathbf{v}_2 = \mathbf{v}_1 \cdot \mathbf{v}_3 = \mathbf{v}_2 \cdot \mathbf{v}_3 = 0$; thus the vectors $\mathbf{v}_1$, $\mathbf{v}_2$, $\mathbf{v}_3$ form an orthogonal set. The corresponding orthonormal set is $\mathbf{q}_1 = (-\frac{2}{\sqrt{6}}, \frac{1}{\sqrt{6}}, \frac{1}{\sqrt{6}})$, $\mathbf{q}_2 = (\frac{1}{\sqrt{5}}, 0, \frac{2}{\sqrt{5}})$, $\mathbf{q}_3 = (-\frac{2}{\sqrt{30}}, -\frac{5}{\sqrt{30}}, \frac{1}{\sqrt{30}})$.
    (d) Although $\mathbf{v}_1 \cdot \mathbf{v}_2 = \mathbf{v}_1 \cdot \mathbf{v}_3 = 0$ we have $\mathbf{v}_2 \cdot \mathbf{v}_3 = (1)(4) + (2)(-3) + (5)(0) = -2 \neq 0$; thus the vectors $\mathbf{v}_1$, $\mathbf{v}_2$, $\mathbf{v}_3$ do not form an orthogonal set.

3. (a) These vectors form an orthonormal set.
    (b) These vectors do not form an orthogonal set since $\mathbf{v}_2 \cdot \mathbf{v}_3 = -\frac{2}{\sqrt{6}} \neq 0$.
    (c) These vectors form an orthogonal set but not an orthonormal set since $\|\mathbf{v}_3\| = \sqrt{3} \neq 1$.

5. (a) $\operatorname{proj}_W \mathbf{x} = (\mathbf{x} \cdot \mathbf{v}_1)\mathbf{v}_1 + (\mathbf{x} \cdot \mathbf{v}_2)\mathbf{v}_2 = (\frac{3}{\sqrt{18}})\mathbf{v}_1 + (\frac{12}{6})\mathbf{v}_2 = (0, \frac{1}{6}, -\frac{2}{3}, -\frac{1}{6}) + (1, \frac{5}{3}, \frac{1}{3}, \frac{1}{3}) = (1, \frac{11}{6}, -\frac{1}{3}, \frac{1}{6})$
    (b) $\operatorname{proj}_W \mathbf{x} = (\mathbf{x} \cdot \mathbf{v}_1)\mathbf{v}_1 + (\mathbf{x} \cdot \mathbf{v}_2)\mathbf{v}_2 + (\mathbf{x} \cdot \mathbf{v}_3)\mathbf{v}_3 = (\frac{3}{\sqrt{18}})\mathbf{v}_1 + (\frac{12}{6})\mathbf{v}_2 + (\frac{5}{\sqrt{18}})\mathbf{v}_3 = (1, \frac{11}{6}, -\frac{1}{3}, \frac{1}{6}) + (\frac{5}{18}, 0, \frac{5}{18}, -\frac{10}{9}) = (\frac{23}{18}, \frac{11}{6} - \frac{1}{18}, -\frac{17}{18})$

7. (a) $\operatorname{proj}_W \mathbf{x} = \frac{(\mathbf{x} \cdot \mathbf{v}_1)}{\|\mathbf{v}_1\|^2}\mathbf{v}_1 + \frac{(\mathbf{x} \cdot \mathbf{v}_2)}{\|\mathbf{v}_2\|^2}\mathbf{v}_2 = (\frac{1}{4})\mathbf{v}_1 + (\frac{5}{4})\mathbf{v}_2 = (\frac{1}{4}, \frac{1}{4}, \frac{1}{4}, \frac{1}{4}) + (\frac{5}{4}, \frac{5}{4}, -\frac{5}{4}, -\frac{5}{4}) = (\frac{3}{2}, \frac{3}{2}, -1, -1)$
    (b) $\operatorname{proj}_W \mathbf{x} = \frac{(\mathbf{x} \cdot \mathbf{v}_1)}{\|\mathbf{v}_1\|^2}\mathbf{v}_1 + \frac{(\mathbf{x} \cdot \mathbf{v}_2)}{\|\mathbf{v}_2\|^2}\mathbf{v}_2 + \frac{(\mathbf{x} \cdot \mathbf{v}_3)}{\|\mathbf{v}_3\|^2}\mathbf{v}_3 = (\frac{1}{4})\mathbf{v}_1 + (\frac{5}{4})\mathbf{v}_2 + (\frac{1}{4})\mathbf{v}_3 = (\frac{7}{4}, \frac{5}{4}, -\frac{3}{4}, -\frac{5}{4})$

9. $\mathbf{w} = (\mathbf{w} \cdot \mathbf{v}_1)\mathbf{v}_1 + (\mathbf{w} \cdot \mathbf{v}_2)\mathbf{v}_2 + (\mathbf{w} \cdot \mathbf{v}_3)\mathbf{v}_3 = (0)\mathbf{v}_1 + (\frac{5}{\sqrt{6}})\mathbf{v}_2 + (\frac{11}{\sqrt{66}})\mathbf{v}_3$

11. $\mathbf{w} = \frac{(\mathbf{w} \cdot \mathbf{v}_1)}{\|\mathbf{v}_1\|^2}\mathbf{v}_1 + \frac{(\mathbf{w} \cdot \mathbf{v}_2)}{\|\mathbf{v}_2\|^2}\mathbf{v}_2 + \frac{(\mathbf{w} \cdot \mathbf{v}_3)}{\|\mathbf{v}_3\|^2}\mathbf{v}_3 + \frac{(\mathbf{w} \cdot \mathbf{v}_4)}{\|\mathbf{v}_4\|^2}\mathbf{v}_4 = (-\frac{2}{18})\mathbf{v}_1 + (\frac{5}{33})\mathbf{v}_3 + (\frac{5}{9})\mathbf{v}_3 + (-\frac{4}{66})\mathbf{v}_4$

13. Using Formula (6), we have $P = \begin{bmatrix} \frac{2}{3} & \frac{1}{3} \\ \frac{1}{3} & \frac{2}{3} \\ \frac{2}{3} & -\frac{2}{3} \end{bmatrix} \begin{bmatrix} \frac{2}{3} & \frac{1}{3} & \frac{2}{3} \\ \frac{1}{3} & \frac{2}{3} & -\frac{2}{3} \end{bmatrix} = \begin{bmatrix} \frac{5}{9} & \frac{4}{9} & \frac{2}{9} \\ \frac{4}{9} & \frac{5}{9} & -\frac{2}{9} \\ \frac{2}{9} & -\frac{2}{9} & \frac{8}{9} \end{bmatrix}$.

15. Using the matrix found in Exercise 13, the orthogonal projection of $\mathbf{w}$ onto $W = \operatorname{span}\{\mathbf{v}_1, \mathbf{v}_2\}$ is

$$ P\mathbf{w} = \begin{bmatrix} \frac{5}{9} & \frac{4}{9} & \frac{2}{9} \\ \frac{4}{9} & \frac{5}{9} & -\frac{2}{9} \\ \frac{2}{9} & -\frac{2}{9} & \frac{8}{9} \end{bmatrix} \begin{bmatrix} 0 \\ 2 \\ -3 \end{bmatrix} = \begin{bmatrix} \frac{2}{9} \\ \frac{16}{9} \\ -\frac{28}{9} \end{bmatrix} $$

On the other hand, using Formula (7), we have

$$ \operatorname{proj}_W \mathbf{w} = (\mathbf{w} \cdot \mathbf{v}_1)\mathbf{v}_1 + (\mathbf{w} \cdot \mathbf{v}_2)\mathbf{v}_2 = (-\frac{4}{3})(\frac{2}{3}, \frac{1}{3}, \frac{2}{3}) + (\frac{10}{3})(\frac{1}{3}, \frac{2}{3}, -\frac{2}{3}) = (\frac{2}{9}, \frac{16}{9}, -\frac{28}{9}) $$

17. We have $\frac{\mathbf{v}_1}{\|\mathbf{v}_1\|} = (\frac{1}{\sqrt{3}}, \frac{1}{\sqrt{3}}, \frac{1}{\sqrt{3}})$ and $\frac{\mathbf{v}_2}{\|\mathbf{v}_2\|} = (\frac{1}{\sqrt{6}}, \frac{1}{\sqrt{6}}, -\frac{2}{\sqrt{6}})$. Thus the standard matrix for the orthogonal projection of R^3 onto $W = \operatorname{span}\{\mathbf{v}_1, \mathbf{v}_2\}$ is given by

$$ P = \begin{bmatrix} \frac{1}{\sqrt{3}} & \frac{1}{\sqrt{6}} \\ \frac{1}{\sqrt{3}} & \frac{1}{\sqrt{6}} \\ \frac{1}{\sqrt{3}} & -\frac{2}{\sqrt{6}} \end{bmatrix} \begin{bmatrix} \frac{1}{\sqrt{3}} & \frac{1}{\sqrt{3}} & \frac{1}{\sqrt{3}} \\ \frac{1}{\sqrt{6}} & \frac{1}{\sqrt{6}} & -\frac{2}{\sqrt{6}} \end{bmatrix} = \begin{bmatrix} \frac{1}{2} & \frac{1}{2} & 0 \\ \frac{1}{2} & \frac{1}{2} & 0 \\ 0 & 0 & 1 \end{bmatrix}. $$

19. Using the matrix found in Exercise 17, the orthogonal projection of $\mathbf{w}$ onto $W = \text{span}\{\mathbf{v}_1, \mathbf{v}_2\}$ is

$$
P\mathbf{w} = \begin{bmatrix} \frac{1}{2} & \frac{1}{2} & 0 \\ \frac{1}{2} & \frac{1}{2} & 0 \\ 0 & 0 & 1 \end{bmatrix} \begin{bmatrix} 4 \\ -5 \\ 0 \end{bmatrix} = \begin{bmatrix} -\frac{1}{2} \\ -\frac{1}{2} \\ 0 \end{bmatrix}
$$

On the other hand, using Formula (8), we have

$$
\text{proj}_W \mathbf{w} = \frac{(\mathbf{w} \cdot \mathbf{v}_1)}{\|\mathbf{v}_1\|^2} \mathbf{v}_1 + \frac{(\mathbf{w} \cdot \mathbf{v}_2)}{\|\mathbf{v}_2\|^2} \mathbf{v}_2 = (-\tfrac{1}{3})(1,1,1) + (-\tfrac{1}{6})(1,1,-2) = (-\tfrac{1}{2}, -\tfrac{1}{2}, 0)
$$

21. From Exercise 17 we have $P = \begin{bmatrix} \frac{1}{2} & \frac{1}{2} & 0 \\ \frac{1}{2} & \frac{1}{2} & 0 \\ 0 & 0 & 1 \end{bmatrix}$; thus $\text{trace}(P) = \tfrac{1}{2} + \tfrac{1}{2} + 1 = 2$.

23. We have $P^T = P$ and it is easy to check that $P^2 = P$; thus P is the standard matrix of an orthogonal projection. The dimension of the range of the projection is equal to $\text{tr}(P) = \tfrac{20}{21} + \tfrac{5}{21} + \tfrac{17}{21} = 2$.

25. We have $P^T = P$ and it is easy to check that $P^2 = P$; thus P is the standard matrix of an orthogonal projection. The dimension of the range of the projection is equal to $\text{tr}(P) = \tfrac{1}{9} + \tfrac{4}{9} + \tfrac{4}{9} = 1$.

27. Let $\mathbf{v}_1 = \mathbf{w}_1 = (1, -3)$ and $\mathbf{v}_2 = \mathbf{w}_2 - \frac{\mathbf{w}_2 \cdot \mathbf{v}_1}{\|\mathbf{v}_1\|^2} \mathbf{v}_1 = (2, 2) - (\tfrac{-4}{10})(1, -3) = (\tfrac{12}{5}, \tfrac{4}{5}) = \tfrac{4}{5}(3, 1)$. Then $\{\mathbf{v}_1, \mathbf{v}_2\}$ is an orthogonal basis for R^2, and the vectors $\mathbf{q}_1 = \frac{\mathbf{v}_1}{\|\mathbf{v}_1\|} = (\tfrac{1}{\sqrt{10}}, \tfrac{-3}{\sqrt{10}})$ and $\mathbf{q}_2 = \frac{\mathbf{v}_2}{\|\mathbf{v}_2\|} = (\tfrac{3}{\sqrt{10}}, \tfrac{1}{\sqrt{10}})$ form an orthonormal basis for R^2.

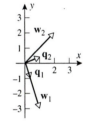

29. Let $\mathbf{v}_1 = \mathbf{w}_1 = (1, 1, 1)$, $\mathbf{v}_2 = \mathbf{w}_2 - \frac{\mathbf{w}_2 \cdot \mathbf{v}_1}{\|\mathbf{v}_1\|^2} \mathbf{v}_1 = (-1, 1, 0) - (\tfrac{0}{3})(1, 1, 1) = (-1, 1, 0)$, and

$$
\mathbf{v}_3 = \mathbf{w}_3 - \frac{\mathbf{w}_3 \cdot \mathbf{v}_1}{\|\mathbf{v}_1\|^2} \mathbf{v}_1 - \frac{\mathbf{w}_3 \cdot \mathbf{v}_2}{\|\mathbf{v}_2\|^2} \mathbf{v}_2 = (1, 2, 1) - (\tfrac{4}{3})(1, 1, 1) - (\tfrac{1}{2})(-1, 1, 0) = (\tfrac{1}{6}, \tfrac{1}{6}, -\tfrac{1}{3})
$$

Then $\{\mathbf{v}_1, \mathbf{v}_2, \mathbf{v}_3\}$ is an orthogonal basis for R^3, and the vectors

$$
\mathbf{q}_1 = \frac{\mathbf{v}_1}{\|\mathbf{v}_1\|} = (\tfrac{1}{\sqrt{3}}, \tfrac{1}{\sqrt{3}}, \tfrac{1}{\sqrt{3}}), \quad \mathbf{q}_2 = \frac{\mathbf{v}_2}{\|\mathbf{v}_2\|} = (-\tfrac{1}{\sqrt{2}}, \tfrac{1}{\sqrt{2}}, 0), \quad \mathbf{q}_3 = \frac{\mathbf{v}_3}{\|\mathbf{v}_3\|} = (\tfrac{1}{\sqrt{6}}, \tfrac{1}{\sqrt{6}}, -\tfrac{2}{\sqrt{6}})
$$

form an orthonormal basis for R^3.

31. Let $\mathbf{v}_1 = \mathbf{w}_1 = (0, 2, 1, 0)$, $\mathbf{v}_2 = \mathbf{w}_2 - \frac{\mathbf{w}_2 \cdot \mathbf{v}_1}{\|\mathbf{v}_1\|^2} \mathbf{v}_1 = (1, -1, 0, 0) - (-\tfrac{2}{5})(0, 2, 1, 0) = (1, -\tfrac{1}{5}, \tfrac{2}{5}, 0)$,

$$
\mathbf{v}_3 = \mathbf{w}_3 - \frac{\mathbf{w}_3 \cdot \mathbf{v}_1}{\|\mathbf{v}_1\|^2} \mathbf{v}_1 - \frac{\mathbf{w}_3 \cdot \mathbf{v}_2}{\|\mathbf{v}_2\|^2} \mathbf{v}_2
$$

$$
= (1, 2, 0, -1) - \left(\frac{4}{5}\right)(0, 2, 1, 0) - \left(\frac{\frac{3}{5}}{\frac{6}{5}}\right)\left(1, -\frac{1}{5}, \frac{2}{5}, 0\right) = \left(\frac{1}{2}, \frac{1}{2}, -1, -1\right)
$$

and

$$\mathbf{v}_4 = \mathbf{w}_4 - \frac{\mathbf{w}_4 \cdot \mathbf{v}_1}{\|\mathbf{v}_1\|^2}\mathbf{v}_1 - \frac{\mathbf{w}_4 \cdot \mathbf{v}_2}{\|\mathbf{v}_2\|^2}\mathbf{v}_2 - \frac{\mathbf{w}_4 \cdot \mathbf{v}_3}{\|\mathbf{v}_3\|^2}\mathbf{v}_3$$

$$= (1,0,0,1) - \left(\frac{0}{5}\right)(0,2,1,0) - \left(\frac{1}{\frac{6}{5}}\right)\left(1,-\frac{1}{5},\frac{2}{5},0\right) - \left(-\frac{\frac{1}{2}}{\frac{5}{2}}\right)\left(\frac{1}{2},\frac{1}{2},-1,-1\right)$$

$$= \left(\frac{4}{15},\frac{4}{15},-\frac{8}{15},\frac{4}{5}\right)$$

Then $\{\mathbf{v}_1, \mathbf{v}_2, \mathbf{v}_3, \mathbf{v}_4\}$ is an orthogonal basis for R^4, and the vectors

$$\mathbf{q}_1 = \frac{\mathbf{v}_1}{\|\mathbf{v}_1\|} = (0, \tfrac{2\sqrt{5}}{5}, \tfrac{\sqrt{5}}{5}, 0), \quad \mathbf{q}_2 = \frac{\mathbf{v}_2}{\|\mathbf{v}_2\|} = (\tfrac{\sqrt{30}}{6}, -\tfrac{\sqrt{30}}{30}, \tfrac{\sqrt{30}}{15}, 0),$$

$$\mathbf{q}_3 = \frac{\mathbf{v}_3}{\|\mathbf{v}_3\|} = (\tfrac{\sqrt{10}}{10}, \tfrac{\sqrt{10}}{10}, -\tfrac{\sqrt{10}}{5}, -\tfrac{\sqrt{10}}{5}), \quad \mathbf{q}_4 = \frac{\mathbf{v}_4}{\|\mathbf{v}_4\|} = (\tfrac{\sqrt{15}}{15}, \tfrac{\sqrt{15}}{15}, -\tfrac{2\sqrt{15}}{15}, \tfrac{\sqrt{15}}{5})$$

form an orthonormal basis for R^4.

33. The vectors $\mathbf{w}_1 = (\tfrac{1}{\sqrt{2}}, \tfrac{1}{\sqrt{2}}, 0)$, $\mathbf{w}_2 = (\tfrac{1}{\sqrt{2}}, -\tfrac{1}{\sqrt{2}}, 0)$, and $\mathbf{w}_3 = (0,0,1)$ form an orthonormal basis for R^3.

35. Note that $\mathbf{w}_3 = \mathbf{w}_1 + \mathbf{w}_2$. Thus the subspace W spanned by the given vectors is 2-dimensional with basis $\{\mathbf{w}_1, \mathbf{w}_2\}$. Let $\mathbf{v}_1 = \mathbf{w}_1 = (0, 1, 2)$ and

$$\mathbf{v}_2 = \mathbf{w}_2 - \frac{\mathbf{w}_2 \cdot \mathbf{v}_1}{\|\mathbf{v}_1\|^2}\mathbf{v}_1 = (-1, 0, 1) - (\tfrac{2}{5})(0,1,2) = (-1, -\tfrac{2}{5}, \tfrac{1}{5})$$

Then $\{\mathbf{v}_1, \mathbf{v}_2\}$ is an orthogonal basis for W, and the vectors

$$\mathbf{u}_1 = \frac{\mathbf{v}_1}{\|\mathbf{v}_1\|} = (0, \tfrac{1}{\sqrt{5}}, \tfrac{2}{\sqrt{5}}), \quad \mathbf{u}_2 = \frac{\mathbf{v}_2}{\|\mathbf{v}_2\|} = (\tfrac{-5}{\sqrt{30}}, \tfrac{-2}{\sqrt{30}}, \tfrac{1}{\sqrt{30}})$$

form an orthonormal basis for W.

37. Note that $\mathbf{u}_1$ and $\mathbf{u}_2$ are orthonormal vectors. Thus the orthogonal projection of $\mathbf{w}$ onto the subspace W spanned by these two vectors is given by

$$\mathbf{w}_1 = \text{proj}_W \mathbf{w} = (\mathbf{w} \cdot \mathbf{u}_1)\mathbf{u}_1 + (\mathbf{w} \cdot \mathbf{u}_2)\mathbf{u}_2 = (-1)(\tfrac{4}{5}, 0, -\tfrac{3}{5}) + (2)(0, 1, 0) = (-\tfrac{4}{5}, 2, \tfrac{3}{5})$$

and the component of $\mathbf{w}$ orthogonal to W is

$$\mathbf{w}_2 = \mathbf{w} - \mathbf{w}_1 = (1, 2, 3) - (-\tfrac{4}{5}, 2, \tfrac{3}{5}) = (\tfrac{9}{5}, 0, \tfrac{12}{5})$$

39. If $\mathbf{w} = (a, b, c)$, then the vector

$$\mathbf{u} = \frac{\mathbf{w}}{\|\mathbf{w}\|} = \left(\frac{a}{\sqrt{a^2 + b^2 + c^2}}, \frac{b}{\sqrt{a^2 + b^2 + c^2}}, \frac{c}{\sqrt{a^2 + b^2 + c^2}}\right)$$

is an orthonormal basis for the 1-dimensional subspace W spanned by $\mathbf{w}$. Thus, using Formula (6), the standard matrix for the orthogonal projection of R^3 onto W is

$$P = \mathbf{u}^T \mathbf{u} = \frac{1}{a^2 + b^2 + c^2}\begin{bmatrix} a \\ b \\ c \end{bmatrix}\begin{bmatrix} a & b & c \end{bmatrix} = \frac{1}{a^2 + b^2 + c^2}\begin{bmatrix} a^2 & ab & ac \\ ab & b^2 & bc \\ ac & bc & c^2 \end{bmatrix}$$

DISCUSSION AND DISCOVERY

D1. If a and b are nonzero, then $\mathbf{u}_1 = (1, 0, a)$ and $\mathbf{u}_2 = (0, 1, b)$ form a basis for the plane $z = ax + by$, and application of the Gram-Schmidt process to these vectors yields an orthonormal basis $\{\mathbf{q}_1, \mathbf{q}_2\}$ where

$$\mathbf{q}_1 = \left(\frac{1}{\sqrt{1 + a^2}}, 0, \frac{a}{\sqrt{1 + a^2}} \right)$$

$$\mathbf{q}_2 = \left(\frac{-ab}{\sqrt{1 + a^2}\sqrt{1 + a^2 + b^2}}, \frac{1 + a^2}{\sqrt{1 + a^2}\sqrt{1 + a^2 + b^2}}, \frac{b}{\sqrt{1 + a^2}\sqrt{1 + a^2 + b^2}} \right)$$

D2. (a) $\text{span}\{\mathbf{v}_1\} = \text{span}\{\mathbf{w}_1\}$, $\text{span}\{\mathbf{v}_1, \mathbf{v}_2\} = \text{span}\{\mathbf{w}_1, \mathbf{w}_2\}$, and $\text{span}\{\mathbf{v}_1, \mathbf{v}_2, \mathbf{v}_3\} = \text{span}\{\mathbf{w}_1, \mathbf{w}_2, \mathbf{w}_3\}$.

(b) $\mathbf{v}_3$ is orthogonal to $\text{span}\{\mathbf{w}_1, \mathbf{w}_2\}$.

D3. If the vectors $\mathbf{w}_1, \mathbf{w}_2, \ldots, \mathbf{w}_k$ are linearly dependent, then at least one of the vectors in the list is a linear combination of the previous ones. If $\mathbf{w}_j$ is a linear combination of $\mathbf{w}_1, \mathbf{w}_2, \ldots, \mathbf{w}_{j-1}$ then, when applying the Gram-Schmidt process at the jth step, the vector $\mathbf{v}_j$ will be $\mathbf{0}$.

D4. If A has orthonormal columns, then AA^T is the standard matrix for the orthogonal projection onto the column space of A.

D5. (a) $\text{col}(M) = \text{col}(P)$

(b) Find an orthonormal basis for $\text{col}(P)$ and use these vectors as the columns of the matrix M.

(c) No. Any orthonormal basis for $\text{col}(P)$ can be used to form the columns of M.

D6. (a) True. Any orthonormal set of vectors is linearly independent.

(b) False. An orthogonal set may contain $\mathbf{0}$. However, it is true that any orthogonal set of nonzero vectors is linearly independent.

(c) False. Strictly speaking, the subspace $\{\mathbf{0}\}$ has no basis, hence no orthonormal basis. However, it is true that any nonzero subspace has an orthonormal basis.

(d) True. The vector $\mathbf{q}_3$ is orthogonal to the subspace $\text{span}\{\mathbf{w}_1, \mathbf{w}_2\}$.

WORKING WITH PROOFS

P1. If $\{\mathbf{v}_1, \mathbf{v}_2, \ldots, \mathbf{v}_k\}$ is an orthogonal basis for W, then $\{\mathbf{v}_1/\|\mathbf{v}_1\|, \ \mathbf{v}_2/\|\mathbf{v}_2\|, \ldots, \mathbf{v}_k/\|\mathbf{v}_k\|\}$ is an orthonormal basis. Thus, using part (a), the orthogonal projection of a vector $\mathbf{x}$ on W can be expressed as

$$\text{proj}_W \mathbf{x} = \left(\mathbf{x} \cdot \frac{\mathbf{v}_1}{\|\mathbf{v}_1\|} \right) \frac{\mathbf{v}_1}{\|\mathbf{v}_1\|} + \left(\mathbf{x} \cdot \frac{\mathbf{v}_2}{\|\mathbf{v}_2\|} \right) \frac{\mathbf{v}_2}{\|\mathbf{v}_2\|} + \cdots + \left(\mathbf{x} \cdot \frac{\mathbf{v}_k}{\|\mathbf{v}_k\|} \right) \frac{\mathbf{v}_k}{\|\mathbf{v}_k\|}$$

$$= \left(\frac{\mathbf{x} \cdot \mathbf{v}_1}{\|\mathbf{v}_1\|^2} \right) \mathbf{v}_1 + \left(\frac{\mathbf{x} \cdot \mathbf{v}_2}{\|\mathbf{v}_2\|^2} \right) \mathbf{v}_2 + \cdots + \left(\frac{\mathbf{x} \cdot \mathbf{v}_k}{\|\mathbf{v}_k\|^2} \right) \mathbf{v}_k$$

P2. If A is symmetric and idempotent, then A is the standard matrix of an orthogonal projection operator; namely the orthogonal projection of R^n onto $W = \text{col}(A)$. Thus $A = UU^T$ where U is any $n \times k$ matrix whose column vectors form an orthonormal basis for W.

P3. We must prove that $\mathbf{v}_j \in \text{span}\{\mathbf{w}_1, \mathbf{w}_2 \ldots, \mathbf{w}_j\}$ for each $j = 1, 2, \ldots$. The proof is by induction on j.

Step 1. Since $\mathbf{v}_1 = \mathbf{w}_1$, we have $\mathbf{v}_1 \in \text{span}\{\mathbf{w}_1\}$; thus the statement is true for $j = 1$.

Step 2 (induction step). Suppose the statement is true for integers k which are less than or equal to j, i.e., for $k = 1, 2, \ldots, j$. Then

$$\mathbf{v}_{j+1} = \mathbf{w}_{j+1} - \frac{\mathbf{w}_{j+1} \cdot \mathbf{v}_1}{\|\mathbf{v}_1\|^2}\mathbf{v}_1 - \frac{\mathbf{w}_{j+1} \cdot \mathbf{v}_2}{\|\mathbf{v}_2\|^2}\mathbf{v}_2 - \cdots - \frac{\mathbf{w}_{j+1} \cdot \mathbf{v}_j}{\|\mathbf{v}_j\|^2}\mathbf{v}_j$$

and since $\mathbf{v}_1 \in \text{span}\{\mathbf{w}_1\}$, $\mathbf{v}_2 \in \text{span}\{\mathbf{w}_1, \mathbf{w}_2\}, \ldots$, and $\mathbf{v}_j \in \text{span}\{\mathbf{w}_1, \mathbf{w}_2, \ldots, \mathbf{w}_j\}$, it follows that $\mathbf{v}_{j+1} \in \text{span}\{\mathbf{w}_1, \mathbf{w}_2, \ldots, \mathbf{w}_j, \mathbf{w}_{j+1}\}$. Thus if the statement is true for each of the integers $k = 1, 2, \ldots, j$ then it is also true for $k = j + 1$.

These two steps complete the proof by induction.

EXERCISE SET 7.10

1. The column vectors of the matrix A are $\mathbf{w}_1 = \begin{bmatrix} 1 \\ 2 \end{bmatrix}$ and $\mathbf{w}_2 = \begin{bmatrix} -1 \\ 3 \end{bmatrix}$. Application of the Gram-Schmidt process to these vector yields

$$\mathbf{q}_1 = \begin{bmatrix} \frac{1}{\sqrt{5}} \\ \frac{2}{\sqrt{5}} \end{bmatrix} \qquad \mathbf{q}_2 = \begin{bmatrix} -\frac{2}{\sqrt{5}} \\ \frac{1}{\sqrt{5}} \end{bmatrix}$$

We have $\mathbf{w}_1 = (\mathbf{w}_1 \cdot \mathbf{q}_1)\mathbf{q}_1 = \sqrt{5}\mathbf{q}_1$ and $\mathbf{w}_2 = (\mathbf{w}_2 \cdot \mathbf{q}_1)\mathbf{q}_1 + (\mathbf{w}_2 \cdot \mathbf{q}_2)\mathbf{q}_2 = \sqrt{5}\mathbf{q}_1 + \sqrt{5}\mathbf{q}_2$. Thus application of Formula (3) yields the following QR-decomposition of A:

$$A = \begin{bmatrix} 1 & -1 \\ 2 & 3 \end{bmatrix} = \begin{bmatrix} \frac{1}{\sqrt{5}} & -\frac{2}{\sqrt{5}} \\ \frac{2}{\sqrt{5}} & \frac{1}{\sqrt{5}} \end{bmatrix} \begin{bmatrix} \sqrt{5} & \sqrt{5} \\ 0 & \sqrt{5} \end{bmatrix} = QR$$

3. Application of the Gram-Schmidt process to the column vectors $\mathbf{w}_1$ and $\mathbf{w}_2$ of A yields

$$\mathbf{q}_1 = \begin{bmatrix} \frac{1}{3} \\ -\frac{2}{3} \\ \frac{2}{3} \end{bmatrix} \qquad \mathbf{q}_2 = \begin{bmatrix} \frac{8}{3\sqrt{26}} \\ \frac{11}{3\sqrt{26}} \\ \frac{7}{3\sqrt{26}} \end{bmatrix}$$

We have $\mathbf{w}_1 = (\mathbf{w}_1 \cdot \mathbf{q}_1)\mathbf{q}_1 = 3\mathbf{q}_1$ and $\mathbf{w}_2 = (\mathbf{w}_2 \cdot \mathbf{q}_1)\mathbf{q}_1 + (\mathbf{w}_2 \cdot \mathbf{q}_2)\mathbf{q}_2 = \frac{1}{3}\mathbf{q}_1 + \frac{\sqrt{26}}{3}\mathbf{q}_2$. This yields the following QR-decomposition of A:

$$A = \begin{bmatrix} 1 & 1 \\ -2 & 1 \\ 2 & 1 \end{bmatrix} = \begin{bmatrix} \frac{1}{3} & \frac{8}{3\sqrt{26}} \\ -\frac{2}{3} & \frac{11}{3\sqrt{26}} \\ \frac{2}{3} & \frac{7}{3\sqrt{26}} \end{bmatrix} \begin{bmatrix} 3 & -\frac{1}{3} \\ 0 & \frac{\sqrt{26}}{3} \end{bmatrix} = QR$$

5. Application of the Gram-Schmidt process to the column vectors $\mathbf{w}_1, \mathbf{w}_2, \mathbf{w}_3$ of A yields

$$\mathbf{q}_1 = \begin{bmatrix} \frac{1}{\sqrt{2}} \\ \frac{1}{\sqrt{2}} \\ 0 \end{bmatrix} \qquad \mathbf{q}_2 = \begin{bmatrix} \frac{1}{\sqrt{38}} \\ -\frac{1}{\sqrt{38}} \\ \frac{6}{\sqrt{38}} \end{bmatrix} \qquad \mathbf{q}_2 = \begin{bmatrix} -\frac{3}{\sqrt{19}} \\ \frac{3}{\sqrt{19}} \\ \frac{1}{\sqrt{19}} \end{bmatrix}$$

We have $\mathbf{w}_1 = \sqrt{2}\mathbf{q}_1$, $\mathbf{w}_2 = \frac{3\sqrt{2}}{2}\mathbf{q}_1 + \frac{\sqrt{38}}{2}\mathbf{q}_2$, and $\mathbf{w}_3 = \sqrt{2}\mathbf{q}_1 + \frac{3\sqrt{38}}{19}\mathbf{q}_2 + \frac{\sqrt{19}}{19}\mathbf{q}_3$. This yields the following QR-decomposition of A:

$$A = \begin{bmatrix} 1 & 2 & 1 \\ 1 & 1 & 1 \\ 0 & 3 & 1 \end{bmatrix} = \begin{bmatrix} \frac{1}{\sqrt{2}} & \frac{1}{\sqrt{38}} & -\frac{3}{\sqrt{19}} \\ \frac{1}{\sqrt{2}} & -\frac{1}{\sqrt{38}} & \frac{3}{\sqrt{19}} \\ 0 & \frac{6}{\sqrt{38}} & \frac{1}{\sqrt{19}} \end{bmatrix} \begin{bmatrix} \sqrt{2} & \frac{3\sqrt{2}}{2} & \sqrt{2} \\ 0 & \frac{\sqrt{38}}{2} & \frac{3\sqrt{38}}{19} \\ 0 & 0 & \frac{\sqrt{19}}{19} \end{bmatrix} = QR$$

7. From Exercise 3, we have $A = \begin{bmatrix} 1 & 1 \\ -2 & 1 \\ 2 & 1 \end{bmatrix} = \begin{bmatrix} \frac{1}{3} & \frac{8}{3\sqrt{26}} \\ -\frac{2}{3} & \frac{11}{3\sqrt{26}} \\ \frac{2}{3} & \frac{7}{3\sqrt{26}} \end{bmatrix} \begin{bmatrix} 3 & \frac{1}{3} \\ 0 & \frac{\sqrt{26}}{3} \end{bmatrix} = QR$. Thus the normal system for

$A\mathbf{x} = \mathbf{b}$ can be expressed as $R\mathbf{x} = Q^T\mathbf{b}$, which is:

$$\begin{bmatrix} 3 & \frac{1}{3} \\ 0 & \frac{\sqrt{26}}{3} \end{bmatrix} \begin{bmatrix} x_1 \\ x_2 \end{bmatrix} = \begin{bmatrix} \frac{1}{3} & -\frac{2}{3} & \frac{2}{3} \\ \frac{8}{3\sqrt{26}} & \frac{11}{3\sqrt{26}} & \frac{7}{3\sqrt{26}} \end{bmatrix} \begin{bmatrix} 1 \\ 1 \\ 0 \end{bmatrix} = \begin{bmatrix} -\frac{1}{3} \\ \frac{19}{3\sqrt{26}} \end{bmatrix}$$

Solving this system by back substitution yields the least squares solution $x_2 = \frac{19}{26}$, $x_1 = -\frac{5}{26}$.

9. From Exercise 5, we have $A = \begin{bmatrix} 1 & 2 & 1 \\ 1 & 1 & 1 \\ 0 & 3 & 1 \end{bmatrix} = \begin{bmatrix} \frac{1}{\sqrt{2}} & \frac{1}{\sqrt{38}} & -\frac{3}{\sqrt{19}} \\ \frac{1}{\sqrt{2}} & -\frac{1}{\sqrt{38}} & \frac{3}{\sqrt{19}} \\ 0 & \frac{6}{\sqrt{38}} & \frac{1}{\sqrt{19}} \end{bmatrix} \begin{bmatrix} \sqrt{2} & \frac{3\sqrt{2}}{2} & \sqrt{2} \\ 0 & \frac{\sqrt{38}}{2} & \frac{3\sqrt{38}}{19} \\ 0 & 0 & \frac{\sqrt{19}}{19} \end{bmatrix} = QR$. Thus the normal

system for $A\mathbf{x} = \mathbf{b}$ can be expressed as $R\mathbf{x} = Q^T\mathbf{b}$, which is:

$$\begin{bmatrix} \sqrt{2} & \frac{3\sqrt{2}}{2} & \sqrt{2} \\ 0 & \frac{\sqrt{38}}{2} & \frac{3\sqrt{38}}{19} \\ 0 & 0 & \frac{\sqrt{19}}{19} \end{bmatrix} \begin{bmatrix} x_1 \\ x_2 \\ x_3 \end{bmatrix} = \begin{bmatrix} \frac{1}{\sqrt{2}} & \frac{1}{\sqrt{2}} & 0 \\ \frac{1}{\sqrt{38}} & -\frac{1}{\sqrt{38}} & \frac{6}{\sqrt{38}} \\ -\frac{3}{\sqrt{19}} & \frac{3}{\sqrt{19}} & \frac{1}{\sqrt{19}} \end{bmatrix} \begin{bmatrix} -2 \\ 3 \\ 1 \end{bmatrix} = \begin{bmatrix} \frac{\sqrt{2}}{2} \\ \frac{\sqrt{38}}{38} \\ \frac{16\sqrt{19}}{19} \end{bmatrix}$$

Solving this system by back substitution yields $x_3 = 16$, $x_2 = -5$, $x_1 = -8$. Note that, in this example, the system $A\mathbf{x} = \mathbf{b}$ is consistent and this is its exact solution.

11. The plane $2x - y + 3z = 0$ corresponds to $\mathbf{a}^{\perp}$ where $\mathbf{a} = (2, -1, 3)$. Thus, writing $\mathbf{a}$ as a column vector, the standard matrix for the reflection of R^3 about the plane is

$$H = I - \frac{2}{\mathbf{a}^T\mathbf{a}}\mathbf{a}\mathbf{a}^T = \begin{bmatrix} 1 & 0 & 0 \\ 0 & 1 & 0 \\ 0 & 0 & 1 \end{bmatrix} - \frac{2}{14}\begin{bmatrix} 4 & -2 & 6 \\ -2 & 1 & -3 \\ 6 & -3 & 9 \end{bmatrix} = \begin{bmatrix} \frac{3}{7} & \frac{2}{7} & -\frac{6}{7} \\ \frac{2}{7} & \frac{6}{7} & \frac{3}{7} \\ -\frac{6}{7} & \frac{3}{7} & -\frac{2}{7} \end{bmatrix}$$

and the reflection of the vector $\mathbf{b} = (1, 2, 2)$ about that plane is given, in column form, by

$$H\mathbf{b} = \begin{bmatrix} \frac{3}{7} & \frac{2}{7} & -\frac{6}{7} \\ \frac{2}{7} & \frac{6}{7} & \frac{3}{7} \\ -\frac{6}{7} & \frac{3}{7} & -\frac{2}{7} \end{bmatrix} \begin{bmatrix} 1 \\ 2 \\ 2 \end{bmatrix} = \begin{bmatrix} -\frac{5}{7} \\ \frac{20}{7} \\ -\frac{4}{7} \end{bmatrix}$$

13. $H = I - \frac{2}{\mathbf{a}^T\mathbf{a}}\mathbf{a}\mathbf{a}^T = \begin{bmatrix} 1 & 0 & 0 \\ 0 & 1 & 0 \\ 0 & 0 & 1 \end{bmatrix} - \frac{2}{3}\begin{bmatrix} 1 & -1 & 1 \\ -1 & 1 & -1 \\ 1 & -1 & 1 \end{bmatrix} = \begin{bmatrix} \frac{1}{3} & \frac{2}{3} & -\frac{2}{3} \\ \frac{2}{3} & \frac{1}{3} & \frac{2}{3} \\ -\frac{2}{3} & \frac{2}{3} & \frac{1}{3} \end{bmatrix}$

15. $H = I - \dfrac{2}{\mathbf{a}^T\mathbf{a}}\mathbf{a}\mathbf{a}^T = \begin{bmatrix} 1 & 0 & 0 & 0 \\ 0 & 1 & 0 & 0 \\ 0 & 0 & 1 & 0 \\ 0 & 0 & 0 & 1 \end{bmatrix} - \dfrac{2}{11}\begin{bmatrix} 0 & 0 & 0 & 0 \\ 0 & 1 & -1 & 3 \\ 0 & -1 & 1 & -3 \\ 0 & 3 & -3 & 9 \end{bmatrix} = \begin{bmatrix} 1 & 0 & 0 & 0 \\ 0 & \frac{9}{11} & \frac{2}{11} & -\frac{6}{11} \\ 0 & \frac{2}{11} & \frac{9}{11} & \frac{6}{11} \\ 0 & -\frac{6}{11} & \frac{6}{11} & -\frac{7}{11} \end{bmatrix}$

17. (a) Let $\mathbf{a} = \mathbf{v} - \mathbf{w} = (3,4) - (5,0) = (-2,4)$, $H = I - \dfrac{2}{\mathbf{a}^T\mathbf{a}}\mathbf{a}\mathbf{a}^T = \begin{bmatrix} 1 & 0 \\ 0 & 1 \end{bmatrix} - \dfrac{2}{20}\begin{bmatrix} 4 & -8 \\ -8 & 16 \end{bmatrix} = \begin{bmatrix} \frac{3}{5} & \frac{4}{5} \\ \frac{4}{5} & -\frac{3}{5} \end{bmatrix}$.

Then H is the Householder matrix for the reflection about $\mathbf{a}^\perp$, and $H\mathbf{v} = \mathbf{w}$.

(b) Let $\mathbf{a} = \mathbf{v} - \mathbf{w} = (3,4) - (0,5) = (3,-1)$, $H = I - \dfrac{2}{\mathbf{a}^T\mathbf{a}}\mathbf{a}\mathbf{a}^T = \begin{bmatrix} 1 & 0 \\ 0 & 1 \end{bmatrix} - \dfrac{2}{10}\begin{bmatrix} 9 & -3 \\ -3 & 1 \end{bmatrix} = \begin{bmatrix} -\frac{4}{5} & \frac{3}{5} \\ \frac{3}{5} & \frac{4}{5} \end{bmatrix}$.

Then H is the Householder matrix for the reflection about $\mathbf{a}^\perp$, and $H\mathbf{v} = \mathbf{w}$.

(c) Let $\mathbf{a} = \mathbf{v} - \mathbf{w} = (3,4) - (\frac{7\sqrt{2}}{2}, -\frac{\sqrt{2}}{2}) = (\frac{6-7\sqrt{2}}{2}, \frac{8+\sqrt{2}}{2})$. Then the appropriate Householder matrix is:

$$H = I - \frac{2}{\mathbf{a}^T\mathbf{a}}\mathbf{a}\mathbf{a}^T = \begin{bmatrix} 1 & 0 \\ 0 & 1 \end{bmatrix} - \frac{2}{50 - 17\sqrt{2}}\begin{bmatrix} \frac{67-42\sqrt{2}}{2} & \frac{17-25\sqrt{2}}{2} \\ \frac{17-25\sqrt{2}}{2} & \frac{33+8\sqrt{2}}{2} \end{bmatrix} = \begin{bmatrix} \frac{1}{\sqrt{2}} & \frac{1}{\sqrt{2}} \\ \frac{1}{\sqrt{2}} & -\frac{1}{\sqrt{2}} \end{bmatrix}$$

19. Let $\mathbf{w} = (\|\mathbf{v}\|, 0, 0) = (3,0,0)$, and $\mathbf{a} = \mathbf{v} - \mathbf{w} = (-5,1,2)$. Then

$$H = I - \frac{2}{\mathbf{a}^T\mathbf{a}}\mathbf{a}\mathbf{a}^T = \begin{bmatrix} 1 & 0 & 0 \\ 0 & 1 & 0 \\ 0 & 0 & 1 \end{bmatrix} - \frac{2}{30}\begin{bmatrix} 25 & -5 & -10 \\ -5 & 1 & 2 \\ -10 & 2 & 4 \end{bmatrix} = \begin{bmatrix} -\frac{2}{3} & \frac{1}{3} & \frac{2}{3} \\ \frac{1}{3} & \frac{14}{15} & -\frac{2}{15} \\ \frac{2}{3} & -\frac{2}{15} & \frac{11}{15} \end{bmatrix}$$

is the standard matrix for the Householder reflection of R^3 about $\mathbf{a}^\perp$, and $H\mathbf{v} = \mathbf{w}$.

21. Let $\mathbf{v} = (1,-1)$, $\mathbf{w} = (\|\mathbf{v}\|, 0) = (\sqrt{2}, 0)$, and $\mathbf{a} = \mathbf{v} - \mathbf{w} = (1 - \sqrt{2}, -1)$. Then the Householder reflection about $\mathbf{a}^\perp$ maps $\mathbf{v}$ into $\mathbf{w}$. The standard matrix for this reflection is

$$H = I - \frac{2}{\mathbf{a}^T\mathbf{a}}\mathbf{a}\mathbf{a}^T = \begin{bmatrix} 1 & 0 \\ 0 & 1 \end{bmatrix} - \frac{2}{4 - 2\sqrt{2}}\begin{bmatrix} (1-\sqrt{2})^2 & -1+\sqrt{2} \\ -1+\sqrt{2} & 1 \end{bmatrix}$$

$$= \begin{bmatrix} 1 & 0 \\ 0 & 1 \end{bmatrix} - \frac{2+\sqrt{2}}{2}\begin{bmatrix} 3-2\sqrt{2} & -1+\sqrt{2} \\ -1+\sqrt{2} & 1 \end{bmatrix}$$

$$= \begin{bmatrix} 1 & 0 \\ 0 & 1 \end{bmatrix} - \begin{bmatrix} \frac{2-\sqrt{2}}{2} & \frac{\sqrt{2}}{2} \\ \frac{\sqrt{2}}{2} & \frac{2+\sqrt{2}}{2} \end{bmatrix} = \begin{bmatrix} \frac{\sqrt{2}}{2} & -\frac{\sqrt{2}}{2} \\ -\frac{\sqrt{2}}{2} & -\frac{\sqrt{2}}{2} \end{bmatrix}$$

We have $HA = \begin{bmatrix} \frac{\sqrt{2}}{2} & -\frac{\sqrt{2}}{2} \\ -\frac{\sqrt{2}}{2} & -\frac{\sqrt{2}}{2} \end{bmatrix}\begin{bmatrix} 1 & 2 \\ -1 & 3 \end{bmatrix} = \begin{bmatrix} \sqrt{2} & -\frac{\sqrt{2}}{2} \\ 0 & -\frac{5\sqrt{2}}{2} \end{bmatrix} = R$ and, setting $Q = H^{-1} = H^T = H$, this yields the following QR-decomposition of the matrix A:

$$A = \begin{bmatrix} 1 & 2 \\ -1 & 3 \end{bmatrix} = \begin{bmatrix} \frac{\sqrt{2}}{2} & -\frac{\sqrt{2}}{2} \\ -\frac{\sqrt{2}}{2} & -\frac{\sqrt{2}}{2} \end{bmatrix}\begin{bmatrix} \sqrt{2} & -\frac{\sqrt{2}}{2} \\ 0 & -\frac{5\sqrt{2}}{2} \end{bmatrix} = QR$$

23. Referring to the construction in Exercise 21, the second entry in the first column of A can be zeroed out by multiplying by the orthogonal matrix Q_1 as indicated below:

$$Q_1 A = \begin{bmatrix} \frac{\sqrt{2}}{2} & -\frac{\sqrt{2}}{2} & 0 \\ -\frac{\sqrt{2}}{2} & -\frac{\sqrt{2}}{2} & 0 \\ 0 & 0 & 1 \end{bmatrix}\begin{bmatrix} 1 & 2 & 1 \\ -1 & -2 & 3 \\ 0 & 4 & 5 \end{bmatrix} = \begin{bmatrix} \sqrt{2} & 2\sqrt{2} & -\sqrt{2} \\ 0 & 0 & -2\sqrt{2} \\ 0 & 4 & 5 \end{bmatrix}$$

Although the matrix on the right is not upper triangular, it can be made so by interchanging the 2nd and 3rd rows, and this can be achieved by interchanging the corresponding rows of Q_1. This yields

$$
Q_2 A = \begin{bmatrix} \frac{\sqrt{2}}{2} & -\frac{\sqrt{2}}{2} & 0 \\ 0 & 0 & 1 \\ -\frac{\sqrt{2}}{2} & -\frac{\sqrt{2}}{2} & 0 \end{bmatrix} \begin{bmatrix} 1 & 2 & 1 \\ -1 & -2 & 3 \\ 0 & 4 & 5 \end{bmatrix} = \begin{bmatrix} \sqrt{2} & 2\sqrt{2} & -\sqrt{2} \\ 0 & 4 & 5 \\ 0 & 0 & -2\sqrt{2} \end{bmatrix} = R
$$

and finally, setting $Q = Q_2^{-1} = Q_2^T$, we obtain the following QR-decomposition of A:

$$
A = \begin{bmatrix} 1 & 2 & 1 \\ -1 & -2 & 3 \\ 0 & 4 & 5 \end{bmatrix} = \begin{bmatrix} \frac{\sqrt{2}}{2} & 0 & -\frac{\sqrt{2}}{2} \\ -\frac{\sqrt{2}}{2} & 0 & -\frac{\sqrt{2}}{2} \\ 0 & 1 & 0 \end{bmatrix} \begin{bmatrix} \sqrt{2} & 2\sqrt{2} & -\sqrt{2} \\ 0 & 4 & 5 \\ 0 & 0 & -2\sqrt{2} \end{bmatrix} = QR
$$

25. Since $A = QR$, the system $A\mathbf{x} = \mathbf{b}$ is equivalent to the upper triangular system $R\mathbf{x} = Q^T\mathbf{b}$ which is:

$$
\begin{bmatrix} \sqrt{3} & -\sqrt{3} & \frac{5}{\sqrt{3}} \\ 0 & \sqrt{2} & 0 \\ 0 & 0 & \frac{4}{\sqrt{6}} \end{bmatrix} \begin{bmatrix} x_1 \\ x_2 \\ x_3 \end{bmatrix} = \begin{bmatrix} \frac{1}{\sqrt{3}} & \frac{1}{\sqrt{3}} & \frac{1}{\sqrt{3}} \\ 0 & \frac{1}{\sqrt{2}} & -\frac{1}{\sqrt{2}} \\ \frac{\sqrt{6}}{3} & -\frac{1}{\sqrt{6}} & -\frac{1}{\sqrt{6}} \end{bmatrix} \begin{bmatrix} 3 \\ 2 \\ 0 \end{bmatrix} = \begin{bmatrix} \frac{5}{\sqrt{3}} \\ \sqrt{2} \\ \frac{4}{\sqrt{6}} \end{bmatrix}
$$

Solving this system by back substitution yields $x_3 = 1, x_2 = 1, x_1 = 1$.

DISCUSSION AND DISCOVERY

D1. The standard matrix for the reflection of R^3 about $\mathbf{e}_1^\perp$ is (as should be expected)

$$
H = I - \frac{2}{\mathbf{e}_1^T \mathbf{e}_1} \mathbf{e}_1 \mathbf{e}_1^T = \begin{bmatrix} 1 & 0 & 0 \\ 0 & 1 & 0 \\ 0 & 0 & 1 \end{bmatrix} - \frac{2}{1}\begin{bmatrix} 1 & 0 & 0 \\ 0 & 0 & 0 \\ 0 & 0 & 0 \end{bmatrix} = \begin{bmatrix} -1 & 0 & 0 \\ 0 & 1 & 0 \\ 0 & 0 & 1 \end{bmatrix}
$$

and similarly for the others.

D2. The standard matrix for the reflection of R^2 about the line $y = mx$ is (taking $\mathbf{a} = (1, m)$) given by

$$
H = I - \frac{2}{\mathbf{a}^T \mathbf{a}} \mathbf{a}\mathbf{a}^T = \begin{bmatrix} 1 & 0 \\ 0 & 1 \end{bmatrix} - \frac{2}{1+m^2}\begin{bmatrix} 1 & m \\ m & m^2 \end{bmatrix} = \begin{bmatrix} -\frac{1-m^2}{1+m^2} & -\frac{2m}{1+m^2} \\ -\frac{2m}{1+m^2} & \frac{1-m^2}{1+m^2} \end{bmatrix}
$$

D3. If $s = \pm\sqrt{53}$, then $\|\mathbf{w}\| = \|\mathbf{v}\|$ and the Householder reflection about $(\mathbf{v} - \mathbf{w})^\perp$ maps $\mathbf{v}$ into $\mathbf{w}$.

D4. Since $\|\mathbf{w}\| = \|\mathbf{v}\|$, the Householder reflection about $(\mathbf{v} - \mathbf{w})^\perp$ maps $\mathbf{v}$ into $\mathbf{w}$. We have $\mathbf{v} - \mathbf{w} = (-8, 12)$, and so $(\mathbf{v} - \mathbf{w})^\perp$ is the line $-8x + 12y = 0$, or $y = \frac{2}{3}x$.

D5. Let $\mathbf{a} = \mathbf{v} - \mathbf{w} = (1, 2, 2) - (0, 0, 3) = (1, 2, -1)$. Then the reflection of R^3 about $\mathbf{a}^\perp$ maps $\mathbf{v}$ into $\mathbf{w}$, and the plane $\mathbf{a}^\perp$ corresponds to $x + 2y - z = 0$ or $z = x + 2y$.

WORKING WITH PROOFS

P1. If $H = I - \frac{2}{\mathbf{a}^T\mathbf{a}}\mathbf{a}\mathbf{a}^T$, then $H^T = (I - \frac{2}{\mathbf{a}^T\mathbf{a}}\mathbf{a}\mathbf{a}^T)^T = I - \frac{2}{\mathbf{a}^T\mathbf{a}}(\mathbf{a}\mathbf{a}^T)^T = I - \frac{2}{\mathbf{a}^T\mathbf{a}}\mathbf{a}^T\mathbf{a} = H$.

P2. To show that $H = I - \frac{2}{\mathbf{a}^T\mathbf{a}}\mathbf{a}\mathbf{a}^T$ is orthogonal we must show that $H^T = H^{-1}$. This follows from

$$HH^T = \left(I - \frac{2}{\mathbf{a}^T\mathbf{a}}\mathbf{a}\mathbf{a}^T\right)\left(I - \frac{2}{\mathbf{a}^T\mathbf{a}}\mathbf{a}\mathbf{a}^T\right)^T = I - \frac{2}{\mathbf{a}^T\mathbf{a}}\mathbf{a}\mathbf{a}^T - \frac{2}{\mathbf{a}^T\mathbf{a}}\mathbf{a}\mathbf{a}^T + \frac{4}{(\mathbf{a}^T\mathbf{a})^2}\mathbf{a}\mathbf{a}^T\mathbf{a}\mathbf{a}^T$$

$$= I - \frac{2}{\mathbf{a}^T\mathbf{a}}\mathbf{a}\mathbf{a}^T - \frac{2}{\mathbf{a}^T\mathbf{a}}\mathbf{a}\mathbf{a}^T + \frac{4}{\mathbf{a}^T\mathbf{a}}\mathbf{a}\mathbf{a}^T = I$$

where we have used the fact that $\mathbf{a}\mathbf{a}^T\mathbf{a}\mathbf{a}^T = \mathbf{a}(\mathbf{a}^T\mathbf{a})\mathbf{a}^T = (\mathbf{a}^T\mathbf{a})\mathbf{a}\mathbf{a}^T$.

P3. One of the features of the Gram-Schmidt process is that $\text{span}\{\mathbf{q}_1, \mathbf{q}_2, \ldots, \mathbf{q}_j\} = \text{span}\{\mathbf{w}_1, \mathbf{w}_2, \ldots, \mathbf{w}_j\}$ for each $j = 1, 2, \ldots, k$. Thus in the expansion

$$\mathbf{w}_j = (\mathbf{w}_j \cdot \mathbf{q}_1)\mathbf{q}_1 + (\mathbf{w}_j \cdot \mathbf{q}_2)\mathbf{q}_2 + \cdots + (\mathbf{w}_j \cdot \mathbf{q}_j)\mathbf{q}_j$$

we must have $\mathbf{w}_j \cdot \mathbf{q}_j \neq 0$, for otherwise $\mathbf{w}_j$ would be in $\text{span}\{\mathbf{q}_1, \mathbf{q}_2, \ldots, \mathbf{q}_{j-1}\} = \text{span}\{\mathbf{w}_1, \mathbf{w}_2, \ldots, \mathbf{w}_{j-1}\}$ which would mean that $\{\mathbf{w}_1, \mathbf{w}_2, \ldots, \mathbf{w}_j\}$ is a linearly dependent set.

P4. If $A = QR$ is a QR-decomposition of A, then $Q = AR^{-1}$. From this it follows that the columns of Q belong to the column space of A. In particular, if $R^{-1} = [s_{ij}]$, then from $Q = AR^{-1}$ it follows that

$$\mathbf{c}_j(Q) = A\mathbf{c}_j(R^{-1}) = s_{1j}\mathbf{c}_1(A) + s_{2j}\mathbf{c}_2(A) + \cdots + s_{kj}\mathbf{c}_k(A)$$

for each $j = 1, 2, \ldots, k$. Finally, since $\dim(\text{col}(A)) = k$ and the vectors $\mathbf{c}_1(Q), \mathbf{c}_2(Q), \ldots, \mathbf{c}_k(Q)$ are linearly independent, it follows that they form a basis for $\text{col}(A)$.

EXERCISE SET 7.11

1. (a) We have $\mathbf{w} = 3\mathbf{v}_1 - 7\mathbf{v}_2$; thus $(\mathbf{w})_B = (3, -7)$ and $[\mathbf{w}]_B = \begin{bmatrix} 3 \\ -7 \end{bmatrix}$.

(b) The vector equation $c_1\mathbf{v}_1 + c_2\mathbf{v}_2 = \mathbf{w}$ is equivalent to the linear system $\begin{bmatrix} 2 & 3 \\ -4 & 8 \end{bmatrix}\begin{bmatrix} c_1 \\ c_2 \end{bmatrix} = \begin{bmatrix} 1 \\ 1 \end{bmatrix}$, and the solution of this system is $c_1 = \frac{5}{28}, c_2 = \frac{3}{14}$. Thus $(\mathbf{w})_B = (\frac{5}{28}, \frac{3}{14})$ and $[\mathbf{w}]_B = \begin{bmatrix} \frac{5}{28} \\ \frac{3}{14} \end{bmatrix}$.

3. The vector equation $c_1\mathbf{v}_1 + c_2\mathbf{v}_2 + c_3\mathbf{v}_3 = \mathbf{w}$ is equivalent to $\begin{bmatrix} 1 & 2 & 3 \\ 0 & 2 & 3 \\ 0 & 0 & 3 \end{bmatrix}\begin{bmatrix} c_1 \\ c_2 \\ c_3 \end{bmatrix} = \begin{bmatrix} 2 \\ -1 \\ 3 \end{bmatrix}$. Solving this system by back substitution yields $c_3 = 1, c_2 = -2, c_1 = 3$. Thus $(\mathbf{w})_B = (3, -2, 1)$ and $[\mathbf{w}]_B = \begin{bmatrix} 3 \\ -2 \\ 1 \end{bmatrix}$.

5. If $(\mathbf{u})_B = (7, -2, 1)$, then $\mathbf{u} = 7\mathbf{v}_1 - 2\mathbf{v}_2 + \mathbf{v}_3 = 7(1, 0, 0) - 2(2, 2, 0) + (3, 3, 3) = (6, -1, 3)$.

7. $(\mathbf{w})_B = (\mathbf{w} \cdot \mathbf{v}_1, \mathbf{w} \cdot \mathbf{v}_2) = (-\frac{4}{\sqrt{2}}, \frac{10}{\sqrt{2}}) = (-2\sqrt{2}, 5\sqrt{2})$

9. $(\mathbf{w})_B = (\mathbf{w} \cdot \mathbf{v}_1, \mathbf{w} \cdot \mathbf{v}_2, \mathbf{w} \cdot \mathbf{v}_3) = (\frac{5}{\sqrt{2}}, -\frac{2}{\sqrt{3}}, \frac{1}{\sqrt{6}}) = (\frac{5\sqrt{2}}{2}, -\frac{2\sqrt{3}}{3}, \frac{\sqrt{6}}{6})$

11. (a) We have $\mathbf{u} = \mathbf{v}_1 + \mathbf{v}_2 = (\frac{7}{5}, -\frac{1}{5})$ and $\mathbf{v} = -\mathbf{v}_1 + 4\mathbf{v}_2 = (\frac{13}{5}, \frac{16}{5})$.

(b) Using Theorem 7.11.2: $\|\mathbf{u}\| = \|(\mathbf{u})_B\| = \sqrt{(1)^2 + (1)^2} = \sqrt{2}$, $\|\mathbf{v}\| = \|(\mathbf{v})_B\| = \sqrt{(-1)^2 + (4)^2} = \sqrt{17}$, and $\mathbf{u} \cdot \mathbf{v} = (\mathbf{u})_B \cdot (\mathbf{v})_B = (1)(-1) + (1)(4) = 3$.

Computing directly: $\|\mathbf{u}\| = \sqrt{(\frac{7}{5})^2 + (-\frac{1}{5})^2} = \sqrt{\frac{50}{25}} = \sqrt{2}$, $\|\mathbf{v}\| = \sqrt{(\frac{13}{5})^2 + (\frac{16}{5})^2} = \sqrt{\frac{425}{25}} = \sqrt{17}$, and $\mathbf{u} \cdot \mathbf{v} = (\frac{7}{5})(\frac{13}{5}) + (-\frac{1}{5})(\frac{16}{5}) = \frac{75}{25} = 3$.

13. $\|\mathbf{u}\| = \|(\mathbf{u})_B\| = \sqrt{(-1)^2 + (2)^2 + (1)^2 + (3)^2} = \sqrt{15}$

$\|\mathbf{v}\| = \|(\mathbf{v})_B\| = \sqrt{(0)^2 + (-3)^2 + (1)^2 + (5)^2} = \sqrt{35}$

$\|\mathbf{w}\| = \|(\mathbf{w})_B\| = \sqrt{(-2)^2 + (-4)^2 + (3)^2 + (1)^2} = \sqrt{30}$

$\|\mathbf{v} + \mathbf{w}\| = \|(\mathbf{v})_B + (\mathbf{w})_B\| = \|(-2, -7, 4, 6)\| = \sqrt{(-2)^2 + (-7)^2 + (4)^2 + (6)^2} = \sqrt{105}$

$\|\mathbf{v} - \mathbf{w}\| = \|(\mathbf{v})_B - (\mathbf{w})_B\| = \|(2, 1, -2, 4)\| = \sqrt{(2)^2 + (1)^2 + (-2)^2 + (4)^2} = 5$

$\mathbf{v} \cdot \mathbf{w} = (\mathbf{v})_B \cdot (\mathbf{w})_B = (0)(-2) + (-3)(-4) + (1)(3) + (5)(1) = 20$

15. Let $B = \{\mathbf{e}_1, \mathbf{e}_2\}$ be the standard basis for R^2, and let $B' = \{\mathbf{v}_1, \mathbf{v}_2\}$ be the basis corresponding to the $x'y'$-system that is described in Figure Ex-15. Then $P_{B' \to B} = [[\mathbf{v}_1]_B \quad [\mathbf{v}_2]_B] = \begin{bmatrix} 1 & \frac{1}{\sqrt{2}} \\ 0 & \frac{1}{\sqrt{2}} \end{bmatrix}$ and so

$P_{B \to B'} = (P_{B' \to B})^{-1} = \begin{bmatrix} 1 & -1 \\ 0 & \sqrt{2} \end{bmatrix}$. It follows that $x'y'$-coordinates are related to xy-coordinates by the equations $x' = x - y$ and $y' = \sqrt{2}y$. In particular:

(a) If $(x, y) = (1, 1)$, then $(x', y') = (0, \sqrt{2})$.   **(b)** If $(x, y) = (1, 0)$, then $(x', y') = (1, 0)$.

(c) If $(x, y) = (0, 1)$, then $(x', y') = (-1, \sqrt{2})$.   **(d)** If $(x, y) = (a, b)$, then $(x', y') = (a - b, \sqrt{2}b)$.

17. (a) We have $\mathbf{v}_1 = 2\mathbf{e}_1 + \mathbf{e}_2$ and $\mathbf{v}_2 = -3\mathbf{e}_1 + 4\mathbf{e}_2$; thus $P_{B \to S} = [(\mathbf{v}_1)_S \quad (\mathbf{v}_2)_S] = \begin{bmatrix} 2 & -3 \\ 1 & 4 \end{bmatrix}$.

(b) The row reduced echelon form of $[B \mid S] = \begin{bmatrix} 2 & -3 & | & 1 & 0 \\ 1 & 4 & | & 0 & 1 \end{bmatrix}$ is $\begin{bmatrix} 1 & 0 & | & \frac{4}{11} & \frac{3}{11} \\ 0 & 1 & | & -\frac{1}{11} & \frac{2}{11} \end{bmatrix}$; thus $P_{S \to B} = \begin{bmatrix} \frac{4}{11} & \frac{3}{11} \\ -\frac{1}{11} & \frac{2}{11} \end{bmatrix}$.

(c) Note that $(P_{B \to S})^{-1} = \begin{bmatrix} 2 & -3 \\ 1 & 4 \end{bmatrix}^{-1} = \frac{1}{11} \begin{bmatrix} 4 & 3 \\ -1 & 2 \end{bmatrix} = P_{S \to B}$.

(d) We have $\mathbf{w} = \mathbf{v}_1 - \mathbf{v}_2$; thus $[\mathbf{w}]_B = \begin{bmatrix} 1 \\ -1 \end{bmatrix}$ and $[\mathbf{w}]_S = P_{B \to S}[\mathbf{w}]_B = \begin{bmatrix} 2 & -3 \\ 1 & 4 \end{bmatrix} \begin{bmatrix} 1 \\ -1 \end{bmatrix} = \begin{bmatrix} 5 \\ -3 \end{bmatrix}$.

(e) We have $\mathbf{w} = 3\mathbf{e}_1 - 5\mathbf{e}_2$; thus $[\mathbf{w}]_S = \begin{bmatrix} 3 \\ -5 \end{bmatrix}$ and $[\mathbf{w}]_B = P_{S \to B}[\mathbf{w}]_S = \begin{bmatrix} \frac{4}{11} & \frac{3}{11} \\ -\frac{1}{11} & \frac{2}{11} \end{bmatrix} \begin{bmatrix} 3 \\ -5 \end{bmatrix} = \begin{bmatrix} -\frac{3}{11} \\ -\frac{13}{11} \end{bmatrix}$.

19. (a) The row reduced echelon form of $[B_1 \mid B_2] = \begin{bmatrix} 2 & 4 & | & 1 & -1 \\ 2 & -1 & | & 3 & -1 \end{bmatrix}$ is $\begin{bmatrix} 1 & 0 & | & \frac{13}{10} & -\frac{1}{2} \\ 0 & 1 & | & -\frac{2}{5} & 0 \end{bmatrix}$; thus $P_{B_2 \to B_1} = \begin{bmatrix} \frac{13}{10} & -\frac{1}{2} \\ -\frac{2}{5} & 0 \end{bmatrix}$.

(b) The row reduced echelon form of $[B_2 \mid B_1] = \begin{bmatrix} 1 & -1 & | & 2 & 4 \\ 3 & -1 & | & 2 & -1 \end{bmatrix}$ is $\begin{bmatrix} 1 & 0 & | & 0 & -\frac{5}{2} \\ 0 & 1 & | & -2 & -\frac{13}{2} \end{bmatrix}$; thus $P_{B_1 \to B_2} = \begin{bmatrix} 0 & -\frac{5}{2} \\ -2 & -\frac{13}{2} \end{bmatrix}$.

(c) Note that $(P_{B_2 \to B_1})^{-1} = \begin{bmatrix} \frac{13}{10} & -\frac{1}{2} \\ -\frac{2}{5} & 0 \end{bmatrix}^{-1} = (-5) \begin{bmatrix} 0 & \frac{1}{2} \\ \frac{2}{5} & \frac{13}{10} \end{bmatrix} = P_{B_1 \to B_2}$.

(d) $\mathbf{w} = -\frac{7}{10}\mathbf{u}_1 + \frac{8}{5}\mathbf{u}_2$; thus $[\mathbf{w}]_{B_1} = \begin{bmatrix} -\frac{7}{10} \\ \frac{8}{5} \end{bmatrix}$ and $[\mathbf{w}]_{B_2} = P_{B_1 \to B_2}[\mathbf{w}]_{B_1} = \begin{bmatrix} 0 & -\frac{5}{2} \\ -2 & -\frac{13}{2} \end{bmatrix} \begin{bmatrix} -\frac{7}{10} \\ \frac{8}{5} \end{bmatrix} = \begin{bmatrix} -4 \\ -9 \end{bmatrix}$.

(e) We have $\mathbf{w} = 4\mathbf{v}_1 - 7\mathbf{v}_2$; thus $[\mathbf{w}]_{B_2} = \begin{bmatrix} -4 \\ -7 \end{bmatrix}$ and $[\mathbf{w}]_{B_1} = P_{B_2 \to B_1}[\mathbf{w}]_{B_2} = \begin{bmatrix} \frac{13}{10} & -\frac{1}{2} \\ -\frac{2}{5} & 0 \end{bmatrix} \begin{bmatrix} -4 \\ -7 \end{bmatrix} = \begin{bmatrix} -\frac{17}{10} \\ \frac{8}{5} \end{bmatrix}$.

21. **(a)** The row reduced echelon form of $[B_2 \mid B_1] = \begin{bmatrix} -6 & -2 & -2 & | & -3 & -3 & 1 \\ -6 & -6 & -3 & | & 0 & 2 & 6 \\ 0 & 4 & 7 & | & -3 & -1 & -1 \end{bmatrix}$ is $\begin{bmatrix} 1 & 0 & 0 & | & \frac{3}{4} & \frac{3}{4} & \frac{1}{12} \\ 0 & 1 & 0 & | & -\frac{3}{4} & -\frac{17}{12} & -\frac{17}{12} \\ 0 & 0 & 1 & | & 0 & \frac{2}{3} & \frac{2}{3} \end{bmatrix}$;

thus $P_{B_1 \to B_2} = \begin{bmatrix} \frac{3}{4} & \frac{3}{4} & \frac{1}{12} \\ -\frac{3}{4} & -\frac{17}{12} & -\frac{17}{12} \\ 0 & \frac{2}{3} & \frac{2}{3} \end{bmatrix}$.

(b) If $\mathbf{w} = (-5, 8, -5)$, we have $(\mathbf{w})_{B_1} = (1, 1, 1)$ and so

$$[\mathbf{w}]_{B_2} = P_{B_1 \to B_2}[\mathbf{w}]_{B_1} = \begin{bmatrix} \frac{3}{4} & \frac{3}{4} & \frac{1}{12} \\ -\frac{3}{4} & -\frac{17}{12} & -\frac{17}{12} \\ 0 & \frac{2}{3} & \frac{2}{3} \end{bmatrix} \begin{bmatrix} 1 \\ 1 \\ 1 \end{bmatrix} = \begin{bmatrix} \frac{19}{12} \\ -\frac{43}{12} \\ \frac{4}{3} \end{bmatrix}$$

(c) The row reduced echelon form of $[B_2 \mid \mathbf{w}] = \begin{bmatrix} -6 & -2 & -2 & | & -5 \\ -6 & -6 & -3 & | & 8 \\ 0 & 4 & 7 & | & -5 \end{bmatrix}$ is $\begin{bmatrix} 1 & 0 & 0 & | & \frac{19}{12} \\ 0 & 1 & 0 & | & -\frac{43}{12} \\ 0 & 0 & 1 & | & \frac{4}{3} \end{bmatrix}$; thus $(\mathbf{w})_{B_2} = $

$(\frac{19}{12}, -\frac{43}{12}, \frac{4}{3})$, which agrees with the computation in part (b).

23. The vector equation $c_1\mathbf{u}_1 + c_2\mathbf{u}_2 + c_3\mathbf{u}_3 = \mathbf{v}_1$ is equivalent to $\begin{bmatrix} \frac{1}{\sqrt{3}} & \frac{1}{\sqrt{2}} & \frac{1}{\sqrt{6}} \\ \frac{1}{\sqrt{3}} & -\frac{1}{\sqrt{2}} & \frac{1}{\sqrt{6}} \\ \frac{1}{\sqrt{3}} & 0 & -\frac{2}{\sqrt{6}} \end{bmatrix} \begin{bmatrix} c_1 \\ c_2 \\ c_3 \end{bmatrix} = \begin{bmatrix} \frac{1}{\sqrt{2}} \\ 0 \\ \frac{1}{\sqrt{2}} \end{bmatrix}$, and the

solution of this system is $c_1 = \frac{2}{\sqrt{6}}$, $c_2 = \frac{1}{2}$, $c_3 = -\frac{1}{2\sqrt{3}}$. Thus $(\mathbf{v}_1)_{B_1} = (\frac{2}{\sqrt{6}}, \frac{1}{2}, -\frac{1}{2\sqrt{3}})$. Similarly, we

have $(\mathbf{v}_2)_{B_1} = (\frac{1}{3}, 0, \frac{4}{3\sqrt{2}})$ and $(\mathbf{v}_3)_{B_1} = (-\frac{2}{3\sqrt{2}}, \frac{3}{2\sqrt{3}}, \frac{1}{6})$. Thus the transition matrix $P_{B_2 \to B_1}$ is

$$P_{B_2 \to B_1} = \begin{bmatrix} \frac{2}{\sqrt{6}} & \frac{1}{3} & -\frac{2}{3\sqrt{2}} \\ \frac{1}{2} & 0 & \frac{3}{2\sqrt{3}} \\ -\frac{1}{2\sqrt{3}} & \frac{4}{3\sqrt{2}} & \frac{1}{6} \end{bmatrix}$$

It is easy to check that

$$(P_{B_2 \to B_1})(P_{B_2 \to B_1})^T = \begin{bmatrix} \frac{2}{\sqrt{6}} & \frac{1}{3} & -\frac{2}{3\sqrt{2}} \\ \frac{1}{2} & 0 & \frac{3}{2\sqrt{3}} \\ -\frac{1}{2\sqrt{3}} & \frac{4}{3\sqrt{2}} & \frac{1}{6} \end{bmatrix} \begin{bmatrix} \frac{2}{\sqrt{6}} & \frac{1}{2} & -\frac{1}{2\sqrt{3}} \\ \frac{1}{3} & 0 & \frac{4}{3\sqrt{2}} \\ -\frac{2}{3\sqrt{2}} & \frac{3}{2\sqrt{3}} & \frac{1}{6} \end{bmatrix} = \begin{bmatrix} 1 & 0 & 0 \\ 0 & 1 & 0 \\ 0 & 0 & 1 \end{bmatrix}$$

Thus $P_{B_2 \to B_1}$ is an orthogonal matrix, and since $P_{B_1 \to B_2} = (P_{B_2 \to B_1})^{-1} = (P_{B_2 \to B_1})^T$, the same is true of $P_{B_1 \to B_2}$.

25. **(a)** We have $\mathbf{v}_1 = (0, 1) = 0\mathbf{e}_1 + 1\mathbf{e}_2$ and $\mathbf{v}_2 = (1, 0) = 1\mathbf{e}_1 + 0\mathbf{e}_2$; thus $P_{B \to S} = \begin{bmatrix} 0 & 1 \\ 1 & 0 \end{bmatrix}$.

(b) If $P = P_{B \to S} = \begin{bmatrix} 0 & 1 \\ 1 & 0 \end{bmatrix}$ then, since P is orthogonal, we have $P^T = P^{-1} = (P_{B \to S})^{-1} = P_{S \to B}$. Geometrically, this corresponds to the fact that reflection about $y = x$ preserves length and thus is an orthogonal transformation.

27. **(a)** If $\begin{bmatrix} x \\ y \end{bmatrix} = \begin{bmatrix} -2 \\ 6 \end{bmatrix}$, then $\begin{bmatrix} x' \\ y' \end{bmatrix} = \begin{bmatrix} \cos(\frac{3\pi}{4}) & \sin(\frac{3\pi}{4}) \\ -\sin(\frac{3\pi}{4}) & \cos(\frac{3\pi}{4}) \end{bmatrix} \begin{bmatrix} x \\ y \end{bmatrix} = \begin{bmatrix} -\frac{1}{\sqrt{2}} & \frac{1}{\sqrt{2}} \\ -\frac{1}{\sqrt{2}} & -\frac{1}{\sqrt{2}} \end{bmatrix} \begin{bmatrix} -2 \\ 6 \end{bmatrix} = \begin{bmatrix} \frac{8}{\sqrt{2}} \\ -\frac{4}{\sqrt{2}} \end{bmatrix} = \begin{bmatrix} 4\sqrt{2} \\ -2\sqrt{2} \end{bmatrix}$.

(b) If $\begin{bmatrix} x' \\ y' \end{bmatrix} = \begin{bmatrix} 5 \\ 2 \end{bmatrix}$, then $\begin{bmatrix} x \\ y \end{bmatrix} = \begin{bmatrix} \cos(\frac{3\pi}{4}) & -\sin(\frac{3\pi}{4}) \\ \sin(\frac{3\pi}{4}) & \cos(\frac{3\pi}{4}) \end{bmatrix} \begin{bmatrix} x' \\ y' \end{bmatrix} = \begin{bmatrix} -\frac{1}{\sqrt{2}} & -\frac{1}{\sqrt{2}} \\ \frac{1}{\sqrt{2}} & -\frac{1}{\sqrt{2}} \end{bmatrix} \begin{bmatrix} 5 \\ 2 \end{bmatrix} = \begin{bmatrix} -\frac{7}{\sqrt{2}} \\ \frac{3}{\sqrt{2}} \end{bmatrix} = \begin{bmatrix} -\frac{7\sqrt{2}}{2} \\ \frac{3\sqrt{2}}{2} \end{bmatrix}$.

29. **(a)** If $\begin{bmatrix} x \\ y \\ z \end{bmatrix} = \begin{bmatrix} -1 \\ 2 \\ 5 \end{bmatrix}$, then $\begin{bmatrix} x' \\ y' \\ z' \end{bmatrix} = \begin{bmatrix} \cos(\frac{\pi}{4}) & \sin(\frac{\pi}{4}) & 0 \\ -\sin(\frac{\pi}{4}) & \cos(\frac{\pi}{4}) & 0 \\ 0 & 0 & 1 \end{bmatrix} \begin{bmatrix} x \\ y \\ z \end{bmatrix} = \begin{bmatrix} \frac{1}{\sqrt{2}} & \frac{1}{\sqrt{2}} & 0 \\ -\frac{1}{\sqrt{2}} & \frac{1}{\sqrt{2}} & 0 \\ 0 & 0 & 1 \end{bmatrix} \begin{bmatrix} -1 \\ 2 \\ 5 \end{bmatrix} = \begin{bmatrix} \frac{1}{\sqrt{2}} \\ \frac{3}{\sqrt{2}} \\ 5 \end{bmatrix}$.

(b) If $\begin{bmatrix} x' \\ y' \\ z' \end{bmatrix} = \begin{bmatrix} 1 \\ 6 \\ -3 \end{bmatrix}$, then $\begin{bmatrix} x \\ y \\ z \end{bmatrix} = \begin{bmatrix} \cos(\frac{\pi}{4}) & -\sin(\frac{\pi}{4}) & 0 \\ \sin(\frac{\pi}{4}) & \cos(\frac{\pi}{4}) & 0 \\ 0 & 0 & 1 \end{bmatrix} \begin{bmatrix} x' \\ y' \\ z' \end{bmatrix} = \begin{bmatrix} \frac{1}{\sqrt{2}} & -\frac{1}{\sqrt{2}} & 0 \\ \frac{1}{\sqrt{2}} & \frac{1}{\sqrt{2}} & 0 \\ 0 & 0 & 1 \end{bmatrix} \begin{bmatrix} 1 \\ 6 \\ -3 \end{bmatrix} = \begin{bmatrix} -\frac{5}{\sqrt{2}} \\ \frac{7}{\sqrt{2}} \\ -3 \end{bmatrix}$.

31. We have $\begin{bmatrix} x' \\ y' \\ z' \end{bmatrix} = \begin{bmatrix} \frac{1}{2} & \frac{\sqrt{3}}{2} & 0 \\ -\frac{\sqrt{3}}{2} & \frac{1}{2} & 0 \\ 0 & 0 & 1 \end{bmatrix} \begin{bmatrix} x \\ y \\ z \end{bmatrix}$ and $\begin{bmatrix} x'' \\ y'' \\ z'' \end{bmatrix} = \begin{bmatrix} \frac{1}{\sqrt{2}} & 0 & -\frac{1}{\sqrt{2}} \\ 0 & 1 & 0 \\ \frac{1}{\sqrt{2}} & 0 & \frac{1}{\sqrt{2}} \end{bmatrix} \begin{bmatrix} x' \\ y' \\ z' \end{bmatrix}$. Thus

$$\begin{bmatrix} x'' \\ y'' \\ z'' \end{bmatrix} = \begin{bmatrix} \frac{1}{\sqrt{2}} & 0 & -\frac{1}{\sqrt{2}} \\ 0 & 1 & 0 \\ \frac{1}{\sqrt{2}} & 0 & \frac{1}{\sqrt{2}} \end{bmatrix} \begin{bmatrix} \frac{1}{2} & \frac{\sqrt{3}}{2} & 0 \\ -\frac{\sqrt{3}}{2} & \frac{1}{2} & 0 \\ 0 & 0 & 1 \end{bmatrix} \begin{bmatrix} x \\ y \\ z \end{bmatrix} = \begin{bmatrix} \frac{\sqrt{2}}{4} & \frac{\sqrt{6}}{4} & -\frac{\sqrt{2}}{2} \\ -\frac{\sqrt{3}}{2} & \frac{1}{2} & 0 \\ \frac{\sqrt{2}}{4} & \frac{\sqrt{6}}{4} & \frac{\sqrt{2}}{2} \end{bmatrix} \begin{bmatrix} x \\ y \\ z \end{bmatrix}$$

DISCUSSION AND DISCOVERY

D1. We have $P_{B_1 \to B_3} = P_{B_2 \to B_3} P_{B_1 \to B_2} = \begin{bmatrix} 7 & 2 \\ 4 & -1 \end{bmatrix} \begin{bmatrix} 3 & 1 \\ 5 & 2 \end{bmatrix} = \begin{bmatrix} 31 & 11 \\ 7 & 2 \end{bmatrix}$ and $P_{B_3 \to B_1} = (P_{B_1 \to B_3})^{-1} = $

$\begin{bmatrix} -\frac{2}{15} & \frac{11}{15} \\ \frac{7}{15} & -\frac{31}{15} \end{bmatrix}$.

D2. **(a)** Let $B = \{\mathbf{v}_1, \mathbf{v}_2, \mathbf{v}_3\}$, where $\mathbf{v}_1 = (1, 1, 0)$, $\mathbf{v}_2 = (1, 0, 2)$, $\mathbf{v}_3 = (0, 2, 1)$ correspond to the column vectors of the matrix P. Then, from Theorem 7.11.8, P is the transition matrix from B to the standard basis $S = \{\mathbf{e}_1, \mathbf{e}_2, \mathbf{e}_3\}$.

(b) If P is the transition matrix from $S = \{\mathbf{e}_1, \mathbf{e}_2, \mathbf{e}_3\}$ to $B = \{\mathbf{w}_1, \mathbf{w}_2, \mathbf{w}_3\}$, then $\mathbf{e}_1 = \mathbf{w}_1 + \mathbf{w}_2$, $\mathbf{e}_2 = \mathbf{w}_1 + 2\mathbf{w}_3$, and $\mathbf{e}_3 = 2\mathbf{w}_2 + \mathbf{w}_3$. Solving these vector equations for $\mathbf{w}_1$, $\mathbf{w}_2$, and $\mathbf{w}_3$ in terms of $\mathbf{e}_1$, $\mathbf{e}_2$, and $\mathbf{e}_3$ results in $\mathbf{w}_1 = \frac{4}{5}\mathbf{e}_1 + \frac{1}{5}\mathbf{e}_2 - \frac{2}{5}\mathbf{e}_3 = (\frac{4}{5}, \frac{1}{5}, -\frac{2}{5})$, $\mathbf{w}_2 = \frac{1}{5}\mathbf{e}_1 - \frac{1}{5}\mathbf{e}_2 + \frac{2}{5}\mathbf{e}_3 = (\frac{1}{5}, -\frac{1}{5}, \frac{2}{5})$, and $\mathbf{w}_3 = -\frac{2}{5}\mathbf{e}_1 + \frac{2}{5}\mathbf{e}_2 + \frac{1}{5}\mathbf{e}_3 = (-\frac{2}{5}, \frac{2}{5}, \frac{1}{5})$. Note that $\mathbf{w}_1$, $\mathbf{w}_2$, $\mathbf{w}_3$ correspond to the column vectors of the matrix P^{-1}.

D3. If $B = \{\mathbf{v}_1, \mathbf{v}_2, \mathbf{v}_3\}$ and if $P = \begin{bmatrix} 1 & 0 & 0 \\ 0 & 3 & 2 \\ 0 & 1 & 1 \end{bmatrix}$ is the transition matrix from B to the given basis, then

$\mathbf{v}_1 = (1, 1, 1)$, $\mathbf{v}_2 = 3(1, 1, 0) + (1, 0, 0) = (4, 3, 0)$, and $\mathbf{v}_3 = 2(1, 1, 0) + (1, 0, 0) = (3, 2, 0)$.

D4. If $[\mathbf{w}]_B = \mathbf{w}$ holds for every $\mathbf{w}$, then the transition matrix from the standard basis S to the basis B is

$$P_{S \to B} = [[\mathbf{e}_1]_B \mid [\mathbf{e}_2]_B \mid [\mathbf{e}_3]_B] = [\mathbf{e}_1 \mid \mathbf{e}_2 \mid \mathbf{e}_3] = I_n$$

and so $B = S = \{\mathbf{e}_1, \mathbf{e}_2, \ldots, \mathbf{e}_n\}$.

D5. If $[\mathbf{x} - \mathbf{y}]_B = \mathbf{0}$, then $[\mathbf{x}]_B = [\mathbf{y}]_B$ and so $\mathbf{x} = \mathbf{y}$.

WORKING WITH PROOFS

P1. If $c_1, c_2, \ldots, c_k$ are scalars, then $(c_1\mathbf{v}_1 + c_2\mathbf{v}_2 + \cdots + c_k\mathbf{v}_k)_B = c_1(\mathbf{v}_1)_B + c_2(\mathbf{v}_2)_B + \cdots + c_k(\mathbf{v}_k)_B$. Note also that $(\mathbf{v})_B = \mathbf{0}$ if and only if $\mathbf{v} = \mathbf{0}$. It follows that $c_1\mathbf{v}_1 + c_2\mathbf{v}_2 + \cdots + c_k\mathbf{v}_k = \mathbf{0}$ if and only if $c_1(\mathbf{v}_1)_B + c_2(\mathbf{v}_2)_B + \cdots + c_k(\mathbf{v}_k)_B = \mathbf{0}$. Thus the vectors $\mathbf{v}_1, \mathbf{v}_2, \ldots, \mathbf{v}_k$ are linearly independent if and only if $(\mathbf{v}_1)_B, (\mathbf{v}_2)_B, \ldots, (\mathbf{v}_k)_B$ are linearly independent.

P2. The vectors $\mathbf{v}_1, \mathbf{v}_2, \ldots, \mathbf{v}_k$ span R^n if and only if every vector $\mathbf{v}$ in R^n can be expressed as a linear combination of them, i.e., there exist scalars $c_1, c_2, \ldots, c_k$ such that $\mathbf{v} = c_1\mathbf{v}_1 + c_2\mathbf{v}_2 + \cdots + c_k\mathbf{v}_k$. Since $(\mathbf{v})_B = c_1(\mathbf{v}_1)_B + c_2(\mathbf{v}_2)_B + \cdots + c_k(\mathbf{v}_k)_B$ and the coordinate mapping $\mathbf{v} \to (\mathbf{v})_B$ is onto, it follows that the vectors $\mathbf{v}_1, \mathbf{v}_2, \ldots, \mathbf{v}_k$ span R^n if and only if $(\mathbf{v}_1)_B, (\mathbf{v}_2)_B, \ldots, (\mathbf{v}_k)_B$ span R^n.

P3. Since the coordinate map $\mathbf{x} \to [\mathbf{x}]_B$ is onto, we have $A[\mathbf{x}]_B = C[\mathbf{x}]_B$ for every $\mathbf{x}$ in R^n if and only if $A\mathbf{y} = C\mathbf{y}$ for every $\mathbf{y}$ in R^n. Thus, using Theorem 3.4.4, we can conclude that $A = C$ if and only if $A[\mathbf{x}]_B = C[\mathbf{x}]_B$ for every $\mathbf{x}$ in R^n.

P4. Suppose $B = \{\mathbf{u}_1, \mathbf{u}_2, \ldots, \mathbf{u}_n\}$ is a basis for R^n. Then if $\mathbf{v} = a_1\mathbf{u}_1 + a_2\mathbf{u}_2 + \cdots + a_n\mathbf{u}_n$ and $\mathbf{w} = b_1\mathbf{u}_1 + b_2\mathbf{u}_2 + \cdots + b_n\mathbf{u}_n$, we have:

$$c\mathbf{v} = c(a_1\mathbf{u}_1 + a_2\mathbf{u}_2 + \cdots + a_n\mathbf{u}_n) = ca_1\mathbf{u}_1 + ca_2\mathbf{u}_2 + \cdots + ca_n\mathbf{u}_n$$

$$\mathbf{v} + \mathbf{w} = (a_1\mathbf{u}_1 + a_2\mathbf{u}_2 + \cdots + a_n\mathbf{u}_n) + (b_1\mathbf{u}_1 + b_2\mathbf{u}_2 + \cdots + b_n\mathbf{u}_n)$$

$$= (a_1 + b_1)\mathbf{u}_1 + (a_2 + b_2)\mathbf{u}_2 + \cdots + (a_n + b_n)\mathbf{u}_n.$$

Thus $(c\mathbf{v})_B = (ca_1, ca_2, \ldots, ca_n) = c(a_1, a_2, \ldots, a_n) = c(\mathbf{v})_B$ and $(\mathbf{v} + \mathbf{w})_B = (a_1 + b_1, \ldots, a_n + b_n) = (a_1, a_2, \ldots, a_n) + (b_1, b_2, \ldots, b_n) = (\mathbf{v})_B + (\mathbf{w})_B$.

CHAPTER 8
Diagonalization

EXERCISE SET 8.1

1. For every vector $\mathbf{x}$ in R^2, we have $\mathbf{x} = \begin{bmatrix} x_1 \\ x_2 \end{bmatrix} = x_2 \begin{bmatrix} 1 \\ 1 \end{bmatrix} + (x_2 - x_1) \begin{bmatrix} -1 \\ 0 \end{bmatrix} = x_2 \mathbf{v}_1 + (x_2 - x_1)\mathbf{v}_2$ and

$$T\mathbf{x} = \begin{bmatrix} 1 & -1 \\ 1 & 1 \end{bmatrix} \begin{bmatrix} x_1 \\ x_2 \end{bmatrix} = \begin{bmatrix} x_1 - x_2 \\ x_1 + x_2 \end{bmatrix} = (x_1 + x_2) \begin{bmatrix} 1 \\ 1 \end{bmatrix} + 2x_2 \begin{bmatrix} -1 \\ 0 \end{bmatrix} = (x_1 + x_2)\mathbf{v}_1 + 2x_2 \mathbf{v}_2$$

Thus $[\mathbf{x}]_B = \begin{bmatrix} x_2 \\ x_2 - x_1 \end{bmatrix}$, $[T\mathbf{x}]_B = \begin{bmatrix} x_1 + x_2 \\ 2x_2 \end{bmatrix}$, and $[T]_B = [[T\mathbf{v}_1]_B \quad [T\mathbf{v}_2]_B] = \begin{bmatrix} 2 & -1 \\ 2 & 0 \end{bmatrix}$. Finally, we note that

$$[T\mathbf{x}]_B = \begin{bmatrix} x_1 + x_2 \\ 2x_2 \end{bmatrix} = \begin{bmatrix} 2 & -1 \\ 2 & 0 \end{bmatrix} \begin{bmatrix} x_2 \\ x_2 - x_1 \end{bmatrix} = [T]_B [\mathbf{x}]_B$$

which is Formula (7).

3. Let $P = P_{S \to B}$ where $S = \{\mathbf{e}_1, \mathbf{e}_2\}$ is the standard basis. Then $P = [[\mathbf{v}_1]_S \quad [\mathbf{v}_2]_S] = \begin{bmatrix} 1 & -1 \\ 1 & 0 \end{bmatrix}$ and

$$P[T]_B P^{-1} = \begin{bmatrix} 1 & -1 \\ 1 & 0 \end{bmatrix} \begin{bmatrix} 2 & -1 \\ 2 & 0 \end{bmatrix} \begin{bmatrix} 0 & 1 \\ -1 & 1 \end{bmatrix} = \begin{bmatrix} 1 & -1 \\ 1 & 1 \end{bmatrix} = [T]$$

5. For every vector $\mathbf{x}$ in R^3, we have

$$\mathbf{x} = \begin{bmatrix} x_1 \\ x_2 \\ x_3 \end{bmatrix} = (\tfrac{1}{2}x_1 - \tfrac{1}{2}x_2 + \tfrac{1}{2}x_3) \begin{bmatrix} 1 \\ 0 \\ 1 \end{bmatrix} + (-\tfrac{1}{2}x_1 + \tfrac{1}{2}x_2 + \tfrac{1}{2}x_3) \begin{bmatrix} 0 \\ 1 \\ 1 \end{bmatrix} + (\tfrac{1}{2}x_1 + \tfrac{1}{2}x_2 - \tfrac{1}{2}x_3) \begin{bmatrix} 1 \\ 1 \\ 0 \end{bmatrix}$$

and

$$T\mathbf{x} = \begin{bmatrix} x_1 - x_2 \\ -x_1 + x_2 \\ x_1 - x_3 \end{bmatrix} = (\tfrac{3}{2}x_1 - x_2 - \tfrac{1}{2}x_3) \begin{bmatrix} 1 \\ 0 \\ 1 \end{bmatrix} + (-\tfrac{1}{2}x_1 + x_2 - \tfrac{1}{2}x_3) \begin{bmatrix} 0 \\ 1 \\ 1 \end{bmatrix} + (-\tfrac{1}{2}x_1 + \tfrac{1}{2}x_3) \begin{bmatrix} 1 \\ 1 \\ 0 \end{bmatrix}$$

Thus $[\mathbf{x}]_B = \begin{bmatrix} \tfrac{1}{2}x_1 - \tfrac{1}{2}x_2 + \tfrac{1}{2}x_3 \\ -\tfrac{1}{2}x_1 + \tfrac{1}{2}x_2 + \tfrac{1}{2}x_3 \\ \tfrac{1}{2}x_1 + \tfrac{1}{2}x_2 - \tfrac{1}{2}x_3 \end{bmatrix}$, $[T\mathbf{x}]_B = \begin{bmatrix} \tfrac{3}{2}x_1 - x_2 - \tfrac{1}{2}x_3 \\ -\tfrac{1}{2}x_1 + x_2 - \tfrac{1}{2}x_3 \\ -\tfrac{1}{2}x_1 + \tfrac{1}{2}x_3 \end{bmatrix}$, and

$$[T]_B = [[T\mathbf{v}_1]_B \quad [T\mathbf{v}_2]_B \quad [T\mathbf{v}_3]_B] = \begin{bmatrix} 1 & -\tfrac{3}{2} & \tfrac{1}{2} \\ -1 & \tfrac{1}{2} & \tfrac{1}{2} \\ 0 & \tfrac{1}{2} & -\tfrac{1}{2} \end{bmatrix}$$

Finally, we note that

$$[T\mathbf{x}]_B = \begin{bmatrix} \tfrac{3}{2}x_1 - x_2 - \tfrac{1}{2}x_3 \\ -\tfrac{1}{2}x_1 + x_2 - \tfrac{1}{2}x_3 \\ -\tfrac{1}{2}x_1 + \tfrac{1}{2}x_3 \end{bmatrix} = \begin{bmatrix} 1 & -\tfrac{3}{2} & \tfrac{1}{2} \\ -1 & \tfrac{1}{2} & \tfrac{1}{2} \\ 0 & \tfrac{1}{2} & -\tfrac{1}{2} \end{bmatrix} \begin{bmatrix} \tfrac{1}{2}x_1 - \tfrac{1}{2}x_2 + \tfrac{1}{2}x_3 \\ -\tfrac{1}{2}x_1 + \tfrac{1}{2}x_2 + \tfrac{1}{2}x_3 \\ \tfrac{1}{2}x_1 + \tfrac{1}{2}x_2 - \tfrac{1}{2}x_3 \end{bmatrix} = [T]_B [\mathbf{x}]_B$$

which is Formula (7).

7. Let $P = P_{S \to B}$ where S is the standard basis. Then $P = [[\mathbf{v}_1]_S \quad [\mathbf{v}_2]_S \quad [\mathbf{v}_3]_S] = \begin{bmatrix} 1 & 0 & 1 \\ 0 & 1 & 1 \\ 1 & 1 & 0 \end{bmatrix}$ and

$$P[T]_B P^{-1} = \begin{bmatrix} 1 & 0 & 1 \\ 0 & 1 & 1 \\ 1 & 1 & 0 \end{bmatrix} \begin{bmatrix} 1 & -\frac{3}{2} & \frac{1}{2} \\ -1 & \frac{1}{2} & \frac{1}{2} \\ 0 & \frac{1}{2} & -\frac{1}{2} \end{bmatrix} \begin{bmatrix} \frac{1}{2} & -\frac{1}{2} & \frac{1}{2} \\ -\frac{1}{2} & \frac{1}{2} & \frac{1}{2} \\ \frac{1}{2} & \frac{1}{2} & -\frac{1}{2} \end{bmatrix} = \begin{bmatrix} 1 & -1 & 0 \\ -1 & 1 & 0 \\ 1 & 0 & -1 \end{bmatrix} = [T]$$

9. We have $T\mathbf{v}_1 = \begin{bmatrix} -11 \\ 5 \end{bmatrix} = -\frac{4}{11} \begin{bmatrix} -1 \\ 5 \end{bmatrix} - \frac{25}{11} \begin{bmatrix} 5 \\ -3 \end{bmatrix} = -\frac{4}{11}\mathbf{v}_1 - \frac{25}{11}\mathbf{v}_2$ and $T\mathbf{v}_2 = \begin{bmatrix} 11 \\ -3 \end{bmatrix} = \frac{9}{11}\mathbf{v}_1 + \frac{26}{11}\mathbf{v}_2$.

Similarly, $T\mathbf{v}_1' = \begin{bmatrix} 0 \\ 1 \end{bmatrix} = \frac{3}{11} \begin{bmatrix} 2 \\ 1 \end{bmatrix} + \frac{2}{11} \begin{bmatrix} -3 \\ 4 \end{bmatrix} = \frac{3}{11}\mathbf{v}_1' + \frac{2}{11}\mathbf{v}_2'$ and $T\mathbf{v}_2' = \begin{bmatrix} -11 \\ 4 \end{bmatrix} = -\frac{32}{11}\mathbf{v}_1' + \frac{19}{11}\mathbf{v}_2'$. Thus

$[T]_B = \begin{bmatrix} -\frac{4}{11} & \frac{9}{11} \\ -\frac{25}{11} & \frac{26}{11} \end{bmatrix}$ and $[T]_{B'} = \begin{bmatrix} \frac{3}{11} & -\frac{32}{11} \\ \frac{2}{11} & \frac{19}{11} \end{bmatrix}$. Since $\mathbf{v}_1 = \mathbf{v}_1' + \mathbf{v}_2'$ and $\mathbf{v}_2 = \mathbf{v}_1' - \mathbf{v}_2'$, we have $P =$

$P_{B \to B'} = \begin{bmatrix} 1 & 1 \\ 1 & -1 \end{bmatrix}$ and

$$P[T]_B P^{-1} = \begin{bmatrix} 1 & 1 \\ 1 & -1 \end{bmatrix} \begin{bmatrix} -\frac{4}{11} & \frac{9}{11} \\ -\frac{25}{11} & \frac{26}{11} \end{bmatrix} \begin{bmatrix} \frac{1}{2} & \frac{1}{2} \\ \frac{1}{2} & -\frac{1}{2} \end{bmatrix} = \begin{bmatrix} \frac{3}{11} & -\frac{32}{11} \\ \frac{2}{11} & \frac{19}{11} \end{bmatrix} = [T]_{B'}$$

11. The equation $P[T]_B P^{-1} = [T]_{B'}$ is equivalent to $[T]_B = P^{-1}[T]_{B'} P$. Thus, from Exercise 9, we have

$$[T]_B = \begin{bmatrix} -\frac{4}{11} & \frac{9}{11} \\ -\frac{25}{11} & \frac{26}{11} \end{bmatrix} = \begin{bmatrix} \frac{1}{2} & \frac{1}{2} \\ \frac{1}{2} & -\frac{1}{2} \end{bmatrix} \begin{bmatrix} \frac{3}{11} & -\frac{32}{11} \\ \frac{2}{11} & \frac{19}{11} \end{bmatrix} \begin{bmatrix} 1 & 1 \\ 1 & -1 \end{bmatrix} = P^{-1}[T]_{B'} P$$

where, as before, $P = P_{B \to B'}$.

13. The standard matrix is $[T] = \begin{bmatrix} 1 & -2 \\ 0 & 1 \end{bmatrix}$ and, from Exercise 9, we have $[T]_B = \begin{bmatrix} -\frac{4}{11} & \frac{9}{11} \\ -\frac{25}{11} & \frac{26}{11} \end{bmatrix}$. These matrices are related by the equation

$$P[T]_B P^{-1} = \begin{bmatrix} -1 & 5 \\ 5 & -3 \end{bmatrix} \begin{bmatrix} -\frac{4}{11} & \frac{9}{11} \\ -\frac{25}{11} & \frac{26}{11} \end{bmatrix} \begin{bmatrix} \frac{3}{22} & \frac{5}{22} \\ \frac{5}{22} & \frac{1}{22} \end{bmatrix} = \begin{bmatrix} 1 & -2 \\ 0 & 1 \end{bmatrix} = [T]$$

where $P = [\mathbf{v}_1 \,|\, \mathbf{v}_2]$.

15. (a) For every $\mathbf{x} = (x, y)$ in R^2, we have

$$\mathbf{x} = \begin{bmatrix} x \\ y \end{bmatrix} = \left(-\tfrac{1}{5}x + \tfrac{2}{5}y\right)\begin{bmatrix} -1 \\ 2 \end{bmatrix} + \left(\tfrac{2}{5}x + \tfrac{1}{5}y\right)\begin{bmatrix} 2 \\ 1 \end{bmatrix} = \left(-\tfrac{1}{5}x + \tfrac{2}{5}y\right)\mathbf{v}_1 + \left(\tfrac{2}{5}x + \tfrac{1}{5}y\right)\mathbf{v}_2$$

and

$$T\mathbf{x} = \begin{bmatrix} 2x + 3y \\ x - 2y \end{bmatrix} = \left(-\tfrac{7}{5}y\right)\begin{bmatrix} -1 \\ 2 \end{bmatrix} + \left(x + \tfrac{4}{5}y\right)\begin{bmatrix} 2 \\ 1 \end{bmatrix} = \left(-\tfrac{7}{5}y\right)\mathbf{v}_1 + \left(x + \tfrac{4}{5}y\right)\mathbf{v}_2$$

thus $[\mathbf{x}]_B = \begin{bmatrix} -\frac{1}{5}x + \frac{2}{5}y \\ \frac{2}{5}x + \frac{1}{5}y \end{bmatrix}$, $[T\mathbf{x}]_B = \begin{bmatrix} -\frac{7}{5}y \\ x + \frac{4}{5}y \end{bmatrix}$, and $[T]_B = [[T\mathbf{v}_1]_B \quad [T\mathbf{v}_2]_B] = \begin{bmatrix} -\frac{14}{5} & -\frac{7}{5} \\ \frac{3}{5} & \frac{14}{5} \end{bmatrix}$.

(b) In agreement with Formula (7), we have

$$[T\mathbf{x}]_B = \begin{bmatrix} -\frac{7}{5}y \\ x + \frac{4}{5}y \end{bmatrix} = \begin{bmatrix} -\frac{14}{5} & -\frac{7}{5} \\ \frac{3}{5} & \frac{14}{5} \end{bmatrix} \begin{bmatrix} -\frac{1}{5}x + \frac{2}{5}y \\ \frac{2}{5}x + \frac{1}{5}y \end{bmatrix} = [T]_B[\mathbf{x}]_B$$

17. For every vector $\mathbf{x}$ in R^2, we have

$$\mathbf{x} = \begin{bmatrix} x_1 \\ x_2 \end{bmatrix} = (\tfrac{2}{5}x_1 + \tfrac{1}{5}x_2)\begin{bmatrix} 1 \\ 3 \end{bmatrix} + (-\tfrac{3}{10}x_1 + \tfrac{1}{10}x_2)\begin{bmatrix} -2 \\ 4 \end{bmatrix} = (\tfrac{2}{5}x_1 + \tfrac{1}{5}x_2)\mathbf{v}_1 + (-\tfrac{3}{10}x_1 + \tfrac{1}{10}x_2)\mathbf{v}_2$$

and

$$T\mathbf{x} = \begin{bmatrix} x_1 + 2x_2 \\ -x_1 \\ 0 \end{bmatrix} = 0\begin{bmatrix} 1 \\ 1 \\ 1 \end{bmatrix} - \tfrac{1}{2}x_1\begin{bmatrix} 2 \\ 2 \\ 0 \end{bmatrix} + (\tfrac{2}{3}x_1 + \tfrac{2}{3}x_2)\begin{bmatrix} 3 \\ 0 \\ 0 \end{bmatrix} = 0\mathbf{v}_1' - \tfrac{1}{2}x_1\mathbf{v}_2' + (\tfrac{2}{3}x_1 + \tfrac{2}{3}x_2)\mathbf{v}_3'$$

thus $[\mathbf{x}]_B = \begin{bmatrix} \frac{2}{5}x_1 + \frac{1}{5}x_2 \\ -\frac{3}{10}x_1 + \frac{1}{10}x_2 \end{bmatrix}$, $[T\mathbf{x}]_{B'} = \begin{bmatrix} 0 \\ -\frac{1}{2}x_1 \\ \frac{2}{3}x_1 + \frac{2}{3}x_2 \end{bmatrix}$, and $[T]_{B'B} = [[T\mathbf{v}_1]_{B'} \quad [T\mathbf{v}_2]_{B'}] = \begin{bmatrix} 0 & 0 \\ -\frac{1}{2} & 1 \\ \frac{8}{3} & \frac{4}{3} \end{bmatrix}$.

Finally, in agreement with Formula (26), we have

$$[T\mathbf{x}]_{B'} = \begin{bmatrix} 0 \\ -\frac{1}{2}x_1 \\ \frac{2}{3}x_1 + \frac{2}{3}x_2 \end{bmatrix} = \begin{bmatrix} 0 & 0 \\ -\frac{1}{2} & 1 \\ \frac{8}{3} & \frac{4}{3} \end{bmatrix} \begin{bmatrix} \frac{2}{5}x_1 + \frac{1}{5}x_2 \\ -\frac{3}{10}x_1 + \frac{1}{10}x_2 \end{bmatrix} = [T]_{B'B}[\mathbf{x}]_B$$

19. For every vector $\mathbf{x}$ in R^3, we have

$$\mathbf{x} = \begin{bmatrix} x_1 \\ x_2 \\ x_3 \end{bmatrix} = (\tfrac{1}{2}x_1 + \tfrac{1}{2}x_2 - \tfrac{1}{2}x_3)\begin{bmatrix} 1 \\ 1 \\ 0 \end{bmatrix} + (\tfrac{1}{2}x_1 - \tfrac{1}{2}x_2 + \tfrac{1}{2}x_3)\begin{bmatrix} 1 \\ 0 \\ 1 \end{bmatrix} + (-\tfrac{1}{2}x_1 + \tfrac{1}{2}x_2 + \tfrac{1}{2}x_3)\begin{bmatrix} 0 \\ 1 \\ 1 \end{bmatrix}$$

and

$$T\mathbf{x} = \begin{bmatrix} 3x_1 + 2x_3 \\ 2x_1 - x_2 \end{bmatrix} = (-\tfrac{5}{7}x_1 - \tfrac{1}{14}x_2 - \tfrac{4}{7}x_3)\begin{bmatrix} -3 \\ 2 \end{bmatrix} + (\tfrac{6}{7}x_1 - \tfrac{3}{14}x_2 + \tfrac{2}{7}x_3)\begin{bmatrix} 1 \\ 4 \end{bmatrix}$$

thus $[\mathbf{x}]_B = \begin{bmatrix} \frac{1}{2}x_1 + \frac{1}{2}x_2 - \frac{1}{2}x_3 \\ \frac{1}{2}x_1 - \frac{1}{2}x_2 + \frac{1}{2}x_3 \\ -\frac{1}{2}x_1 + \frac{1}{2}x_2 + \frac{1}{2}x_3 \end{bmatrix}$, $[T\mathbf{x}]_{B'} = \begin{bmatrix} -\frac{5}{7}x_1 - \frac{1}{14}x_2 - \frac{4}{7}x_3 \\ \frac{6}{7}x_1 - \frac{3}{14}x_2 + \frac{2}{7}x_3 \end{bmatrix}$, and $[T]_{B'B} = \begin{bmatrix} -\frac{11}{14} & -\frac{9}{7} & -\frac{9}{14} \\ \frac{9}{14} & \frac{8}{7} & \frac{1}{14} \end{bmatrix}$.

Finally, in agreement with Formula (26), we have

$$[T\mathbf{x}]_{B'} = \begin{bmatrix} -\frac{5}{7}x_1 - \frac{1}{14}x_2 - \frac{4}{7}x_3 \\ \frac{6}{7}x_1 - \frac{3}{14}x_2 + \frac{2}{7}x_3 \end{bmatrix} = \begin{bmatrix} -\frac{11}{14} & -\frac{9}{7} & -\frac{9}{14} \\ \frac{9}{14} & \frac{8}{7} & \frac{1}{14} \end{bmatrix} \begin{bmatrix} \frac{1}{2}x_1 + \frac{1}{2}x_2 - \frac{1}{2}x_3 \\ \frac{1}{2}x_1 - \frac{1}{2}x_2 + \frac{1}{2}x_3 \\ -\frac{1}{2}x_1 + \frac{1}{2}x_2 + \frac{1}{2}x_3 \end{bmatrix} = [T]_{B'B}[\mathbf{x}]_B$$

21. (a) $[T\mathbf{v}_1]_B = \begin{bmatrix} 1 \\ -2 \end{bmatrix}$ and $[T\mathbf{v}_2]_B = \begin{bmatrix} 3 \\ 5 \end{bmatrix}$.

(b) $T\mathbf{v}_1 = \mathbf{v}_1 - 2\mathbf{v}_2 = \begin{bmatrix} -1 \\ 2 \end{bmatrix} - 2\begin{bmatrix} 2 \\ 1 \end{bmatrix} = \begin{bmatrix} -5 \\ 0 \end{bmatrix}$ and $T\mathbf{v}_2 = 3\mathbf{v}_1 + 5\mathbf{v}_2 = 3\begin{bmatrix} -1 \\ 2 \end{bmatrix} + 5\begin{bmatrix} 2 \\ 1 \end{bmatrix} = \begin{bmatrix} 7 \\ 11 \end{bmatrix}$.

(c) For every vector $\mathbf{x}$ in R^2, we have

$$\mathbf{x} = \begin{bmatrix} x_1 \\ x_2 \end{bmatrix} = (-\tfrac{1}{5}x_1 + \tfrac{2}{5}x_2) \begin{bmatrix} -1 \\ 2 \end{bmatrix} + (\tfrac{2}{5}x_1 + \tfrac{1}{5}x_2) \begin{bmatrix} 2 \\ 1 \end{bmatrix} = (-\tfrac{1}{5}x_1 + \tfrac{2}{5}x_2)\mathbf{v}_1 + (\tfrac{2}{5}x_1 + \tfrac{1}{5}x_2)\mathbf{v}_2$$

Thus, using the linearity of T, it follows that

$$T\mathbf{x} = (-\tfrac{1}{5}x_1 + \tfrac{2}{5}x_2) \begin{bmatrix} -5 \\ 0 \end{bmatrix} + (\tfrac{2}{5}x_1 + \tfrac{1}{5}x_2) \begin{bmatrix} 7 \\ 11 \end{bmatrix} = \begin{bmatrix} \tfrac{19}{5}x_1 - \tfrac{3}{5}x_2 \\ \tfrac{22}{5}x_1 + \tfrac{11}{5}x_2 \end{bmatrix}$$

or, in the comma delimited form, $T(x_1, x_2) = (\tfrac{19}{5}x_1 - \tfrac{3}{5}x_2, \tfrac{22}{5}x_1 + \tfrac{11}{5}x_2)$.

(d) Using the formula obtained in part (c), we have $T(1,1) = (\tfrac{19}{5} - \tfrac{3}{5}, \tfrac{22}{5} + \tfrac{11}{5}) = (\tfrac{16}{5}, \tfrac{33}{5})$.

23. If T is the identity operator then, since $T\mathbf{e}_1 = \mathbf{e}_1$ and $T\mathbf{e}_2 = \mathbf{e}_2$, we have $[T] = \begin{bmatrix} 1 & 0 \\ 0 & 1 \end{bmatrix}$. Similarly,

$[T]_B = \begin{bmatrix} 1 & 0 \\ 0 & 1 \end{bmatrix}$ and $[T]_{B'} = \begin{bmatrix} 1 & 0 \\ 0 & 1 \end{bmatrix}$. On the other hand, $T\mathbf{v}_1 = \begin{bmatrix} 2 \\ 3 \end{bmatrix} = \tfrac{12}{17} \begin{bmatrix} 1 \\ 7 \end{bmatrix} + \tfrac{11}{51} \begin{bmatrix} 6 \\ -9 \end{bmatrix} = \tfrac{12}{17}\mathbf{v}_1' + \tfrac{11}{51}\mathbf{v}_2'$

and $T\mathbf{v}_2 = \begin{bmatrix} -1 \\ 4 \end{bmatrix} = \tfrac{5}{17}\mathbf{v}_1' - \tfrac{11}{51}\mathbf{v}_2'$; thus $[T]_{B',B} = \begin{bmatrix} \tfrac{12}{17} & \tfrac{5}{17} \\ \tfrac{11}{51} & -\tfrac{11}{51} \end{bmatrix}$.

25. Let $B = \{\mathbf{v}_1, \mathbf{v}_2, \ldots, \mathbf{v}_n\}$ and $B' = \{\mathbf{u}_1, \mathbf{u}_2, \ldots, \mathbf{u}_m\}$ be bases for R^n and R^m respectively. Then, if T is the zero transformation, we have

$$T\mathbf{v}_i = \mathbf{0} = 0\mathbf{u}_1 + 0\mathbf{u}_2 + \cdots + 0\mathbf{u}_m$$

for each $i = 1, 2, \ldots, n$. Thus $[T]_{B',B} = [\mathbf{0}\,|\,\mathbf{0}\,|\,\cdots\,|\,\mathbf{0}]$ is the zero matrix.

27. $[T] = \begin{bmatrix} 0 & 1 & 0 \\ 0 & 0 & 1 \\ 1 & 0 & 0 \end{bmatrix}$

29. We have $T\mathbf{v}_1 = \begin{bmatrix} 1 & 5 \\ 5 & 1 \end{bmatrix} \begin{bmatrix} \tfrac{1}{\sqrt{2}} \\ -\tfrac{1}{\sqrt{2}} \end{bmatrix} = \begin{bmatrix} -\tfrac{4}{\sqrt{2}} \\ \tfrac{4}{\sqrt{2}} \end{bmatrix} = -4\mathbf{v}_1$ and $T\mathbf{v}_2 = \begin{bmatrix} 1 & 5 \\ 5 & 1 \end{bmatrix} \begin{bmatrix} \tfrac{1}{\sqrt{2}} \\ \tfrac{1}{\sqrt{2}} \end{bmatrix} = \begin{bmatrix} \tfrac{6}{\sqrt{2}} \\ \tfrac{6}{\sqrt{2}} \end{bmatrix} = 6\mathbf{v}_2$; thus $[T]_B = \begin{bmatrix} -4 & 0 \\ 0 & 6 \end{bmatrix}$. From this we see that the effect of the operator T is to stretch the $\mathbf{v}_1$ component of a vector by a factor of 4 and reverse its direction, and to stretch the $\mathbf{v}_2$ component by a factor of 6. If the xy-coordinate axes are rotated 45 degrees clockwise to produce an $x'y'$-coordinate system whose axes are aligned with the directions of the vectors $\mathbf{v}_1$ and $\mathbf{v}_2$, then the effect is to stretch by a factor of 4 in the x'-direction, reflect about the y'-axis, and stretch by a factor of 6 in the y'-direction.

DISCUSSION AND DISCOVERY

D1. Since $T\mathbf{v}_1 = \mathbf{v}_2$ and $T\mathbf{v}_2 = \mathbf{v}_1$, the matrix of T with respect to the basis $B = \{\mathbf{v}_1, \mathbf{v}_2\}$ is $[T]_B = \begin{bmatrix} 0 & 1 \\ 1 & 0 \end{bmatrix}$. On the other hand, since $\mathbf{e}_1 = \tfrac{1}{2}\mathbf{v}_1$ and $\mathbf{e}_2 = \tfrac{1}{4}\mathbf{v}_2$, we have $T\mathbf{e}_1 = \tfrac{1}{2}T\mathbf{v}_1 = \tfrac{1}{2}\mathbf{v}_2 = 2\mathbf{e}_2$ and $T\mathbf{e}_2 = \tfrac{1}{4}T\mathbf{v}_2 = \tfrac{1}{4}\mathbf{v}_1 = \tfrac{1}{2}\mathbf{e}_1$; thus the standard matrix for T is $[T] = \begin{bmatrix} 0 & \tfrac{1}{2} \\ 2 & 0 \end{bmatrix}$.

D2. The appropriate diagram is

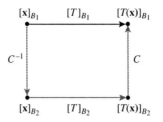

where $C = P_{B_2 \to B_1}$. Thus $[T]_{B_1} = C[T]_{B_2}C^{-1}$, and $[T]_{B_2} = C^{-1}[T]_{B_1}C$.

D3. The appropriate diagram is

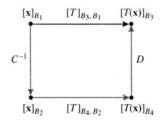

where $C = P_{B_2 \to B_1}$ and $D = P_{B_4 \to B_3}$. Thus $[T]_{B_3, B_1} = D[T]_{B_4, B_2}C^{-1}$.

D4. (a) True. We have $[T_1(\mathbf{x})]_{B'} = [T_1]_{B', B}[\mathbf{x}]_B = [T_2]_{B', B}[\mathbf{x}]_B = [T_2(\mathbf{x})]_{B'}$; thus $T_1(\mathbf{x}) = T_2(\mathbf{x})$.

    (b) False. For example, the zero operator has the same matrix (the zero matrix) with respect to any basis for R^2.

    (c) True. If $B = \{\mathbf{v}_1, \mathbf{v}_2, \ldots, \mathbf{v}_n\}$ and $[T]_B = I$, then $T(\mathbf{v}_k) = \mathbf{v}_k$ for each $k = 1, 2, \ldots, n$ and it follows from this that $T(\mathbf{x}) = \mathbf{x}$ for all $\mathbf{x}$.

    (d) False. For example, let $B = \{\mathbf{e}_1, \mathbf{e}_2\}$, $B' = \{\mathbf{e}_2, \mathbf{e}_1\}$, and $T(x, y) = (y, x)$. Then $[T]_{B', B} = I_2$, but T is not the identity operator.

D5. One reason is that the representation of the operator in some other basis may more clearly reflect the geometric effect of the operator.

WORKING WITH PROOFS

P1. If $\mathbf{x}$ and $\mathbf{y}$ are vectors and c is a scalar then, since T is linear, we have

$$c[\mathbf{x}]_B = [c\mathbf{x}]_B \to [T(c\mathbf{x})]_B = [cT(\mathbf{x})]_B = c[T(\mathbf{x})]_B$$

and

$$[\mathbf{x}]_B + [\mathbf{y}]_B = [\mathbf{x} + \mathbf{y}]_B \to [T(\mathbf{x} + \mathbf{y})]_B = [T(\mathbf{x}) + T(\mathbf{y})]_B = [T(\mathbf{x})]_B + [T(\mathbf{y})]_B$$

This shows that the mapping $[\mathbf{x}]_B \to [T(\mathbf{x})]_B$ is linear.

P2. If $\mathbf{x}$ is in R^n and $\mathbf{y}$ is in R^k, then we have $[T_1(\mathbf{x})]_{B'} = [T_1]_{B', B}[\mathbf{x}]_B$ and $[T_2(\mathbf{y})]_{B''} = [T_2]_{B'', B'}[\mathbf{y}]_{B'}$. Thus

$$[T_2(T_1(\mathbf{x}))]_{B''} = [T_2]_{B'', B'}[T_1(\mathbf{x})]_{B'} = [T_2]_{B'', B'}[T_1]_{B', B}[\mathbf{x}]_B$$

and from this it follows that $[T_2 \circ T_1]_{B'', B} = [T_2]_{B'', B'}[T_1]_{B', B}$.

P3. If $\mathbf{x}$ is a vector in R^n, then $[T]_B[\mathbf{x}]_B = [T\mathbf{x}]_B = \mathbf{0}$ if and only if $T\mathbf{x} = \mathbf{0}$. Thus, if T is one-to-one, it follows that $[T]_B[\mathbf{x}]_B = \mathbf{0}$ if and only if $[\mathbf{x}]_B = \mathbf{0}$, i.e., that $[T]_B$ is an invertible matrix. Furthermore, since $[T^{-1}]_B[T]_B = [T^{-1} \circ T]_B = [I]_B = I$, we have $[T^{-1}]_B = [T_B]^{-1}$.

P4. $[T]_B = [T(\mathbf{v}_1)_B \,|\, T(\mathbf{v}_2)_B \,|\, \cdots \,|\, T(\mathbf{v}_n)_B] = [T]_{B,B}$

EXERCISE SET 8.2

1. We have $\mathrm{tr}(A) = 3$ and $\mathrm{tr}(B) = -1$; thus A and B are not similar.

3. We have $\mathrm{rank}(A) = 3$ and $\mathrm{rank}(B) = 2$; thus A and B are not similar.

5. **(a)** The size of the matrix corresponds to the degree of its characteristic polynomial; so in this case we have a 5×5 matrix. The eigenvalues of the matrix with their algebraic multiplicities are $\lambda = 0$ (multiplicity 1), $\lambda = -1$ (multiplicity 2), and $\lambda = 1$ (multiplicity 2). The eigenspace corresponding to $\lambda = 0$ has dimension 1, and the eigenspaces corresponding to $\lambda = -1$ or $\lambda = 1$ have dimension 1 or 2.

   (b) The matrix is 11×11 with eigenvalues $\lambda = -3$ (multiplicity 1), $\lambda = -1$ (multiplicity 3), and $\lambda = 8$ (multiplicity 7). The eigenspace corresponding to $\lambda = -3$ has dimension 1; the eigenspace corresponding to $\lambda = -1$ has dimension 1, 2, or 3; and the eigenspace corresponding to $\lambda = 8$ have dimension 1, 2, 3, 4, 5, 6, 7, or 8.

7. Since A is triangular, its characteristic polynomial is $p(\lambda) = (\lambda - 1)(\lambda - 1)(\lambda - 2) = (\lambda - 1)^2(\lambda - 2)$. Thus the eigenvalues of A are $\lambda = 1$ and $\lambda = 2$, with algebraic multiplicities 2 and 1 respectively. The eigenspace corresponding to $\lambda = 1$ is the solution space of the system $(I - A)\mathbf{x} = \mathbf{0}$, which is

$$\begin{bmatrix} 0 & -1 & -4 \\ 0 & 0 & -1 \\ 0 & 0 & -1 \end{bmatrix} \begin{bmatrix} x_1 \\ x_2 \\ x_3 \end{bmatrix} = \begin{bmatrix} 0 \\ 0 \\ 0 \end{bmatrix}$$

The general solution of this system is $\mathbf{x} = \begin{bmatrix} t \\ 0 \\ 0 \end{bmatrix} = t \begin{bmatrix} 1 \\ 0 \\ 0 \end{bmatrix}$; thus the eigenspace is 1-dimensional and so $\lambda = 1$ has geometric multiplicity 1. The eigenspace corresponding to $\lambda = 2$ is the solution space of the system $(2I - A)\mathbf{x} = \mathbf{0}$ which is

$$\begin{bmatrix} 1 & -1 & -4 \\ 0 & 1 & -1 \\ 0 & 0 & 0 \end{bmatrix} \begin{bmatrix} x_1 \\ x_2 \\ x_3 \end{bmatrix} = \begin{bmatrix} 0 \\ 0 \\ 0 \end{bmatrix}$$

The solution space of this system is $\mathbf{x} = \begin{bmatrix} 5s \\ s \\ s \end{bmatrix} = s \begin{bmatrix} 5 \\ 1 \\ 1 \end{bmatrix}$; thus the eigenspace is 1-dimensional and so $\lambda = 2$ also has geometric multiplicity 1.

9. The characteristic polynomial of A is $p(\lambda) = \det(\lambda I - A) = (\lambda - 5)^2(\lambda - 3)$. Thus the eigenvalues of A are $\lambda = 5$ and $\lambda = 3$, with algebraic multiplicities 2 and 1 respectively. The eigenspace corresponding to $\lambda = 5$ is the solution space of the system $(5I - A)\mathbf{x} = \mathbf{0}$, which is

$$\begin{bmatrix} 0 & 0 & 0 \\ -1 & 0 & -1 \\ 1 & 0 & 2 \end{bmatrix} \begin{bmatrix} x_1 \\ x_2 \\ x_3 \end{bmatrix} = \begin{bmatrix} 0 \\ 0 \\ 0 \end{bmatrix}$$

The general solution of this system is $\mathbf{x} = \begin{bmatrix} 0 \\ t \\ 0 \end{bmatrix} = t \begin{bmatrix} 0 \\ 1 \\ 0 \end{bmatrix}$; thus the eigenspace is 1-dimensional and so $\lambda = 5$ has geometric multiplicity 1. The eigenspace corresponding to $\lambda = 3$ is the solution space of the system $(3I - A)\mathbf{x} = \mathbf{0}$ which is

$$\begin{bmatrix} -2 & 0 & 0 \\ -1 & -2 & -1 \\ 1 & 0 & 0 \end{bmatrix} \begin{bmatrix} x_1 \\ x_2 \\ x_3 \end{bmatrix} = \begin{bmatrix} 0 \\ 0 \\ 0 \end{bmatrix}$$

The solution space of this system is $\mathbf{x} = \begin{bmatrix} 0 \\ s \\ -2s \end{bmatrix} = s \begin{bmatrix} 0 \\ 1 \\ -2 \end{bmatrix}$; thus the eigenspace is 1-dimensional and so $\lambda = 3$ also has geometric multiplicity 1.

11. The characteristic polynomial of A is $p(\lambda) = \lambda^3 + 3\lambda^2 = \lambda^2(\lambda + 3)$; thus the eigenvalues are $\lambda = 0$ and $\lambda = -3$, with algebraic multiplicities 2 and 1 respectively. The rank of the matrix

$$0I - A = -A = \begin{bmatrix} 1 & 1 & -1 \\ 1 & 1 & -1 \\ -1 & -1 & 1 \end{bmatrix}$$

is clearly 1 since each of the rows is a scalar multiple of the 1st row. Thus nullity$(0I - A) = 3 - 1 = 2$, and this is the geometric multiplicity of $\lambda = 0$. On the other hand, the matrix

$$-3I - A = \begin{bmatrix} -2 & 1 & -1 \\ 1 & -2 & -1 \\ -1 & -1 & -2 \end{bmatrix}$$

has rank 2 since its reduced row echelon form is $\begin{bmatrix} 1 & 0 & 1 \\ 0 & 1 & 1 \\ 0 & 0 & 0 \end{bmatrix}$. Thus nullity$(-3I - A) = 3 - 2 = 1$, and this is the geometric multiplicity of $\lambda = -3$.

13. The characteristic polynomial of A is $p(\lambda) = \lambda^3 - 11\lambda^2 + 39\lambda - 45 = (\lambda - 5)(\lambda - 3)^2$; thus the eigenvalues are $\lambda = 5$ and $\lambda = 3$, with algebraic multiplicities 1 and 2 respectively. The matrix

$$5I - A = \begin{bmatrix} 1 & 0 & -1 \\ -2 & 2 & -2 \\ -1 & 0 & 1 \end{bmatrix}$$

has rank 2, since its reduced row echelon form is $\begin{bmatrix} 1 & 0 & 1 \\ 0 & 1 & 1 \\ 0 & 0 & 0 \end{bmatrix}$. Thus nullity$(5I - A) = 3 - 2 = 1$, and this is the geometric multiplicity of $\lambda = 5$. On the other hand, the matrix

$$3I - A = \begin{bmatrix} -1 & 0 & -1 \\ -2 & 0 & -2 \\ -1 & 0 & -1 \end{bmatrix}$$

has rank 1 since each of the rows is a scalar multiple of the 1st row. Thus nullity$(3I - A) = 3 - 1 = 2$, and this is the geometric multiplicity of $\lambda = 3$. It follows that A has $1 + 2 = 3$ linearly independent eigenvectors and thus is diagonalizable.

15. The characteristic polynomial of A is $p(\lambda) = \lambda^2 - 3\lambda + 2 = (\lambda - 1)(\lambda - 2)$; thus A has two distinct eigenvalues, $\lambda = 1$ and $\lambda = 2$. The eigenspace corresponding to $\lambda = 1$ is obtained by solving the system $(I - A)\mathbf{x} = \mathbf{0}$, which is $\begin{bmatrix} 15 & -12 \\ 20 & -16 \end{bmatrix} \begin{bmatrix} x_1 \\ x_2 \end{bmatrix} = \begin{bmatrix} 0 \\ 0 \end{bmatrix}$. The general solution of this system is $\mathbf{x} = t \begin{bmatrix} \frac{4}{5} \\ 1 \end{bmatrix}$. Thus, taking $t = 5$, we see that $\mathbf{p}_1 = \begin{bmatrix} 4 \\ 5 \end{bmatrix}$ is an eigenvector for $\lambda = 1$. Similarly, $\mathbf{p}_2 = \begin{bmatrix} 3 \\ 4 \end{bmatrix}$ is an eigenvector for $\lambda = 2$. Finally, the matrix $P = [\mathbf{p}_1 \quad \mathbf{p}_2] = \begin{bmatrix} 4 & 3 \\ 5 & 4 \end{bmatrix}$ has the property that

$$P^{-1}AP = \begin{bmatrix} 4 & -3 \\ -5 & 4 \end{bmatrix} \begin{bmatrix} -14 & 12 \\ -20 & 17 \end{bmatrix} \begin{bmatrix} 4 & 3 \\ 5 & 4 \end{bmatrix} = \begin{bmatrix} 1 & 0 \\ 0 & 2 \end{bmatrix}$$

17. The characteristic polynomial of A is $p(\lambda) = \lambda(\lambda - 1)(\lambda - 2)$; thus A has three distinct eigenvalues, $\lambda = 0$, $\lambda = 1$, and $\lambda = 2$. The eigenspace corresponding to $\lambda = 0$ is obtained by solving the system $(0I - A)\mathbf{x} = \mathbf{0}$, which is $\begin{bmatrix} -1 & 0 & 0 \\ 0 & -1 & -1 \\ 0 & -1 & -1 \end{bmatrix} \begin{bmatrix} x_1 \\ x_2 \\ x_3 \end{bmatrix} = \begin{bmatrix} 0 \\ 0 \\ 0 \end{bmatrix}$. The general solution of this system is $\mathbf{x} = r \begin{bmatrix} 0 \\ -1 \\ 1 \end{bmatrix}$. Similarly, the general solution of $(I - A)\mathbf{x} = \mathbf{0}$ is $\mathbf{x} = s \begin{bmatrix} 1 \\ 0 \\ 0 \end{bmatrix}$, and the general solution of $(2I - A)\mathbf{x} = \mathbf{0}$ is $\mathbf{x} = t \begin{bmatrix} 0 \\ 1 \\ 1 \end{bmatrix}$. Thus the matrix $P = \begin{bmatrix} 0 & 1 & 0 \\ -1 & 0 & 1 \\ 1 & 0 & 1 \end{bmatrix}$ has the property that

$$P^{-1}AP = \frac{1}{2} \begin{bmatrix} 0 & -1 & 1 \\ 2 & 0 & 0 \\ 0 & 1 & 1 \end{bmatrix} \begin{bmatrix} 1 & 0 & 0 \\ 0 & 1 & 1 \\ 0 & 1 & 1 \end{bmatrix} \begin{bmatrix} 0 & 1 & 0 \\ -1 & 0 & 1 \\ 1 & 0 & 1 \end{bmatrix} = \begin{bmatrix} 0 & 0 & 0 \\ 0 & 1 & 0 \\ 0 & 0 & 2 \end{bmatrix}$$

19. The characteristic polynomial of A is $p(\lambda) = \lambda^3 - 6\lambda^2 + 11\lambda - 6 = (\lambda - 1)(\lambda - 2)(\lambda - 3)$; thus A has three distinct eigenvalues $\lambda_1 = 1$, $\lambda_2 = 2$ and $\lambda_3 = 3$. Corresponding eigenvectors are $\mathbf{v}_1 = \begin{bmatrix} 1 \\ 1 \\ 1 \end{bmatrix}$, $\mathbf{v}_2 = \begin{bmatrix} 2 \\ 3 \\ 3 \end{bmatrix}$, and $\mathbf{v}_3 = \begin{bmatrix} 1 \\ 3 \\ 4 \end{bmatrix}$. Thus A is diagonalizable and $P = \begin{bmatrix} 1 & 2 & 1 \\ 1 & 3 & 3 \\ 1 & 3 & 4 \end{bmatrix}$ has the property that

$$P^{-1}AP = \begin{bmatrix} 3 & -5 & 3 \\ -1 & 3 & -2 \\ 0 & -1 & 1 \end{bmatrix} \begin{bmatrix} -1 & 4 & -2 \\ -3 & 4 & 0 \\ -3 & 1 & 3 \end{bmatrix} \begin{bmatrix} 1 & 2 & 1 \\ 1 & 3 & 3 \\ 1 & 3 & 4 \end{bmatrix} = \begin{bmatrix} 1 & 0 & 0 \\ 0 & 2 & 0 \\ 0 & 0 & 3 \end{bmatrix}$$

Note. The diagonalizing matrix P is not unique; it depends on the choice (and the order) of the eigenvectors. This is just one possibility.

21. The characteristic polynomial of A is $p(\lambda) = (\lambda - 5)^3$; thus A has one eigenvalue, $\lambda = 5$, which has algebraic multiplicity 3. The eigenspace corresponding to $\lambda = 5$ is obtained by solving the system $(5I - A)\mathbf{x} = \mathbf{0}$, which is $\begin{bmatrix} 0 & 0 & 0 \\ -1 & 0 & 0 \\ 0 & -1 & 0 \end{bmatrix} \begin{bmatrix} x_1 \\ x_2 \\ x_3 \end{bmatrix} = \begin{bmatrix} 0 \\ 0 \\ 0 \end{bmatrix}$. The general solution of this system is $\mathbf{x} = t \begin{bmatrix} 1 \\ 0 \\ 0 \end{bmatrix}$, which shows that the eigenspace has dimension 1, i.e., the eigenvalue has geometric multiplicity 1. It follows that A is not diagonalizable since the sum of the geometric multiplicities of its eigenvalues is less than 3.

23. The characteristic polynomial of A is $p(\lambda) = (\lambda + 2)^2(\lambda - 3)^2$; thus A has two eigenvalues, $\lambda = -2$ and $\lambda = 3$, each of which has algebraic multiplicity 2. The eigenspace corresponding to $\lambda = -2$ is obtained by solving the system $(-2I - A)\mathbf{x} = \mathbf{0}$, which is

$$\begin{bmatrix} 0 & 0 & 0 & 0 \\ 0 & 0 & 0 & 0 \\ 0 & 0 & -5 & 0 \\ 0 & 0 & -1 & -5 \end{bmatrix} \begin{bmatrix} x_1 \\ x_1 \\ x_3 \\ x_4 \end{bmatrix} = \begin{bmatrix} 0 \\ 0 \\ 0 \\ 0 \end{bmatrix}$$

The general solution of this system is $\mathbf{x} = r \begin{bmatrix} 1 \\ 0 \\ 0 \\ 0 \end{bmatrix} + s \begin{bmatrix} 0 \\ 1 \\ 0 \\ 0 \end{bmatrix}$, which shows that the eigenspace has dimension

2, i.e., that the eigenvalue $\lambda = -2$ has geometric multiplicity 2. On the other hand, the general

solution of $(3I - A)\mathbf{x} = \mathbf{0}$ is $\mathbf{x} = t \begin{bmatrix} 0 \\ 0 \\ 0 \\ 1 \end{bmatrix}$, and so $\lambda = 3$ has geometric multiplicity 1. It follows that A is

not diagonalizable since the sum of the geometric multiplicities of its eigenvalues is less than 4.

25. If the matrix A is upper triangular with 1's on the main diagonal, then its characteristic polynomial is $p(\lambda) = (\lambda - 1)^n$ and $\lambda = 1$ is the only eigenvalue. Thus, in order for A to be diagonalizable, the system $(I - A)\mathbf{x} = \mathbf{0}$ must have n linearly independent solutions. But, if this is true, then $(I - A)\mathbf{x} = \mathbf{0}$ for every vector $\mathbf{x}$ in R^n and so $I - A$ is the zero matrix, i.e., $A = I$.

27. If C is similar to A then there is an invertible matrix P such that $C = P^{-1}AP$. It follows that if A is invertible, then C is invertible since it is a product of invertible matrices. Similarly, since $PCP^{-1} = A$, the invertibility of C implies the invertibility of A.

29. The standard matrix of the linear operator T is $A = \begin{bmatrix} -2 & 1 & -1 \\ 1 & -2 & -1 \\ -1 & -1 & -2 \end{bmatrix}$, and the characteristic polynomial

of A is $p(\lambda) = \lambda^3 + 6\lambda^2 + 9\lambda = \lambda(\lambda + 3)^2$. Thus the eigenvalues of T are $\lambda = 0$ and $\lambda = -3$, with algebraic multiplicities 1 and 2 respectively. Since $\lambda = 0$ has algebraic multiplicity 1, its geometric multiplicity is also 1. The eigenspace associated with $\lambda = -3$ is found by solving $(-3I - A)\mathbf{x} = \mathbf{0}$ which is

$$\begin{bmatrix} -1 & -1 & 1 \\ -1 & -1 & 1 \\ 1 & 1 & -1 \end{bmatrix} \begin{bmatrix} x_1 \\ x_2 \\ x_3 \end{bmatrix} = \begin{bmatrix} 0 \\ 0 \\ 0 \end{bmatrix}$$

The general solution of this system is $x = s \begin{bmatrix} 1 \\ 0 \\ 1 \end{bmatrix} + t \begin{bmatrix} 0 \\ 1 \\ 1 \end{bmatrix}$; thus $\lambda = -3$ has geometric multiplicity 2. It

follows that T is diagonalizable since the sum of the geometric multiplicities of its eigenvalues is 3.

31. If $\mathbf{x}$ is a vector in R^n and λ is scalar, then $[T\mathbf{x}]_B = [T]_B[\mathbf{x}]_B$ and $[\lambda\mathbf{x}]_B = \lambda[\mathbf{x}]_B$. It follows that $T\mathbf{x} = \lambda\mathbf{x}$ if and only if $[T]_B[\mathbf{x}]_B = [T\mathbf{x}]_B = [\lambda\mathbf{x}]_B = \lambda[\mathbf{x}]_B$; thus $\mathbf{x}$ is an eigenvector of T corresponding to λ if and only if $[\mathbf{x}]_B$ is an eigenvector of $[T]_B$ corresponding to λ.

DISCUSSION AND DISCOVERY

D1. Similar matrices must have the same rank; thus all one needs to do is to produce two matrices with different ranks. For example, in the 3×3 case, the matrices $A = \begin{bmatrix} 1 & 0 & 0 \\ 0 & 0 & 0 \\ 0 & 0 & 0 \end{bmatrix}$ and $B = \begin{bmatrix} 1 & 0 & 0 \\ 0 & 1 & 0 \\ 0 & 0 & 0 \end{bmatrix}$ are not similar since $\text{rank}(A) = 1$ and $\text{rank}(B) = 2$.

D2. (a) True. We have $A = P^{-1}AP$ where $P = I$.

   (b) True. If A is similar to B and B is similar to C, then there are invertible matrices P_1 and P_2 such that $A = P_1^{-1}BP_1$ and $B = P_2^{-1}CP_2$. It follows that $A = P_1^{-1}(P_2^{-1}CP_2)P_1 = (P_2P_1)^{-1}C(P_2P_1)$; thus A is similar to C.

   (c) True. If $A = P^{-1}BP$, then $A^{-1} = (P^{-1}BP)^{-1} = P^{-1}B^{-1}(P^{-1})^{-1} = P^{-1}B^{-1}P$.

   (d) False. This statement does not guarantee that there are enough linearly independent eigenvectors. For example, the matrix $A = \begin{bmatrix} 1 & 0 & 0 \\ 0 & 0 & 1 \\ 0 & -1 & 0 \end{bmatrix}$ has only one (real) eigenvalue, $\lambda = 1$, which has algebraic multiplicity 1, but A is not diagonalizable.

D3. (a) False. For example, $A = \begin{bmatrix} 1 & 0 \\ 0 & 0 \end{bmatrix}$ is diagonalizable.

   (b) False. For example, if $P^{-1}AP$ is a diagonal matrix then so is $Q^{-1}AQ$ where $Q = 2P$. The diagonalizing matrix (if it exists) is not unique!

   (c) True. Vectors from different eigenspaces correspond to different eigenvalues and are therefore linearly independent. In the situation described $\{\mathbf{v}_1, \mathbf{v}_2, \mathbf{v}_3\}$ is a linearly independent set.

   (d) True. If an invertible matrix A is similar to a diagonal matrix D, then D must also be invertible; thus D has nonzero diagonal entries and D^{-1} is the diagonal matrix whose diagonal entries are the reciprocals of the corresponding entries of D. Finally, if P is an invertible matrix such that $P^{-1}AP = D$, we have $P^{-1}A^{-1}P = (P^{-1}AP)^{-1} = D^{-1}$ and so A^{-1} is similar to D^{-1}.

   (e) True. The vectors in a basis are linearly independent; thus A has n linear independent eigenvectors.

D4. (a) A is a 6×6 matrix.

   (b) The eigenspace corresponding to $\lambda = 1$ has dimension 1. The eigenspace corresponding to $\lambda = 3$ has dimension 1 or 2. The eigenspace corresponding to $\lambda = 4$ has dimension 1, 2 or 3.

   (c) If A is diagonalizable, then the eigenspaces corresponding to $\lambda = 1$, $\lambda = 3$, and $\lambda = 4$ have dimensions 1, 2, and 3 respectively.

   (d) These vectors must correspond to the eigenvalue $\lambda = 4$.

D5. (a) If λ_1 has geometric multiplicity 2 and λ_2 has geometric multiplicity 3, then λ_3 must have multiplicity 1. Thus the sum of the geometric multiplicities is 6 and so A is diagonalizable.

   (b) In this case the matrix is not diagonalizable since the sum of the geometric multiplicities of the eigenvalues is less than 6.

   (c) The matrix may or may not be diagonalizable. The geometric multiplicity of λ_3 must be 1 or 2. If the geometric multiplicity of λ_3 is 2, then the matrix is diagonalizable. If the geometric multiplicity of λ_3 is 1, then the matrix is not diagonalizable.

WORKING WITH PROOFS

P1. If A and B are similar, then there is an invertible matrix P such that $A = P^{-1}BP$. Thus $PA = BP$ and so, using the result of the cited Exercise, we have $\text{rank}(A) = \text{rank}(PA) = \text{rank}(BP) = \text{rank}(B)$ and $\text{nullity}(A) = \text{nullity}(PA) = \text{nullity}(BP) = \text{nullity}(B)$.

P2. If A and B are similar, then there is an invertible matrix P such that $A = P^{-1}BP$. Thus, using part (e) of Theorem 3.2.12, we have $\text{tr}(A) = \text{tr}(P^{-1}BP) = \text{tr}(P^{-1}(BP)) = \text{tr}((BP)P^{-1}) = \text{tr}(B)$.

P3. If $\mathbf{x} \neq \mathbf{0}$ and $A\mathbf{x} = \lambda\mathbf{x}$ then, since P is invertible and $CP^{-1} = P^{-1}A$, we have

$$CP^{-1}\mathbf{x} = P^{-1}A\mathbf{x} = P^{-1}(\lambda\mathbf{x}) = \lambda P^{-1}\mathbf{x}$$

with $P^{-1}\mathbf{x} \neq \mathbf{0}$. Thus $P^{-1}\mathbf{x}$ is an eigenvector of C corresponding to λ.

P4. If A and B are similar, then there is an invertible matrix P such that $A = P^{-1}BP$. We will prove, by induction, that $A^k = P^{-1}B^kP$ (thus A^k and B^k are similar) for every positive integer k.

Step 1. The fact that $A^1 = A = P^{-1}BP = P^{-1}B^1P$ is given.

Step 2. If $A^k = P^{-1}B^kP$, where k is a fixed integer ≥ 1, then we have

$$A^{k+1} = AA^k = (P^{-1}BP)(P^{-1}B^kP) = P^{-1}B(PP^{-1})B^kP = P^{-1}B^{k+1}P$$

These two steps complete the proof by induction.

P5. If A is diagonalizable, then there is an invertible matrix P and a diagonal matrix D such that $P^{-1}AP = D$. We will prove, by induction, that $P^{-1}A^kP = D^k$ for every positive integer k. Since D^k is diagonal this shows that A^k is diagonalizable.

Step 1. The fact that $P^{-1}A^1P = P^{-1}AP = D = D^1$ is given.

Step 2. If $P^{-1}A^kP = D^k$, where k is a fixed integer ≥ 1, then we have

$$P^{-1}A^{k+1}P = P^{-1}AA^kP = (P^{-1}AP)(P^{-1}A^kP) = DD^k = D^{k+1}$$

These two steps complete the proof by induction.

P6. **(a)** Let W be the eigenspace corresponding to λ_0. Choose a basis $\{\mathbf{u}_1, \mathbf{u}_2, \ldots, \mathbf{u}_k\}$ for W, then extend it to obtain a basis $B = \{\mathbf{u}_1, \mathbf{u}_2, \ldots, \mathbf{u}_k, \mathbf{u}_{k+1}, \ldots, \mathbf{u}_n\}$ for R^n.

(b) If $P = [\mathbf{u}_1 \mid \mathbf{u}_2 \mid \cdots \mid \mathbf{u}_k \mid \mathbf{u}_{k+1} \mid \cdots \mid \mathbf{u}_n] = [B_1 \mid B_2]$ then the product AP has the form

$$AP = [\lambda_0\mathbf{u}_1 \mid \lambda_0\mathbf{u}_2 \mid \cdots \mid \lambda_0\mathbf{u}_k \mid AB_2]$$

On the other hand, if C is an $n \times n$ matrix of the form $C = \begin{bmatrix} \lambda_0 I_k & X \\ 0 & Y \end{bmatrix}$, then PC has the form

$$PC = [\lambda_0\mathbf{u}_1 \mid \lambda_0\mathbf{u}_2 \mid \cdots \mid \lambda_0\mathbf{u}_k \mid PZ]$$

where $Z = \begin{bmatrix} X \\ Y \end{bmatrix}$. Thus if $Z = P^{-1}AB_2$, we have $AP = PC$.

(c) Since $AP = PC$, we have $P^{-1}AP = C$. Thus A is similar to $C = \begin{bmatrix} \lambda_0 I_k & X \\ 0 & Y \end{bmatrix}$, and so A and C have the same characteristic polynomial.

(d) Due to the special block structure of C, its characteristic polynomial of has the form

$$\det(\lambda I - C) = \det \begin{bmatrix} (\lambda - \lambda_0)I_k & -X \\ 0 & \lambda I_{n-k} - Y \end{bmatrix} = (\lambda - \lambda_0)^k \det(\lambda I_{n-k} - Y)$$

Thus the algebraic multiplicity of λ_0 as an eigenvalue of C, and of A, is greater than or equal to k.

EXERCISE SET 8.3

1. The characteristic polynomial of A is $p(\lambda) = \lambda^2 - 5\lambda = \lambda(\lambda - 5)$. Thus the eigenvalues of A are $\lambda = 0$ and $\lambda = 5$, and each of the eigenspaces has dimension 1.

3. The characteristic polynomial of A is $p(\lambda) = \lambda^3 - 3\lambda^2 = \lambda^2(\lambda - 3)$. Thus the eigenvalues of A are $\lambda = 0$ and $\lambda = 3$. The eigenspace corresponding to $\lambda = 0$ has dimension 2, and the eigenspace corresponding to $\lambda = 3$ has dimension 1.

5. The general solution of the system $(0I - A)\mathbf{x} = \mathbf{0}$ is $\mathbf{x} = r \begin{bmatrix} -1 \\ 0 \\ 1 \end{bmatrix} + s \begin{bmatrix} 0 \\ -1 \\ 1 \end{bmatrix}$; thus the vectors $\mathbf{v}_1 = \begin{bmatrix} -1 \\ 0 \\ 1 \end{bmatrix}$

and $\mathbf{v}_2 = \begin{bmatrix} 0 \\ -1 \\ 1 \end{bmatrix}$ form a basis for the eigenspace corresponding to $\lambda = 0$. Similarly the vector $\mathbf{v}_3 = \begin{bmatrix} 1 \\ 1 \\ 1 \end{bmatrix}$

forms a basis for the eigenspace corresponding to $\lambda = 3$. Since $\mathbf{v}_3$ is orthogonal to both $\mathbf{v}_1$ and $\mathbf{v}_2$ it follows that the two eigenspaces are orthogonal.

7. The characteristic polynomial of A is $p(\lambda) = \lambda^2 - 6\lambda + 8 = (\lambda - 2)(\lambda - 4)$; thus the eigenvalues of A are $\lambda = 2$ and $\lambda = 4$. The vector $\mathbf{v}_1 = \begin{bmatrix} -1 \\ 1 \end{bmatrix}$ forms a basis for the eigenspace corresponding to $\lambda = 2$, and the vector $\mathbf{v}_2 = \begin{bmatrix} 1 \\ 1 \end{bmatrix}$ forms a basis for the eigenspace corresponding to $\lambda = 4$. These vectors are orthogonal to each other, and the orthogonal matrix $P = [\frac{\mathbf{v}_1}{\|\mathbf{v}_1\|} \quad \frac{\mathbf{v}_2}{\|\mathbf{v}_2\|}] = \begin{bmatrix} -\frac{1}{\sqrt{2}} & \frac{1}{\sqrt{2}} \\ \frac{1}{\sqrt{2}} & \frac{1}{\sqrt{2}} \end{bmatrix}$ has the property that

$$P^T A P = \begin{bmatrix} -\frac{1}{\sqrt{2}} & \frac{1}{\sqrt{2}} \\ \frac{1}{\sqrt{2}} & \frac{1}{\sqrt{2}} \end{bmatrix} \begin{bmatrix} 3 & 1 \\ 1 & 3 \end{bmatrix} \begin{bmatrix} -\frac{1}{\sqrt{2}} & \frac{1}{\sqrt{2}} \\ \frac{1}{\sqrt{2}} & \frac{1}{\sqrt{2}} \end{bmatrix} = \begin{bmatrix} 2 & 0 \\ 0 & 4 \end{bmatrix} = D$$

9. The characteristic polynomial of A is $p(\lambda) = \lambda^3 + 6\lambda^2 - 32 = (\lambda - 2)(\lambda + 4)^2$; thus the eigenvalues of A are $\lambda = 2$ and $\lambda = -4$. The general solution of $(2I - A)\mathbf{x} = \mathbf{0}$ is $\mathbf{x} = r \begin{bmatrix} 1 \\ 1 \\ 2 \end{bmatrix}$, and the general solution

of $(-4I - A)\mathbf{x} = \mathbf{0}$ is $\mathbf{x} = s \begin{bmatrix} -1 \\ 1 \\ 0 \end{bmatrix} + t \begin{bmatrix} -2 \\ 0 \\ 1 \end{bmatrix}$. Thus the vector $\mathbf{v}_1 = \begin{bmatrix} 1 \\ 1 \\ 2 \end{bmatrix}$ forms a basis for the eigenspace

corresponding to $\lambda = 2$, and the vectors $\mathbf{v}_2 = \begin{bmatrix} -1 \\ 1 \\ 0 \end{bmatrix}$ and $\mathbf{v}_3 = \begin{bmatrix} -2 \\ 0 \\ 1 \end{bmatrix}$ form a basis for the eigenspace

corresponding to $\lambda = -4$. Application of the Gram-Schmidt process to $\{\mathbf{v}_1\}$ and to $\{\mathbf{v}_2, \mathbf{v}_3\}$ yield orthonormal bases $\{\mathbf{u}_1\}$ and $\{\mathbf{u}_2, \mathbf{u}_3\}$ for the eigenspaces, and the orthogonal matrix

$$P = [\mathbf{u}_1 \quad \mathbf{u}_2 \quad \mathbf{u}_3] = \begin{bmatrix} \frac{1}{\sqrt{6}} & -\frac{1}{\sqrt{2}} & -\frac{1}{\sqrt{3}} \\ \frac{1}{\sqrt{6}} & \frac{1}{\sqrt{2}} & -\frac{1}{\sqrt{3}} \\ \frac{2}{\sqrt{6}} & 0 & \frac{1}{\sqrt{3}} \end{bmatrix}$$

has the property that

$$P^T A P = \begin{bmatrix} \frac{1}{\sqrt{6}} & \frac{1}{\sqrt{6}} & \frac{2}{\sqrt{6}} \\ -\frac{1}{\sqrt{2}} & \frac{1}{\sqrt{2}} & 0 \\ -\frac{1}{\sqrt{3}} & -\frac{1}{\sqrt{3}} & \frac{1}{\sqrt{3}} \end{bmatrix} \begin{bmatrix} -3 & 1 & 2 \\ 1 & -3 & 2 \\ 2 & 2 & 0 \end{bmatrix} \begin{bmatrix} \frac{1}{\sqrt{6}} & -\frac{1}{\sqrt{2}} & -\frac{1}{\sqrt{3}} \\ \frac{1}{\sqrt{6}} & \frac{1}{\sqrt{2}} & -\frac{1}{\sqrt{3}} \\ \frac{2}{\sqrt{6}} & 0 & \frac{1}{\sqrt{3}} \end{bmatrix} = \begin{bmatrix} 2 & 0 & 0 \\ 0 & -4 & 0 \\ 0 & 0 & -4 \end{bmatrix} = D$$

Note. The diagonalizing matrix P is not unique. It depends on the choice of bases for the eigenspaces. This is just one possibility.

11. The characteristic polynomial of A is $p(\lambda) = \lambda^3 - 2\lambda^2 = \lambda^2(\lambda - 2)$; thus the eigenvalues of A are $\lambda = 0$ and $\lambda = 2$. The general solution of $(0I - A)\mathbf{x} = \mathbf{0}$ is $\mathbf{x} = r\begin{bmatrix} 0 \\ 0 \\ 1 \end{bmatrix} + s\begin{bmatrix} -1 \\ 1 \\ 0 \end{bmatrix}$, and the general solution of

$(2I - A)\mathbf{x} = \mathbf{0}$ is $\mathbf{x} = t\begin{bmatrix} 1 \\ 1 \\ 0 \end{bmatrix}$. Thus the vectors $\mathbf{v}_1 = \begin{bmatrix} 0 \\ 0 \\ 1 \end{bmatrix}$ and $\mathbf{v}_2 = \begin{bmatrix} -1 \\ 1 \\ 0 \end{bmatrix}$ form a basis for the eigenspace

corresponding to $\lambda = 0$, and the vector $\mathbf{v}_3 = \begin{bmatrix} 1 \\ 1 \\ 0 \end{bmatrix}$ forms a basis for the eigenspace corresponding to

$\lambda = 2$. These vectors are mutually orthogonal, and the orthogonal matrix

$$P = \begin{bmatrix} \dfrac{\mathbf{v}_1}{\|\mathbf{v}_1\|} & \dfrac{\mathbf{v}_2}{\|\mathbf{v}_2\|} & \dfrac{\mathbf{v}_3}{\|\mathbf{v}_3\|} \end{bmatrix} = \begin{bmatrix} 0 & -\frac{1}{\sqrt{2}} & \frac{1}{\sqrt{2}} \\ 0 & \frac{1}{\sqrt{2}} & \frac{1}{\sqrt{2}} \\ 1 & 0 & 0 \end{bmatrix}$$

has the property that

$$P^T A P = \begin{bmatrix} 0 & 0 & 1 \\ -\frac{1}{\sqrt{2}} & \frac{1}{\sqrt{2}} & 0 \\ \frac{1}{\sqrt{2}} & \frac{1}{\sqrt{2}} & 0 \end{bmatrix} \begin{bmatrix} 1 & 1 & 0 \\ 1 & 1 & 0 \\ 0 & 0 & 0 \end{bmatrix} \begin{bmatrix} 0 & -\frac{1}{\sqrt{2}} & \frac{1}{\sqrt{2}} \\ 0 & \frac{1}{\sqrt{2}} & \frac{1}{\sqrt{2}} \\ 1 & 0 & 0 \end{bmatrix} = \begin{bmatrix} 0 & 0 & 0 \\ 0 & 0 & 0 \\ 0 & 0 & 2 \end{bmatrix} = D$$

13. The characteristic polynomial of A is $p(\lambda) = \lambda^4 - 6\lambda^3 + 8\lambda^2 = \lambda^2(\lambda - 2)(\lambda - 4)$; thus the eigenvalues

of A are $\lambda = 0$, $\lambda = 2$, and $\lambda = 4$. The general solution of $(0I - A)\mathbf{x} = \mathbf{0}$ is $\mathbf{x} = r\begin{bmatrix} 0 \\ 0 \\ 0 \\ 1 \end{bmatrix} + s\begin{bmatrix} 0 \\ 0 \\ 1 \\ 0 \end{bmatrix}$, the

general solution of $(2I - A)\mathbf{x} = \mathbf{0}$ is $\mathbf{x} = t\begin{bmatrix} -1 \\ 1 \\ 0 \\ 0 \end{bmatrix}$, and the general solution of $(4I - A)\mathbf{x} = \mathbf{0}$ is $\mathbf{x} = u\begin{bmatrix} 1 \\ 1 \\ 0 \\ 0 \end{bmatrix}$.

Thus the vectors $\mathbf{v}_1 = \begin{bmatrix} 0 \\ 0 \\ 0 \\ 1 \end{bmatrix}$ and $\mathbf{v}_2 = \begin{bmatrix} 0 \\ 0 \\ 1 \\ 0 \end{bmatrix}$ form a basis for the eigenspace corresponding to $\lambda = 0$,

$\mathbf{v}_3 = \begin{bmatrix} -1 \\ 1 \\ 0 \\ 0 \end{bmatrix}$ forms a basis for the eigenspace corresponding to $\lambda = 2$, and $\mathbf{v}_4 = \begin{bmatrix} 1 \\ 1 \\ 0 \\ 0 \end{bmatrix}$ forms a basis

for the eigenspace corresponding to $\lambda = 2$. These four vectors are mutually orthogonal, and the orthogonal matrix $P = \begin{bmatrix} \frac{\mathbf{v}_1}{\|\mathbf{v}_1\|} & \frac{\mathbf{v}_2}{\|\mathbf{v}_2\|} & \frac{\mathbf{v}_3}{\|\mathbf{v}_3\|} & \frac{\mathbf{v}_4}{\|\mathbf{v}_4\|} \end{bmatrix} = \begin{bmatrix} 0 & 0 & -\frac{1}{\sqrt{2}} & \frac{1}{\sqrt{2}} \\ 0 & 0 & \frac{1}{\sqrt{2}} & \frac{1}{\sqrt{2}} \\ 0 & 1 & 0 & 0 \\ 1 & 0 & 0 & 0 \end{bmatrix}$ has the property that

$$P^T A P = \begin{bmatrix} 0 & 0 & 0 & 1 \\ 0 & 0 & 1 & 0 \\ -\frac{1}{\sqrt{2}} & \frac{1}{\sqrt{2}} & 0 & 0 \\ \frac{1}{\sqrt{2}} & \frac{1}{\sqrt{2}} & 0 & 0 \end{bmatrix} \begin{bmatrix} 3 & 1 & 0 & 0 \\ 1 & 3 & 0 & 0 \\ 0 & 0 & 0 & 0 \\ 0 & 0 & 0 & 0 \end{bmatrix} \begin{bmatrix} 0 & 0 & -\frac{1}{\sqrt{2}} & \frac{1}{\sqrt{2}} \\ 0 & 0 & \frac{1}{\sqrt{2}} & \frac{1}{\sqrt{2}} \\ 0 & 1 & 0 & 0 \\ 1 & 0 & 0 & 0 \end{bmatrix} = \begin{bmatrix} 0 & 0 & 0 & 0 \\ 0 & 0 & 0 & 0 \\ 0 & 0 & 2 & 0 \\ 0 & 0 & 0 & 4 \end{bmatrix}$$

15. The eigenvalues of the matrix $A = \begin{bmatrix} 3 & 1 \\ 1 & 3 \end{bmatrix}$ are $\lambda_1 = 2$ and $\lambda_2 = 4$, with corresponding normalized eigenvectors $\mathbf{u}_1 = \begin{bmatrix} \frac{1}{\sqrt{2}} \\ -\frac{1}{\sqrt{2}} \end{bmatrix}$ and $\mathbf{u}_2 = \begin{bmatrix} \frac{1}{\sqrt{2}} \\ \frac{1}{\sqrt{2}} \end{bmatrix}$. Thus the spectral decomposition of A is

$$\begin{bmatrix} 3 & 1 \\ 1 & 3 \end{bmatrix} = (2) \begin{bmatrix} \frac{1}{\sqrt{2}} \\ -\frac{1}{\sqrt{2}} \end{bmatrix} \begin{bmatrix} \frac{1}{\sqrt{2}} & -\frac{1}{\sqrt{2}} \end{bmatrix} + (4) \begin{bmatrix} \frac{1}{\sqrt{2}} \\ \frac{1}{\sqrt{2}} \end{bmatrix} \begin{bmatrix} \frac{1}{\sqrt{2}} & \frac{1}{\sqrt{2}} \end{bmatrix} = 2 \begin{bmatrix} \frac{1}{2} & -\frac{1}{2} \\ -\frac{1}{2} & \frac{1}{2} \end{bmatrix} + 4 \begin{bmatrix} \frac{1}{2} & \frac{1}{2} \\ \frac{1}{2} & \frac{1}{2} \end{bmatrix}$$

17. The eigenvalues of the matrix A are $\lambda_1 = 2$ and $\lambda_2 = -4$, with corresponding orthonormal eigenbases $\{\mathbf{u}_1\}$ and $\{\mathbf{u}_2, \mathbf{u}_3\}$ where $\mathbf{u}_1 = \begin{bmatrix} \frac{1}{\sqrt{6}} \\ \frac{1}{\sqrt{6}} \\ \frac{2}{\sqrt{6}} \end{bmatrix}$, $\mathbf{u}_2 = \begin{bmatrix} -\frac{1}{\sqrt{2}} \\ \frac{1}{\sqrt{2}} \\ 0 \end{bmatrix}$, $\mathbf{u}_3 = \begin{bmatrix} -\frac{1}{\sqrt{3}} \\ -\frac{1}{\sqrt{3}} \\ \frac{1}{\sqrt{3}} \end{bmatrix}$. Thus a spectral decomposition of A is

$$\begin{bmatrix} -3 & 1 & 2 \\ 1 & -3 & 2 \\ 2 & 2 & 0 \end{bmatrix} = 2 \begin{bmatrix} \frac{1}{\sqrt{6}} \\ \frac{1}{\sqrt{6}} \\ \frac{2}{\sqrt{6}} \end{bmatrix} \begin{bmatrix} \frac{1}{\sqrt{6}} & \frac{1}{\sqrt{6}} & \frac{2}{\sqrt{6}} \end{bmatrix} - 4 \begin{bmatrix} -\frac{1}{\sqrt{2}} \\ \frac{1}{\sqrt{2}} \\ 0 \end{bmatrix} \begin{bmatrix} -\frac{1}{\sqrt{2}} & \frac{1}{\sqrt{2}} & 0 \end{bmatrix} - 4 \begin{bmatrix} -\frac{1}{\sqrt{3}} \\ -\frac{1}{\sqrt{3}} \\ \frac{1}{\sqrt{3}} \end{bmatrix} \begin{bmatrix} -\frac{1}{\sqrt{3}} & -\frac{1}{\sqrt{3}} & \frac{1}{\sqrt{3}} \end{bmatrix}$$

$$= 2 \begin{bmatrix} \frac{1}{6} & \frac{1}{6} & \frac{1}{3} \\ \frac{1}{6} & \frac{1}{6} & \frac{1}{3} \\ \frac{1}{3} & \frac{1}{3} & \frac{2}{3} \end{bmatrix} - 4 \begin{bmatrix} \frac{1}{2} & -\frac{1}{2} & 0 \\ -\frac{1}{2} & \frac{1}{2} & 0 \\ 0 & 0 & 0 \end{bmatrix} - 4 \begin{bmatrix} \frac{1}{3} & \frac{1}{3} & -\frac{1}{3} \\ \frac{1}{3} & \frac{1}{3} & -\frac{1}{3} \\ -\frac{1}{3} & -\frac{1}{3} & \frac{1}{3} \end{bmatrix}$$

Note. The spectral decomposition is not unique. It depends on the choice of bases for the eigenspaces. This is just one possibility.

19. The matrix A has eigenvalues $\lambda = -1$ and $\lambda = 2$, with corresponding eigenvectors $\begin{bmatrix} -1 \\ 1 \end{bmatrix}$ and $\begin{bmatrix} -3 \\ 2 \end{bmatrix}$. Thus the matrix $P = \begin{bmatrix} -1 & -3 \\ 1 & 2 \end{bmatrix}$ has the property that $P^{-1} A P = D = \begin{bmatrix} -1 & 0 \\ 0 & 2 \end{bmatrix}$. It follows that

$$A^{10} = PD^{10}P^{-1} = \begin{bmatrix} -1 & -3 \\ 1 & 2 \end{bmatrix} \begin{bmatrix} 1 & 0 \\ 0 & 1024 \end{bmatrix} \begin{bmatrix} 2 & 3 \\ -1 & -1 \end{bmatrix} = \begin{bmatrix} 3070 & 3069 \\ -2046 & -2045 \end{bmatrix}$$

21. The matrix A has eigenvalues $\lambda = -1$ and $\lambda = 1$. The vector $\begin{bmatrix} 6 \\ 3 \\ 4 \end{bmatrix}$ forms a basis for the eigenspace corresponding to $\lambda = -1$, and the vectors $\begin{bmatrix} 1 \\ 0 \\ 1 \end{bmatrix}$ and $\begin{bmatrix} 0 \\ 1 \\ 0 \end{bmatrix}$ form a basis for the eigenspace corresponding to $\lambda = 1$. Thus the matrix $P = \begin{bmatrix} 6 & 1 & 0 \\ 3 & 0 & 1 \\ 4 & 1 & 0 \end{bmatrix}$ has the property that $P^{-1} A P = D = \begin{bmatrix} -1 & 0 & 0 \\ 0 & 1 & 0 \\ 0 & 0 & 1 \end{bmatrix}$, and it follows that

$$A^{1000} = PD^{1000}P^{-1} = \begin{bmatrix} 6 & 1 & 0 \\ 3 & 0 & 1 \\ 4 & 1 & 0 \end{bmatrix} \begin{bmatrix} 1 & 0 & 0 \\ 0 & 1 & 0 \\ 0 & 0 & 1 \end{bmatrix} \left(\frac{1}{2}\right) \begin{bmatrix} 1 & 0 & -1 \\ -4 & 0 & 6 \\ -3 & 2 & 3 \end{bmatrix} = \begin{bmatrix} 1 & 0 & 0 \\ 0 & 1 & 0 \\ 0 & 0 & 1 \end{bmatrix}$$

23. (a) The characteristic polynomial of A is $p(\lambda) = \lambda^3 - 6\lambda^2 + 12\lambda - 8$. Computing successive powers

of A, we have $A^2 = \begin{bmatrix} 8 & -8 & 4 \\ 8 & -12 & 8 \\ 12 & -24 & 16 \end{bmatrix}$ and $A^3 = \begin{bmatrix} 20 & -24 & 12 \\ 24 & -40 & 24 \\ 36 & -72 & 44 \end{bmatrix}$; thus

$$A^3 - 6A^2 + 12A = \begin{bmatrix} 20 & -24 & 12 \\ 24 & -40 & 24 \\ 36 & -72 & 44 \end{bmatrix} - 6 \begin{bmatrix} 8 & -8 & 4 \\ 8 & -12 & 8 \\ 12 & -24 & 16 \end{bmatrix} + 12 \begin{bmatrix} 3 & -2 & 1 \\ 2 & -2 & 2 \\ 3 & -6 & 5 \end{bmatrix} = 8I$$

which shows that A satisfies its characteristic equation, i.e., that $p(A) = 0$.

(b) Since $A^3 = 6A^2 - 12A + 8I$, we have $A^4 = 6A^3 - 12A^2 + 8A = 24A^2 - 64A + 48I$.

(c) Since $A^3 - 6A^2 + 12A - 8I = 0$, we have $A(A^2 - 6A + 12I) = 8I$ and $A^{-1} = \frac{1}{8}(A^2 - 6A + 12I)$.

25. From Exercise 7 we have $P^T A P = \begin{bmatrix} \frac{1}{\sqrt{2}} & -\frac{1}{\sqrt{2}} \\ \frac{1}{\sqrt{2}} & \frac{1}{\sqrt{2}} \end{bmatrix} \begin{bmatrix} 3 & 1 \\ 1 & 3 \end{bmatrix} \begin{bmatrix} \frac{1}{\sqrt{2}} & \frac{1}{\sqrt{2}} \\ -\frac{1}{\sqrt{2}} & \frac{1}{\sqrt{2}} \end{bmatrix} = \begin{bmatrix} 2 & 0 \\ 0 & 4 \end{bmatrix} = D$. Thus $A = PDP^T$ and

$$e^{tA} = Pe^{tD}P^T = \begin{bmatrix} \frac{1}{\sqrt{2}} & \frac{1}{\sqrt{2}} \\ -\frac{1}{\sqrt{2}} & \frac{1}{\sqrt{2}} \end{bmatrix} \begin{bmatrix} e^{2t} & 0 \\ 0 & e^{4t} \end{bmatrix} \begin{bmatrix} \frac{1}{\sqrt{2}} & -\frac{1}{\sqrt{2}} \\ \frac{1}{\sqrt{2}} & \frac{1}{\sqrt{2}} \end{bmatrix} = \frac{1}{2} \begin{bmatrix} e^{2t} + e^{4t} & -e^{2t} + e^{4t} \\ -e^{2t} + e^{4t} & e^{2t} + e^{4t} \end{bmatrix}.$$

27. From Exercise 9 we have

$$P^T A P = \begin{bmatrix} \frac{1}{\sqrt{6}} & \frac{1}{\sqrt{6}} & \frac{2}{\sqrt{6}} \\ -\frac{1}{\sqrt{2}} & \frac{1}{\sqrt{2}} & 0 \\ -\frac{1}{\sqrt{3}} & -\frac{1}{\sqrt{3}} & \frac{1}{\sqrt{3}} \end{bmatrix} \begin{bmatrix} -3 & 1 & 2 \\ 1 & -3 & 2 \\ 2 & 2 & 0 \end{bmatrix} \begin{bmatrix} \frac{1}{\sqrt{6}} & -\frac{1}{\sqrt{2}} & -\frac{1}{\sqrt{3}} \\ \frac{1}{\sqrt{6}} & \frac{1}{\sqrt{2}} & -\frac{1}{\sqrt{3}} \\ \frac{2}{\sqrt{6}} & 0 & \frac{1}{\sqrt{3}} \end{bmatrix} = \begin{bmatrix} 2 & 0 & 0 \\ 0 & -4 & 0 \\ 0 & 0 & -4 \end{bmatrix} = D$$

Thus $A = PDP^T$ and

$$e^{tA} = Pe^{tD}P^T = \begin{bmatrix} \frac{1}{\sqrt{6}} & -\frac{1}{\sqrt{2}} & -\frac{1}{\sqrt{3}} \\ \frac{1}{\sqrt{6}} & \frac{1}{\sqrt{2}} & -\frac{1}{\sqrt{3}} \\ \frac{2}{\sqrt{6}} & 0 & \frac{1}{\sqrt{3}} \end{bmatrix} \begin{bmatrix} e^{2t} & 0 & 0 \\ 0 & e^{-4t} & 0 \\ 0 & 0 & e^{-4t} \end{bmatrix} \begin{bmatrix} \frac{1}{\sqrt{6}} & \frac{1}{\sqrt{6}} & \frac{2}{\sqrt{6}} \\ -\frac{1}{\sqrt{2}} & \frac{1}{\sqrt{2}} & 0 \\ -\frac{1}{\sqrt{3}} & -\frac{1}{\sqrt{3}} & \frac{1}{\sqrt{3}} \end{bmatrix}$$

$$= \frac{1}{6} \begin{bmatrix} e^{2t} + 5e^{-4t} & e^{2t} - e^{-4t} & 2e^{2t} - 2e^{-4t} \\ e^{2t} - e^{-4t} & e^{2t} + 5e^{-4t} & 2e^{2t} - 2e^{-4t} \\ 2e^{2t} - 2e^{-4t} & 2e^{2t} - 2e^{-4t} & 4e^{2t} + 2e^{-4t} \end{bmatrix}$$

29. Note that $\begin{bmatrix} \sin(2\pi) & 0 & 0 \\ 0 & \sin(-4\pi) & 0 \\ 0 & 0 & \sin(-4\pi) \end{bmatrix} = \begin{bmatrix} 0 & 0 & 0 \\ 0 & 0 & 0 \\ 0 & 0 & 0 \end{bmatrix}$. Thus, proceeding as in Exercise 27:

$$\sin(\pi A) = P \sin(\pi D) P^T = \begin{bmatrix} \frac{1}{\sqrt{6}} & -\frac{1}{\sqrt{2}} & -\frac{1}{\sqrt{3}} \\ \frac{1}{\sqrt{6}} & \frac{1}{\sqrt{2}} & -\frac{1}{\sqrt{3}} \\ \frac{2}{\sqrt{6}} & 0 & \frac{1}{\sqrt{3}} \end{bmatrix} \begin{bmatrix} 0 & 0 & 0 \\ 0 & 0 & 0 \\ 0 & 0 & 0 \end{bmatrix} \begin{bmatrix} \frac{1}{\sqrt{6}} & \frac{1}{\sqrt{6}} & \frac{2}{\sqrt{6}} \\ -\frac{1}{\sqrt{2}} & \frac{1}{\sqrt{2}} & 0 \\ -\frac{1}{\sqrt{3}} & -\frac{1}{\sqrt{3}} & \frac{1}{\sqrt{3}} \end{bmatrix} = \begin{bmatrix} 0 & 0 & 0 \\ 0 & 0 & 0 \\ 0 & 0 & 0 \end{bmatrix}$$

31. If $A = \begin{bmatrix} 0 & 0 & 0 \\ 1 & 0 & 0 \\ 2 & 1 & 0 \end{bmatrix}$, then $A^2 = \begin{bmatrix} 0 & 0 & 0 \\ 0 & 0 & 0 \\ 1 & 0 & 0 \end{bmatrix}$ and $A^3 = \begin{bmatrix} 0 & 0 & 0 \\ 0 & 0 & 0 \\ 0 & 0 & 0 \end{bmatrix}$. Thus A is nilpotent and

$$e^A = e^0 I + e^0 A + \frac{1}{2!} e^0 A^2 = 1 \begin{bmatrix} 1 & 0 & 0 \\ 0 & 1 & 0 \\ 0 & 0 & 1 \end{bmatrix} + 1 \begin{bmatrix} 0 & 0 & 0 \\ 1 & 0 & 0 \\ 2 & 1 & 0 \end{bmatrix} + \frac{1}{2} \begin{bmatrix} 0 & 0 & 0 \\ 0 & 0 & 0 \\ 1 & 0 & 0 \end{bmatrix} = \begin{bmatrix} 1 & 0 & 0 \\ 1 & 1 & 0 \\ \frac{5}{2} & 1 & 1 \end{bmatrix}$$

33. If P is symmetric and orthogonal, then $P^T = P$ and $P^T P = I$; thus $P^2 = P^T P = I$. If λ is an eigenvalue of P, then there is a nonzero vector $\mathbf{x}$ such that $P\mathbf{x} = \lambda\mathbf{x}$. Since $P^2 = I$ it follows that $\lambda^2 \mathbf{x} = P^2 \mathbf{x} = I\mathbf{x} = \mathbf{x}$; thus $\lambda^2 = 1$ and so $\lambda = \pm 1$.

DISCUSSION AND DISCOVERY

D1. **(a)** True. The matrix AA^T is symmetric and hence is orthogonally diagonalizable.

(b) False. If A is diagonalizable but not symmetric (therefore not orthogonally diagonalizable), then there is a basis for R^n (but not an orthogonal basis) consisting of eigenvectors of A.

(c) False. An orthogonal matrix need not be symmetric; for example $A = \begin{bmatrix} \frac{1}{\sqrt{2}} & -\frac{1}{\sqrt{2}} \\ \frac{1}{\sqrt{2}} & \frac{1}{\sqrt{2}} \end{bmatrix}$.

(d) True. If A is an invertible orthogonally diagonalizable matrix, then there is an orthogonal matrix P such that $P^T A P = D$ where D is a diagonal matrix with nonzero entries (the eigenvalues of A) on the main diagonal. It follows that $P^T A^{-1} P = (P^T A P)^{-1} = D^{-1}$ and that D^{-1} is a diagonal matrix with nonzero entries (the reciprocals of the eigenvalues) on the main diagonal. Thus the matrix A^{-1} is orthogonally diagonalizable.

(e) True. If A is orthogonally diagonalizable, then A is symmetric and thus has real eigenvalues.

D2. **(a)** $A = PDP^T = \begin{bmatrix} 0 & 1 & 0 \\ \frac{1}{\sqrt{2}} & 0 & \frac{1}{\sqrt{2}} \\ -\frac{1}{\sqrt{2}} & 0 & \frac{1}{\sqrt{2}} \end{bmatrix} \begin{bmatrix} -1 & 0 & 0 \\ 0 & 3 & 0 \\ 0 & 0 & 7 \end{bmatrix} \begin{bmatrix} 0 & \frac{1}{\sqrt{2}} & -\frac{1}{\sqrt{2}} \\ 1 & 0 & 0 \\ 0 & \frac{1}{\sqrt{2}} & \frac{1}{\sqrt{2}} \end{bmatrix} = \begin{bmatrix} 3 & 0 & 0 \\ 0 & 3 & 4 \\ 0 & 4 & 3 \end{bmatrix}$

(b) No. The vectors $\mathbf{v}_2$ and $\mathbf{v}_3$ correspond to different eigenvalues, but are not orthogonal. Therefore they cannot be eigenvectors of a symmetric matrix.

D3. Yes. Since A is diagonalizable and the eigenspaces are mutually orthogonal, there is an orthonormal basis for R^n consisting of eigenvectors of A. Thus A is orthogonally diagonalizable and therefore must be symmetric.

WORKING WITH PROOFS

P1. We first show that if A and C are orthogonally similar, then there exist orthonormal bases with respect to which they represent the same linear operator. For this purpose, let T be the operator defined by $T(x) = Ax$. Then $A = [T]$, i.e., A is the matrix of T relative to the standard basis $B = \{\mathbf{e}_1, \mathbf{e}_2, \ldots, \mathbf{e}_n\}$. Since A and C are orthogonally similar, there is an orthogonal matrix P such that $C = P^T A P$. Let $B' = \{\mathbf{v}_1, \mathbf{v}_2, \ldots, \mathbf{v}_n\}$ where $\mathbf{v}_1, \mathbf{v}_2, \ldots, \mathbf{v}_n$ are the column vectors of P. Then B' is an orthonormal basis for R^n, and $P = P_{B' \to B}$. Thus $[T]_B = P[T]_{B'} P^T$ and $[T]_{B'} = P^T [T]_B P = P^T A P = C$. This shows that there exist orthonormal bases with respect to which A and C represent the same linear operator.

Conversely, suppose that $A = [T]_B$ and $C = [T]_{B'}$ where $T : R^n \to R^n$ is a linear operator and B, B' are bases for R^n. If $P = P_{B' \to B}$ then P is an orthogonal matrix and $C = [T]_{B'} = P^T [T]_B P = P^T A P$. Thus A and C are orthogonally similar.

P2. Suppose $A = c_1 \mathbf{u}_1 \mathbf{u}_1^T + c_2 \mathbf{u}_2 \mathbf{u}_2^T + \cdots + c_n \mathbf{u}_n \mathbf{u}_n^T$ where $\{\mathbf{u}_1, \mathbf{u}_2, \ldots, \mathbf{u}_n\}$ is an orthonormal basis for R^n. Since $(\mathbf{u}_j \mathbf{u}_j^T)^T = \mathbf{u}_j^{TT} \mathbf{u}_j^T = \mathbf{u}_j \mathbf{u}_j^T$ it follows that $A^T = A$; thus A is symmetric. Furthermore, since $\mathbf{u}_i^T \mathbf{u}_j = \mathbf{u}_i \cdot \mathbf{u}_j = \delta_{ij}$, we have

$$Au_j = (c_1 \mathbf{u}_1 \mathbf{u}_1^T + c_2 \mathbf{u}_2 \mathbf{u}_2^T + \cdots + c_n \mathbf{u}_n \mathbf{u}_n^T) \mathbf{u}_j = \sum_{i=1}^{n} c_i \mathbf{u}_i \mathbf{u}_i^T \mathbf{u}_j = c_j \mathbf{u}_j$$

for each $j = 1, 2, \ldots, n$. Thus $c_1, c_2, \ldots, c_n$ are eigenvalues of A.

P3. The spectral decomposition $A = \lambda_1 \mathbf{u}_1 \mathbf{u}_1^T + \lambda_2 \mathbf{u}_2 \mathbf{u}_2^T + \cdots + \lambda_n \mathbf{u}_n \mathbf{u}_n^T$ is equivalent to $A = PDP^T$ where $P = [\mathbf{u}_1 \,|\, \mathbf{u}_2 \,|\, \cdots \,|\, \mathbf{u}_n]$ and $D = \operatorname{diag}(\lambda_1, \lambda_2, \ldots, \lambda_n)$; thus

$$f(A) = Pf(D)P^T = P\operatorname{diag}(f(\lambda_1), f(\lambda_2), \ldots, f(\lambda_n))P^T$$
$$= f(\lambda_1) \mathbf{u}_1 \mathbf{u}_1^T + f(\lambda_2) \mathbf{u}_2 \mathbf{u}_2^T + \cdots + f(\lambda_n) \mathbf{u}_n \mathbf{u}_n^T$$

P4. **(a)** Suppose A is a symmetric matrix, and λ_0 is an eigenvalue of A having geometric multiplicity k. Let W be the eigenspace corresponding to λ_0. Choose an orthonormal basis $\{\mathbf{u}_1, \mathbf{u}_2, \ldots, \mathbf{u}_k\}$ for W, extend it to an orthonormal basis $B = \{\mathbf{u}_1, \mathbf{u}_2, \ldots, \mathbf{u}_k, \mathbf{u}_{k+1}, \ldots, \mathbf{u}_n\}$ for R^n, and let P be the orthogonal matrix having the vectors of B as its columns. Then, as shown in Exercise P6(b) of Section 8.2, the product AP can be written as $AP = P\begin{bmatrix} \lambda_0 I_k & X \\ 0 & Y \end{bmatrix}$. Since P is orthogonal, we have

$$P^T AP = P^T P \begin{bmatrix} \lambda_0 I_k & X \\ 0 & Y \end{bmatrix} = \begin{bmatrix} \lambda_0 I_k & X \\ 0 & Y \end{bmatrix}$$

and since $P^T AP$ is a symmetric matrix, it follows that $X = 0$.

(b) Since A is similar to $C = \begin{bmatrix} \lambda_0 I_k & 0 \\ 0 & Y \end{bmatrix}$, it has the same characteristic polynomial as C, namely $(\lambda - \lambda_0)^k \det(\lambda I_{n-k} - Y) = (\lambda - \lambda_0)^k p_Y(\lambda)$ where $p_Y(\lambda)$ is the characteristic polynomial of Y. We will now prove that $p_Y(\lambda_0) \neq 0$ and thus that the algebraic multiplicity of λ_0 is exactly k. The proof is by contradiction:

Suppose $p_Y(\lambda_0) = 0$, i.e., that λ_0 is an eigenvalue of the matrix Y. Then there is a nonzero vector $\mathbf{y}$ in R^{n-k} such that $Y\mathbf{y} = \lambda_0 \mathbf{y}$. Let $\mathbf{x} = \begin{bmatrix} 0 \\ \mathbf{y} \end{bmatrix}$ be the vector in R^n whose first k components are 0 and whose last $n - k$ components are those of $\mathbf{y}$. Then

$$C\mathbf{x} = \begin{bmatrix} \lambda_0 I_{n-k} & 0 \\ \hline 0 & Y \end{bmatrix} \begin{bmatrix} 0 \\ \mathbf{y} \end{bmatrix} = \begin{bmatrix} 0 \\ \lambda_0 \mathbf{y} \end{bmatrix} = \lambda_0 \mathbf{x}$$

and so $\mathbf{x}$ is an eigenvector of C corresponding to λ_0. Since $AP = PC$, it follows that $P\mathbf{x}$ is an eigenvector of A corresponding to λ_0. But note that $\mathbf{e}_1, \ldots, \mathbf{e}_k$ are also eigenvectors of C corresponding to λ_0, and that $\{\mathbf{e}_1, \ldots, \mathbf{e}_k, \mathbf{x}\}$ is a linearly independent set. It follows that $\{P\mathbf{e}_1, \ldots, P\mathbf{e}_k, P\mathbf{x}\}$ is a linearly independent set of eigenvectors of A corresponding to λ_0. But this implies that the geometric multiplicity of λ_0 is greater than k, a contradiction!

(c) It follows from part (b) that the sum of the dimensions of the eigenspaces of A is equal to n; thus A is diagonalizable. Furthermore, since A is symmetric, the eigenspaces corresponding to different eigenvalues are orthogonal. Thus we can form an orthonormal basis for R^n by choosing an orthonormal basis for each of the eigenspaces and joining them together. Since the sum of the dimensions is n, this will be an orthonormal basis consisting of eigenvectors of A. Thus A is orthogonally diagonalizable.

EXERCISE SET 8.4

1. (a) $3x_1^2 + 7x_2^2 = [x_1 \ \ x_2] \begin{bmatrix} 3 & 0 \\ 0 & 7 \end{bmatrix} \begin{bmatrix} x_1 \\ x_2 \end{bmatrix}$

    (b) $4x_1^2 - 9x_2^2 - 6x_1x_2 = [x_1 \ \ x_2] \begin{bmatrix} 4 & -3 \\ -3 & -9 \end{bmatrix} \begin{bmatrix} x_1 \\ x_2 \end{bmatrix}$

    (c) $9x_1^2 - x_2^2 + 4x_3^2 + 6x_1x_2 - 8x_1x_3 + x_2x_3 = [x_1 \ \ x_2 \ \ x_3] \begin{bmatrix} 9 & 3 & -4 \\ 3 & -1 & \frac{1}{2} \\ -4 & \frac{1}{2} & 4 \end{bmatrix} \begin{bmatrix} x_1 \\ x_2 \\ x_3 \end{bmatrix}$

3. $[x \ \ y] \begin{bmatrix} 2 & -3 \\ -3 & 5 \end{bmatrix} \begin{bmatrix} x \\ y \end{bmatrix} = 2x^2 + 5y^2 - 6xy$

5. The quadratic form $Q = 2x_1^2 + 2x_2^2 - 2x_1x_2$ can be expressed in matrix notation as

$$Q = \mathbf{x}^T A \mathbf{x} = [x_1 \ \ x_2] \begin{bmatrix} 2 & -1 \\ -1 & 2 \end{bmatrix} \begin{bmatrix} x_1 \\ x_2 \end{bmatrix}$$

The matrix A has eigenvalues $\lambda_1 = 1$ and $\lambda_2 = 3$, with corresponding eigenvectors $\mathbf{v}_1 = \begin{bmatrix} 1 \\ 1 \end{bmatrix}$ and $\mathbf{v}_2 = \begin{bmatrix} -1 \\ 1 \end{bmatrix}$ respectively. Thus the matrix $P = \begin{bmatrix} \frac{1}{\sqrt{2}} & -\frac{1}{\sqrt{2}} \\ \frac{1}{\sqrt{2}} & \frac{1}{\sqrt{2}} \end{bmatrix}$ orthogonally diagonalizes A, and the change of variable $\begin{bmatrix} x_1 \\ x_2 \end{bmatrix} = \mathbf{x} = P\mathbf{y} = \begin{bmatrix} \frac{1}{\sqrt{2}} & -\frac{1}{\sqrt{2}} \\ \frac{1}{\sqrt{2}} & \frac{1}{\sqrt{2}} \end{bmatrix} \begin{bmatrix} y_1 \\ y_2 \end{bmatrix}$ eliminates the cross product terms in Q:

$$Q = \mathbf{x}^T A \mathbf{x} = \mathbf{y}^T (P^T A P) \mathbf{y} = [y_1 \ \ y_2] \begin{bmatrix} 1 & 0 \\ 0 & 3 \end{bmatrix} \begin{bmatrix} y_1 \\ y_2 \end{bmatrix} = y_1^2 + 3y_2^2$$

Note that the inverse relationship between $\mathbf{x}$ and $\mathbf{y}$ is $\begin{bmatrix} y_1 \\ y_2 \end{bmatrix} = \mathbf{y} = P^T \mathbf{x} = \begin{bmatrix} \frac{1}{\sqrt{2}} & \frac{1}{\sqrt{2}} \\ -\frac{1}{\sqrt{2}} & \frac{1}{\sqrt{2}} \end{bmatrix} \begin{bmatrix} x_1 \\ x_2 \end{bmatrix}$.

7. The given quadratic form can be expressed in matrix notation as $Q = \mathbf{x}^T A \mathbf{x}$ where $A = \begin{bmatrix} 3 & 2 & 0 \\ 2 & 4 & -2 \\ 0 & -2 & 5 \end{bmatrix}$.

The matrix A has eigenvalues $\lambda_1 = 1$, $\lambda_2 = 4$, $\lambda_3 = 7$, with corresponding (orthogonal) eigenvectors

$\mathbf{v}_1 = \begin{bmatrix} -2 \\ 2 \\ 1 \end{bmatrix}$, $\mathbf{v}_2 = \begin{bmatrix} 2 \\ 1 \\ 2 \end{bmatrix}$, $\mathbf{v}_3 = \begin{bmatrix} 1 \\ 2 \\ -2 \end{bmatrix}$. Thus the matrix $P = \begin{bmatrix} -\frac{2}{3} & \frac{2}{3} & \frac{1}{3} \\ \frac{2}{3} & \frac{1}{3} & \frac{2}{3} \\ \frac{1}{3} & \frac{2}{3} & -\frac{2}{3} \end{bmatrix}$ orthogonally diagonalizes A,

and the change of variable $\mathbf{x} = P\mathbf{y}$ eliminates the cross product terms in Q:

$$Q = \mathbf{x}^T A \mathbf{x} = \mathbf{y}^T (P^T A P) \mathbf{y} = [y_1 \ \ y_2 \ \ y_3] \begin{bmatrix} 1 & 0 & 0 \\ 0 & 4 & 0 \\ 0 & 0 & 7 \end{bmatrix} \begin{bmatrix} y_1 \\ y_2 \\ y_3 \end{bmatrix} = y_1^2 + 4y_2^2 + 7y_3^2$$

Note that the diagonalizing matrix P is symmetric and so the inverse relationship between $\mathbf{x}$ and $\mathbf{y}$ is

$$\begin{bmatrix} y_1 \\ y_2 \\ y_3 \end{bmatrix} = \mathbf{y} = P^T \mathbf{x} = P\mathbf{x} = \begin{bmatrix} -\frac{2}{3} & \frac{2}{3} & \frac{1}{3} \\ \frac{2}{3} & \frac{1}{3} & \frac{2}{3} \\ \frac{1}{3} & \frac{2}{3} & -\frac{2}{3} \end{bmatrix} \begin{bmatrix} x_1 \\ x_2 \\ x_3 \end{bmatrix}$$

9. (a) $[x \ \ y] \begin{bmatrix} 2 & \frac{1}{2} \\ \frac{1}{2} & 0 \end{bmatrix} \begin{bmatrix} x \\ y \end{bmatrix} + [1 \ \ -6] \begin{bmatrix} x \\ y \end{bmatrix} + 2 = 0$

(b) $[x \ \ y] \begin{bmatrix} 0 & 0 \\ 0 & 1 \end{bmatrix} \begin{bmatrix} x \\ y \end{bmatrix} + [7 \ \ -8] \begin{bmatrix} x \\ y \end{bmatrix} - 5 = 0$

11. (a) Ellipse   (b) Hyperbola   (c) Parabola   (d) Circle

13. The equation can be written in matrix form as $\mathbf{x}^T A\mathbf{x} = -8$ where $A = \begin{bmatrix} 2 & -2 \\ -2 & -1 \end{bmatrix}$. The eigenvalues of A
are $\lambda_1 = 3$ and $\lambda_2 = -2$, with corresponding eigenvectors $\mathbf{v}_1 = \begin{bmatrix} 2 \\ -1 \end{bmatrix}$ and $\mathbf{v}_2 = \begin{bmatrix} 1 \\ 2 \end{bmatrix}$ respectively. Thus
the matrix $P = \begin{bmatrix} \frac{2}{\sqrt{5}} & \frac{1}{\sqrt{5}} \\ -\frac{1}{\sqrt{5}} & \frac{2}{\sqrt{5}} \end{bmatrix}$ orthogonally diagonalizes A. Note that $\det(P) = 1$, so P is a rotation
matrix. The equation of the conic in the rotated $x'y'$-coordinate system is

$$[x' \ \ y'] \begin{bmatrix} 3 & 0 \\ 0 & -2 \end{bmatrix} \begin{bmatrix} x' \\ y' \end{bmatrix} = -8$$

which can be written as $2y'^2 - 3x'^2 = 8$; thus the conic is a hyperbola. The angle through which the
axes have been rotated is $\theta = \tan^{-1}(-\frac{1}{2}) \approx -26.6°$.

15. The equation can be written in matrix form as $\mathbf{x}^T A\mathbf{x} = 15$ where $A = \begin{bmatrix} 11 & 12 \\ 12 & 4 \end{bmatrix}$. The eigenvalues of
A are $\lambda_1 = 20$ and $\lambda_2 = -5$, with corresponding eigenvectors $\mathbf{v}_1 = \begin{bmatrix} 4 \\ 3 \end{bmatrix}$ and $\mathbf{v}_2 = \begin{bmatrix} -3 \\ 4 \end{bmatrix}$ respectively.
Thus the matrix $P = \begin{bmatrix} \frac{4}{5} & -\frac{3}{5} \\ \frac{3}{5} & \frac{4}{5} \end{bmatrix}$ orthogonally diagonalizes A. Note that $\det(P) = 1$, so P is a rotation
matrix. The equation of the conic in the rotated $x'y'$-coordinate system is

$$[x' \ \ y'] \begin{bmatrix} 20 & 0 \\ 0 & -5 \end{bmatrix} \begin{bmatrix} x' \\ y' \end{bmatrix} = 15$$

which we can write as $4x'^2 - y'^2 = 3$; thus the conic is a hyperbola. The angle through which the
axes have been rotated is $\theta = \tan^{-1}(\frac{3}{4}) \approx 36.9°$.

17. (a) The eigenvalues of $A = \begin{bmatrix} 1 & 0 \\ 0 & 2 \end{bmatrix}$ and $\lambda = 1$ and $\lambda = 2$; thus A is positive definite.
   (b) negative definite   (c) indefinite
   (d) positive semidefinite   (e) negative semidefinite

19. We have $Q = x_1^2 + x_2^2 > 0$ for $(x_1, x_2) \neq (0,0)$; thus Q is positive definite.

21. We have $Q = (x_1 - x_2)^2 > 0$ for $x_1 \neq x_2$ and $Q = 0$ for $x_1 = x_2$; thus Q is positive semidefinite.

23. We have $Q = x_1^2 - x_2^2 > 0$ for $x_1 \neq 0, x_2 = 0$ and $Q < 0$ for $x_1 = 0, x_2 \neq 0$; thus Q is indefinite.

25. (a) The eigenvalues of the matrix $A = \begin{bmatrix} 5 & -2 \\ -2 & 5 \end{bmatrix}$ are $\lambda = 3$ and $\lambda = 7$; thus A is positive definite.
   Since $|5| = 5$ and $\begin{vmatrix} 5 & -2 \\ -2 & 5 \end{vmatrix} = 21$ are positive, we reach the same conclusion using Theorem 8.4.5.

(b) The eigenvalues of $A = \begin{bmatrix} 2 & -1 & 0 \\ -1 & 2 & 0 \\ 0 & 0 & 5 \end{bmatrix}$ are $\lambda = 1$, $\lambda = 3$, and $\lambda = 5$; thus A is positive definite.

The determinants of the principal submatrices are $|2| = 2$, $\begin{vmatrix} 2 & -1 \\ -1 & 2 \end{vmatrix} = 3$, and $\begin{vmatrix} 2 & -1 & 0 \\ -1 & 2 & 0 \\ 0 & 0 & 5 \end{vmatrix} = 15$;

thus we reach the same conclusion using Theorem 8.4.5.

27. (a) The matrix A has eigenvalues $\lambda_1 = 3$ and $\lambda_2 = 7$, with corresponding eigenvectors $\mathbf{v}_1 = \begin{bmatrix} 1 \\ 1 \end{bmatrix}$ and

$\mathbf{v}_2 = \begin{bmatrix} -1 \\ 1 \end{bmatrix}$. Thus the matrix $P = \begin{bmatrix} \frac{1}{\sqrt{2}} & -\frac{1}{\sqrt{2}} \\ \frac{1}{\sqrt{2}} & \frac{1}{\sqrt{2}} \end{bmatrix}$ orthogonally diagonalizes A, and the matrix

$$B = \begin{bmatrix} \frac{1}{\sqrt{2}} & -\frac{1}{\sqrt{2}} \\ \frac{1}{\sqrt{2}} & \frac{1}{\sqrt{2}} \end{bmatrix} \begin{bmatrix} \sqrt{3} & 0 \\ 0 & \sqrt{7} \end{bmatrix} \begin{bmatrix} \frac{1}{\sqrt{2}} & \frac{1}{\sqrt{2}} \\ -\frac{1}{\sqrt{2}} & \frac{1}{\sqrt{2}} \end{bmatrix} = \begin{bmatrix} \frac{\sqrt{3}}{2} + \frac{\sqrt{7}}{2} & \frac{\sqrt{3}}{2} - \frac{\sqrt{7}}{2} \\ \frac{\sqrt{3}}{2} - \frac{\sqrt{7}}{2} & \frac{\sqrt{3}}{2} + \frac{\sqrt{7}}{2} \end{bmatrix}$$

has the property that $B^2 = A$.

(b) The matrix A has eigenvalues $\lambda_1 = 1$, $\lambda_2 = 3$, $\lambda_3 = 5$, with corresponding eigenvectors $\mathbf{v}_1 = \begin{bmatrix} 1 \\ 1 \\ 1 \end{bmatrix}$,

$\mathbf{v}_2 = \begin{bmatrix} -1 \\ 1 \\ 0 \end{bmatrix}$, and $\mathbf{v}_3 = \begin{bmatrix} 0 \\ 0 \\ 1 \end{bmatrix}$. Thus $P = \begin{bmatrix} \frac{1}{\sqrt{2}} & -\frac{1}{\sqrt{2}} & 0 \\ \frac{1}{\sqrt{2}} & \frac{1}{\sqrt{2}} & 0 \\ 0 & 0 & 1 \end{bmatrix}$ orthogonally diagonalizes A, and

$$B = \begin{bmatrix} \frac{1}{\sqrt{2}} & -\frac{1}{\sqrt{2}} & 0 \\ \frac{1}{\sqrt{2}} & \frac{1}{\sqrt{2}} & 0 \\ 0 & 0 & 1 \end{bmatrix} \begin{bmatrix} 1 & 0 & 0 \\ 0 & \sqrt{3} & 0 \\ 0 & 0 & \sqrt{5} \end{bmatrix} \begin{bmatrix} \frac{1}{\sqrt{2}} & \frac{1}{\sqrt{2}} & 0 \\ -\frac{1}{\sqrt{2}} & \frac{1}{\sqrt{2}} & 0 \\ 0 & 0 & 1 \end{bmatrix} = \begin{bmatrix} \frac{1}{2} + \frac{\sqrt{3}}{2} & \frac{1}{2} - \frac{\sqrt{3}}{2} & 0 \\ \frac{1}{2} - \frac{\sqrt{3}}{2} & \frac{1}{2} + \frac{\sqrt{3}}{2} & 0 \\ 0 & 0 & \sqrt{5} \end{bmatrix}$$

has the property that $B^2 = A$.

29. The quadratic form $Q = 5x_1^2 + x_2^2 + kx_3^2 + 4x_1x_2 - 2x_1x_3 - 2x_2x_3$ can be expressed in matrix nota-

tion as $Q = \mathbf{x}^T A \mathbf{x}$ where $A = \begin{bmatrix} 5 & 2 & -1 \\ 2 & 1 & -1 \\ -1 & -1 & k \end{bmatrix}$. The determinants of the principal submatrices of A are

$|5| = 5$, $\begin{vmatrix} 5 & 2 \\ 2 & 1 \end{vmatrix} = 1$, and $\begin{vmatrix} 5 & 2 & -1 \\ 2 & 1 & -1 \\ -1 & -1 & k \end{vmatrix} = k - 2$. Thus Q is positive definite if and only if $k > 2$.

31. (a) The matrix A has eigenvalues $\lambda_1 = 3$ and $\lambda_2 = 15$, with corresponding eigenvectors $\mathbf{v}_1 = \begin{bmatrix} -1 \\ 1 \end{bmatrix}$

and $\mathbf{v}_2 = \begin{bmatrix} 1 \\ 1 \end{bmatrix}$. Thus A is positive definite, the matrix $P = \begin{bmatrix} -\frac{1}{\sqrt{2}} & \frac{1}{\sqrt{2}} \\ \frac{1}{\sqrt{2}} & \frac{1}{\sqrt{2}} \end{bmatrix}$ orthogonally diagonalizes

A, and the matrix

$$B = \begin{bmatrix} -\frac{1}{\sqrt{2}} & \frac{1}{\sqrt{2}} \\ \frac{1}{\sqrt{2}} & \frac{1}{\sqrt{2}} \end{bmatrix} \begin{bmatrix} \sqrt{3} & 0 \\ 0 & \sqrt{15} \end{bmatrix} \begin{bmatrix} -\frac{1}{\sqrt{2}} & \frac{1}{\sqrt{2}} \\ \frac{1}{\sqrt{2}} & \frac{1}{\sqrt{2}} \end{bmatrix} = \begin{bmatrix} \frac{\sqrt{3}}{2} + \frac{\sqrt{15}}{2} & -\frac{\sqrt{3}}{2} + \frac{\sqrt{15}}{2} \\ -\frac{\sqrt{3}}{2} + \frac{\sqrt{15}}{2} & \frac{\sqrt{3}}{2} + \frac{\sqrt{15}}{2} \end{bmatrix}$$

has the property that $B^2 = A$.

(b) The LDU-decomposition (p159-160) of the matrix A is

$$A = \begin{bmatrix} 9 & 6 \\ 6 & 9 \end{bmatrix} = \begin{bmatrix} 1 & 0 \\ \frac{2}{3} & 1 \end{bmatrix} \begin{bmatrix} 9 & 0 \\ 0 & 5 \end{bmatrix} \begin{bmatrix} 1 & \frac{2}{3} \\ \frac{2}{3} & 1 \end{bmatrix} = LDU$$

and, since $L = U^T$, this can be written as

$$A = \begin{bmatrix} 9 & 6 \\ 6 & 9 \end{bmatrix} = \left(\begin{bmatrix} 1 & 0 \\ \frac{2}{3} & 1 \end{bmatrix} \begin{bmatrix} 3 & 0 \\ 0 & \sqrt{5} \end{bmatrix} \right) \left(\begin{bmatrix} 3 & 0 \\ 0 & \sqrt{5} \end{bmatrix} \begin{bmatrix} 1 & \frac{2}{3} \\ \frac{2}{3} & 1 \end{bmatrix} \right) = \begin{bmatrix} 3 & 0 \\ 2 & \sqrt{5} \end{bmatrix} \begin{bmatrix} 3 & 2 \\ 0 & \sqrt{5} \end{bmatrix} = C^T C$$

which is a factorization of the required type.

33. We have $(c_1 x_1 + c_2 x_2 + \cdots + c_n x_n)^2 = \sum\limits_{i=1}^{n} c_i^2 x_i^2 + \sum\limits_{i=1}^{n} \sum\limits_{j=i+1}^{n} 2 c_i c_j x_i x_j = \mathbf{x}^T A \mathbf{x}$ where

$$A = \begin{bmatrix} c_1^2 & c_1 c_2 & \cdots & c_1 c_n \\ c_1 c_2 & c_2^2 & \cdots & c_2 c_n \\ \vdots & \vdots & \ddots & \vdots \\ c_1 c_n & c_2 c_n & \cdots & c_n^2 \end{bmatrix}$$

35. (a) The quadratic form $Q = \frac{4}{3} x^2 + \frac{4}{3} y^2 + \frac{4}{3} z^2 + \frac{4}{3} xy + \frac{4}{3} xz + \frac{4}{3} yz$ can be expressed in matrix no-

tation as $Q = \mathbf{x}^T A \mathbf{x}$ where $A = \begin{bmatrix} \frac{4}{3} & \frac{2}{3} & \frac{2}{3} \\ \frac{2}{3} & \frac{4}{3} & \frac{2}{3} \\ \frac{2}{3} & \frac{2}{3} & \frac{4}{3} \end{bmatrix}$. The matrix A has eigenvalues $\lambda = \frac{2}{3}$ and $\lambda = \frac{8}{3}$.

The vectors $\mathbf{v}_1 = \begin{bmatrix} 0 \\ -1 \\ 1 \end{bmatrix}$ and $\mathbf{v}_2 = \begin{bmatrix} 1 \\ -1 \\ 0 \end{bmatrix}$ form a basis for the eigenspace corresponding to $\lambda = \frac{2}{3}$,

and $\mathbf{v}_3 = \begin{bmatrix} 1 \\ 1 \\ 1 \end{bmatrix}$ forms a basis for the eigenspace corresponding to $\lambda = \frac{8}{3}$. Application of the

Gram-Schmidt process to $\{\mathbf{v}_1, \mathbf{v}_2, \mathbf{v}_3\}$ produces orthonormal eigenvectors $\{\mathbf{p}_1, \mathbf{p}_2, \mathbf{p}_3\}$, and the matrix

$$P = [\mathbf{p}_1 \quad \mathbf{p}_2 \quad \mathbf{p}_3] = \begin{bmatrix} 0 & \frac{2}{\sqrt{6}} & \frac{1}{\sqrt{3}} \\ -\frac{1}{\sqrt{2}} & -\frac{1}{\sqrt{6}} & \frac{1}{\sqrt{3}} \\ \frac{1}{\sqrt{2}} & -\frac{1}{\sqrt{6}} & \frac{1}{\sqrt{3}} \end{bmatrix}$$

orthogonally diagonalizes A. Thus the change of variable $\mathbf{x} = P\mathbf{x}'$ converts Q into a quadratic form in the variables $\mathbf{x}' = (x', y', z')$ without cross product terms:

$$Q = \frac{2}{3} x'^2 + \frac{2}{3} y'^2 + \frac{8}{3} z'^2$$

From this we conclude that the equation $Q = 1$ corresponds to an ellipsoid with axis lengths $2\sqrt{\frac{3}{2}} = \sqrt{6}$ in the x' and y' directions, and $2\sqrt{\frac{3}{8}} = \sqrt{\frac{3}{2}}$ in the z' direction.

(b) The matrix A must be positive definite.

DISCUSSION AND DISCOVERY

D1. (a) False. For example the matrix $A = \begin{bmatrix} 1 & 2 \\ 2 & 1 \end{bmatrix}$ has eigenvalues -1 and 3; thus A is indefinite.

(b) False. The term $4x_1 x_2 x_3$ is not quadratic in the variables x_1, x_2, x_3.

(c) True. When expanded, each of the terms of the resulting expression is quadratic (of degree 2) in the variables.

(d) True. The eigenvalues of a positive definite matrix A are strictly positive; in particular, 0 is not an eigenvalue of A and so A is invertible.

(e) False. For example the matrix $A = \begin{bmatrix} 1 & 0 \\ 0 & 0 \end{bmatrix}$ is positive semi-definite.

(f) True. If the eigenvalues of A are positive, then the eigenvalues of $-A$ are negative.

D2. (a) True. When written in matrix form, we have $\mathbf{x} \cdot \mathbf{x} = \mathbf{x}^T A \mathbf{x}$ where $A = I$.

(b) True. If A has positive eigenvalues, then so does A^{-1}.

(c) True. See Theorem 8.4.3(a).

(d) True. Both of the principal submatrices of A will have a positive determinant.

(e) False. If $A = \begin{bmatrix} 1 & 1 \\ -1 & 1 \end{bmatrix}$, then $\mathbf{x}^T A \mathbf{x} = x^2 + y^2$. On the other hand, the statement is true if A is assumed to be symmetric.

(f) False. If $c > 0$ the graph is an ellipse. If $c < 0$ the graph is empty.

D3. The eigenvalues of A must be positive and equal to each other; in other words, A must have a positive eigenvalue of multiplicity 2.

WORKING WITH PROOFS

P1. Rotating the coordinate axes through an angle θ corresponds to the change of variable $\mathbf{x} = P\mathbf{x}'$ where $P = \begin{bmatrix} \cos\theta & -\sin\theta \\ \sin\theta & \cos\theta \end{bmatrix}$, i.e., $x = x'\cos\theta - y'\sin\theta$ and $y = x'\sin\theta + y'\cos\theta$. Substituting these expressions into the quadratic form $ax^2 + 2bxy + cy^2$ leads to $Ax'^2 + Bx'y' + Cy'^2$, where the coefficient of the cross product term is

$$B = -2a\cos\theta\sin\theta + 2b(\cos^2\theta - \sin^2\theta) + 2c\cos\theta\sin\theta = (-a+c)\sin 2\theta + 2b\cos 2\theta$$

Thus the resulting quadratic form in the variables x' and y' has no cross product term if and only if $(-a+c)\sin 2\theta + 2b\cos 2\theta = 0$, or (equivalently) $\cot 2\theta = \frac{a-c}{2b}$.

P2. From the Principal Axis Theorem (8.4.1), there is an orthogonal change of variable $x = Py$ for which $\mathbf{x}^T A \mathbf{x} = \mathbf{y}^T D \mathbf{y} = \lambda_1 y_1^2 + \lambda_2 y_2^2$, where λ_1 and λ_2 are the eigenvalues of A. Since λ_1 and λ_2 are nonnegative, it follows that $\mathbf{x}^T A \mathbf{x} \geq 0$ for every vector $\mathbf{x}$ in R^n.

EXERCISE SET 8.5

1. (a) The first partial derivatives of f are $f_x(x,y) = 4y - 4x^3$ and $f_y(x,y) = 4x - 4y^3$. To find the critical points we set f_x and f_y equal to zero. This yields the equations $y = x^3$ and $x = y^3$. From this we conclude that $y = y^9$ and so $y = 0$ or $y = \pm 1$. Since $x = y^3$ the corresponding values of x are $x = 0$ and $x = \pm 1$ respectively. Thus there are three critical points: $(0,0)$, $(1,1)$, and $(-1,-1)$.

(b) The Hessian matrix is $H(x,y) = \begin{bmatrix} f_{xx}(x,y) & f_{xy}(x,y) \\ f_{yx}(x,y) & f_{yy}(x,y) \end{bmatrix} = \begin{bmatrix} -12x^2 & 4 \\ 4 & -12y^2 \end{bmatrix}$. Evaluating this matrix at the critical points of f yields

$$H(0,0) = \begin{bmatrix} 0 & 4 \\ 4 & 0 \end{bmatrix}, \quad H(1,1) = \begin{bmatrix} -12 & 4 \\ 4 & -12 \end{bmatrix}, \quad H(-1,-1) = \begin{bmatrix} -12 & 4 \\ 4 & -12 \end{bmatrix}$$

The eigenvalues of $\begin{bmatrix} 0 & 4 \\ 4 & 0 \end{bmatrix}$ are $\lambda = \pm 4$; thus the matrix $H(0,0)$ is indefinite and so f has a saddle point at $(0,0)$. The eigenvalues of $\begin{bmatrix} -12 & 4 \\ 4 & -12 \end{bmatrix}$ are $\lambda = -8$ and $\lambda = -16$; thus the matrix

$H(1, 1) = H(-1, -1)$ is negative definite and so f has a relative maximum at $(1, 1)$ and at $(-1, -1)$.

3. The first partial derivatives of f are $f_x(x, y) = 3x^2 - 3y$ and $f_y(x, y) = -3x - 3y^2$. To find the critical points we set f_x and f_y equal to zero. This yields the equations $y = x^2$ and $x = -y^2$. From this we conclude that $y = y^4$ and so $y = 0$ or $y = 1$. The corresponding values of x are $x = 0$ and $x = -1$ respectively. Thus there are two critical points: $(0, 0)$ and $(-1, 1)$.

The Hessian matrix is $H(x, y) = \begin{bmatrix} f_{xx}(x, y) & f_{xy}(x, y) \\ f_{yx}(x, y) & f_{yy}(x, y) \end{bmatrix} = \begin{bmatrix} 6x & -3 \\ -3 & -6y \end{bmatrix}$. The eigenvalues of $H(0, 0) = \begin{bmatrix} 0 & -3 \\ -3 & 0 \end{bmatrix}$ are $\lambda = \pm 3$; this matrix is indefinite and so f has a saddle point at $(0, 0)$. The eigenvalues of $H(-1, 1) = \begin{bmatrix} -6 & -3 \\ -3 & -6 \end{bmatrix}$ are $\lambda = -3$ and $\lambda = -9$; this matrix is negative definite and so f has a relative maximum at $(-1, 1)$.

5. The first partial derivatives of f are $f_x(x, y) = 2x - 2xy$ and $f_y(x, y) = 4y - x^2$. To find the critical points we set f_x and f_y equal to zero. This yields the equations $x = xy$ and $y = \frac{1}{4}x^2$. From this we conclude that $x = \frac{1}{4}x^3$ and so $x = 0$, or $x = \pm 2$. The corresponding values of y are $y = 0$, or $y = 1$ respectively. Thus there are three critical points: $(0, 0)$, $(2, 1)$, and $(-2, 1)$.

The Hessian matrix is $H(x, y) = \begin{bmatrix} f_{xx}(x, y) & f_{xy}(x, y) \\ f_{yx}(x, y) & f_{yy}(x, y) \end{bmatrix} = \begin{bmatrix} 2 - 2y & -2x \\ -2x & 4 \end{bmatrix}$. The eigenvalues of the matrix $H(0, 0) = \begin{bmatrix} 2 & 0 \\ 0 & 4 \end{bmatrix}$ are $\lambda = 2$ and $\lambda = 4$; this matrix is positive definite and so f has a relative minimum at $(0, 0)$. The eigenvalues of $H(2, 1) = \begin{bmatrix} 0 & -4 \\ -4 & 4 \end{bmatrix}$ are $\lambda = 2 \pm 2\sqrt{5}$. One of these is positive and one is negative; thus this matrix is indefinite and f has a saddle point at $(2, 1)$. Similarly, the eigenvalues of $H(-2, 1) = \begin{bmatrix} 0 & 4 \\ 4 & 4 \end{bmatrix}$ are $\lambda = 2 \pm 2\sqrt{5}$; thus f has a saddle point at $(-2, 1)$.

7. The quadratic form $z = 5x^2 - y^2$ can be written in matrix notation as $z = \mathbf{x}^T A \mathbf{x}$ where $A = \begin{bmatrix} 5 & 0 \\ 0 & -1 \end{bmatrix}$. The eigenvalues of A are $\lambda_1 = 5$ and $\lambda_2 = -1$, with corresponding eigenvectors $\mathbf{v}_1 = \begin{bmatrix} 1 \\ 0 \end{bmatrix}$ and $\mathbf{v}_2 = \begin{bmatrix} 0 \\ 1 \end{bmatrix}$. Thus the constrained maximum is $z = 5$ occurring at $(x, y) = (\pm 1, 0)$, and the constrained minimum is $z = -1$ occurring at $(x, y) = (0, \pm 1)$.

9. The quadratic form $z = 3x^2 + 7y^2$ can be written in matrix notation as $z = \mathbf{x}^T A \mathbf{x}$ where $A = \begin{bmatrix} 3 & 0 \\ 0 & 7 \end{bmatrix}$. The eigenvalues of A are $\lambda_1 = 7$ and $\lambda_2 = 3$, with corresponding unit eigenvectors $\mathbf{v}_1 = \begin{bmatrix} 0 \\ 1 \end{bmatrix}$ and $\mathbf{v}_2 = \begin{bmatrix} 1 \\ 0 \end{bmatrix}$. Thus the constrained maximum is $z = 7$ occurring at $(x, y) = (0, \pm 1)$, and the constrained minimum is $z = 3$ occurring at $(x, y) = (\pm 1, 0)$.

11. The quadratic form $w = 9x^2 + 4y^2 + 3z^2$ can be expressed as $w = \mathbf{x}^T A \mathbf{x}$ where $A = \begin{bmatrix} 9 & 0 & 0 \\ 0 & 4 & 0 \\ 0 & 0 & 3 \end{bmatrix}$. The eigenvalues of A are $\lambda_1 = 9$, $\lambda_2 = 4$, $\lambda_3 = 3$, with corresponding eigenvectors $\mathbf{v}_1 = \begin{bmatrix} 1 \\ 0 \\ 0 \end{bmatrix}$, $\mathbf{v}_2 = \begin{bmatrix} 0 \\ 1 \\ 0 \end{bmatrix}$, $\mathbf{v}_3 = \begin{bmatrix} 0 \\ 0 \\ 1 \end{bmatrix}$. Thus the constrained maximum is $w = 9$ occurring at $(x, y, z) = (\pm 1, 0, 0)$, and the constrained minimum is $w = 3$ occurring at $(x, y, z) = (0, 0, \pm 1)$.

13. The constraint equation $4x^2 + 8y^2 = 16$ can be rewritten as $(\frac{x}{2})^2 + (\frac{y}{\sqrt{2}})^2 = 1$. Thus, with the change of variable $(x, y) = (2x', \sqrt{2}y')$, the problem is to find the extreme values of $z = xy = 2\sqrt{2}x'y'$ subject to $x'^2 + y'^2 = 1$. Note that $z = xy = 2\sqrt{2}x'y'$ can be expressed as $z = \mathbf{x}'^T A\mathbf{x}'$ where $A = \begin{bmatrix} 0 & \sqrt{2} \\ \sqrt{2} & 0 \end{bmatrix}$.
The eigenvalues of A are $\lambda_1 = \sqrt{2}$ and $\lambda_2 = -\sqrt{2}$, with corresponding (normalized) eigenvectors $\mathbf{v}_1 = \begin{bmatrix} \frac{1}{\sqrt{2}} \\ \frac{1}{\sqrt{2}} \end{bmatrix}$ and $\mathbf{v}_2 = \begin{bmatrix} -\frac{1}{\sqrt{2}} \\ \frac{1}{\sqrt{2}} \end{bmatrix}$. Thus the constrained maximum is $z = \sqrt{2}$ occurring at $(x', y') = (\frac{1}{\sqrt{2}}, \frac{1}{\sqrt{2}})$ or $(x, y) = (\sqrt{2}, 1)$. Similarly, the constrained minimum is $z = -\sqrt{2}$ occurring at $(x', y') = (-\frac{1}{\sqrt{2}}, \frac{1}{\sqrt{2}})$ or $(x, y) = (-\sqrt{2}, 1)$.

15. The level curve corresponding to the constrained maximum is the hyperbola $5x^2 - y^2 = 5$; it touches the unit circle at $(x, y) = (\pm 1, 0)$. The level curve corresponding to the constrained minimum is the hyperbola $5x^2 - y^2 = -1$; it touches the unit circle at $(x, y) = (0, \pm 1)$.

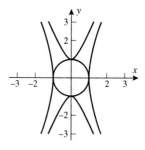

17. The area of the inscribed rectangle is $z = 4xy$, where (x, y) is the corner point that lies in the first quadrant. Our problem is to find the maximum value of $z = 4xy$ subject to the constraints $x \geq 0$, $y \geq 0$, $x^2 + 25y^2 = 25$. The constraint equation can be rewritten as $x'^2 + y'^2 = 1$ where $x = 5x'$ and $y = y'$. In terms of the variables x' and y', our problem is to find the maximum value of $z = 20x'y'$ subject to $x'^2 + y'^2 = 1$, $x' \geq 0$, $y' \geq 0$. Note that $z = \mathbf{x}'^T A\mathbf{x}'$ where $A = \begin{bmatrix} 0 & 10 \\ 10 & 0 \end{bmatrix}$. The largest eigenvalue of A is $\lambda = 10$ with corresponding (normalized) eigenvector $\begin{bmatrix} \frac{1}{\sqrt{2}} \\ \frac{1}{\sqrt{2}} \end{bmatrix}$. Thus the maximum area is $z = 10$, and this occurs when $(x', y') = (\frac{1}{\sqrt{2}}, \frac{1}{\sqrt{2}})$ or $(x, y) = (\frac{5}{\sqrt{2}}, \frac{1}{\sqrt{2}})$.

DISCUSSION AND DISCOVERY

D1. (a) We have $f_x(x, y) = 4x^3$ and $f_y(x, y) = 4y^3$; thus f has a critical point at $(0, 0)$. Similarly, $g_x(x, y) = 4x^3$ and $g_y(x, y) = -4y^3$, and so g has a critical point at $(0, 0)$. The Hessian matrices for f and g are $H_f(x, y) = \begin{bmatrix} 12x^2 & 0 \\ 0 & 12y^2 \end{bmatrix}$ and $H_g(x, y) = \begin{bmatrix} 12x^2 & 0 \\ 0 & -12y^2 \end{bmatrix}$ respectively. Since $H_f(0, 0) = H_g(0, 0) = \begin{bmatrix} 0 & 0 \\ 0 & 0 \end{bmatrix}$, the second derivative test is inconclusive in both cases.

(b) It is clear that f has a relative minimum at $(0, 0)$ since $f(0, 0) = 0$ and $f(x, y) = x^4 + y^4$ is strictly positive at all other points (x, y). In contrast, we have $g(0, 0) = 0$, $g(x, 0) = x^4 > 0$ for $x \neq 0$, and $g(0, y) = -y^4 < 0$ for $y \neq 0$. Thus g has a saddle point at $(0, 0)$.

D2. The eigenvalues of $H = \begin{bmatrix} 2 & 4 \\ 4 & 2 \end{bmatrix}$ are $\lambda = 6$ and $\lambda = -2$. Thus H is indefinite and so the critical points of f (if any) are saddle points. Starting from $f_{xx}(x, y) = f_{yy}(x, y) = 2$ and $f_{yx}(x, y) = f_{xy}(xy) = 4$ it follows, using partial integration, that the quadratic form f is $f(x, y) = x^2 + 4xy + y^2$. This function has one critical point (a saddle) which is located at the origin.

D3. If $\mathbf{x}$ is a unit eigenvector corresponding to λ, then $q(\mathbf{x}) = \mathbf{x}^T A\mathbf{x} = \mathbf{x}^T (\lambda \mathbf{x}) = \lambda(\mathbf{x}^T \mathbf{x}) = \lambda(1) = \lambda$.

WORKING WITH PROOFS

P1. First note that, as in D3, we have $u_m^T A u_m = m$ and $u_M^T A u_M = M$. On the other hand, since u_m and u_M are orthogonal, we have $u_m^T A u_M = u_m^T (M u_M) = M(u_m^T u_M) = M(0) = 0$ and $u_M^T A u_m = 0$. It follows that if $x_c = \sqrt{\frac{M-c}{M-m}} u_m + \sqrt{\frac{c-m}{M-m}} u_M$, then

$$x_c^T A x_c = \left(\frac{M-c}{M-m}\right) u_m^T A u_m + 0 + 0 + \left(\frac{c-m}{M-m}\right) u_M^T A u_M = \left(\frac{M-c}{M-m}\right) m + \left(\frac{c-m}{M-m}\right) M = c$$

EXERCISE SET 8.6

1. The characteristic polynomial of $A^T A = \begin{bmatrix} 1 \\ 2 \\ 0 \end{bmatrix} \begin{bmatrix} 1 & 2 & 0 \end{bmatrix} = \begin{bmatrix} 1 & 2 & 0 \\ 2 & 4 & 0 \\ 0 & 0 & 0 \end{bmatrix}$ is $\lambda^2(\lambda - 5)$; thus the eigenvalues of $A^T A$ are $\lambda_1 = 5$ and $\lambda_2 = 0$, and $\sigma_1 = \sqrt{5}$ is a singular value of A.

3. The eigenvalues of $A^T A = \begin{bmatrix} 1 & 2 \\ -2 & 1 \end{bmatrix} \begin{bmatrix} 1 & -2 \\ 2 & 1 \end{bmatrix} = \begin{bmatrix} 5 & 0 \\ 0 & 5 \end{bmatrix}$ are $\lambda_1 = 5$ and $\lambda_2 = 5$ (i.e., $\lambda = 5$ is an eigenvalue of multiplicity 2); thus the singular values of A are $\sigma_1 = \sqrt{5}$ and $\sigma_2 = \sqrt{5}$.

5. The only eigenvalue of $A^T A = \begin{bmatrix} 1 & 1 \\ -1 & 1 \end{bmatrix} \begin{bmatrix} 1 & -1 \\ 1 & 1 \end{bmatrix} = \begin{bmatrix} 2 & 0 \\ 0 & 2 \end{bmatrix}$ is $\lambda = 2$ (multiplicity 2), and the vectors $v_1 = \begin{bmatrix} 1 \\ 0 \end{bmatrix}$ and $v_2 = \begin{bmatrix} 0 \\ 1 \end{bmatrix}$ form an orthonormal basis for the eigenspace (which is all of R^2). The singular values of A are $\sigma_1 = \sqrt{2}$ and $\sigma_2 = \sqrt{2}$. We have $u_1 = \frac{1}{\sigma_1} A v_1 = \frac{1}{\sqrt{2}} \begin{bmatrix} 1 & -1 \\ 1 & 1 \end{bmatrix} \begin{bmatrix} 1 \\ 0 \end{bmatrix} = \begin{bmatrix} \frac{1}{\sqrt{2}} \\ \frac{1}{\sqrt{2}} \end{bmatrix}$, and $u_2 = \frac{1}{\sigma_2} A v_2 = \frac{1}{\sqrt{2}} \begin{bmatrix} 1 & -1 \\ 1 & 1 \end{bmatrix} \begin{bmatrix} 0 \\ 1 \end{bmatrix} = \begin{bmatrix} -\frac{1}{\sqrt{2}} \\ \frac{1}{\sqrt{2}} \end{bmatrix}$. This results in the following singular value decomposition of A:

$$A = \begin{bmatrix} 1 & -1 \\ 1 & 1 \end{bmatrix} = \begin{bmatrix} \frac{1}{\sqrt{2}} & -\frac{1}{\sqrt{2}} \\ \frac{1}{\sqrt{2}} & \frac{1}{\sqrt{2}} \end{bmatrix} \begin{bmatrix} \sqrt{2} & 0 \\ 0 & \sqrt{2} \end{bmatrix} \begin{bmatrix} 1 & 0 \\ 0 & 1 \end{bmatrix} = U\Sigma V^T$$

7. The eigenvalues of $A^T A = \begin{bmatrix} 4 & 0 \\ 6 & 4 \end{bmatrix} \begin{bmatrix} 4 & 6 \\ 0 & 4 \end{bmatrix} = \begin{bmatrix} 16 & 24 \\ 24 & 52 \end{bmatrix}$ are $\lambda_1 = 64$ and $\lambda_2 = 4$, with corresponding unit eigenvectors $v_1 = \begin{bmatrix} \frac{1}{\sqrt{5}} \\ \frac{2}{\sqrt{5}} \end{bmatrix}$ and $v_2 = \begin{bmatrix} -\frac{2}{\sqrt{5}} \\ \frac{1}{\sqrt{5}} \end{bmatrix}$ respectively. The singular values of A are $\sigma_1 = 8$ and $\sigma_2 = 2$. We have $u_1 = \frac{1}{\sigma_1} A v_1 = \frac{1}{8} \begin{bmatrix} 4 & 6 \\ 0 & 4 \end{bmatrix} \begin{bmatrix} \frac{1}{\sqrt{5}} \\ \frac{2}{\sqrt{5}} \end{bmatrix} = \begin{bmatrix} \frac{2}{\sqrt{5}} \\ \frac{1}{\sqrt{5}} \end{bmatrix}$, and $u_2 = \frac{1}{\sigma_2} A v_2 = \frac{1}{2} \begin{bmatrix} 4 & 6 \\ 0 & 4 \end{bmatrix} \begin{bmatrix} -\frac{2}{\sqrt{5}} \\ \frac{1}{\sqrt{5}} \end{bmatrix} = \begin{bmatrix} -\frac{1}{\sqrt{5}} \\ \frac{2}{\sqrt{5}} \end{bmatrix}$. This results in the following singular value decomposition:

$$A = \begin{bmatrix} 4 & 6 \\ 0 & 4 \end{bmatrix} = \begin{bmatrix} \frac{2}{\sqrt{5}} & -\frac{1}{\sqrt{5}} \\ \frac{1}{\sqrt{5}} & \frac{2}{\sqrt{5}} \end{bmatrix} \begin{bmatrix} 8 & 0 \\ 0 & 2 \end{bmatrix} \begin{bmatrix} \frac{1}{\sqrt{5}} & \frac{2}{\sqrt{5}} \\ -\frac{2}{\sqrt{5}} & \frac{1}{\sqrt{5}} \end{bmatrix} = U\Sigma V^T$$

9. The eigenvalues of $A^T A = \begin{bmatrix} -2 & -1 & 2 \\ 2 & 1 & -2 \end{bmatrix} \begin{bmatrix} -2 & 2 \\ -1 & 1 \\ 2 & -2 \end{bmatrix} = \begin{bmatrix} 9 & -9 \\ -9 & 9 \end{bmatrix}$ are $\lambda_1 = 18$ and $\lambda_2 = 0$, with corresponding unit eigenvectors $v_1 = \begin{bmatrix} -\frac{1}{\sqrt{2}} \\ \frac{1}{\sqrt{2}} \end{bmatrix}$ and $v_2 = \begin{bmatrix} \frac{1}{\sqrt{2}} \\ \frac{1}{\sqrt{2}} \end{bmatrix}$ respectively. The only singular value of

A is $\sigma_1 = \sqrt{18} = 3\sqrt{2}$, and we have $\mathbf{u}_1 = \frac{1}{\sigma_1}A\mathbf{v}_1 = \frac{1}{3\sqrt{2}}\begin{bmatrix} -2 & 2 \\ -1 & 1 \\ 2 & -2 \end{bmatrix}\begin{bmatrix} -\frac{1}{\sqrt{2}} \\ \frac{1}{\sqrt{2}} \end{bmatrix} = \begin{bmatrix} \frac{2}{3} \\ \frac{1}{3} \\ -\frac{2}{3} \end{bmatrix}$. We must choose

the vectors $\mathbf{u}_2$ and $\mathbf{u}_3$ so that $\{\mathbf{u}_1, \mathbf{u}_2, \mathbf{u}_3\}$ is an orthonormal basis for R^3, e.g., $\mathbf{u}_2 = \begin{bmatrix} \frac{1}{3} \\ \frac{2}{3} \\ \frac{2}{3} \end{bmatrix}$ and

$\mathbf{u}_3 = \begin{bmatrix} \frac{2}{3} \\ -\frac{2}{3} \\ \frac{1}{3} \end{bmatrix}$. This results in the following singular value decomposition:

$$A = \begin{bmatrix} -2 & 2 \\ -1 & 1 \\ 2 & -2 \end{bmatrix} = \begin{bmatrix} \frac{2}{3} & \frac{1}{3} & \frac{2}{3} \\ \frac{1}{3} & \frac{2}{3} & -\frac{2}{3} \\ -\frac{2}{3} & \frac{2}{3} & \frac{1}{3} \end{bmatrix}\begin{bmatrix} 3\sqrt{2} & 0 \\ 0 & 0 \\ 0 & 0 \end{bmatrix}\begin{bmatrix} -\frac{1}{\sqrt{2}} & \frac{1}{\sqrt{2}} \\ \frac{1}{\sqrt{2}} & \frac{1}{\sqrt{2}} \end{bmatrix} = U\Sigma V^T$$

Note. The singular value decomposition is not unique. It depends on the choice of the (extended) orthonormal basis for R^3. This is just one possibility.

11. The eigenvalues of $A^T A = \begin{bmatrix} 1 & 1 & -1 \\ 0 & 1 & 1 \end{bmatrix}\begin{bmatrix} 1 & 0 \\ 1 & 1 \\ -1 & 1 \end{bmatrix} = \begin{bmatrix} 3 & 0 \\ 0 & 2 \end{bmatrix}$ are $\lambda_1 = 3$ and $\lambda_2 = 2$, with corresponding

unit eigenvectors $\mathbf{v}_1 = \begin{bmatrix} 1 \\ 0 \end{bmatrix}$ and $\mathbf{v}_2 = \begin{bmatrix} 0 \\ 1 \end{bmatrix}$ respectively. The singular values of A are $\sigma_1 = \sqrt{3}$ and

$\sigma_2 = \sqrt{2}$. We have $\mathbf{u}_1 = \frac{1}{\sigma_1}A\mathbf{v}_1 = \frac{1}{\sqrt{3}}\begin{bmatrix} 1 & 0 \\ 1 & 1 \\ -1 & 1 \end{bmatrix}\begin{bmatrix} 1 \\ 0 \end{bmatrix} = \begin{bmatrix} \frac{1}{\sqrt{3}} \\ \frac{1}{\sqrt{3}} \\ -\frac{1}{\sqrt{3}} \end{bmatrix}$ and $\mathbf{u}_2 = \frac{1}{\sqrt{2}}\begin{bmatrix} 1 & 0 \\ 1 & 1 \\ -1 & 1 \end{bmatrix}\begin{bmatrix} 0 \\ 1 \end{bmatrix} = \begin{bmatrix} 0 \\ \frac{1}{\sqrt{2}} \\ \frac{1}{\sqrt{2}} \end{bmatrix}$. We

choose $\mathbf{u}_3 = \begin{bmatrix} \frac{2}{\sqrt{6}} \\ -\frac{1}{\sqrt{6}} \\ \frac{1}{\sqrt{6}} \end{bmatrix}$ so that $\{\mathbf{u}_1, \mathbf{u}_2, \mathbf{u}_3\}$ is an orthonormal basis for R^3. This results in the following

singular value decomposition:

$$A = \begin{bmatrix} 1 & 0 \\ 1 & 1 \\ -1 & 1 \end{bmatrix} = \begin{bmatrix} \frac{1}{\sqrt{3}} & 0 & \frac{2}{\sqrt{6}} \\ \frac{1}{\sqrt{3}} & \frac{1}{\sqrt{2}} & -\frac{1}{\sqrt{6}} \\ -\frac{1}{\sqrt{3}} & \frac{1}{\sqrt{2}} & \frac{1}{\sqrt{6}} \end{bmatrix}\begin{bmatrix} \sqrt{3} & 0 \\ 0 & \sqrt{2} \\ 0 & 0 \end{bmatrix}\begin{bmatrix} 1 & 0 \\ 0 & 1 \end{bmatrix} = U\Sigma V^T$$

13. Using the singular value decomposition $A = \begin{bmatrix} 4 & 6 \\ 0 & 4 \end{bmatrix} = \begin{bmatrix} \frac{2}{\sqrt{5}} & -\frac{1}{\sqrt{5}} \\ \frac{1}{\sqrt{5}} & \frac{2}{\sqrt{5}} \end{bmatrix}\begin{bmatrix} 8 & 0 \\ 0 & 2 \end{bmatrix}\begin{bmatrix} \frac{1}{\sqrt{5}} & \frac{2}{\sqrt{5}} \\ -\frac{2}{\sqrt{5}} & \frac{1}{\sqrt{5}} \end{bmatrix} = U\Sigma V^T$ found

in Exercise 7, we have the following polar decomposition of A:

$$A = (U\Sigma U^T)(UV^T) = \left(\begin{bmatrix} \frac{2}{\sqrt{5}} & -\frac{1}{\sqrt{5}} \\ \frac{1}{\sqrt{5}} & \frac{2}{\sqrt{5}} \end{bmatrix}\begin{bmatrix} 8 & 0 \\ 0 & 2 \end{bmatrix}\begin{bmatrix} \frac{2}{\sqrt{5}} & \frac{1}{\sqrt{5}} \\ -\frac{1}{\sqrt{5}} & \frac{2}{\sqrt{5}} \end{bmatrix}\right)\left(\begin{bmatrix} \frac{2}{\sqrt{5}} & -\frac{1}{\sqrt{5}} \\ \frac{1}{\sqrt{5}} & \frac{2}{\sqrt{5}} \end{bmatrix}\begin{bmatrix} \frac{1}{\sqrt{5}} & \frac{2}{\sqrt{5}} \\ -\frac{2}{\sqrt{5}} & \frac{1}{\sqrt{5}} \end{bmatrix}\right)$$

$$= \begin{bmatrix} \frac{34}{5} & \frac{12}{5} \\ \frac{12}{5} & \frac{16}{5} \end{bmatrix}\begin{bmatrix} \frac{4}{5} & \frac{3}{5} \\ -\frac{3}{5} & \frac{4}{5} \end{bmatrix} = PQ$$

15. In Exercise 11 we found the following singular value decomposition:

$$A = \begin{bmatrix} 1 & 0 \\ 1 & 1 \\ -1 & 1 \end{bmatrix} = \begin{bmatrix} \frac{1}{\sqrt{3}} & 0 & \frac{2}{\sqrt{6}} \\ \frac{1}{\sqrt{3}} & \frac{1}{\sqrt{2}} & -\frac{1}{\sqrt{6}} \\ -\frac{1}{\sqrt{3}} & \frac{1}{\sqrt{2}} & \frac{1}{\sqrt{6}} \end{bmatrix}\begin{bmatrix} \sqrt{3} & 0 \\ 0 & \sqrt{2} \\ 0 & 0 \end{bmatrix}\begin{bmatrix} 1 & 0 \\ 0 & 1 \end{bmatrix} = U\Sigma V^T$$

Since A has rank 2, the corresponding reduced singular value decomposition is

$$A = \begin{bmatrix} 1 & 0 \\ 1 & 1 \\ -1 & 1 \end{bmatrix} = \begin{bmatrix} \frac{1}{\sqrt{3}} & 0 \\ \frac{1}{\sqrt{3}} & \frac{1}{\sqrt{2}} \\ -\frac{1}{\sqrt{3}} & \frac{1}{\sqrt{2}} \end{bmatrix} \begin{bmatrix} \sqrt{3} & 0 \\ 0 & \sqrt{2} \end{bmatrix} \begin{bmatrix} 1 & 0 \\ 0 & 1 \end{bmatrix} = U_1 \Sigma_1 V_1^T$$

and the reduced singular value expansion is

$$A = \begin{bmatrix} 1 & 0 \\ 1 & 1 \\ -1 & 1 \end{bmatrix} = \sqrt{3} \begin{bmatrix} \frac{1}{\sqrt{3}} \\ \frac{1}{\sqrt{3}} \\ -\frac{1}{\sqrt{3}} \end{bmatrix} \begin{bmatrix} 1 & 0 \end{bmatrix} + \sqrt{2} \begin{bmatrix} 0 \\ \frac{1}{\sqrt{2}} \\ \frac{1}{\sqrt{2}} \end{bmatrix} \begin{bmatrix} 0 & 1 \end{bmatrix} = \sigma_1 \mathbf{u}_1 \mathbf{v}_1^T + \sigma_2 \mathbf{u}_2 \mathbf{v}_2^T$$

17. The characteristic polynomial of A is $(\lambda + 1)(\lambda - 3)^2$; thus $\lambda = -1$ is an eigenvalue of multiplicity 1 and $\lambda = 3$ is an eigenvalue of multiplicity 2. The vector $\mathbf{v}_1 = \begin{bmatrix} -1 \\ 1 \\ 0 \end{bmatrix}$ forms a basis for the eigenspace corresponding to $\lambda = -1$, and the vectors $\mathbf{v}_2 = \begin{bmatrix} 0 \\ 0 \\ 1 \end{bmatrix}$ and $\mathbf{v}_3 = \begin{bmatrix} 1 \\ 1 \\ 0 \end{bmatrix}$ form an (orthogonal) basis for the eigenspace corresponding to $\lambda = 3$. Thus the matrix $P = \begin{bmatrix} \frac{\mathbf{v}_1}{\|\mathbf{v}_1\|} & \frac{\mathbf{v}_2}{\|\mathbf{v}_2\|} & \frac{\mathbf{v}_3}{\|\mathbf{v}_3\|} \end{bmatrix}$ orthogonally diagonalizes A and the eigenvalue decomposition of A is

$$A = \begin{bmatrix} 1 & 2 & 0 \\ 2 & 1 & 0 \\ 0 & 0 & 3 \end{bmatrix} = \begin{bmatrix} -\frac{1}{\sqrt{2}} & 0 & \frac{1}{\sqrt{2}} \\ \frac{1}{\sqrt{2}} & 0 & \frac{1}{\sqrt{2}} \\ 0 & 1 & 0 \end{bmatrix} \begin{bmatrix} -1 & 0 & 0 \\ 0 & 3 & 0 \\ 0 & 0 & 3 \end{bmatrix} \begin{bmatrix} -\frac{1}{\sqrt{2}} & \frac{1}{\sqrt{2}} & 0 \\ 0 & 0 & 1 \\ \frac{1}{\sqrt{2}} & \frac{1}{\sqrt{2}} & 0 \end{bmatrix} = PDP^T$$

The corresponding singular value decomposition of A is obtained by shifting the negative sign from the diagonal factor to the second orthogonal factor:

$$A = \begin{bmatrix} 1 & 2 & 0 \\ 2 & 1 & 0 \\ 0 & 0 & 3 \end{bmatrix} = \begin{bmatrix} -\frac{1}{\sqrt{2}} & 0 & \frac{1}{\sqrt{2}} \\ \frac{1}{\sqrt{2}} & 0 & \frac{1}{\sqrt{2}} \\ 0 & 1 & 0 \end{bmatrix} \begin{bmatrix} 1 & 0 & 0 \\ 0 & 3 & 0 \\ 0 & 0 & 3 \end{bmatrix} \begin{bmatrix} \frac{1}{\sqrt{2}} & -\frac{1}{\sqrt{2}} & 0 \\ 0 & 0 & 1 \\ \frac{1}{\sqrt{2}} & \frac{1}{\sqrt{2}} & 0 \end{bmatrix} = U\Sigma V^T$$

19. **(a)** We have $\text{rank}(A) = 2$; thus $\mathbf{u}_1 = \begin{bmatrix} \frac{1}{2} \\ \frac{1}{2} \\ \frac{1}{2} \\ \frac{1}{2} \end{bmatrix}$ and $\mathbf{u}_2 = \begin{bmatrix} \frac{1}{2} \\ -\frac{1}{2} \\ -\frac{1}{2} \\ \frac{1}{2} \end{bmatrix}$ form a basis for col(A), $\mathbf{u}_3 = \begin{bmatrix} \frac{1}{2} \\ -\frac{1}{2} \\ \frac{1}{2} \\ -\frac{1}{2} \end{bmatrix}$ and

$\mathbf{u}_4 = \begin{bmatrix} \frac{1}{2} \\ \frac{1}{2} \\ -\frac{1}{2} \\ -\frac{1}{2} \end{bmatrix}$ form a basis for col$(A)^\perp = $ null(A^T), $\mathbf{v}_1^T = \begin{bmatrix} \frac{2}{3} & -\frac{1}{3} & \frac{2}{3} \end{bmatrix}$ and $\mathbf{v}_2^T = \begin{bmatrix} \frac{2}{3} & \frac{2}{3} & -\frac{1}{3} \end{bmatrix}$

form a basis for row(A), and $\mathbf{v}_3 = \begin{bmatrix} -\frac{1}{3} \\ \frac{2}{3} \\ \frac{2}{3} \end{bmatrix}$ forms a basis for row$(A)^\perp = $ null(A).

(b) $A = \begin{bmatrix} 12 & 0 & 6 \\ 4 & -8 & 10 \\ 4 & -8 & 10 \\ 12 & 0 & 6 \end{bmatrix} = \begin{bmatrix} \frac{1}{2} & \frac{1}{2} \\ \frac{1}{2} & -\frac{1}{2} \\ \frac{1}{2} & -\frac{1}{2} \\ \frac{1}{2} & \frac{1}{2} \end{bmatrix} \begin{bmatrix} 24 & 0 \\ 0 & 12 \end{bmatrix} \begin{bmatrix} \frac{2}{3} & -\frac{1}{3} & \frac{2}{3} \\ \frac{2}{3} & \frac{2}{3} & -\frac{1}{3} \end{bmatrix} = U_1 \Sigma_1 V_1^T$

21. Since $A^T A$ is positive semidefinite and symmetric its eigenvalues are nonnegative, and its singular values are the same as its nonzero eigenvalues. On the other hand, the singular values of A are the square roots of the nonzero eigenvalues of $A^T A$. Thus the singular values of $A^T A$ are the squares of the singular values of A.

23. We have $Q = \begin{bmatrix} \frac{\sqrt{3}}{2} & \frac{1}{2} \\ -\frac{1}{2} & \frac{\sqrt{3}}{3} \end{bmatrix} = \begin{bmatrix} \cos\theta & -\sin\theta \\ \sin\theta & \cos\theta \end{bmatrix}$ where $\theta = 330°$; thus multiplication by Q corresponds to rotation about the origin through an angle of $330°$. The symmetric matrix P has eigenvalues $\lambda = 3$ and $\lambda = 1$, with corresponding unit eigenvectors $\mathbf{u}_1 = \begin{bmatrix} \frac{\sqrt{3}}{2} \\ \frac{1}{2} \end{bmatrix}$ and $\mathbf{u}_2 = \begin{bmatrix} -\frac{1}{2} \\ \frac{\sqrt{3}}{2} \end{bmatrix}$. Thus $V = [\mathbf{u}_1 \ \mathbf{u}_2]$ is a diagonalizing matrix for P:

$$V^T PV = \begin{bmatrix} \frac{\sqrt{3}}{2} & \frac{1}{2} \\ -\frac{1}{2} & \frac{\sqrt{3}}{2} \end{bmatrix} \begin{bmatrix} \sqrt{3} & 2 \\ 0 & \sqrt{3} \end{bmatrix} \begin{bmatrix} \frac{\sqrt{3}}{2} & -\frac{1}{2} \\ \frac{1}{2} & \frac{\sqrt{3}}{2} \end{bmatrix} = \begin{bmatrix} 3 & 0 \\ 0 & 1 \end{bmatrix}$$

From this we conclude that multiplication by P stretches R^2 by a factor of 3 in the direction of $\mathbf{u}_1$ and by a factor of 1 in the direction of $\mathbf{u}_2$.

DISCUSSION AND DISCOVERY

D1. (a) If $A = U\Sigma V^T$ is a singular value decomposition of an $m \times n$ matrix of rank k, then U has size $m \times m$, Σ has size $m \times n$ and V has size $n \times n$.

(b) If $A = U_1 \Sigma_1 V_1^T$ is a reduced singular value decomposition of an $m \times n$ matrix of rank k, then U has size $m \times k$, Σ_1 has size $k \times k$ and V has size $n \times k$.

D2. If A is an invertible matrix, then its eigenvalues are nonzero. Thus if $A = U\Sigma V^T$ is a singular value decomposition of A, then Σ is invertible and $A^{-1} = (V^T)^{-1}\Sigma^{-1}U^{-1} = V\Sigma^{-1}U^T$. Note also that the diagonal entries of Σ^{-1} are the reciprocals of the diagonal entries of Σ, which are the singular values of A^{-1}. Thus $A^{-1} = V\Sigma^{-1}U^T$ is a singular value decomposition of A^{-1}.

D3. If A and B are orthogonally similar matrices, then there is an orthogonal matrix P such that $B = PAP^T$. Thus, if $A = U\Sigma V^T$ is a singular value decomposition of A, then

$$B = PAP^T = P(U\Sigma V^T)P^T = (PU)\Sigma(PV)^T$$

is a singular value decomposition of B. It follows that B and A have the same singular values (the nonzero diagonal entries of Σ).

D4. If P is the matrix for the orthogonal projection of R^n onto a subspace W of dimension k, then $P^2 = P$ and the eigenvalues of P are $\lambda = 1$ (with multiplicity k) and $\lambda = 0$ (with multiplicity $n - k$). Thus the singular values of P are $\sigma_1 = 1, \sigma_2 = 1, \ldots, \sigma_k = 1$.

EXERCISE SET 8.7

1. We have $A^T A = [3 \ 4]\begin{bmatrix} 3 \\ 4 \end{bmatrix} = 25$; thus $A^+ = (A^T A)^{-1}A^T = \frac{1}{25}[3 \ 4] = [\frac{3}{25} \ \frac{4}{25}]$.

3. We have $A^T A = \begin{bmatrix} 7 & 0 & 5 \\ 1 & 0 & 5 \end{bmatrix}\begin{bmatrix} 7 & 1 \\ 0 & 0 \\ 5 & 5 \end{bmatrix} = \begin{bmatrix} 74 & 32 \\ 32 & 26 \end{bmatrix}$; thus the pseudoinverse of A is

$$A^+ = (A^T A)^{-1}A^T = \frac{1}{900}\begin{bmatrix} 26 & -32 \\ -32 & 74 \end{bmatrix}\begin{bmatrix} 7 & 0 & 5 \\ 1 & 0 & 5 \end{bmatrix} = \begin{bmatrix} \frac{1}{6} & 0 & -\frac{1}{30} \\ -\frac{1}{6} & 0 & \frac{7}{30} \end{bmatrix}$$

5. (a) $AA^+ A = \begin{bmatrix} 3 \\ 4 \end{bmatrix} \begin{bmatrix} \frac{3}{25} & \frac{4}{25} \end{bmatrix} \begin{bmatrix} 3 \\ 4 \end{bmatrix} = \begin{bmatrix} 3 \\ 4 \end{bmatrix} [1] = \begin{bmatrix} 3 \\ 4 \end{bmatrix} = A$

(b) $A^+ AA^+ = \begin{bmatrix} \frac{3}{25} & \frac{4}{25} \end{bmatrix} \begin{bmatrix} 3 \\ 4 \end{bmatrix} \begin{bmatrix} \frac{3}{25} & \frac{4}{25} \end{bmatrix} = [1] \begin{bmatrix} \frac{3}{25} & \frac{4}{25} \end{bmatrix} = \begin{bmatrix} \frac{3}{25} & \frac{4}{25} \end{bmatrix} = A^+$

(c) $AA^+ = \begin{bmatrix} 3 \\ 4 \end{bmatrix} \begin{bmatrix} \frac{3}{25} & \frac{4}{25} \end{bmatrix} = \begin{bmatrix} \frac{9}{25} & \frac{12}{25} \\ \frac{12}{25} & \frac{16}{26} \end{bmatrix}$ is symmetric; thus $(AA^+)^T = AA^+$.

(d) $A^+ A = \begin{bmatrix} \frac{3}{25} & \frac{4}{25} \end{bmatrix} \begin{bmatrix} 3 \\ 4 \end{bmatrix} = [1]$ is symmetric; thus $(A^+ A)^T = A^+ A$.

(e) The eigenvalues of $AA^T = \begin{bmatrix} 9 & 12 \\ 12 & 16 \end{bmatrix}$ are $\lambda_1 = 25$ and $\lambda_2 = 0$, with corresponding unit eigen-

vectors $\mathbf{v}_1 = \begin{bmatrix} \frac{3}{5} \\ \frac{4}{5} \end{bmatrix}$ and $\mathbf{v}_2 = \begin{bmatrix} -\frac{4}{5} \\ \frac{3}{5} \end{bmatrix}$ respectively. The only singular value of A^T is $\sigma_1 = 5$, and

we have $\mathbf{u}_1 = \frac{1}{\sigma_1} A^T \mathbf{v}_1 = \frac{1}{5} [3 \ 4] \begin{bmatrix} \frac{3}{5} \\ \frac{4}{5} \end{bmatrix} = [1]$. This results in the singular value decomposition

$$A^T = [3 \ 4] = [1] [5 \ 0] \begin{bmatrix} \frac{3}{5} & \frac{4}{5} \\ -\frac{4}{5} & \frac{3}{5} \end{bmatrix} = U\Sigma V^T$$

The corresponding reduced singular value decomposition is

$$A^T = [3 \ 4] = [1] [5] \begin{bmatrix} \frac{3}{5} & \frac{4}{5} \end{bmatrix} = U_1 \Sigma_1 V_1^T$$

and from this we obtain

$$(A^T)^+ = V_1 \Sigma_1^{-1} U_1^T = \begin{bmatrix} \frac{3}{5} \\ \frac{4}{5} \end{bmatrix} \begin{bmatrix} \frac{1}{5} \end{bmatrix} [1] = \begin{bmatrix} \frac{3}{25} \\ \frac{4}{25} \end{bmatrix} = (A^+)^T$$

(f) The eigenvalues of $(A^+)^T A^+ = \frac{1}{625} \begin{bmatrix} 9 & 12 \\ 12 & 16 \end{bmatrix}$ are $\lambda_1 = \frac{1}{25}$ and $\lambda_2 = 0$, with corresponding unit

eigenvectors $\mathbf{v}_1 = \begin{bmatrix} \frac{3}{5} \\ \frac{4}{5} \end{bmatrix}$ and $\mathbf{v}_2 = \begin{bmatrix} -\frac{4}{5} \\ \frac{3}{5} \end{bmatrix}$ respectively. The only singular value of A^+ is $\sigma_1 = \frac{1}{5}$,

and we have $\mathbf{u}_1 = \frac{1}{\sigma_1} A^+ \mathbf{v}_1 = 5 \begin{bmatrix} \frac{3}{25} & \frac{4}{25} \end{bmatrix} \begin{bmatrix} \frac{3}{5} \\ \frac{4}{5} \end{bmatrix} = [1]$. This results in the singular value decompo-

sition

$$A^+ = \begin{bmatrix} \frac{3}{25} & \frac{4}{25} \end{bmatrix} = [1] \begin{bmatrix} \frac{1}{5} & 0 \end{bmatrix} \begin{bmatrix} \frac{3}{5} & \frac{4}{5} \\ -\frac{4}{5} & \frac{3}{5} \end{bmatrix} = U\Sigma V^T$$

The corresponding reduced singular value decomposition is

$$A^+ = \begin{bmatrix} \frac{3}{25} & \frac{4}{25} \end{bmatrix} = [1] \begin{bmatrix} \frac{1}{5} \end{bmatrix} \begin{bmatrix} \frac{3}{5} & \frac{4}{5} \end{bmatrix} = U_1 \Sigma_1 V_1^T$$

and from this we obtain

$$(A^+)^+ = V_1 \Sigma_1^{-1} U_1^T = \begin{bmatrix} \frac{3}{5} \\ \frac{4}{5} \end{bmatrix} [5] [1] = \begin{bmatrix} 3 \\ 4 \end{bmatrix} = A$$

7. The only eigenvalue of $A^T A = [25]$ is $\lambda_1 = 25$, with corresponding unit eigenvector $\mathbf{v}_1 = [1]$. The only singular value of A is $\sigma_1 = 5$. We have $\mathbf{u}_1 = \frac{1}{\sigma_1} A \mathbf{v}_1 = \frac{1}{5} \begin{bmatrix} 3 \\ 4 \end{bmatrix} [1] = \begin{bmatrix} \frac{3}{5} \\ \frac{4}{5} \end{bmatrix}$, and we choose $\mathbf{u}_2 = \begin{bmatrix} -\frac{4}{5} \\ \frac{3}{5} \end{bmatrix}$ so that $\{\mathbf{u}_1, \mathbf{u}_2\}$ is an orthonormal basis for R^2. This results in the singular value decomposition

$$A = \begin{bmatrix} 3 \\ 4 \end{bmatrix} = \begin{bmatrix} \frac{3}{5} & -\frac{4}{5} \\ \frac{4}{5} & \frac{3}{5} \end{bmatrix} \begin{bmatrix} 5 \\ 0 \end{bmatrix} [1] = U \Sigma V^T$$

The corresponding reduced singular value decomposition is

$$A = \begin{bmatrix} 3 \\ 4 \end{bmatrix} = \begin{bmatrix} \frac{3}{5} \\ \frac{4}{5} \end{bmatrix} [5][1] = U_1 \Sigma_1 V_1^T$$

and from this we obtain

$$A^+ = V_1 \Sigma_1^{-1} U_1^T = [1] \begin{bmatrix} \frac{1}{5} \end{bmatrix} \begin{bmatrix} \frac{3}{5} & \frac{4}{5} \end{bmatrix} = \begin{bmatrix} \frac{3}{25} & \frac{4}{25} \end{bmatrix}$$

9. The eigenvalues of $A^T A = \begin{bmatrix} 74 & 32 \\ 32 & 26 \end{bmatrix}$ are $\lambda_1 = 90$ and $\lambda_2 = 10$, with corresponding unit eigenvectors $\mathbf{v}_1 = \begin{bmatrix} \frac{2}{\sqrt{5}} \\ \frac{1}{\sqrt{5}} \end{bmatrix}$ and $\mathbf{v}_2 = \begin{bmatrix} \frac{1}{\sqrt{5}} \\ -\frac{2}{\sqrt{5}} \end{bmatrix}$. The singular values of A are $\sigma_1 = \sqrt{90} = 3\sqrt{10}$ and $\sigma_2 = \sqrt{10}$. We have $\mathbf{u}_1 = \frac{1}{\sigma_1} A \mathbf{v}_1 = \frac{1}{3\sqrt{10}} \begin{bmatrix} 7 & 1 \\ 0 & 0 \\ 5 & 5 \end{bmatrix} \begin{bmatrix} \frac{2}{\sqrt{5}} \\ \frac{1}{\sqrt{5}} \end{bmatrix} = \begin{bmatrix} \frac{1}{\sqrt{2}} \\ 0 \\ \frac{1}{\sqrt{2}} \end{bmatrix}$, $\mathbf{u}_2 = \frac{1}{\sigma_2} A \mathbf{v}_2 = \frac{1}{3\sqrt{10}} \begin{bmatrix} 7 & 1 \\ 0 & 0 \\ 5 & 5 \end{bmatrix} \begin{bmatrix} \frac{1}{\sqrt{5}} \\ -\frac{2}{\sqrt{5}} \end{bmatrix} = \begin{bmatrix} \frac{1}{\sqrt{2}} \\ 0 \\ -\frac{1}{\sqrt{2}} \end{bmatrix}$, and we choose $\mathbf{u}_3 = \begin{bmatrix} 0 \\ 1 \\ 0 \end{bmatrix}$ so that $\{\mathbf{u}_1, \mathbf{u}_2, \mathbf{u}_3\}$ is an orthonormal basis for R^3. This yields the singular value decomposition

$$A = \begin{bmatrix} 7 & 1 \\ 0 & 0 \\ 5 & 5 \end{bmatrix} = \begin{bmatrix} \frac{1}{\sqrt{2}} & \frac{1}{\sqrt{2}} & 0 \\ 0 & 0 & 1 \\ \frac{1}{\sqrt{2}} & -\frac{1}{\sqrt{2}} & 0 \end{bmatrix} \begin{bmatrix} 3\sqrt{10} & 0 \\ 0 & \sqrt{10} \\ 0 & 0 \end{bmatrix} \begin{bmatrix} \frac{2}{\sqrt{5}} & \frac{1}{\sqrt{5}} \\ \frac{1}{\sqrt{5}} & -\frac{2}{\sqrt{5}} \end{bmatrix} = U \Sigma V^T$$

The corresponding reduced singular value decomposition is

$$A = \begin{bmatrix} 7 & 1 \\ 0 & 0 \\ 5 & 5 \end{bmatrix} = \begin{bmatrix} \frac{1}{\sqrt{2}} & \frac{1}{\sqrt{2}} \\ 0 & 0 \\ \frac{1}{\sqrt{2}} & -\frac{1}{\sqrt{2}} \end{bmatrix} \begin{bmatrix} 3\sqrt{10} & 0 \\ 0 & \sqrt{10} \end{bmatrix} \begin{bmatrix} \frac{2}{\sqrt{5}} & \frac{1}{\sqrt{5}} \\ \frac{1}{\sqrt{5}} & -\frac{2}{\sqrt{5}} \end{bmatrix} = U_1 \Sigma_1 V_1^T$$

and from this we obtain

$$A^+ = V_1 \Sigma_1^{-1} U_1^T = \begin{bmatrix} \frac{2}{\sqrt{5}} & \frac{1}{\sqrt{5}} \\ \frac{1}{\sqrt{5}} & -\frac{2}{\sqrt{5}} \end{bmatrix} \begin{bmatrix} \frac{1}{3\sqrt{10}} & 0 \\ 0 & \frac{1}{\sqrt{10}} \end{bmatrix} \begin{bmatrix} \frac{1}{\sqrt{2}} & 0 & \frac{1}{\sqrt{2}} \\ \frac{1}{\sqrt{2}} & 0 & -\frac{1}{\sqrt{2}} \end{bmatrix} = \begin{bmatrix} \frac{1}{6} & 0 & -\frac{1}{30} \\ -\frac{1}{6} & 0 & \frac{7}{30} \end{bmatrix}$$

11. Since $A = \begin{bmatrix} 2 & 2 \\ -1 & 1 \end{bmatrix}$ has full column rank, we have $A^+ = (A^T A)^{-1} A^T$; thus

$$A^+ = \begin{bmatrix} 5 & 3 \\ 3 & 5 \end{bmatrix}^{-1} \begin{bmatrix} 2 & -1 \\ 2 & 1 \end{bmatrix} = \frac{1}{16} \begin{bmatrix} 5 & -3 \\ -3 & 5 \end{bmatrix} \begin{bmatrix} 2 & -1 \\ 2 & 1 \end{bmatrix} = \frac{1}{4} \begin{bmatrix} 1 & -2 \\ 1 & 1 \end{bmatrix} = A^{-1}$$

13. The matrix $A = \begin{bmatrix} 1 & 1 \\ 1 & 1 \end{bmatrix}$ does not have full column rank, so Formula (3) does not apply. The eigenvalues of $A^T A = \begin{bmatrix} 2 & 2 \\ 2 & 2 \end{bmatrix}$ are $\lambda_1 = 4$ and $\lambda_2 = 0$, with corresponding unit eigenvectors $\mathbf{v}_1 = \begin{bmatrix} \frac{1}{\sqrt{2}} \\ \frac{1}{\sqrt{2}} \end{bmatrix}$

and $\mathbf{v}_2 = \begin{bmatrix} -\frac{1}{\sqrt{2}} \\ \frac{1}{\sqrt{2}} \end{bmatrix}$. The only singular value of A is $\sigma_1 = 2$. We have $\mathbf{u}_1 = \frac{1}{\sigma_1} A\mathbf{v}_1 = \frac{1}{2} \begin{bmatrix} 1 & 1 \\ 1 & 1 \end{bmatrix} \begin{bmatrix} \frac{1}{\sqrt{2}} \\ \frac{1}{\sqrt{2}} \end{bmatrix} =$

$\begin{bmatrix} \frac{1}{\sqrt{2}} \\ \frac{1}{\sqrt{2}} \end{bmatrix}$ and we choose $\mathbf{u}_1 = \begin{bmatrix} -\frac{1}{\sqrt{2}} \\ \frac{1}{\sqrt{2}} \end{bmatrix}$ so that $\{\mathbf{u}_1, \mathbf{u}_2\}$ is an orthonormal basis for R^2. This results in the singular value decomposition

$$A = \begin{bmatrix} 1 & 1 \\ 1 & 1 \end{bmatrix} = \begin{bmatrix} \frac{1}{\sqrt{2}} & -\frac{1}{\sqrt{2}} \\ \frac{1}{\sqrt{2}} & \frac{1}{\sqrt{2}} \end{bmatrix} \begin{bmatrix} 2 & 0 \\ 0 & 0 \end{bmatrix} \begin{bmatrix} \frac{1}{\sqrt{2}} & \frac{1}{\sqrt{2}} \\ -\frac{1}{\sqrt{2}} & \frac{1}{\sqrt{2}} \end{bmatrix} = U\Sigma V^T$$

The corresponding reduced singular value decomposition is

$$A = \begin{bmatrix} 1 & 1 \\ 1 & 1 \end{bmatrix} = \begin{bmatrix} \frac{1}{\sqrt{2}} \\ \frac{1}{\sqrt{2}} \end{bmatrix} [2] \begin{bmatrix} \frac{1}{\sqrt{2}} & \frac{1}{\sqrt{2}} \end{bmatrix} = U_1\Sigma_1 V_1^T$$

and from this we obtain

$$A^+ = V_1\Sigma_1^{-1}U_1^T = \begin{bmatrix} \frac{1}{\sqrt{2}} \\ \frac{1}{\sqrt{2}} \end{bmatrix} [\frac{1}{2}] \begin{bmatrix} \frac{1}{\sqrt{2}} & \frac{1}{\sqrt{2}} \end{bmatrix} = \begin{bmatrix} \frac{1}{4} & \frac{1}{4} \\ \frac{1}{4} & \frac{1}{4} \end{bmatrix}$$

15. The standard matrix for the orthogonal projection of R^3 onto $\text{col}(A)$ is

$$AA^+ = \begin{bmatrix} 1 & 1 \\ 1 & 3 \\ 2 & 1 \end{bmatrix} \begin{bmatrix} \frac{1}{6} & -\frac{7}{30} & \frac{8}{15} \\ 0 & \frac{2}{5} & -\frac{1}{5} \end{bmatrix} = \begin{bmatrix} \frac{1}{6} & \frac{1}{6} & \frac{1}{3} \\ \frac{1}{6} & \frac{29}{30} & -\frac{1}{15} \\ \frac{1}{3} & -\frac{1}{15} & \frac{13}{15} \end{bmatrix}$$

17. The given system can be written in matrix form as $A\mathbf{x} = \mathbf{b}$ where $A = \begin{bmatrix} 1 & 1 \\ 2 & 2 \\ 2 & 2 \end{bmatrix}$ and $\mathbf{b} = \begin{bmatrix} 1 \\ 0 \\ -1 \end{bmatrix}$. The matrix A has the following reduced singular value decomposition and pseudoinverse:

$$A = U_1\Sigma_1 V_1^T = \begin{bmatrix} \frac{1}{3} \\ \frac{2}{3} \\ \frac{2}{3} \end{bmatrix} [3\sqrt{2}] \begin{bmatrix} \frac{1}{\sqrt{2}} & \frac{1}{\sqrt{2}} \end{bmatrix}$$

$$A^+ = V_1\Sigma_1^{-1}U_1^T = \begin{bmatrix} \frac{1}{\sqrt{2}} \\ \frac{1}{\sqrt{2}} \end{bmatrix} [\frac{1}{3\sqrt{2}}] \begin{bmatrix} \frac{1}{3} & \frac{2}{3} & \frac{2}{3} \end{bmatrix} = \begin{bmatrix} \frac{1}{18} & \frac{1}{9} & \frac{1}{9} \\ \frac{1}{18} & \frac{1}{9} & \frac{1}{9} \end{bmatrix}$$

Thus the least squares solution of minimum norm for the system $A\mathbf{x} = \mathbf{b}$ is:

$$\mathbf{x} = A^+\mathbf{b} = \begin{bmatrix} \frac{1}{18} & \frac{1}{9} & \frac{1}{9} \\ \frac{1}{18} & \frac{1}{9} & \frac{1}{9} \end{bmatrix} \begin{bmatrix} 1 \\ 0 \\ -1 \end{bmatrix} = \begin{bmatrix} -\frac{1}{18} \\ -\frac{1}{18} \end{bmatrix}$$

19. Since A^T has full column rank, we have $(A^+)^T = (A^T)^+ = (AA^T)^{-1}A = [14]^{-1} [1 \ 2 \ 3] = [\frac{1}{14} \ \frac{2}{14} \ \frac{3}{14}]$

and so $A^+ = \begin{bmatrix} \frac{1}{14} \\ \frac{2}{14} \\ \frac{3}{14} \end{bmatrix}$.

21. If A is invertible, then A has full column rank and $A^+ = (A^T A)^{-1} A^T = A^{-1}(A^T)^{-1} A^T = A^{-1}$.

DISCUSSION AND DISCOVERY

D1. If $A = [\mathbf{c}_1 \quad \mathbf{c}_2 \quad \cdots \quad \mathbf{c}_n]$ is an $m \times n$ matrix with orthogonal (nonzero) column vectors, then

$$A^+ = (A^T A)^{-1} A^T = \begin{bmatrix} \|\mathbf{c}_1\|^2 & 0 & \cdots & 0 \\ 0 & \|\mathbf{c}_2\|^2 & \cdots & 0 \\ \vdots & \vdots & \ddots & \vdots \\ 0 & 0 & \cdots & \|\mathbf{c}_n\|^2 \end{bmatrix}^{-1} A^T = \begin{bmatrix} \frac{1}{\|\mathbf{c}_1\|^2} & 0 & \cdots & 0 \\ 0 & \frac{1}{\|\mathbf{c}_2\|^2} & \cdots & 0 \\ \vdots & \vdots & \ddots & \vdots \\ 0 & 0 & \cdots & \frac{1}{\|\mathbf{c}_n\|^2} \end{bmatrix} A^T$$

Note. If the columns of A are orthonormal, then $A^+ = A^T$.

D2. (a) If $A = \sigma \mathbf{u} \mathbf{v}^T$, then $A^+ = \frac{1}{\sigma} \mathbf{v} \mathbf{u}^T$.

(b) $A^+ A = (\frac{1}{\sigma} \mathbf{v} \mathbf{u}^T)(\sigma \mathbf{u} \mathbf{v}^T) = \mathbf{v}(\mathbf{u}^T \mathbf{u})\mathbf{v}^T = \mathbf{v}\mathbf{v}^T$ and $AA^+ = (\sigma \mathbf{u} \mathbf{v}^T)(\frac{1}{\sigma} \mathbf{v} \mathbf{u}^T) = \mathbf{u}\mathbf{u}^T$.

D3. If c is a nonzero scalar, then $(cA)^+ = \frac{1}{c} A^+$.

D4. We have $A^+ = V\Sigma^{-1} U^T = \begin{bmatrix} -\frac{2}{3} \\ -\frac{1}{3} \\ \frac{2}{3} \end{bmatrix} [\frac{1}{3}] [-1] = \begin{bmatrix} \frac{2}{9} \\ \frac{1}{9} \\ -\frac{2}{9} \end{bmatrix}$, $AA^+ = [2 \quad 1 \quad -2] \begin{bmatrix} \frac{2}{9} \\ \frac{1}{9} \\ -\frac{2}{9} \end{bmatrix} = [1]$, and $A^+ A = $

$\begin{bmatrix} \frac{2}{9} \\ \frac{1}{9} \\ -\frac{2}{9} \end{bmatrix} [2 \quad 1 \quad -2] = \begin{bmatrix} \frac{4}{9} & \frac{2}{9} & -\frac{4}{9} \\ \frac{2}{9} & \frac{1}{9} & -\frac{2}{9} \\ -\frac{4}{9} & -\frac{2}{9} & \frac{4}{9} \end{bmatrix}$.

D5. (a) AA^+ and $A^+ A$ are the standard matrices of orthogonal projection operators.

(b) Using parts (a) and (b) of Theorem 8.7.2, we have $(AA^+)(AA^+) = A(A^+ AA^+) = AA^+$ and $(A^+ A)(A^+ A) = A^+(AA^+ A) = A^+ A$; thus AA^+ and $A^+ A$ are idempotent.

WORKING WITH PROOFS

P1. If A has rank k, then we have $U_1^T U_1 = V_1^T V_1 = I_k$; thus

$$AA^+ A = (U_1 \Sigma_1 V_1^T)(V_1 \Sigma_1^{-1} U_1^T)(U_1 \Sigma_1 V_1^T) = U_1 \Sigma_1 \Sigma_1^{-1} \Sigma_1 V_1^T = U_1 \Sigma_1 V_1^T = A$$

P2. Since $V_1^T V_1 = I_k$, we have $AA^+ = (U_1 \Sigma_1 V_1^T)(V_1 \Sigma_1^{-1} U_1^T) = U_1 U_1^T$; thus AA^+ is symmetric.

P3. If $A = U_1 \Sigma_1 V_1^T$ then, since Σ_1 is diagonal, we have $A^T = V_1 \Sigma_1 U_1^T$; thus

$$(A^T)^+ = U_1 \Sigma_1^{-1} V_1^T = (V_1 \Sigma_1^{-1} U_1^T)^T = (A^+)^T$$

P4. If $A = U_1 \Sigma_1 V_1^T$ then $A^+ = V_1 \Sigma_1^{-1} U_1^T$; thus $A^{++} = (A^+)^+ = U_1^{TT}(\Sigma_1^{-1})^{-1} V_1^T = U_1 \Sigma_1 V_1^T = A$.

P5. Using P4 and P1, we have $A^+ AA^+ = A^+ A^{++} A^+ = A^+$.

P6. Using P4 and P2, we have $(AA^+)^T = (A^{++} A^+)^T = A^{++} A^+ = AA^+$.

P7. First note that, as in Exercise P2, we have $AA^+ = (U_1 \Sigma_1 V_1^T)(V_1 \Sigma_1^{-1} U_1^T) = U_1 U_1^T$. Thus, since the columns of U_1 form an orthonormal basis for $\text{col}(A)$, the matrix $AA^+ = U_1 U_1^T$ is the standard matrix of the orthogonal projection of R^n onto $\text{col}(A)$.

P8. It follows from Exercise P7 that $A^T(A^T)^+$ is the standard matrix of the orthogonal projection of R^n onto $\text{col}(A^T) = \text{row}(A)$. Furthermore, using parts (d), (e), and (f) of Theorem 8.7.2, we have

$$A^T(A^T)^+ = (A^{++})^T(A^+)^T = (A^+A^{++})^T = (A^+A)^T = A^+A$$

and so A^+A is the matrix of the orthogonal projection of R^n onto $\text{row}(A)$.

EXERCISE SET 8.8

1. If $u = (2 - i, 4i, 1 + i)$, then $\bar{u} = (2 + i, -4i, 1 - i)$, $\text{Re}(u) = (2, 0, 1)$, $\text{Im}(u) = (-1, 4, 1)$, and

$$\|u\| = \sqrt{|2 - i|^2 + |4i|^2 + |1 + i|^2} = \sqrt{5 + 16 + 2} = \sqrt{23}$$

3. (a) We have $\bar{u} = (3 + 4i, 2 - i, 6i)$; thus $\bar{\bar{u}} = (3 - 4i, 2 + i, -6i) = u$.

    (b) We have $ku = i(3 - 4i, 2 + i, -6i) = (4 + 3i, -1 + 2i, 6)$; thus $\overline{ku} = (4 - 3i, -1 - 2i, 6)$. On the other hand, $\bar{k}\bar{u} = -i(3 + 4i, 2 - i, 6i) = (4 - 3i, -1 - 2i, 6)$, and so $\overline{ku} = \bar{k}\bar{u}$.

    (c) We have $u + v = (4 - 3i, 4, 4 - 6i)$; thus $\overline{u + v} = (4 + 3i, 4, 4 + 6i)$. On the other hand, $\bar{u} + \bar{v} = (3 + 4i, 2 - i, 6i) + (1 - i, 2 + i, 4) = (4 + 3i, 4, 4 + 6i)$, and so $\overline{u + v} = \bar{u} + \bar{v}$.

    (d) We have $u - v = (2 - 5i, 2i, -4 - 6i)$; thus $\overline{u - v} = (2 + 5i, -2i, -4 + 6i)$. On the other hand, $\bar{u} - \bar{v} = (3 + 4i, 2 - i, 6i) - (1 - i, 2 + i, 4) = (2 + 5i, -2i, -4 + 6i)$, and so $\overline{u - v} = \bar{u} - \bar{v}$.

5. If $ix - 3v = \bar{u}$, then $ix = \bar{u} + 3v = (3 + 4i, 2 - i, 6i) + 3(1 + i, 2 - i, 4) = (6 + 7i, 8 - 4i, 12 + 6i)$; thus $x = \frac{1}{i}(\bar{u} + 3v) = -i(\bar{u} + 3v) = -i(6 + 7i, 8 - 4i, 12 + 6i) = (7 - 6i, -4 - 8i, 6 - 12i)$.

7. If $A = \begin{bmatrix} -5i & 4 \\ 2 - i & 1 + 5i \end{bmatrix}$, then we have $\bar{A} = \begin{bmatrix} 5i & 4 \\ 2 + i & 1 - 5i \end{bmatrix}$, $\text{Re}(A) = \begin{bmatrix} 0 & 4 \\ 2 & 1 \end{bmatrix}$, $\text{Im}(A) = \begin{bmatrix} -5 & 0 \\ -1 & 5 \end{bmatrix}$, $\det(A) = (-5i)(1 + 5i) - 4(2 - i) = 17 - i$, and $\text{tr}(A) = (-5i) + (1 + 5i) = 1$.

9. (a) We have $\bar{A} = \begin{bmatrix} 5i & 4 \\ 2 + i & 1 - 5i \end{bmatrix}$; thus $\bar{\bar{A}} = \begin{bmatrix} -5i & 4 \\ 2 - i & 1 + 5i \end{bmatrix} = A$.

    (b) We have $A^T = \begin{bmatrix} -5i & 2 - i \\ 4 & 1 + 5i \end{bmatrix}$; thus $\overline{(A^T)} = \begin{bmatrix} 5i & 2 + i \\ 4 & 1 - 5i \end{bmatrix} = (\bar{A})^T$.

    (c) We have $AB = \begin{bmatrix} -5i & 4 \\ 2 - i & 1 + 5i \end{bmatrix}\begin{bmatrix} 1 - i \\ 2i \end{bmatrix} = \begin{bmatrix} (-5i)(1 - i) + 4(2i) \\ (2 - i)(1 - i) + (1 + 5i)(2i) \end{bmatrix} = \begin{bmatrix} -5 + 3i \\ -9 - i \end{bmatrix}$ and $\bar{A}\bar{B} = \begin{bmatrix} 5i & 4 \\ 2 + i & 1 - 5i \end{bmatrix}\begin{bmatrix} 1 + i \\ -2i \end{bmatrix} = \begin{bmatrix} (5i)(1 + i) + 4(-2i) \\ (2 + i)(1 + i) + (1 - 5i)(-2i) \end{bmatrix} = \begin{bmatrix} -5 - 3i \\ -9 + i \end{bmatrix}$; thus $\overline{AB} = \bar{A}\bar{B}$.

11. If $u = (i, 2i, 3)$, $v = (4, -2i, 1 + i)$, $w = (2 - i, 2i, 5 + 3i)$, and $k = 2i$, then:

$$u \cdot v = (i)(4) + (2i)(2i) + (3)(1 - i) = -1 + i$$
$$u \cdot w = (i)(2 + i) + (2i)(-2i) + (3)(5 - 3i) = 18 - 7i$$
$$v \cdot w = (4)(2 + i) + (-2i)(-2i) + (1 + i)(5 - 3i) = 12 + 6i$$
$$v \cdot u = (4)(-i) + (-2i)(-2i) + (1 + i)(3) = -1 - i = \overline{u \cdot v}$$
$$u \cdot (v + w) = (i)(6 + i) + (2i)(0) + (3)(6 - 4i) = 17 - 6i = (-1 + i) + (18 - 7i) = u \cdot v + u \cdot w$$

$(k\mathbf{u}) \cdot \mathbf{v} = (-2, -4, 6i) \cdot \mathbf{v} = (-2)(4) + (-4)(2i) + (6i)(1 - i) = -2 - 2i = (2i)(-1 + i) = k(\mathbf{u} \cdot \mathbf{v})$

$\mathbf{u}^T \bar{\mathbf{v}} = \begin{bmatrix} i & 2i & 3 \end{bmatrix} \begin{bmatrix} 4 \\ 2i \\ 1 - i \end{bmatrix} = 4i - 4 + 3(1 - i) = -1 + i = \mathbf{u} \cdot \mathbf{v}$

$\bar{\mathbf{v}}^T \mathbf{u} = \begin{bmatrix} 4 & 2i & 1 - i \end{bmatrix} \begin{bmatrix} i \\ 2i \\ 3 \end{bmatrix} = 4i - 4 + 3(1 - i) = -1 + i = \mathbf{u} \cdot \mathbf{v}$

13. We have $\mathbf{u} \cdot \bar{\mathbf{v}} = (i)(4) + (2i)(-2i) + (3)(1 + i) = 7 + 7i$ and $\mathbf{w} \cdot \mathbf{u} = 18 + 7i$; thus

$$\overline{(\mathbf{u} \cdot \bar{\mathbf{v}}) - \mathbf{w} \cdot \mathbf{u}} = \overline{-11 + 14i} = -11 - 14i$$

15. The characteristic polynomial of the matrix $A = \begin{bmatrix} 4 & -5 \\ 1 & 0 \end{bmatrix}$ is $p(\lambda) = \lambda^2 - 4\lambda + 5$; thus the eigenvalues of A (roots of $p(\lambda) = 0$) are $\lambda = 2 \pm i$. The corresponding eigenspaces are found by solving the system $(\lambda I - A)\mathbf{x} = \mathbf{0}$ with $\lambda = 2 + i$ and then with $\lambda = 2 - i$. With $\lambda_1 = 2 + i$ the augmented matrix of this system becomes

$$\begin{bmatrix} -2 + i & 5 & | & 0 \\ -1 & 2 + i & | & 0 \end{bmatrix}$$

Multiplying the first row by $\frac{1}{-2+i} = \frac{-2-i}{5}$ results in

$$\begin{bmatrix} 1 & -2 - i & | & 0 \\ -1 & 2 + i & | & 0 \end{bmatrix}$$

and from this we see that the solutions consist of all scalar multiples of the vector $\mathbf{v}_1 = \begin{bmatrix} 2 + i \\ 1 \end{bmatrix}$.

Similarly, the eigenspace corresponding to $\lambda_2 = 2 - i$ consists of all scalar multiples of $\mathbf{v}_2 = \begin{bmatrix} 2 - i \\ 1 \end{bmatrix}$.

17. The characteristic polynomial of the matrix $A = \begin{bmatrix} 5 & -2 \\ 1 & 3 \end{bmatrix}$ is $p(\lambda) = \lambda^2 - 8\lambda + 17$; thus the eigenvalues of A are $\lambda = 4 \pm i$. The corresponding eigenspaces are found by solving $(\lambda I - A)\mathbf{x} = \mathbf{0}$ with $\lambda = 4 + i$ and $\lambda = 4 - i$. With $\lambda_1 = 4 + i$ the augmented matrix of this system becomes

$$\begin{bmatrix} -1 + i & 2 & | & 0 \\ -1 & 1 + i & | & 0 \end{bmatrix}$$

Multiplying the first row by $\frac{1}{-1+i} = \frac{-1-i}{2}$ results in

$$\begin{bmatrix} 1 & -1 - i & | & 0 \\ -1 & 1 + i & | & 0 \end{bmatrix}$$

and from this we see that the solutions consist of all scalar multiples of the vector $\mathbf{v}_1 = \begin{bmatrix} 1 + i \\ 1 \end{bmatrix}$.

Similarly, the eigenspace corresponding to $\lambda_2 = 4 - i$ consists of all scalar multiples of $\mathbf{v}_2 = \begin{bmatrix} 1 - i \\ 1 \end{bmatrix}$.

19. The eigenvalues of $C = \begin{bmatrix} 1 & -1 \\ 1 & 1 \end{bmatrix}$ are $\lambda = 1 \pm i$ and, taking $\phi = \frac{\pi}{4}$, we have:

$$C = \begin{bmatrix} 1 & -1 \\ 1 & 1 \end{bmatrix} = \begin{bmatrix} \sqrt{2} & 0 \\ 0 & \sqrt{2} \end{bmatrix} \begin{bmatrix} \frac{1}{\sqrt{2}} & -\frac{1}{\sqrt{2}} \\ \frac{1}{\sqrt{2}} & \frac{1}{\sqrt{2}} \end{bmatrix} = \begin{bmatrix} |\lambda| & 0 \\ 0 & |\lambda| \end{bmatrix} \begin{bmatrix} \cos\phi & -\sin\phi \\ \sin\phi & \cos\phi \end{bmatrix}$$

21. The eigenvalues of $C = \begin{bmatrix} 1 & \sqrt{3} \\ -\sqrt{3} & 1 \end{bmatrix}$ are $\lambda = 1 \pm \sqrt{3}i$ and, taking $\phi = -\frac{\pi}{3}$, we have:

$$C = \begin{bmatrix} 1 & \sqrt{3} \\ -\sqrt{3} & 1 \end{bmatrix} = \begin{bmatrix} 2 & 0 \\ 0 & 2 \end{bmatrix} \begin{bmatrix} \frac{1}{2} & \frac{\sqrt{3}}{2} \\ -\frac{3}{2} & \frac{1}{2} \end{bmatrix} = \begin{bmatrix} |\lambda| & 0 \\ 0 & |\lambda| \end{bmatrix} \begin{bmatrix} \cos\phi & -\sin\phi \\ \sin\phi & \cos\phi \end{bmatrix}$$

23. The characteristic polynomial of A is $p(\lambda) = \lambda^2 - 6\lambda + 13$; thus the eigenvalues of A are $\lambda = 3 \pm 2i$. With $\lambda = 3 - 2i$ the augmented matrix of the system $(\lambda I - A)\mathbf{x} = \mathbf{0}$ becomes

$$\begin{bmatrix} 4 - 2i & 5 & | & 0 \\ -4 & -4 - 2i & | & 0 \end{bmatrix}$$

Multiplying the first row by $\frac{1}{4-2i} = \frac{4+2i}{20}$ and the second row by $\frac{1}{4}$ results in

$$\begin{bmatrix} 1 & 1 + \frac{1}{2}i & | & 0 \\ -1 & -1 - \frac{1}{2}i & | & 0 \end{bmatrix}$$

and from this we see that $\mathbf{x} = \begin{bmatrix} -1 - \frac{1}{2}i \\ 1 \end{bmatrix}$ is an eigenvector corresponding to $\lambda = 3 - 2i$. Proceeding as outlined in Theorem 8.8.9, this results in the following factorization of A:

$$A = \begin{bmatrix} -1 & -5 \\ 4 & 7 \end{bmatrix} = \begin{bmatrix} -1 & -\frac{1}{2} \\ 1 & 0 \end{bmatrix} \begin{bmatrix} 3 & -2 \\ 2 & 3 \end{bmatrix} \begin{bmatrix} 0 & 1 \\ -2 & -2 \end{bmatrix} = PCP^{-1}$$

Note. The matrix P is not unique. It depends on the choice of the eigenvector $\mathbf{x}$. This is just one possibility.

25. The characteristic polynomial of A is $p(\lambda) = \lambda^2 - 10\lambda + 34$; thus the eigenvalues are $\lambda = 5 \pm 3i$. With $\lambda = 5 - 3i$ the augmented matrix of the system $(\lambda I - A)\mathbf{x} = \mathbf{0}$ becomes

$$\begin{bmatrix} -3 - 3i & -6 & | & 0 \\ 3 & 3 - 3i & | & 0 \end{bmatrix}$$

Multiplying the first row by $\frac{1}{-3-3i} = \frac{-1+i}{6}$ and the second row by $\frac{1}{3}$ results in

$$\begin{bmatrix} 1 & 1 - i & | & 0 \\ 1 & 1 - i & | & 0 \end{bmatrix}$$

and from this we see that $\mathbf{x} = \begin{bmatrix} -1 + i \\ 1 \end{bmatrix}$ is an eigenvector corresponding to $\lambda = 5 - 3i$. Proceeding as outlined in Theorem 8.8.9, this leads to the following factorization of A:

$$A = \begin{bmatrix} 8 & 6 \\ -3 & 2 \end{bmatrix} = \begin{bmatrix} -1 & 1 \\ 1 & 0 \end{bmatrix} \begin{bmatrix} 5 & -3 \\ 3 & 5 \end{bmatrix} \begin{bmatrix} 0 & 1 \\ 1 & 1 \end{bmatrix} = PCP^{-1}$$

27. (a) We have $\mathbf{u} \cdot \mathbf{v} = (2i)(-i) + (i)(-6i) + (3i)(\bar{k}) = 8 + 3i\bar{k}$; thus $\mathbf{u} \cdot \mathbf{v} = 0$ (so $\mathbf{u}$ and $\mathbf{v}$ are orthogonal) if and only if $\bar{k} = -\frac{8}{3i} = \frac{8}{3}i$, i.e., $k = -\frac{8}{3}i$.

(b) We have $\mathbf{u} \cdot \mathbf{v} = (k)(1) + (k)(-1) + (1 + i)(1 + i) = 2i \neq 0$; thus there is no value of k for which $\mathbf{u}$ and $\mathbf{v}$ are orthogonal.

29. (a) We have $\beta^2 = \begin{bmatrix} I_2 & 0 \\ 0 & -I_2 \end{bmatrix} \begin{bmatrix} I_2 & 0 \\ 0 & -I_2 \end{bmatrix} = \begin{bmatrix} I_2^2 & 0 \\ 0 & (-I_2)^2 \end{bmatrix} = \begin{bmatrix} I_2 & 0 \\ 0 & I_2 \end{bmatrix} = I_4$. It is easy to verify that

$\sigma_1^2 = \sigma_2^2 = \sigma_3^2 = I_2$; thus $\alpha_x^2 = \begin{bmatrix} 0 & \sigma_1 \\ \sigma_1 & 0 \end{bmatrix} \begin{bmatrix} 0 & \sigma_1 \\ \sigma_1 & 0 \end{bmatrix} = \begin{bmatrix} \sigma_1^2 & 0 \\ 0 & \sigma_1^2 \end{bmatrix} = \begin{bmatrix} I_2 & 0 \\ 0 & I_2 \end{bmatrix} = I_4$ and, similarly, $\alpha_y^2 = \alpha_z^2 = I_4$.

(b) We have $\beta\alpha_x = \begin{bmatrix} I_2 & 0 \\ 0 & -I_2 \end{bmatrix} \begin{bmatrix} 0 & \sigma_1 \\ \sigma_1 & 0 \end{bmatrix} = \begin{bmatrix} 0 & \sigma_1 \\ -\sigma_1 & 0 \end{bmatrix}$ and $\alpha_x\beta = \begin{bmatrix} 0 & \sigma_1 \\ \sigma_1 & 0 \end{bmatrix} \begin{bmatrix} I_2 & 0 \\ 0 & -I_2 \end{bmatrix} = \begin{bmatrix} 0 & -\sigma_1 \\ \sigma_1 & 0 \end{bmatrix}$;

thus $\alpha_x\beta = -\beta\alpha_x$. Similarly, $\alpha_y\beta = -\beta\alpha_y$ and $\alpha_z\beta = -\beta\alpha_z$.

We have $\sigma_1\sigma_2 = \begin{bmatrix} 0 & 1 \\ 1 & 0 \end{bmatrix} \begin{bmatrix} 0 & -i \\ i & 0 \end{bmatrix} = \begin{bmatrix} i & 0 \\ 0 & -i \end{bmatrix}$ and $\sigma_2\sigma_1 = \begin{bmatrix} 0 & -i \\ i & 0 \end{bmatrix} \begin{bmatrix} 0 & 1 \\ 1 & 0 \end{bmatrix} = \begin{bmatrix} -i & 0 \\ 0 & i \end{bmatrix} = -\sigma_1\sigma_2$. Similarly, $\sigma_3\sigma_1 = -\sigma_1\sigma_3$ and $\sigma_3\sigma_2 = -\sigma_2\sigma_3$. From this it follows that

$$\alpha_y\alpha_x = \begin{bmatrix} 0 & \sigma_2 \\ \sigma_2 & 0 \end{bmatrix} \begin{bmatrix} 0 & \sigma_1 \\ \sigma_1 & 0 \end{bmatrix} = \begin{bmatrix} \sigma_2\sigma_1 & 0 \\ 0 & \sigma_2\sigma_1 \end{bmatrix} = -\begin{bmatrix} \sigma_1\sigma_2 & 0 \\ 0 & \sigma_1\sigma_2 \end{bmatrix} = -\alpha_x\alpha_y$$

and, similarly, $\alpha_z\alpha_x = -\alpha_x\alpha_z$ and $\alpha_z\alpha_y = -\alpha_y\alpha_z$.

DISCUSSION AND DISCOVERY

D1. If $\mathbf{u} \cdot \mathbf{v} = a + bi$, then $(i\mathbf{u}) \cdot \mathbf{v} = i(\mathbf{u} \cdot \mathbf{v}) = -b + ai$, $\mathbf{u} \cdot (i\mathbf{v}) = \bar{i}(\mathbf{u} \cdot \mathbf{v}) = -i(\mathbf{u} \cdot \mathbf{v}) = b - ai$, and $\mathbf{v} \cdot (i\mathbf{u}) = \overline{(i\mathbf{u}) \cdot \mathbf{v}} = -b - ai$.

D2. Yes. We have $\|\mathbf{v}\|^2 = \mathbf{v} \cdot \mathbf{v}$; thus $\|k\mathbf{v}\|^2 = (k\mathbf{v}) \cdot (k\mathbf{v}) = k\bar{k}(\mathbf{v} \cdot \mathbf{v}) = |k|^2 \|\mathbf{v}\|^2$ and $\|k\mathbf{v}\| = |k| \|\mathbf{v}\|$.

D3. $\cdots$, then $\bar{\lambda} = a - bi$ is also an eigenvalue of A and $\bar{\mathbf{x}} = (u_1 - v_1 i, u_2 - v_2 i)$ is a corresponding eigenvector.

D4. The characteristic polynomial of A is $p(\lambda) = \lambda^2 - 4\lambda + 19$; thus the eigenvalues are $\lambda = 2 \pm i\sqrt{15}$. This does not contradict Theorem 8.8.7, which applies only to real symmetric matrices.

WORKING WITH PROOFS

P1. $\overline{\mathbf{u} + \mathbf{v}} = \overline{(u_1 + v_1, u_2 + v_2, \ldots, u_n + v_n)} = (\overline{u_1 + v_1}, \overline{u_2 + v_2}, \ldots, \overline{u_n + v_n}) = (\bar{u}_1 + \bar{v}_1, \bar{u}_2 + \bar{v}_2, \ldots, \bar{u}_n + \bar{v}_n) = (\bar{u}_1, \bar{u}_2, \ldots, \bar{u}_n) + (\bar{v}_1, \bar{v}_2, \ldots, \bar{v}_n) = \bar{\mathbf{u}} + \bar{\mathbf{v}}$

P2. (a) If $A = [a_{ij}]$, then $\bar{A} = [\bar{a}_{ij}]$ and so $\bar{\bar{A}} = [\bar{\bar{a}}_{ij}] = [a_{ij}] = A$.

(b) If $A = [a_{ij}]$, then $A^T = [a_{ji}]$ and so $(\overline{A^T}) = [\bar{a}_{ji}]$. On the other hand, we have $\bar{A} = [\bar{a}_{ij}]$ and so $(\bar{A})^T = [\bar{a}_{ji}]$. Thus $(\overline{A^T}) = (\bar{A})^T$.

(c) If $A = [a_{ir}]_{m \times k}$ and $B = [b_{rj}]_{k \times n}$, then $\overline{AB} = [\bar{c}_{ij}]_{m \times n}$ where $c_{ij} = \sum_{r=1}^{k} a_{ir}b_{rj}$. On the other hand, $\bar{A}\bar{B} = [d_{ij}]$ where $d_{ij} = \sum_{r=1}^{k} \bar{a}_{ir}\bar{b}_{rj} = \overline{\sum_{r=1}^{k} a_{ir}b_{rj}} = \bar{c}_{ij}$. Thus $\overline{AB} = \bar{A}\bar{B}$.

P3. From basic properties of the inner product, Theorem 8.8.5 and Formula (11), we have

$$\|\mathbf{u} + \mathbf{v}\|^2 = (\mathbf{u} + \mathbf{v}) \cdot (\mathbf{u} + \mathbf{v}) = \mathbf{u} \cdot \mathbf{u} + \mathbf{u} \cdot \mathbf{v} + \mathbf{v} \cdot \mathbf{u} + \mathbf{v} \cdot \mathbf{v} = \|\mathbf{u}\|^2 + 2\mathrm{Re}(\mathbf{u} \cdot \mathbf{v}) + \|\mathbf{v}\|^2$$

$$\|\mathbf{u} - \mathbf{v}\|^2 = \|\mathbf{u}\|^2 - 2\mathrm{Re}(\mathbf{u} \cdot \mathbf{v}) + \|\mathbf{v}\|^2$$

$$\|\mathbf{u} + i\mathbf{v}\|^2 = (\mathbf{u} + i\mathbf{v}) \cdot (\mathbf{u} + i\mathbf{v}) = \mathbf{u} \cdot \mathbf{u} + \mathbf{u} \cdot (i\mathbf{v}) + (i\mathbf{v}) \cdot \mathbf{u} + (i\mathbf{v}) \cdot (i\mathbf{v})$$

$$= \|\mathbf{u}\|^2 + \bar{i}(\mathbf{u} \cdot \mathbf{v}) + i(\overline{\mathbf{u} \cdot \mathbf{v}}) - \|\mathbf{v}\|^2 = \|\mathbf{u}\|^2 + 2\mathrm{Im}(\mathbf{u} \cdot \mathbf{v}) - \|\mathbf{v}\|^2$$

and $\|u - iv\|^2 = \|u\|^2 - 2\mathrm{Im}(u \cdot v) - \|v\|^2$. Putting the pieces together, it follows that

$$\frac{1}{4}\|u + v\|^2 - \frac{1}{4}\|u - v\|^2 + \frac{i}{4}\|u + iv\|^2 - \frac{i}{4}\|u - iv\|^2 = \mathrm{Re}(u \cdot v) + i\mathrm{Im}(u \cdot v) = u \cdot v$$

P4. If $\lambda = \cos\phi + i\sin\phi$, the augmented matrix of the system $(\lambda I - R_\phi)x = 0$ is

$$\begin{bmatrix} i\sin\phi & \sin\phi & \vdots & 0 \\ -\sin\phi & i\sin\phi & \vdots & 0 \end{bmatrix}$$

which, upon multiplying the first row by $\frac{1}{i} = -i$, reduces to

$$\begin{bmatrix} \sin\phi & -i\sin\phi & \vdots & 0 \\ -\sin\phi & i\sin\phi & \vdots & 0 \end{bmatrix}$$

From this we see that the eigenvectors of R_ϕ corresponding to $\lambda = \cos\phi + i\sin\phi$ are of the form $x = z\begin{bmatrix} i \\ 1 \end{bmatrix}$ where z is a complex scalar. If $z = x + iy$ where x and y are real, then

$$x = z\begin{bmatrix} i \\ 1 \end{bmatrix} = (x + iy)\begin{bmatrix} i \\ 1 \end{bmatrix} = \begin{bmatrix} -y + ix \\ x + iy \end{bmatrix} = \begin{bmatrix} -y \\ x \end{bmatrix} + i\begin{bmatrix} x \\ y \end{bmatrix}$$

Thus the vectors $\mathrm{Re}(x) = \begin{bmatrix} -y \\ x \end{bmatrix}$ and $\mathrm{Im}(x) = \begin{bmatrix} x \\ y \end{bmatrix}$ are orthogonal, and each has length $|z| = \sqrt{x^2 + y^2}$. Similarly, the eigenvectors of R_ϕ corresponding to $\lambda = \cos\phi - i\sin\phi$ are of the form

$$x = z\begin{bmatrix} -i \\ 1 \end{bmatrix} = (x + iy)\begin{bmatrix} -i \\ 1 \end{bmatrix} = \begin{bmatrix} y - ix \\ x + iy \end{bmatrix} = \begin{bmatrix} y \\ x \end{bmatrix} + i\begin{bmatrix} -x \\ y \end{bmatrix}$$

P5. **(a)** Using the notation suggested, we have

$$Au + iAv = A(u + iv) = Ax = \lambda x = (a - bi)(u + iv) = (au + bv) + i(-bu + av)$$

and, equating real and imaginary parts, this leads to $Au = au + bv$ and $Av = -bu + av$. When written in matrix form, these equations take the form

$$AP = A\begin{bmatrix} u & v \end{bmatrix} = \begin{bmatrix} Au & Av \end{bmatrix} = \begin{bmatrix} au + bv & -bu + av \end{bmatrix} = \begin{bmatrix} u & v \end{bmatrix}\begin{bmatrix} a & -b \\ b & a \end{bmatrix} = PM$$

(b) If the matrix $P = \begin{bmatrix} u & v \end{bmatrix}$ is not invertible then one of the vectors u and v is a (real) scalar multiple of the other, say $v = cu$. Substituting this into the equations obtained in part (a), we have $Au = au + bcu$ and $cAu = Av = -bu + acu$. Multiplying the first equation by c and adding to the second equation leads to $(1 + c^2)bu = 0$ and, since $1 + c^2 \neq 0$ and $u \neq 0$, we conclude that $b = 0$. But this contradicts the hypothesis that the eigenvalues of A are complex. Thus the matrix P is invertible, and from part (a) it follows that $A = PMP^{-1}$.

P6. **(a)** $(u - kv) \cdot (u - kv) = u \cdot u - u \cdot (kv) - (kv) \cdot u + (kv) \cdot (kv) =$
$\|u\|^2 - \bar{k}(u \cdot v) - k\overline{(u \cdot v)} + |k|^2\|v\|^2$

(b) Since $(u - kv) \cdot (u - kv) = \|u - kv\|^2 \geq 0$, we have $0 \leq \|u\|^2 - \bar{k}(u \cdot v) - k\overline{(u \cdot v)} + |k|^2\|v\|^2$ for every complex scalar k.

(c) Setting $k = \frac{u \cdot v}{\|v\|^2}$ in part (b), we have $0 \leq \|u\|^2 - \frac{\overline{(u \cdot v)}}{\|v\|^2}(u \cdot v) - \frac{(u \cdot v)}{\|v\|^2}\overline{(u \cdot v)} + \frac{|(u \cdot v)|^2}{\|v\|^4}\|v\|^2 =$
$\|u\|^2 - 2\frac{|(u \cdot v)|^2}{\|v\|^2} + \frac{|(u \cdot v)|^2}{\|v\|^2} = \|u\|^2 - \frac{|(u \cdot v)|^2}{\|v\|^2}$, and from this it follows that $|u \cdot v|^2 \leq \|u\|^2\|v\|^2$
or $|u \cdot v| \leq \|u\|\|v\|$.

EXERCISE SET 8.9

1. $A^* = \begin{bmatrix} -2i & 4 & 5-i \\ 1+i & 3-i & 0 \end{bmatrix}$

3. $A = \begin{bmatrix} 1 & i & 2-3i \\ -i & -3 & 1 \\ 2+3i & 1 & 2 \end{bmatrix}$

5. **(a)** A is not Hermitian since $a_{31} \neq \bar{a}_{13}$ **(b)** A is not Hermitian since a_{22} is not real

7. The characteristic polynomial of the matrix $A = \begin{bmatrix} 3 & 2-3i \\ 2+3i & -1 \end{bmatrix}$ is

$$\det(\lambda I - A) = \begin{vmatrix} \lambda - 3 & -2+3i \\ -2-3i & \lambda+1 \end{vmatrix} = (\lambda-3)(\lambda+1) - (-2-3i)(-2+3i) = \lambda^2 - 2\lambda - 16$$

Thus the eigenvalues of A are $\lambda = 1 \pm \sqrt{17}$. For $\lambda = 1 + \sqrt{17}$ the augmented matrix of the system $(\lambda I - A)\mathbf{x} = \mathbf{0}$ is

$$\begin{bmatrix} -2+\sqrt{17} & -2+3i & \vdots & 0 \\ -2-3i & 2+\sqrt{17} & \vdots & 0 \end{bmatrix}$$

Multiplying the 1st row by $\frac{1}{-2+\sqrt{17}} = \frac{-2-\sqrt{17}}{4-17} = \frac{2+\sqrt{17}}{13}$ and the 2nd row by $\frac{1}{-2-3i} = \frac{-2+3i}{13}$ results in

$$\begin{bmatrix} 1 & \frac{(2+\sqrt{17})(-2+3i)}{13} & \vdots & 0 \\ 1 & \frac{(-2+3i)(2+\sqrt{17})}{13} & \vdots & 0 \end{bmatrix}$$

and from this we conclude that $\mathbf{v}_1 = \begin{bmatrix} \frac{(2-3i)(2+\sqrt{17})}{13} \\ 1 \end{bmatrix}$ forms a basis for the eigenspace corresponding to $\lambda = 1 + \sqrt{17}$. A similar computation shows that $\mathbf{v}_2 = \begin{bmatrix} \frac{(2-3i)(2-\sqrt{17})}{13} \\ 1 \end{bmatrix}$ forms a basis for the eigenspace corresponding to $\lambda = 1 - \sqrt{17}$. Finally, since

$$\mathbf{v}_1 \cdot \mathbf{v}_2 = \frac{(2-3i)(2+\sqrt{17})}{13} \frac{(2+3i)(2-\sqrt{17})}{13} + 1 = -1 + 1 = 0$$

it follows that the two eigenspaces are orthogonal.

9. The following computations show that the row vectors of A are orthonormal:

$$\|\mathbf{r}_1\| = \sqrt{\left|\tfrac{3}{5}\right|^2 + \left|\tfrac{4}{5}i\right|^2} = \sqrt{\tfrac{9}{25} + \tfrac{16}{25}} = 1$$

$$\|\mathbf{r}_2\| = \sqrt{\left|-\tfrac{4}{5}\right|^2 + \left|\tfrac{3}{5}i\right|^2} = \sqrt{\tfrac{16}{25} + \tfrac{9}{25}} = 1$$

$$\mathbf{r}_1 \cdot \mathbf{r}_2 = (\tfrac{3}{5})(-\tfrac{4}{5}) + (\tfrac{4}{5}i)(-\tfrac{3}{5}i) = -\tfrac{12}{5} + \tfrac{12}{5} = 0$$

Thus A is unitary, and $A^{-1} = A^* = \begin{bmatrix} \tfrac{3}{5} & -\tfrac{4}{5} \\ -\tfrac{4}{5}i & -\tfrac{3}{5}i \end{bmatrix}$.

11. The following computations show that the column vectors of A are orthonormal:

$$\|\mathbf{c}_1\| = \sqrt{\left|\tfrac{1}{2\sqrt{2}}(\sqrt{3}+i)\right|^2 + \left|\tfrac{1}{2\sqrt{2}}(1+i\sqrt{3})\right|^2} = \sqrt{\tfrac{4}{8}+\tfrac{4}{8}} = 1$$

$$\|\mathbf{c}_2\| = \sqrt{\left|\tfrac{1}{2\sqrt{2}}(1-i\sqrt{3})\right|^2 + \left|\tfrac{1}{2\sqrt{2}}(i-\sqrt{3})\right|^2} = \sqrt{\tfrac{4}{8}+\tfrac{4}{8}} = 1$$

$$\mathbf{c}_1 \cdot \mathbf{c}_2 = \tfrac{1}{2\sqrt{2}}(\sqrt{3}+i)\tfrac{1}{2\sqrt{2}}(1+i\sqrt{3}) + \tfrac{1}{2\sqrt{2}}(1+i\sqrt{3})\tfrac{1}{2\sqrt{2}}(-i-\sqrt{3}) = 0$$

Thus A is unitary, and $A^{-1} = A^* = \begin{bmatrix} \tfrac{1}{2\sqrt{2}}(\sqrt{3}-i) & \tfrac{1}{2\sqrt{2}}(1-i\sqrt{3}) \\ \tfrac{1}{2\sqrt{2}}(1+i\sqrt{3}) & \tfrac{1}{2\sqrt{2}}(-i-\sqrt{3}) \end{bmatrix}$.

13. The characteristic polynomial of A is $\lambda^2 - 9\lambda + 18 = (\lambda - 3)(\lambda - 6)$; thus the eigenvalues of A are $\lambda = 3$ and $\lambda = 6$. The augmented matrix of the system $(3I - A)\mathbf{x} = \mathbf{0}$ is $\begin{bmatrix} -1 & -1+i & \vert & 0 \\ -1-i & 2 & \vert & 0 \end{bmatrix}$, which reduces to $\begin{bmatrix} 1 & 1-i & \vert & 0 \\ 1 & 1-i & \vert & 0 \end{bmatrix}$. Thus $\mathbf{v}_1 = \begin{bmatrix} -1+i \\ 1 \end{bmatrix}$ is a basis for the eigenspace corresponding to $\lambda = 3$, and $\mathbf{p}_1 = \begin{bmatrix} \tfrac{-1+i}{\sqrt{3}} \\ \tfrac{1}{\sqrt{3}} \end{bmatrix}$ is a unit eigenvector. A similar computation shows that, $\mathbf{p}_2 = \begin{bmatrix} \tfrac{1-i}{\sqrt{6}} \\ \tfrac{2}{\sqrt{6}} \end{bmatrix}$ is a unit eigenvector corresponding to $\lambda = 6$. The vectors $\mathbf{p}_1$ and $\mathbf{p}_2$ are orthogonal, and the unitary matrix $P = [\mathbf{p}_1 \quad \mathbf{p}_2]$ diagonalizes the matrix A:

$$P^* AP = \begin{bmatrix} \tfrac{-1-i}{\sqrt{3}} & \tfrac{1}{\sqrt{3}} \\ \tfrac{1+i}{\sqrt{6}} & \tfrac{2}{\sqrt{6}} \end{bmatrix} \begin{bmatrix} 4 & 1-i \\ 1+i & 5 \end{bmatrix} \begin{bmatrix} \tfrac{-1+i}{\sqrt{3}} & \tfrac{1-i}{\sqrt{6}} \\ \tfrac{1}{\sqrt{3}} & \tfrac{2}{\sqrt{6}} \end{bmatrix} = \begin{bmatrix} 3 & 0 \\ 0 & 6 \end{bmatrix}$$

15. The characteristic polynomial of A is $\lambda^2 - 10\lambda + 16 = (\lambda - 2)(\lambda - 8)$; thus the eigenvalues of A are $\lambda = 2$ and $\lambda = 8$. The augmented matrix of the system $(2I - A)\mathbf{x} = \mathbf{0}$ is $\begin{bmatrix} -4 & -2-2i & \vert & 0 \\ -2+2i & -2 & \vert & 0 \end{bmatrix}$, which reduces to $\begin{bmatrix} 2 & 1+i & \vert & 0 \\ 2 & 1+i & \vert & 0 \end{bmatrix}$. Thus $\mathbf{v}_1 = \begin{bmatrix} -1-i \\ 2 \end{bmatrix}$ is a basis for the eigenspace corresponding to $\lambda = 2$, and $\mathbf{p}_1 = \begin{bmatrix} \tfrac{-1-i}{\sqrt{6}} \\ \tfrac{2}{\sqrt{6}} \end{bmatrix}$ is a unit eigenvector. A similar computation shows that $\mathbf{p}_2 = \begin{bmatrix} \tfrac{1+i}{\sqrt{3}} \\ \tfrac{1}{\sqrt{3}} \end{bmatrix}$ is a unit eigenvector corresponding to $\lambda = 8$. The vectors $\mathbf{p}_1$ and $\mathbf{p}_2$ are orthogonal, and the unitary matrix $P = [\mathbf{p}_1 \quad \mathbf{p}_2]$ diagonalizes the matrix A:

$$P^* AP = \begin{bmatrix} \tfrac{-1+i}{\sqrt{6}} & \tfrac{2}{\sqrt{6}} \\ \tfrac{1-i}{\sqrt{3}} & \tfrac{1}{\sqrt{3}} \end{bmatrix} \begin{bmatrix} 6 & 2+2i \\ 2-2i & 4 \end{bmatrix} \begin{bmatrix} \tfrac{-1-i}{\sqrt{6}} & \tfrac{1+i}{\sqrt{3}} \\ \tfrac{2}{\sqrt{6}} & \tfrac{1}{\sqrt{3}} \end{bmatrix} = \begin{bmatrix} 2 & 0 \\ 0 & 8 \end{bmatrix}$$

17. The characteristic polynomial of A is $(\lambda - 5)(\lambda^2 + \lambda - 2) = (\lambda + 2)(\lambda - 1)(\lambda - 5)$; thus the eigenvalues of A are $\lambda_1 = -2$, $\lambda_2 = 1$, and $\lambda_3 = 5$. The augmented matrix of the system $(-2I - A)\mathbf{x} = \mathbf{0}$ is $\begin{bmatrix} -7 & 0 & 0 & \vert & 0 \\ 0 & -1 & 1-i & \vert & 0 \\ 0 & 1+i & -2 & \vert & 0 \end{bmatrix}$, which can be reduced to $\begin{bmatrix} 1 & 0 & 0 & \vert & 0 \\ 0 & 1 & -1+i & \vert & 0 \\ 0 & 1 & -1+i & \vert & 0 \end{bmatrix}$. Thus $\mathbf{v}_1 = \begin{bmatrix} 0 \\ 1-i \\ 1 \end{bmatrix}$ is a basis for the eigenspace corresponding to $\lambda_2 = -2$, and $\mathbf{p}_1 = \begin{bmatrix} 0 \\ \tfrac{1-i}{\sqrt{3}} \\ \tfrac{1}{\sqrt{3}} \end{bmatrix}$ is a unit eigenvector. Similar computations show that $\mathbf{p}_2 = \begin{bmatrix} 0 \\ \tfrac{-1+i}{\sqrt{6}} \\ \tfrac{2}{\sqrt{6}} \end{bmatrix}$ is a unit eigenvector corresponding to $\lambda_2 = 1$, and $\mathbf{p}_3 = \begin{bmatrix} 1 \\ 0 \\ 0 \end{bmatrix}$ is

a unit eigenvector corresponding to $\lambda_3 = 5$. The vectors $\{\mathbf{p}_1, \mathbf{p}_2, \mathbf{p}_3\}$ form an orthogonal set, and the unitary matrix $P = [\mathbf{p}_1 \quad \mathbf{p}_2 \quad \mathbf{p}_3]$ diagonalizes the matrix A:

$$P^*AP = \begin{bmatrix} 0 & \frac{1+i}{\sqrt{3}} & \frac{1}{\sqrt{3}} \\ 0 & \frac{-1-i}{\sqrt{6}} & \frac{2}{\sqrt{6}} \\ 1 & 0 & 0 \end{bmatrix} \begin{bmatrix} 5 & 0 & 0 \\ 0 & -1 & -1+i \\ 0 & -1-i & 0 \end{bmatrix} \begin{bmatrix} 0 & 0 & 1 \\ \frac{1-i}{\sqrt{3}} & \frac{-1+i}{\sqrt{6}} & 0 \\ \frac{1}{\sqrt{3}} & \frac{2}{\sqrt{6}} & 0 \end{bmatrix} = \begin{bmatrix} -2 & 0 & 0 \\ 0 & 1 & 0 \\ 0 & 0 & 5 \end{bmatrix}$$

Note. The matrix P is not unique. It depends on the particular choice of the unit eigenvectors. This is just one possibility.

19. $A = \begin{bmatrix} 0 & i & 2-3i \\ i & 0 & 1 \\ -2-3i & -1 & 4i \end{bmatrix}$

21. (a) $a_{21} = -i \neq i = -\bar{a}_{12}$ and $a_{31} \neq -\bar{a}_{13}$ 　　　　(b) $a_{11} \neq -\bar{a}_{11}$, $a_{31} \neq -\bar{a}_{13}$, and $a_{32} \neq -\bar{a}_{23}$

23. The characteristic polynomial of $A = \begin{bmatrix} 0 & -1+i \\ 1+i & i \end{bmatrix}$ is $\lambda^2 - i\lambda + 2 = (\lambda - 2i)(\lambda + i)$; thus the eigenvalues of A are $\lambda = 2i$ and $\lambda = -i$.

25. We have $AA^* = \begin{bmatrix} 1+2i & 2+i & -2-i \\ 2+i & 1+i & -i \\ -2-i & -i & 1+i \end{bmatrix} \begin{bmatrix} 1-2i & 2-i & -2+i \\ 2-i & 1-i & i \\ -2+i & i & 1-i \end{bmatrix} = \begin{bmatrix} 15 & 8 & -8 \\ 8 & 8 & -7 \\ -8 & -7 & 8 \end{bmatrix}$ and

$A^*A = \begin{bmatrix} 1-2i & 2-i & -2+i \\ 2-i & 1-i & i \\ -2+i & i & 1-i \end{bmatrix} \begin{bmatrix} 1+2i & 2+i & -2-i \\ 2+i & 1+i & -i \\ -2-i & -i & 1+i \end{bmatrix} = \begin{bmatrix} 15 & 8 & -8 \\ 8 & 8 & -7 \\ -8 & -7 & 8 \end{bmatrix}$; thus A is normal.

27. We have $A^*A = \frac{1}{\sqrt{2}} \begin{bmatrix} e^{-i\theta} & -ie^{-i\theta} \\ e^{i\theta} & ie^{i\theta} \end{bmatrix} \frac{1}{\sqrt{2}} \begin{bmatrix} e^{i\theta} & e^{-i\theta} \\ ie^{i\theta} & -ie^{-i\theta} \end{bmatrix} = \frac{1}{2} \begin{bmatrix} 1+1 & e^{-2i\theta}-e^{-2i\theta} \\ e^{2i\theta}-e^{2i\theta} & 1+1 \end{bmatrix} = \begin{bmatrix} 1 & 0 \\ 0 & 1 \end{bmatrix}$; thus $A^* = A^{-1}$ and A is unitary.

29. (a) If $B = \frac{1}{2}(A + A^*)$, then $B^* = \frac{1}{2}(A + A^*)^* = \frac{1}{2}(A^* + A^{**}) = \frac{1}{2}(A^* + A) = B$. 　Similarly, $C^* = C$.

　　(b) We have $B + iC = \frac{1}{2}(A + A^*) + \frac{1}{2}(A - A^*) = A$ and $B - iC = \frac{1}{2}(A + A^*) - \frac{1}{2}(A - A^*) = A^*$.

　　(c) $AA^* = (B + iC)(B - iC) = B^2 - iBC + iCB + C^2$ and $A^*A = B^2 + iBC - iCB + C^2$. Thus $AA^* = A^*A$ if and only if $-iBC + iCB = iBC - iCB$, or $2iCB = 2iBC$. Thus A is normal if and only if B and C commute, i.e., $CB = BC$.

31. If A is unitary, then $A^{-1} = A^*$ and so $(A^*)^{-1} = (A^{-1})^* = (A^*)^*$; thus A^* is also unitary.

33. A unitary matrix A has the property that $\|A\mathbf{x}\| = \|\mathbf{x}\|$ for all x in C^n. Thus if A is unitary and $A\mathbf{x} = \lambda\mathbf{x}$ where $\mathbf{x} \neq \mathbf{0}$, we must have $|\lambda| \, \|\mathbf{x}\| = \|A\mathbf{x}\| = \|\mathbf{x}\|$ and so $|\lambda| = 1$.

35. If $H = I - 2\mathbf{u}\mathbf{u}^*$, then $H^* = (I - 2\mathbf{u}\mathbf{u}^*)^* = I^* - 2\mathbf{u}^{**}\mathbf{u}^* = I - 2\mathbf{u}\mathbf{u}^* = H$; thus H is Hermitian. Furthermore, if $\|\mathbf{u}\| = 1$ then

$$HH^* = (I - 2\mathbf{u}\mathbf{u}^*)(I - 2\mathbf{u}\mathbf{u}^*) = I - 2\mathbf{u}\mathbf{u}^* - 2\mathbf{u}\mathbf{u}^* + 4\mathbf{u}\mathbf{u}^*\mathbf{u}\mathbf{u}^* = I - 4\mathbf{u}\mathbf{u}^* + 4\mathbf{u}\|\mathbf{u}\|^2\mathbf{u}^* = I$$

and so H is unitary.

DISCUSSION AND DISCOVERY

D1. If A is both unitary and Hermitian, then $A^{-1} = A^*$ and $A^* = A$; thus $A^{-1} = A$.

D2. $A = \begin{bmatrix} \frac{1}{\sqrt{2}} & -\frac{i}{\sqrt{2}} \\ \frac{i}{\sqrt{2}} & -\frac{1}{\sqrt{2}} \end{bmatrix}$ is both Hermitian and unitary.

D3. We have $AA^* = \begin{bmatrix} a & 0 & 0 \\ 0 & 0 & c \\ 0 & b & 0 \end{bmatrix} \begin{bmatrix} \bar{a} & 0 & 0 \\ 0 & 0 & \bar{b} \\ 0 & \bar{c} & 0 \end{bmatrix} = \begin{bmatrix} |a|^2 & 0 & 0 \\ 0 & |c|^2 & 0 \\ 0 & 0 & |b|^2 \end{bmatrix}$ and $A^*A = \begin{bmatrix} |a|^2 & 0 & 0 \\ 0 & |b|^2 & 0 \\ 0 & 0 & |c|^2 \end{bmatrix}$; thus $AA^* = A^*A$ if

and only if $|b| = |c|$.

D4. If $P = uu^*$, then $Px = (uu^*)x = (u\bar{u}^T)x = u(\bar{u}^Tx) = (x \cdot u)u$. Thus multiplication of x by P corresponds to $\|u\|^2$ times the orthogonal projection of x onto $W = \text{span}\{u\}$. If $\|u\| = 1$, then multiplication of x by $H = I - 2uu^*$ corresponds to reflection of x about the hyperplane $u^\perp$.

WORKING WITH PROOFS

P1. If A is invertible, then $A^*(A^{-1})^* = (A^{-1}A)^* = I^* = I$; thus A^* is invertible and $(A^*)^{-1} = (A^{-1})^*$.

P2. (a) Using the cited formula, we have $\det(\bar{A}) = \sum \pm \bar{a}_{1j_1}\bar{a}_{2j_2}\cdots\bar{a}_{nj_n} = \overline{\sum \pm a_{1j_1}a_{2j_2}\cdots a_{nj_n}} = \overline{\det(A)}$.

(b) We have $A^* = (\bar{A})^T$; thus $\det(A^*) = \det((\bar{A})^T) = \det(\bar{A}) = \overline{\det(A)}$.

P3. (a) If $A = A^*$, then $\det(A) = \det(A^*) = \overline{\det(A)}$; thus $\det(A)$ is real.

(b) If $AA^* = I$, then $|\det(A)|^2 = \det(A)\overline{\det(A)} = \det(A)\det(A^*) = \det(AA^*) = \det(I) = 1$; thus $|\det(A)| = 1$.

P4. (a) $A^{**} = (\overline{A^*})^T = (\overline{(\bar{A})^T})^T = (\bar{\bar{A}})^{TT} = \bar{\bar{A}} = A$

(e) $(AB)^* = (\overline{AB})^T = (\bar{A}\bar{B})^T = (\bar{B})^T(\bar{A})^T = B^*A^*$

P5. (b) $(A + B)^* = (\overline{A + B})^T = (\bar{A} + \bar{B})^T = (\bar{A})^T + (\bar{B})^T = A^* + B^*$

(d) $(kA)^* = (\overline{kA})^T = (\bar{k}\bar{A})^T = \bar{k}(\bar{A})^T = \bar{k}A^*$

P6. If $A = [c_1 \quad c_2 \quad \cdots \quad c_n]$, then $A^*A = \begin{bmatrix} c_1^* \\ c_2^* \\ \vdots \\ c_n^* \end{bmatrix} [c_1 \quad c_2 \quad \cdots \quad c_n] = \begin{bmatrix} c_1 \cdot c_1 & c_2 \cdot c_1 & \cdots & c_n \cdot c_1 \\ c_1 \cdot c_2 & c_2 \cdot c_2 & \cdots & c_n \cdot c_2 \\ \vdots & \vdots & \ddots & \vdots \\ c_1 \cdot c_n & c_2 \cdot c_n & \cdots & c_n \cdot c_n \end{bmatrix}$; thus

$A^*A = I$ if and only if $c_i \cdot c_j = \delta_{ij}$, i.e., if and only if the vectors $c_1, c_2, \cdots, c_n$ form an orthonormal set.

P7. If λ is an eigenvalue of A and x is a corresponding eigenvector then, as in the proof of Theorem 8.8.7, we have $\lambda = \frac{x^*Ax}{\|x\|^2}$; thus to show that λ is real it suffices to show that $\overline{x^*Ax} = x^*Ax$. This results from the following computation:

$$\overline{x^*Ax} = \overline{x}^*\overline{Ax} = x^T\overline{Ax} = (\overline{Ax})^Tx = (\bar{A}\bar{x})^Tx = (\bar{x}^T\bar{A}^T)x = x^*A^*x = x^*Ax$$

EXERCISE SET 8.10

1. The general solution of the equation $y' = -3y$ is $y = ce^{-3t}$, and in order that $y(0) = 5$ we must have $c = 5$. Thus the solution of the given initial value problem is $y = 5e^{-3t}$.

3. Since the matrix is diagonal the given system "decouples" into a set of three equations which can be solved individually: $y_1' = y_1$, $y_2' = 4y_2$, and $y_3' = -2y_3$. The general solutions of these equations are $y_1 = c_1e^t$, $y_2 = c_2e^{4t}$, and $y_3 = c_3e^{-2t}$ respectively. Thus the general solution of the given system can be written in vector form as

$$\mathbf{y} = \begin{bmatrix} y_1 \\ y_2 \\ y_3 \end{bmatrix} = \begin{bmatrix} c_1e^t \\ c_2e^{4t} \\ c_3e^{-2t} \end{bmatrix} = c_1 \begin{bmatrix} e^t \\ 0 \\ 0 \end{bmatrix} + c_2 \begin{bmatrix} 0 \\ e^{4t} \\ 0 \end{bmatrix} + c_3 \begin{bmatrix} 0 \\ 0 \\ e^{-2t} \end{bmatrix}$$

    and $\mathbf{y}_1 = \begin{bmatrix} e^t \\ 0 \\ 0 \end{bmatrix}$, $\mathbf{y}_2 = \begin{bmatrix} 0 \\ e^{4t} \\ 0 \end{bmatrix}$, $\mathbf{y}_3 = \begin{bmatrix} 0 \\ 0 \\ e^{-2t} \end{bmatrix}$ form a fundamental set of solutions. The solution which satisfies the given initial condition is:

$$\mathbf{y} = \begin{bmatrix} y_1 \\ y_2 \\ y_3 \end{bmatrix} = \begin{bmatrix} e^t \\ e^{4t} \\ e^{-2t} \end{bmatrix} = \begin{bmatrix} e^t \\ 0 \\ 0 \end{bmatrix} + \begin{bmatrix} 0 \\ e^{4t} \\ 0 \end{bmatrix} + \begin{bmatrix} 0 \\ 0 \\ e^{-2t} \end{bmatrix}$$

5. The given system can be written in matrix form as $\mathbf{y}' = A\mathbf{y}$ where $A = \begin{bmatrix} 1 & 4 \\ 2 & 3 \end{bmatrix}$. The characteristic polynomial of A is $\lambda^2 - 4\lambda - 5 = (\lambda + 1)(\lambda - 5)$; thus the eigenvalues of A are $\lambda_1 = -1$ and $\lambda_2 = 5$. Corresponding eigenvectors are $\mathbf{v}_1 = \begin{bmatrix} -2 \\ 1 \end{bmatrix}$ and $\mathbf{v}_2 = \begin{bmatrix} 1 \\ 1 \end{bmatrix}$ respectively; thus a general solution is:

$$\mathbf{y} = \begin{bmatrix} y_1 \\ y_2 \end{bmatrix} = c_1 \begin{bmatrix} -2 \\ 1 \end{bmatrix} e^{-t} + c_2 \begin{bmatrix} 1 \\ 1 \end{bmatrix} e^{5t} = \begin{bmatrix} -2c_1e^{-t} + c_2e^{5t} \\ c_1e^{-t} + c_2e^{5t} \end{bmatrix}$$

    In order to satisfy $\mathbf{y}(0) = \begin{bmatrix} 0 \\ 0 \end{bmatrix}$ the coefficients c_1 and c_2 above must satisfy the linear system

$$-2c_1 + c_2 = 0$$
$$c_1 + c_2 = 0$$

    which has only the trivial solution $c_1 = c_2 = 0$. Thus the solution of the initial value problem is

$$\mathbf{y} = \begin{bmatrix} y_1 \\ y_2 \end{bmatrix} = \begin{bmatrix} 0 \\ 0 \end{bmatrix}$$

7. The system can be written in matrix form as $\mathbf{y}' = A\mathbf{y}$ where $A = \begin{bmatrix} 4 & 0 & 1 \\ -2 & 1 & 0 \\ -2 & 0 & 1 \end{bmatrix}$. The characteristic polynomial of A is $\lambda^3 - 6\lambda^2 + 11\lambda - 6 = (\lambda - 1)(\lambda - 2)(\lambda - 3)$; thus the eigenvalues are $\lambda_1 = 1$, $\lambda_2 = 2$, and $\lambda_3 = 3$. Corresponding eigenvectors are $\mathbf{v}_1 = \begin{bmatrix} 0 \\ 1 \\ 0 \end{bmatrix}$, $\mathbf{v}_2 = \begin{bmatrix} -1 \\ 2 \\ 2 \end{bmatrix}$ and $\mathbf{v}_3 = \begin{bmatrix} -1 \\ 1 \\ 1 \end{bmatrix}$; thus a general solution of the system is:

$$\mathbf{y} = \begin{bmatrix} y_1 \\ y_2 \\ y_3 \end{bmatrix} = c_1 \begin{bmatrix} 0 \\ 1 \\ 0 \end{bmatrix} e^t + c_2 \begin{bmatrix} -1 \\ 2 \\ 2 \end{bmatrix} e^{2t} + c_3 \begin{bmatrix} -1 \\ 1 \\ 1 \end{bmatrix} e^{3t} = \begin{bmatrix} -c_2e^{2t} - c_3e^{3t} \\ c_1e^t + 2c_2e^{2t} + c_3e^{3t} \\ 2c_2e^{2t} + c_3e^{3t} \end{bmatrix}$$

In order to satisfy the condition $\mathbf{y}(0) = \begin{bmatrix} -1 \\ 1 \\ 0 \end{bmatrix}$ the coefficients c_1, c_2, c_3 must satisfy the linear system

$$-c_2 - c_3 = -1$$
$$c_1 + 2c_2 + c_3 = 1$$
$$2c_2 + c_3 = 0$$

which has the solution $c_1 = 1$, $c_2 = -1$, $c_3 = 2$. Thus the solution of the initial value problem is

$$\mathbf{y} = \begin{bmatrix} y_1 \\ y_2 \\ y_3 \end{bmatrix} = \begin{bmatrix} 0 \\ 1 \\ 0 \end{bmatrix} e^t - \begin{bmatrix} -1 \\ 2 \\ 2 \end{bmatrix} e^{2t} + 2 \begin{bmatrix} -1 \\ 1 \\ 1 \end{bmatrix} e^{3t} = \begin{bmatrix} e^{2t} - 2e^{3t} \\ e^t - 2e^{2t} + 2e^{3t} \\ -2e^{2t} + 2e^{3t} \end{bmatrix}$$

9. Let $y_1(t) =$ the amount of salt (kg) in tank 1 at time t, and $y_2(t) =$ the amount of salt in tank 2 at time t. The rates of change of $y_1(t)$ and $y_2(t)$ are given by

$$y_1'(t) = \text{rate in} - \text{rate out} = \left(10\, \frac{L}{\min}\right)\left(\frac{y_2(t)}{120}\, \frac{\text{kg}}{L}\right) - \left(90\, \frac{L}{\min}\right)\left(\frac{y_1(t)}{120}\, \frac{\text{kg}}{L}\right) = -\frac{3}{4}y_1 + \frac{1}{12}y_2$$

$$y_2'(t) = \text{rate in} - \text{rate out} = \left(90\, \frac{L}{\min}\right)\left(\frac{y_1(t)}{120}\, \frac{\text{kg}}{L}\right) - \left(90\, \frac{L}{\min}\right)\left(\frac{y_2(t)}{120}\, \frac{\text{kg}}{L}\right) = \frac{3}{4}y_1 - \frac{3}{4}y_2$$

Thus the amount of salt in the tanks at time t can be found by solving the initial value problem

$$\begin{bmatrix} y_1' \\ y_2' \end{bmatrix} = \begin{bmatrix} -\frac{3}{4} & \frac{1}{12} \\ \frac{3}{4} & -\frac{3}{4} \end{bmatrix} \begin{bmatrix} y_1 \\ y_2 \end{bmatrix}; \qquad \begin{bmatrix} y_1(0) \\ y_2(0) \end{bmatrix} = \begin{bmatrix} 30 \\ 40 \end{bmatrix}$$

The characteristic polynomial of the matrix $A = \begin{bmatrix} -\frac{3}{4} & \frac{1}{12} \\ \frac{3}{4} & -\frac{3}{4} \end{bmatrix}$ is $\lambda^2 + \frac{3}{2}\lambda + \frac{1}{2} = \frac{1}{2}(\lambda+1)(2\lambda+1)$; thus the eigenvalues of A are $\lambda_1 = -1$ and $\lambda_2 = -\frac{1}{2}$. Corresponding eigenvectors are $\mathbf{v}_1 = \begin{bmatrix} 1 \\ -3 \end{bmatrix}$ and $\mathbf{v}_2 = \begin{bmatrix} 1 \\ 3 \end{bmatrix}$ respectively. Thus a general solution of the system is given by

$$\mathbf{y} = \begin{bmatrix} y_1 \\ y_2 \end{bmatrix} = c_1 e^{-t} \begin{bmatrix} 1 \\ -3 \end{bmatrix} + c_2 e^{-t/2} \begin{bmatrix} 1 \\ 3 \end{bmatrix} = \begin{bmatrix} c_1 e^{-t} + c_2 e^{-t/2} \\ -3c_1 e^{-t} + 3c_2 e^{-t/2} \end{bmatrix}$$

In order to satisfy the condition $\mathbf{y}(0) = \begin{bmatrix} 30 \\ 40 \end{bmatrix}$ the coefficients c_1 and c_2 must satisfy the linear system

$$c_1 + c_2 = 30$$
$$-3c_1 + 3c_2 = 40$$

which has the solution $c_1 = \frac{25}{3}$, $c_2 = \frac{65}{3}$. Thus the solution of the initial value problem is

$$\mathbf{y} = \begin{bmatrix} y_1 \\ y_2 \end{bmatrix} = \frac{25}{3} e^{-t} \begin{bmatrix} 1 \\ -3 \end{bmatrix} + \frac{65}{3} e^{-t/2} \begin{bmatrix} 1 \\ 3 \end{bmatrix} = \begin{bmatrix} \frac{25}{3} e^{-t} + \frac{65}{3} e^{-t/2} \\ -25c_1 e^{-t} + 65c_2 e^{-t/2} \end{bmatrix}$$

11. The eigenvalues of the matrix $A = \begin{bmatrix} 4 & -2 \\ 1 & 1 \end{bmatrix}$ are $\lambda_1 = 2$ and $\lambda_2 = 3$, with corresponding eigenvectors $\mathbf{v}_1 = \begin{bmatrix} 1 \\ 1 \end{bmatrix}$ and $\mathbf{v}_2 = \begin{bmatrix} 2 \\ 1 \end{bmatrix}$; thus the matrix $P = \begin{bmatrix} 1 & 2 \\ 1 & 1 \end{bmatrix}$ has the property that $P^{-1}AP = D = \begin{bmatrix} 2 & 0 \\ 0 & 3 \end{bmatrix}$. From this it follows that $A = PDP^{-1}$ and that

$$e^{tA} = Pe^{tD}P^{-1} = \begin{bmatrix} 1 & 2 \\ 1 & 1 \end{bmatrix} \begin{bmatrix} e^{2t} & 0 \\ 0 & e^{3t} \end{bmatrix} \begin{bmatrix} -1 & 2 \\ 1 & -1 \end{bmatrix} = \begin{bmatrix} -e^{2t} + 2e^{3t} & 2e^{2t} - 2e^{3t} \\ -e^{2t} + e^{3t} & 2e^{2t} - e^{3t} \end{bmatrix}$$

Thus the solution of the given initial value problem is:

$$\begin{bmatrix} y_1 \\ y_2 \end{bmatrix} = \mathbf{y} = e^{tA}\mathbf{y}_0 = \begin{bmatrix} -e^{2t} + 2e^{3t} & 2e^{2t} - 2e^{3t} \\ -e^{2t} + e^{3t} & 2e^{2t} - e^{3t} \end{bmatrix} \begin{bmatrix} 3 \\ -4 \end{bmatrix} = \begin{bmatrix} -11e^{2t} + 14e^{3t} \\ -11e^{2t} + 7e^{3t} \end{bmatrix}$$

13. The eigenvalues of $A = \begin{bmatrix} 1 & -1 & -1 \\ 1 & 3 & 1 \\ -3 & 1 & -1 \end{bmatrix}$ are $\lambda_1 = -2$, $\lambda_2 = 3$, and $\lambda_3 = 2$, with corresponding eigen-

vectors $\mathbf{v}_1 = \begin{bmatrix} 1 \\ -1 \\ 4 \end{bmatrix}$, $\mathbf{v}_2 = \begin{bmatrix} -1 \\ 1 \\ 1 \end{bmatrix}$, and $\mathbf{v}_3 = \begin{bmatrix} -1 \\ 0 \\ 1 \end{bmatrix}$; thus the matrix $P = \begin{bmatrix} 1 & -1 & -1 \\ -1 & 1 & 0 \\ 4 & 1 & 1 \end{bmatrix}$ has the property

that $P^{-1}AP = D = \begin{bmatrix} -2 & 0 & 0 \\ 0 & 3 & 0 \\ 0 & 0 & 2 \end{bmatrix}$. From this it follows that $A = PDP^{-1}$ and that

$$e^{tA} = Pe^{tD}P^{-1} = \begin{bmatrix} 1 & -1 & -1 \\ -1 & 1 & 0 \\ 4 & 1 & 1 \end{bmatrix} \begin{bmatrix} e^{-2t} & 0 & 0 \\ 0 & e^{3t} & 0 \\ 0 & 0 & e^{2t} \end{bmatrix} \begin{bmatrix} \frac{1}{5} & 0 & \frac{1}{5} \\ \frac{1}{5} & 1 & \frac{1}{5} \\ -1 & -1 & 0 \end{bmatrix}$$

$$= \begin{bmatrix} \frac{1}{5}e^{-2t} - \frac{1}{5}e^{3t} + e^{2t} & -e^{3t} + e^{2t} & \frac{1}{5}e^{-2t} - \frac{1}{5}e^{3t} \\ -\frac{1}{5}e^{-2t} + \frac{1}{5}e^{3t} & e^{3t} & -\frac{1}{5}e^{-2t} + \frac{1}{5}e^{3t} \\ \frac{4}{5}e^{-2t} + \frac{1}{5}e^{3t} - e^{2t} & e^{3t} - e^{2t} & \frac{4}{5}e^{-2t} + \frac{1}{5}e^{3t} \end{bmatrix}$$

Thus the solution of the given initial value problem is:

$$\begin{bmatrix} y_1 \\ y_2 \\ y_3 \end{bmatrix} = \begin{bmatrix} \frac{1}{5}e^{-2t} - \frac{1}{5}e^{3t} + e^{2t} & -e^{3t} + e^{2t} & \frac{1}{5}e^{-2t} - \frac{1}{5}e^{3t} \\ -\frac{1}{5}e^{-2t} + \frac{1}{5}e^{3t} & e^{3t} & -\frac{1}{5}e^{-2t} + \frac{1}{5}e^{3t} \\ \frac{4}{5}e^{-2t} + \frac{1}{5}e^{3t} - e^{2t} & e^{3t} - e^{2t} & \frac{4}{5}e^{-2t} + \frac{1}{5}e^{3t} \end{bmatrix} \begin{bmatrix} 2 \\ 0 \\ -1 \end{bmatrix} = \begin{bmatrix} \frac{1}{5}e^{-2t} - \frac{1}{5}e^{3t} + 2e^{2t} \\ -\frac{1}{5}e^{-2t} + \frac{1}{5}e^{3t} \\ \frac{4}{5}e^{-2t} + \frac{1}{5}e^{3t} - 2e^{2t} \end{bmatrix}$$

15. The coefficient matrix $A = \begin{bmatrix} 0 & 1 & 2 \\ 0 & 0 & -1 \\ 0 & 0 & 0 \end{bmatrix}$ is nilpotent with $A^2 = \begin{bmatrix} 0 & 0 & -1 \\ 0 & 0 & 0 \\ 0 & 0 & 0 \end{bmatrix}$ and $A^3 = 0$. Thus

$$e^{tA} = I + tA + \tfrac{1}{2}t^2A^2 = \begin{bmatrix} 1 & t & 2t - \frac{1}{2}t^2 \\ 0 & 1 & -t \\ 0 & 0 & 1 \end{bmatrix}$$

and the solution of the given initial value problem is

$$\begin{bmatrix} y_1 \\ y_2 \\ y_3 \end{bmatrix} = \mathbf{y} = e^{tA}\mathbf{y}_0 = \begin{bmatrix} 1 & t & 2t - \frac{1}{2}t^2 \\ 0 & 1 & -t \\ 0 & 0 & 1 \end{bmatrix} \begin{bmatrix} -1 \\ 4 \\ 2 \end{bmatrix} = \begin{bmatrix} -1 + 8t - t^2 \\ 4 - 2t \\ 2 \end{bmatrix}$$

17. If $A = \begin{bmatrix} 0 & -1 \\ 1 & 0 \end{bmatrix}$, then $A^2 = -I$, $A^3 = -A$, $A^4 = I$, $A^5 = A$, etc. Thus

$$e^{tA} = I + tA + \tfrac{1}{2!}t^2A^2 + \tfrac{1}{3!}t^3A^3 + \tfrac{1}{4!}t^4A^4 + \tfrac{1}{5!}t^5A^5 + \cdots$$

$$= I(1 - \tfrac{1}{2!}t^2 + \tfrac{1}{4!}t^4 + \cdots) + A(t - \tfrac{1}{3!}t^3 + \tfrac{1}{5!}t^5 + \cdots) = I\cos t + A\sin t = \begin{bmatrix} \cos t & -\sin t \\ \sin t & \cos t \end{bmatrix}$$

and the solution of the given initial value problem is

$$\begin{bmatrix} y_1 \\ y_2 \end{bmatrix} = \mathbf{y} = e^{tA}\mathbf{y}_0 = \begin{bmatrix} \cos t & -\sin t \\ \sin t & \cos t \end{bmatrix}\begin{bmatrix} -1 \\ 2 \end{bmatrix} = \begin{bmatrix} -\cos t - 2\sin t \\ -\sin t + 2\cos t \end{bmatrix}$$

19. (a) Let $y_1 = y$ and $y_2 = y'$. Then $y_2' = y''$ and so the equation $y'' - 6y' - 6y = 0$ is equivalent to the system $y_1' = y_2$, $y_2' = 6y_1 + y_2$; or

$$\begin{bmatrix} y_1' \\ y_2' \end{bmatrix} = \mathbf{y}' = A\mathbf{y} = \begin{bmatrix} 0 & 1 \\ 6 & 6 \end{bmatrix}\begin{bmatrix} y_1 \\ y_2 \end{bmatrix}$$

(b) The eigenvalues of A are $\lambda_1 = 3 + \sqrt{15}$ and $\lambda_2 = 3 - \sqrt{15}$, with corresponding eigenvectors

$$\mathbf{v}_1 = \begin{bmatrix} 1 \\ 3 + \sqrt{15} \end{bmatrix} \text{ and } \mathbf{v}_2 = \begin{bmatrix} 1 \\ 3 - \sqrt{15} \end{bmatrix}. \text{ Thus a general solution of the system } \mathbf{y}' = A\mathbf{y} \text{ is}$$

$$\begin{bmatrix} y \\ y' \end{bmatrix} = \begin{bmatrix} y_1 \\ y_2 \end{bmatrix} = c_1 e^{(3+\sqrt{15})t}\begin{bmatrix} 1 \\ 3 + \sqrt{15} \end{bmatrix} + c_2 e^{(3-\sqrt{15})t}\begin{bmatrix} 1 \\ 3 - \sqrt{15} \end{bmatrix}$$

$$= \begin{bmatrix} c_1 e^{(3+\sqrt{15})t} + c_2 e^{(3-\sqrt{15})t} \\ c_1(3 + \sqrt{15})e^{(3+\sqrt{15})t} + c_2(3 - \sqrt{15})e^{(3-\sqrt{15})t} \end{bmatrix}$$

These solutions do in fact satisfy the original equation since

$$y'' - 6y' - 6y = c_1[(3 + \sqrt{15})^2 - 6(3 + \sqrt{15}) - 6]e^{(3+\sqrt{15})t}$$
$$+ c_2[(3 - \sqrt{15})^2 - 6(3 - \sqrt{15}) - 6]e^{(3-\sqrt{15})t}$$
$$= c_1[0]e^{(3+\sqrt{15})t} + c_2[0]e^{(3-\sqrt{15})t} = 0$$

21. Let $y_1(t) = $ the amount of salt (kg) in tank 1 at time t, and $y_2(t) = $ the amount of salt in tank 2 at time t. The rates of change of $y_1(t)$ and $y_2(t)$ are given by

$$y_1'(t) = \text{rate in} - \text{rate out} = \left(30\frac{L}{\min}\right)\left(0\frac{kg}{L}\right) - \left(30\frac{L}{\min}\right)\left(\frac{y_1(t)}{60}\frac{kg}{L}\right) = -\frac{1}{2}y_1$$

$$y_2'(t) = \text{rate in} - \text{rate out} = \left(10\frac{L}{\min}\right)\left(\frac{y_1(t)}{60}\frac{kg}{L}\right) - \left(10\frac{L}{\min}\right)\left(\frac{y_2(t)}{60}\frac{kg}{L}\right) = \frac{1}{6}y_1 - \frac{1}{6}y_2$$

Thus the amount of salt in the tanks at time t can be found by solving the initial value problem

$$\mathbf{y}' = \begin{bmatrix} y_1' \\ y_2' \end{bmatrix} = \begin{bmatrix} -\frac{1}{2} & 0 \\ \frac{1}{6} & -\frac{1}{6} \end{bmatrix}\begin{bmatrix} y_1 \\ y_2 \end{bmatrix} = A\mathbf{y}; \quad \mathbf{y}(0) = \begin{bmatrix} y_1(0) \\ y_2(0) \end{bmatrix} = \begin{bmatrix} 10 \\ 7 \end{bmatrix}$$

The eigenvalues of A are $\lambda_1 = -\frac{1}{2}$ and $\lambda_2 = -\frac{1}{6}$, with corresponding eigenvectors $\mathbf{v}_1 = \begin{bmatrix} -2 \\ 1 \end{bmatrix}$ and $\mathbf{v}_2 = \begin{bmatrix} 0 \\ 1 \end{bmatrix}$; thus the matrix $P = \begin{bmatrix} -2 & 0 \\ 1 & 1 \end{bmatrix}$ has the property that $P^{-1}AP = D = \begin{bmatrix} -\frac{1}{2} & 0 \\ 0 & -\frac{1}{6} \end{bmatrix}$. From this it follows that $A = PDP^{-1}$ and that

$$e^{tA} = Pe^{tD}P^{-1} = \begin{bmatrix} -2 & 0 \\ 1 & 1 \end{bmatrix}\begin{bmatrix} e^{-t/2} & 0 \\ 0 & e^{-t/6} \end{bmatrix}\begin{bmatrix} -\frac{1}{2} & 0 \\ \frac{1}{2} & 1 \end{bmatrix} = \begin{bmatrix} e^{-t/2} & 0 \\ -\frac{1}{2}e^{-t/2} + \frac{1}{2}e^{-t/6} & e^{-t/6} \end{bmatrix}$$

Thus the solution of the initial value problem is:

$$\begin{bmatrix} y_1 \\ y_2 \end{bmatrix} = \mathbf{y} = e^{tA}\mathbf{y}_0 = \begin{bmatrix} e^{-t/2} & 0 \\ -\frac{1}{2}e^{-t/2} + \frac{1}{2}e^{-t/6} & e^{-t/6} \end{bmatrix}\begin{bmatrix} 10 \\ 7 \end{bmatrix} = \begin{bmatrix} 10e^{-t/2} \\ -5e^{-t/2} + 12e^{-t/6} \end{bmatrix}$$

WORKING WITH PROOFS

P1. If $y = f(t)$ is a solution of $y' = ay$, then $f'(t) = dy/dt = ay = af(t)$; thus

$$\frac{d}{dt}[f(t)e^{-at}] = f(t)(-ae^{-at}) + f'(t)e^{at} = -af(t)e^{-at} + af(t)e^{at} = 0$$

From this it follows that $f(t)e^{-at}$ is constant, i.e., that $y = f(t) = ce^{at}$ for some constant c.

CHAPTER 9
General Vector Spaces

EXERCISE SET 9.1

1. **(a)** If $\mathbf{u} = (-1, 2)$, $\mathbf{v} = (3, 4)$, and $k = 3$, then $\mathbf{u} + \mathbf{v} = (-1 + 3, 2 + 4) = (2, 6)$ and $k\mathbf{u} = (-3, 0)$.
   (b) The result of either operation is an ordered pair of real numbers.
   (c) Axioms 1-5.
   (d) If $\mathbf{u} = (u_1, u_2)$ and $\mathbf{v} = (v_1, v_2)$, then

   $$k(\mathbf{u} + \mathbf{v}) = k(u_1 + u_2, v_1 + v_2) = (ku_1 + ku_2, 0) = (ku_1, 0) + (ku_2, 0) = k\mathbf{u} + k\mathbf{v}$$
   $$(k + l)\mathbf{u} = ((k + l)u_1, 0) = (ku_1 + lu_1, 0) = (ku_1, 0) + (lu_1, 0) = k\mathbf{u} + l\mathbf{u}$$
   $$k(l\mathbf{u}) = k(lu_1, 0) = (klu_1, 0) = ((kl)u_1, 0) = (kl)\mathbf{u}$$

   thus Axioms 7, 8 and 9 hold.
   (e) If $\mathbf{u} = (u_1, u_2)$ with $u_2 \neq 0$, then $1\mathbf{u} = (u_1, 0) \neq \mathbf{u}$; thus Axiom 10 fails.

3. **(a)** If $\mathbf{u} = (0, 4)$, $\mathbf{v} = (1, -3)$, and $k = 2$, then $\mathbf{u} + \mathbf{v} = (0 + 1 + 1, 4 - 3 + 1) = (2, 2)$ and $k\mathbf{u} = (0, 8)$.
   (b) We have $(u_1, u_2) + (0, 0) = (u_1 + 1, u_2 + 1)$; thus $\mathbf{u} + (0, 0) \neq \mathbf{u}$ for every $\mathbf{u}$ in V.
   (c) We have $(u_1, u_2) + (-1, -1) = (u_1 - 1 + 1, u_2 - 1 + 1) = (u_1, u_2)$; thus $\mathbf{u} + (-1, -1) = \mathbf{u}$ for every $\mathbf{u}$ in V.
   (d) If $\mathbf{u} = (u_1, u_2)$, then $-\mathbf{u} = (-u_1 - 2, -u_2 - 2)$ satisfies $\mathbf{u} + (-\mathbf{u}) = (-1, -1) = \mathbf{0}$; thus Axiom 5 holds.
   (e) Axioms 7 and 8 fail to hold. For example:

   $$k(\mathbf{u} + \mathbf{v}) = k(u_1 + v_1 + 1, u_2 + v_2 + 1) = (ku_1 + kv_1 + k, ku_2 + kv_2 + k)$$
   $$k\mathbf{u} + k\mathbf{v} = (ku_1, ku_2) + (kv_1, kv_2) = (ku_1 + kv_1 + 1, ku_2 + kv_2 + 1)$$

5. (a) and (c) are subspaces of $F(-\infty, \infty)$; (b) and (d) are not subspaces.

7. All except (a) are subspaces of M_{nn}.

9. (a) and (b) are subspaces of P_2; (c) is not a subspace.

11. **(a)** The set V of 2×2 upper triangular matrices consists of those matrices of the form $A = \begin{bmatrix} x & y \\ 0 & z \end{bmatrix}$
   where x, y, z can be any real numbers. Clearly this set of matrices is closed under addition and scalar multiplication. Thus V is a subspace of M_{22}.
   (b) The matrices $A_1 = \begin{bmatrix} 1 & 0 \\ 0 & 0 \end{bmatrix}$, $A_2 = \begin{bmatrix} 0 & 1 \\ 0 & 0 \end{bmatrix}$, and $A_3 = \begin{bmatrix} 0 & 0 \\ 0 & 1 \end{bmatrix}$ form a basis for V; thus $\dim(V) = 3$.

13. **(a)** $\cos 4x = \cos^2 2x - \sin^2 2x$; thus $\sin^2 2x - \cos^2 2x + \cos 4x = 0$, i.e., $f_1(x) - f_2(x) + f_3(x) = 0$.
   (b) $\sin^2(\frac{1}{2}x) = \frac{1}{2}(1 - \cos x)$; thus $\sin^2(\frac{1}{2}x) - (\frac{1}{2})1 + \frac{1}{2}\cos x = 0$, i.e., $f_1(x) - \frac{1}{2}f_2(x) + \frac{1}{2}f_3(x) = 0$.

15. If $f_1(x) = x$ and $f_2(x) = \cos x$, then $W(x) = \begin{vmatrix} f_1(x) & f_2(x) \\ f_1'(x) & f_2'(x) \end{vmatrix} = \begin{vmatrix} x & \cos x \\ 1 & -\sin x \end{vmatrix} = -x \sin x - \cos x$ is not
   identically zero on $(-\infty, \infty)$; thus $f_1(x)$ and $f_2(x)$ are linearly independent in $F(-\infty, \infty)$.

17. If $f_1(x) = e^x$, $f_2(x) = xe^x$, and $f_3(x) = x^2 e^x$, then

$$W(x) = \begin{vmatrix} f_1(x) & f_2(x) & f_3(x) \\ f_1'(x) & f_2'(x) & f_3'(x) \\ f_1''(x) & f_2''(x) & f_3''(x) \end{vmatrix} = \begin{vmatrix} e^x & xe^x & x^2 e^x \\ e^x & xe^x + e^x & x^2 e^x + 2xe^x \\ e^x & xe^x + 2e^x & x^2 e^x + 4xe^x + 2e^x \end{vmatrix} = 2e^{3x} \neq 0$$

Thus the functions $f_1(x)$, $f_2(x)$, $f_3(x)$ are linearly independent and span a three dimensional subspace of $F(-\infty, \infty)$.

19. From Example 13, we know that $\dim(M_{22}) = 4$. Thus to show that the matrices A_1, A_2, A_3, A_4 form a basis for M_{22} we have only to show that they are linearly independent.

   If $c_1 A_1 + c_2 A_2 + c_3 A_3 + c_4 A_4 = 0$ then, equating corresponding matrix entries, the scalars c_1, c_2, c_3, c_4 must satisfy the linear system

$$\begin{aligned} c_1 + c_2 + c_3 \quad\quad &= 0 \\ c_2 \quad\quad &= 0 \\ c_1 \quad\quad\quad + c_4 &= 0 \\ c_3 \quad\quad &= 0 \end{aligned}$$

which has only the trivial solution $c_1 = c_2 = c_3 = c_4 = 0$. From this we conclude that A_1, A_2, A_3, A_4 are linearly independent and thus form a basis for M_{22}.

   In order that $c_1 A_1 + c_2 A_2 + c_3 A_3 + c_4 A_4 = A$, the scalars c_1, c_2, c_3, c_4 must satisfy the linear system

$$\begin{aligned} c_1 + c_2 + c_3 \quad\quad &= 6 \\ c_2 \quad\quad &= 2 \\ c_1 \quad\quad\quad + c_4 &= 5 \\ c_3 \quad\quad &= 3 \end{aligned}$$

which has the solution $c_1 = 1$, $c_2 = 2$, $c_3 = 3$, $c_4 = 4$. Thus $A = A_1 + 2A_2 + 3A_3 + 4A_4$.

21. If $c_1 \mathbf{p_1} + c_2 \mathbf{p_2} + c_3 \mathbf{p_3} = \mathbf{0}$, then the scalars c_1, c_2, c_3 must satisfy the linear system

$$\begin{aligned} c_1 + 2c_2 + 3c_3 &= 0 \\ 2c_1 + 9c_2 + 3c_3 &= 0 \\ c_1 \quad\quad + 4c_3 &= 0 \end{aligned}$$

which has only the trivial solution $c_1 = c_2 = c_3 = 0$. From this we conclude that $\mathbf{p_1}, \mathbf{p_2}, \mathbf{p_3}$ are linearly independent and thus form a basis for P_2.

   In order that $c_1 \mathbf{p_1} + c_2 \mathbf{p_2} + c_3 \mathbf{p_3} = \mathbf{p}$, the scalars c_1, c_2, c_3 must satisfy the linear system

$$\begin{aligned} c_1 + 2c_2 + 3c_3 &= 2 \\ 2c_1 + 9c_2 + 3c_3 &= 17 \\ c_1 \quad\quad + 4c_3 &= -3 \end{aligned}$$

which has the solution $c_1 = 1$, $c_2 = 2$, $c_3 = -1$. Thus $\mathbf{p} = \mathbf{p_1} + 2\mathbf{p_2} - \mathbf{p_3}$.

23. The Lagrange interpolating polynomials corresponding to $x_1 = 0$, $x_2 = 1$, $x_3 = 2$ are

$$p_1(x) = \frac{(x-1)(x-2)}{(0-1)(0-2)} = \frac{1}{2}x^2 - \frac{3}{2}x + 1$$

$$p_2(x) = \frac{(x-0)(x-2)}{(1-0)(1-2)} = -x^2 + 2x$$

$$p_3(x) = \frac{(x-0)(x-1)}{(2-0)(2-1)} = \frac{1}{2}x^2 - \frac{1}{2}x$$

and we have $p(x) = 1p_1(x) + 3p_2(x) + 7p_3(x)$.

25. The Lagrange interpolating polynomials corresponding to $x_1 = 2$, $x_2 = 3$, $x_3 = 4$, $x_4 = 5$ are

$$p_1(x) = \frac{(x-3)(x-4)(x-5)}{(2-3)(2-4)(2-5)} = -\frac{1}{6}x^3 + 2x^2 - \frac{47}{6}x + 10$$

$$p_2(x) = \frac{(x-2)(x-4)(x-5)}{(3-2)(3-4)(3-5)} = \frac{1}{2}x^3 - \frac{11}{2}x^2 + 19x - 20$$

$$p_3(x) = \frac{(x-2)(x-3)(x-5)}{(4-2)(4-3)(4-5)} = -\frac{1}{2}x^3 + 5x^2 - \frac{31}{2}x + 15$$

$$p_4(x) = \frac{(x-2)(x-3)(x-4)}{(5-2)(5-3)(5-4)} = \frac{1}{6}x^3 - \frac{3}{2}x^2 + \frac{13}{3}x - 4$$

and the cubic polynomial passing through the given points is

$$p(x) = p_1(x) + p_2(x) - 3p_3(x) = \frac{11}{6}x^3 - \frac{37}{2}x^2 + \frac{137}{3}x - 55$$

27. We have $\begin{bmatrix} a & a-b \\ a-b & b \end{bmatrix} = a\begin{bmatrix} 1 & 1 \\ 1 & 0 \end{bmatrix} + b\begin{bmatrix} 0 & -1 \\ -1 & 0 \end{bmatrix}$; thus $V = \text{span}(\{A, B\})$ where $A = \begin{bmatrix} 1 & 1 \\ 1 & 0 \end{bmatrix}$ and $B = \begin{bmatrix} 0 & -1 \\ -1 & 0 \end{bmatrix}$. This shows that V is a subspace of M_{22}. It is also easy to see that $aA + bB = 0$ if and only if $a = b = 0$. Thus the matrices A and B form a basis for V, and $\dim(V) = 2$.

29. (a) $1 \oplus 1 = 1 + 1 - 1 = 1$　　　　　　(b) $0 \otimes 2 = 0 + (1 - 0) = 1$
(c) Axioms 1 and 6 are automatic, and Axioms 2, 3, and 10 are easy to verify. We omit these details. The following computations serve to verify the remaining Axioms.

Axiom 4. We have $\mathbf{u} \oplus 1 = u + 1 - 1 = \mathbf{u}$ for every $\mathbf{u} = u$ in V; thus the number 1 plays the role of the zero vector in V; i.e., $\mathbf{0} = 1$.

Axiom 5. For each $\mathbf{u} = u$ in V, we have $\mathbf{u} \oplus (-u + 2) = u + (-u + 2) - 1 = 1 = \mathbf{0}$; thus the number $-\mathbf{u} = -u + 2$ serves as the negative of $\mathbf{u} = u$ in V.

Axiom 7. $k \otimes (\mathbf{u} \oplus \mathbf{v}) = k \otimes (u + v - 1) = k(u + v - 1) + (1 - k) = ku + kv + 1 - 2k$, and $(k \otimes \mathbf{u}) \oplus (k \otimes \mathbf{v}) = (ku + (1 - k)) \oplus (kv + (1 - k)) = ku + (1 - k) + kv + (1 - k) - 1 = ku + kv + 1 - 2k$.

Axiom 8. $(k + l) \otimes \mathbf{u} = (k + l)u + (1 - (k + l)) = ku + lu + 1 - k - l$, and $(k \otimes \mathbf{u}) \oplus (l \otimes \mathbf{u}) = (ku + (1 - k)) \oplus (lu + (1 - l)) = ku + (1 - k) + lu + (1 - l) - 1 = ku + lu + 1 - k - l$.

Axiom 9. $k \otimes (l \otimes \mathbf{u}) = k \otimes (lu + (1 - l)) = k(lu + (1 - l)) + (1 - k) = (kl)u + (1 - kl) = (kl) \otimes \mathbf{u}$.

31. The set V with these operations is not a vector space; in particular, Axiom 4 fails to hold.

DISCUSSION AND DISCOVERY

D1. Yes. The element Jupiter acts as the $\mathbf{0}$ vector and all ten axioms hold in a trivial way.

D2. If $\mathbf{v} \neq \mathbf{0}$, then $t\mathbf{v} = \mathbf{0}$ if and only if $t = 0$; thus $t_1\mathbf{v} = t_2\mathbf{v}$ if and only if $t_1 = t_2$. It follows that if $\mathbf{v}$ is a nonzero element of V, then $\text{span}\{\mathbf{v}\}$ is an infinite set.

D3. Yes. If A and C commute with B, then we have $(A + C)B = AB + CB = BA + BC = B(A + C)$ and $(kA)B = kAB = kBA = B(kA)$; thus $A + C$ and kA commute with B. This shows that the set of matrices which commute with B is a subspace of M_{22}.

D4. The figures below illustrate the fact that if the terminal points of $\mathbf{u}$ and $\mathbf{v}$ lie on a line L not passing through the origin, then the terminal points of $\mathbf{u} + \mathbf{v}$ and $k\mathbf{u}$ $(k \neq 1)$ do not lie on L. Thus the line L does not correspond to a subspace of R^2.

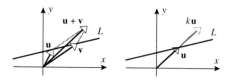

D5. We have $f_2(x) = 3\cos(\frac{1}{2}\pi - x) = 3[\cos(\frac{1}{2}\pi)\cos x + \sin(\frac{1}{2}\pi)\sin x] = 3\sin x = 3f_1(x)$; thus the subspace of $F(-\infty, \infty)$ that is spanned by f_1 and f_2 is 1-dimensional.

D6. Axiom 8 does not hold.

D7. For each $j = 1, 2, \ldots$, let $\mathbf{v}_j$ be the infinite sequence which has 1 in its j-th component and 0 in all other components. Then the set $S = \{\mathbf{v}_1, \mathbf{v}_2, \ldots, \mathbf{v}_j, \ldots\}$ is a linearly independent subset of R^∞.

WORKING WITH PROOFS

P1. If $\mathbf{0}_1$ and $\mathbf{0}_2$ are two zero vectors, then $\mathbf{0}_1 = \mathbf{0}_1 + \mathbf{0}_2 = \mathbf{0}_2 + \mathbf{0}_1 = \mathbf{0}_2$.

P2.
$$\mathbf{u}_1 + (\mathbf{u} + \mathbf{u}_2) = (\mathbf{u}_1 + \mathbf{u}) + \mathbf{u}_2 \qquad \text{Axiom 3}$$
$$\mathbf{u}_1 + \mathbf{0} = (\mathbf{u}_1 + \mathbf{u}) + \mathbf{u}_2 \qquad \text{Axiom 5}$$
$$\mathbf{u}_1 = (\mathbf{u}_1 + \mathbf{u}) + \mathbf{u}_2 \qquad \text{Axiom 4}$$
$$\mathbf{u}_1 = (\mathbf{u} + \mathbf{u}_1) + \mathbf{u}_2 \qquad \text{Axiom 2}$$
$$\mathbf{u}_1 = \mathbf{0} + \mathbf{u}_2 \qquad \text{Axiom 5}$$
$$\mathbf{u}_1 = \mathbf{u}_2 + \mathbf{0} \qquad \text{Axiom 2}$$
$$\mathbf{u}_1 = \mathbf{u}_2 \qquad \text{Axiom 4}$$

P3.
$$\mathbf{u} + \mathbf{w} = \mathbf{v} + \mathbf{w} \qquad \text{Hypothesis}$$
$$(\mathbf{u} + \mathbf{w}) + (-\mathbf{w}) = (\mathbf{v} + \mathbf{w}) + (-\mathbf{w}) \qquad \text{Add } -\mathbf{w} \text{ to both sides.}$$
$$\mathbf{u} + [\mathbf{w} + (-\mathbf{w})] = \mathbf{v} + [\mathbf{w} + (-\mathbf{w})] \qquad \text{Axiom 3}$$
$$\mathbf{u} + \mathbf{0} = \mathbf{v} + \mathbf{0} \qquad \text{Axiom 5}$$
$$\mathbf{u} = \mathbf{v} \qquad \text{Axiom 4}$$

P4. If $k\mathbf{u} = \mathbf{0}$ and $k \neq 0$, then $\mathbf{u} = 1\mathbf{u} = (\frac{1}{k}k)\mathbf{u} = \frac{1}{k}(k\mathbf{u}) = \frac{1}{k}\mathbf{0} = \mathbf{0}$.

P5. Let $W = W_1 \cap W_2$. If $\mathbf{u}$ and $\mathbf{v}$ belong to W then these vectors belong to both W_1 and W_2 and, since W_1 and W_2 are subspaces, it follows that $\mathbf{u} + \mathbf{v}$ belongs to both W_1 and W_2. This shows that W is closed under addition. Similarly, if $\mathbf{u}$ belongs to W and k is a scalar, then $k\mathbf{u}$ belongs to both W_1 and W_2; thus W is closed under scalar multiplication.

EXERCISE SET 9.2

1. **(a)** $\langle \mathbf{u}, \mathbf{v} \rangle = 3(2)(1) + 2(-3)(4) = -18$

(b) $\|\mathbf{u}\| = \sqrt{3(2)^2 + 2(-3)^2} = \sqrt{30}, \|\mathbf{v}\| = \sqrt{35}$

(c) $\quad \cos \theta = \dfrac{\langle \mathbf{u}, \mathbf{v} \rangle}{\|\mathbf{u}\| \, \|\mathbf{v}\|} = -\dfrac{18}{\sqrt{30}\sqrt{35}}$

(d) $\quad d(\mathbf{u}, \mathbf{v}) = \sqrt{3(2-1)^2 + 2(-3-4)^2} = \sqrt{101}$

3. We have $\langle \mathbf{u}, \mathbf{v} \rangle = 3(1)(2) + 2(1)(-3) = 0$, $\|\mathbf{u}\|^2 = 3(1)^2 + 2(1)^2 = 5$, $\|\mathbf{v}\|^2 = 3(2)^2 + 2(-3)^2 = 30$, and $\|\mathbf{u} + \mathbf{v}\|^2 = 3(1+2)^2 + 2(1-3)^2 = 35$. Thus $\mathbf{u}$ and $\mathbf{v}$ are orthogonal with respect to the given inner product and $\|\mathbf{u} + \mathbf{v}\|^2 = \|\mathbf{u}\|^2 + \|\mathbf{v}\|^2$.

5. (a) Axiom 4 $\qquad\qquad$ (b) Axioms 1 and 4 $\qquad\qquad$ (c) Axiom 4

7. We have $\langle \mathbf{u}, \mathbf{v} \rangle = -18$, $\|\mathbf{u}\| = \sqrt{30}$, $\|\mathbf{v}\| = \sqrt{35}$, and $\|\mathbf{u} + \mathbf{v}\| = \sqrt{3(2+1)^2 + 2(-3+4)^2} = \sqrt{29}$. Thus $\langle \mathbf{u}, \mathbf{v} \rangle^2 = (-18)^2 = 324 \le 1025 = (30)(35) = \|\mathbf{u}\|^2 \|\mathbf{v}\|^2$, and $\|\mathbf{u} + \mathbf{v}\| = \sqrt{29} \le \sqrt{30} + \sqrt{35} = \|\mathbf{u}\| + \|\mathbf{v}\|$.

9. We have $\langle \mathbf{v}_1, \mathbf{v}_2 \rangle = 1(1)(1) + 3(1)(-1) + 2(1)(1) = 0$, $\langle \mathbf{v}_1, \mathbf{v}_3 \rangle = 1(1)(2) + 3(1)(0) + 2(1)(-1) = 0$, and $\langle \mathbf{v}_2, \mathbf{v}_3 \rangle = 1(1)(2) + 3(-1)(0) + 2(1)(-1) = 0$; thus $S = \{\mathbf{v}_1, \mathbf{v}_2, \mathbf{v}_3\}$ is an orthogonal set relative to the given inner product. We have $\|\mathbf{v}_1\| = \sqrt{1(1)^2 + 3(1)^2 + 2(1)^2} = \sqrt{6}$ and, similarly, $\|\mathbf{v}_2\| = \|\mathbf{v}_3\| = \sqrt{6}$. Thus the corresponding orthonormal set is

$$S' = \left\{ \frac{\mathbf{v}_1}{\|\mathbf{v}_1\|}, \frac{\mathbf{v}_2}{\|\mathbf{v}_2\|}, \frac{\mathbf{v}_3}{\|\mathbf{v}_3\|} \right\} = \left\{ \left(\frac{1}{\sqrt{6}}, \frac{1}{\sqrt{6}}, \frac{1}{\sqrt{6}} \right), \left(\frac{1}{\sqrt{6}}, -\frac{1}{\sqrt{6}}, \frac{1}{\sqrt{6}} \right), \left(\frac{2}{\sqrt{6}}, 0, -\frac{1}{\sqrt{6}} \right) \right\}.$$

11. We have $\|\mathbf{x}\| = \sqrt{\frac{1}{16}x^2 + \frac{1}{4}y^2}$; thus the equation of the unit circle is $\frac{1}{16}x^2 + \frac{1}{4}y^2 = 1$.

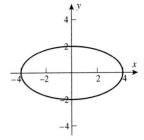

13. $\langle \mathbf{u}, \mathbf{v} \rangle = \frac{1}{25}u_1 v_1 + \frac{1}{16}u_2 v_2$

15. (a) $\quad \langle f, g \rangle = \displaystyle\int_0^1 (x)(x^2)\,dx = \dfrac{1}{4}$

(b) $\quad \|f\| = \sqrt{\displaystyle\int_0^1 (x)(x)\,dx} = \dfrac{1}{\sqrt{3}}$ and $\|g\| = \sqrt{\int_0^1 (x^2)(x^2)\,dx} = \dfrac{1}{\sqrt{5}}$

(c) $\quad \cos\theta = \dfrac{\langle f, g \rangle}{\|f\| \, \|g\|} = \dfrac{\frac{1}{4}}{\frac{1}{\sqrt{3}} \frac{1}{\sqrt{5}}} = \dfrac{\sqrt{15}}{4}$

(d) $\quad d(f, g) = \|f - g\| = \sqrt{\displaystyle\int_0^1 (x - x^2)^2\,dx} = \dfrac{1}{\sqrt{30}}$

17. (a) $\quad \langle f_1, f_2 \rangle = \int_0^1 (1)(\frac{1}{2} - x)\,dx = 0$; thus f_1 and f_2 are orthogonal in $C[0, 1]$.

(b) $\quad \|f_1\|^2 = \int_0^1 (1)^2\,dx = 1$, $\|f_2\|^2 = \int_0^1 (\frac{1}{2} - x)^2\,dx = \frac{1}{12}$, and $\|f_1 + f_2\|^2 = \int_0^1 [1 + (\frac{1}{2} - x)]^2\,dx = \frac{13}{12}$; thus $\|f_1 + f_2\|^2 = \|f_1\|^2 + \|f_2\|^2$.

(c) The functions $g_1(x) = 1$ and $g_2(x) = \sqrt{12}f_2(x) = 2\sqrt{3}(\frac{1}{2} - x)$ form an orthonormal basis for span$\{f_1, f_2\}$. We have $\langle f, g_1 \rangle = \int_0^1 (x^2)(1)dx = \frac{1}{3}$ and $\langle f, g_2 \rangle = \int_0^1 (x^2)(2\sqrt{3}(\frac{1}{2} - x))dx = -\frac{\sqrt{3}}{6}$. Thus the best mean square approximation of $f(x) = x^2$ by functions in $W = \text{span}\{f_1, f_2\}$ is

$$\hat{f}(x) = \langle f, g_1 \rangle g_1(x) + \langle f, g_2 \rangle g_2(x) = \frac{1}{3} - \left(\frac{1}{2} - x\right) = x - \frac{1}{6}$$

19. Using the trigonometric identity $\sin(A)\sin(B) = \frac{1}{2}[\cos(A - B) - \cos(A + B)]$ we have

$$\langle f, g \rangle = \int_0^{2\pi} \sin px \sin qx\, dx = \frac{1}{2}\int_0^{2\pi}[\cos(p - q)x - \cos(p + q)x]dx$$

$$= \frac{\sin 2\pi(p - q)}{2(p - q)} - \frac{\sin 2\pi(p + q)}{2(p + q)} = 0 - 0 = 0$$

Thus the functions $f(x) = \sin px$ and $g(x) = \sin qx$ are orthogonal.

21. $a_0 = \frac{1}{\pi}\int_0^{2\pi} e^x\, dx = \frac{e^{2\pi} - 1}{\pi},\quad a_1 = \frac{1}{\pi}\int_0^{2\pi} e^x \cos x\, dx = \frac{1}{2}\frac{e^{2\pi} - 1}{\pi},\quad a_2 = \frac{1}{\pi}\int_0^{2\pi} e^x \cos 2x\, dx =$

$\frac{1}{5}\frac{e^{2\pi} - 1}{\pi},\quad b_1 = \frac{1}{\pi}\int_0^{2\pi} e^x \sin x\, dx = -\frac{1}{2}\frac{e^{2\pi} - 1}{\pi},\quad$ and $\quad b_2 = \frac{1}{\pi}\int_0^{2\pi} e^x \sin 2x\, dx = -\frac{2}{5}\frac{e^{2\pi} - 1}{\pi}.$

Thus the second-order Fourier approximation of $f(x) = e^x$ is

$$e^x \approx \frac{e^{2\pi} - 1}{\pi}\left(\frac{1}{2} + \frac{1}{2}\cos x - \frac{1}{2}\sin x + \frac{1}{5}\cos 2x - \frac{2}{5}\sin 2x\right)$$

23. We have $a_0 = \frac{1}{\pi}\int_0^{2\pi} 3x\, dx = 6\pi$ and, using integration by parts,

$$a_k = \frac{1}{\pi}\int_0^{2\pi} 3x \cos kx\, dx = \frac{3}{\pi}\left[\frac{1}{k^2}\cos kx + \frac{x}{k}\sin kx\right]_0^{2\pi} = 0$$

$$b_k = \frac{1}{\pi}\int_0^{2\pi} 3x \sin kx\, dx = \frac{3}{\pi}\left[\frac{1}{k^2}\sin kx - \frac{x}{k}\cos kx\right]_0^{2\pi} = -\frac{6}{k}$$

for $k = 1, 2, \ldots, n$. Thus the nth-order Fourier approximation to $f(x) = 3x$ is

$$3x \approx 3\pi - 6\left(\sin x + \frac{\sin 2x}{2} + \frac{\sin 3x}{3} + \cdots + \frac{\sin nx}{n}\right)$$

25. We have $\langle u, v \rangle = u^T A v$ where $A = \begin{bmatrix} 2 & -1 \\ -1 & 4 \end{bmatrix}$. Since $a_{11} = 2 > 0$ and $\det(A) = 7 > 0$, the matrix A is positive definite. Thus, from Theorem 9.2.8, $\langle u, v \rangle = u^T A v$ is an inner product on R^2.

27. We have $\|u + v\|^2 = \langle u + v, u + v \rangle = \langle u, u \rangle + \langle u, v \rangle + \langle v, u \rangle + \langle v, v \rangle = \|u\|^2 + 2\langle u, v \rangle + \|v\|^2$ and, similarly, $\|u - v\|^2 = \|u\|^2 - 2\langle u, v \rangle + \|v\|^2$. Thus $\|u + v\|^2 + \|u - v\|^2 = 2(\|u\|^2 + \|v\|^2)$.

29. (a) $\langle u, v \rangle = 3u_1\bar{v}_1 + 2u_2\bar{v}_2 = \overline{3v_1\bar{u}_1 + 2v_2\bar{u}_2} = \overline{\langle v, u \rangle}$

$\langle u + v, w \rangle = 3(u_1 + v_1)\bar{w}_1 + 2(u_2 + v_2)\bar{w}_2 = (3u_1\bar{w}_1 + 2u_2\bar{w}_2) + (3v_1\bar{w}_1 + 2v_2\bar{w}_2) =$
$\langle u, w \rangle + \langle v, w \rangle$

$k\langle u, v \rangle = 3(ku_1)\bar{v}_1 + 2(ku_2)\bar{v}_2 = k(3u_1\bar{v}_1 + 2u_2\bar{v}_2) = k\langle u, v \rangle$

$\langle v, v \rangle = 3v_1\bar{v}_1 + 2v_2\bar{v}_2 = 3|v_1|^2 + 2|v_2|^2 \geq 0$, and $\langle v, v \rangle = 0$ if and only if $v = (0, 0) = 0$

(b) We have $\langle \mathbf{u}, \mathbf{v} \rangle = 3(i)(-3i) + 2(-2i)(i) = 13$, $\|\mathbf{u}\| = \sqrt{3(i)(-i) + 2(-2i)(2i)} = \sqrt{11}$,
$\|\mathbf{v}\| = \sqrt{3(3i)(-3i) + 2(-i)(i)} = \sqrt{29}$, and $d(\mathbf{u}, \mathbf{v}) = \|\mathbf{u} - \mathbf{v}\| = \sqrt{3(-2i)(2i) + 2(-i)(i)} = \sqrt{14}$.

DISCUSSION AND DISCOVERY

D1. $\|\mathbf{u} - \mathbf{v}\| = \sqrt{\langle \mathbf{u} - \mathbf{v}, \mathbf{u} - \mathbf{v} \rangle} = \sqrt{\langle \mathbf{u}, \mathbf{u} \rangle - \langle \mathbf{u}, \mathbf{v} \rangle - \langle \mathbf{v}, \mathbf{u} \rangle + \langle \mathbf{v}, \mathbf{v} \rangle} = \sqrt{1 - 0 - 0 + 1} = \sqrt{2}$

D2. The given expression can be written as $\langle \mathbf{u}, \mathbf{v} \rangle = \mathbf{u}^T A \mathbf{v}$ where $A = \begin{bmatrix} 5 & -3 \\ -3 & c \end{bmatrix}$. This defines an inner product on R^2 if and only if A is positive definite, and this occurs if and only if $\det(A) = 5c - 9 > 0$.

D3. If U and V are 2×2 matrices, then $\operatorname{tr}(U^T V)$ is the same as $\mathbf{u} \cdot \mathbf{v}$ where $\mathbf{u} = (u_{11}, u_{12}, u_{21}, u_{22})$ and $\mathbf{v} = (v_{11}, v_{12}, v_{21}, v_{22})$. Thus $\langle U, V \rangle = \operatorname{tr}(U^T V)$ defines an inner product on the vector space M_{22}.

D4. If $p(x) = c_0 + c_1 x + \cdots + c_n x^n$ and $q(x) = d_0 + d_1 x + \cdots + d_n x^n$, then

$$\langle p, q \rangle = c_0 d_0 + c_1 d_1 + \cdots + c_n d_n = \mathbf{u} \cdot \mathbf{v}$$

where $\mathbf{u} = (c_0, c_1, \ldots, c_n)$ and $\mathbf{v} = (d_0, d_1, \ldots, d_n)$ in the vector space R^{n+1}. Thus $\langle p, q \rangle$ defines an inner product on the vector space P_n.

D5. We have $\langle \mathbf{u}, \mathbf{v} \rangle = (r\mathbf{u}) \cdot (s\mathbf{v}) = (rs)(\mathbf{u} \cdot \mathbf{v})$; thus $\langle \mathbf{u}, \mathbf{v} \rangle$ is an inner product on R^n if and only if $rs > 0$.

WORKING WITH PROOFS

P1. The inner product axioms are verified as follows:
(1) $\langle \mathbf{u}, \mathbf{v} \rangle = A\mathbf{u} \cdot A\mathbf{v} = A\mathbf{v} \cdot A\mathbf{u} = \langle \mathbf{v}, \mathbf{u} \rangle$
(2) $\langle \mathbf{u} + \mathbf{v}, \mathbf{w} \rangle = A(\mathbf{u} + \mathbf{v}) \cdot A\mathbf{w} = (A\mathbf{u} + A\mathbf{v}) \cdot A\mathbf{w} = A\mathbf{u} \cdot A\mathbf{w} + A\mathbf{v} \cdot A\mathbf{w} = \langle \mathbf{u}, \mathbf{w} \rangle + \langle \mathbf{v}, \mathbf{w} \rangle$
(3) $\langle k\mathbf{u}, \mathbf{v} \rangle = A(k\mathbf{u}) \cdot A\mathbf{v} = (kA\mathbf{u}) \cdot A\mathbf{v} = k(A\mathbf{u} \cdot A\mathbf{v}) = k \langle \mathbf{u}, \mathbf{v} \rangle$
(4) $\langle \mathbf{v}, \mathbf{v} \rangle = A\mathbf{v} \cdot A\mathbf{v} = \|A\mathbf{v}\|^2 \geq 0$ and, since A is invertible, $\langle \mathbf{v}, \mathbf{v} \rangle = \|A\mathbf{v}\|^2 = 0$ if and only if $\mathbf{v} = \mathbf{0}$.

P2. The inner product axioms are verified as follows:
(1) $\langle \mathbf{x}, \mathbf{y} \rangle = w_1 x_1 y_1 + w_2 x_2 y_2 + \cdots + w_n x_n y_n = w_1 y_1 x_1 + w_2 y_2 x_2 + \cdots + w_n y_n x_n = \langle \mathbf{y}, \mathbf{x} \rangle$
(2) $\langle \mathbf{x} + \mathbf{y}, \mathbf{z} \rangle = w_1(x_1 + y_1)z_1 + w_2(x_2 + y_2)z_2 + \cdots + w_n(x_n + y_n)z_n$
$= (w_1 x_1 z_1 + w_2 x_2 z_2 + \cdots + w_n x_n z_n) + (w_1 y_1 z_1 + w_2 y_2 z_2 + \cdots + w_n y_n z_n) = \langle \mathbf{x}, \mathbf{z} \rangle + \langle \mathbf{y}, \mathbf{z} \rangle$
(3) $\langle k\mathbf{u}, \mathbf{v} \rangle = w_1(ku_1)v_1 + w_2(ku_2)v_2 + \cdots + w_n(ku_n)v_n = k(w_1 u_1 v_1 + w_2 u_2 v_2 + \cdots + w_n u_n v_n) = k \langle \mathbf{u}, \mathbf{v} \rangle$
(4) $\langle \mathbf{v}, \mathbf{v} \rangle = w_1 v_1^2 + w_2 v_2^2 + \cdots + w_n v_n^2 \geq 0$ and, since $w_1, w_2, \ldots, w_n$ are positive, we have $\langle \mathbf{v}, \mathbf{v} \rangle = 0$ if and only if $\mathbf{v} = (0, 0, \ldots, 0) = \mathbf{0}$.

P3. If $L(\mathbf{u}) = \mathbf{u}^T A\mathbf{y} = \mathbf{u} \cdot \mathbf{y}$, we have $L(\mathbf{u} + \mathbf{v}) = (\mathbf{u} + \mathbf{v})^T A\mathbf{y} = (\mathbf{u}^T + \mathbf{v}^T)A\mathbf{y} = \mathbf{u}^T A\mathbf{y} + \mathbf{v}^T A\mathbf{y} = L(\mathbf{u}) + L(\mathbf{v})$ and $L(k\mathbf{u}) = (k\mathbf{u})^T A\mathbf{y} = k\mathbf{u}^T A\mathbf{y} = kL(\mathbf{u})$; thus $L: R^n \to R$ is a linear transformation.

P4. If **u** and **v** are vectors in R^n then, using the properties of an inner product, we have

$$\langle \mathbf{u}, \mathbf{v} \rangle = \langle u_1\mathbf{e}_1 + u_2\mathbf{e}_2 + \cdots + u_n\mathbf{e}_n, v_1\mathbf{e}_1 + v_2\mathbf{e}_2 + \cdots + v_n\mathbf{e}_n \rangle = \sum_{i=1}^{n}\sum_{j=1}^{n} u_i v_j \langle \mathbf{e}_i, \mathbf{e}_j \rangle = \mathbf{u}^T A \mathbf{v}$$

where $A = \begin{bmatrix} \langle \mathbf{e}_1, \mathbf{e}_1 \rangle & \langle \mathbf{e}_1, \mathbf{e}_2 \rangle & \cdots & \langle \mathbf{e}_1, \mathbf{e}_n \rangle \\ \langle \mathbf{e}_2, \mathbf{e}_1 \rangle & \langle \mathbf{e}_2, \mathbf{e}_2 \rangle & \cdots & \langle \mathbf{e}_2, \mathbf{e}_n \rangle \\ \vdots & \vdots & \ddots & \vdots \\ \langle \mathbf{e}_n, \mathbf{e}_1 \rangle & \langle \mathbf{e}_n, \mathbf{e}_2 \rangle & \cdots & \langle \mathbf{e}_n, \mathbf{e}_n \rangle \end{bmatrix}$. Since $\langle \mathbf{e}_i, \mathbf{e}_j \rangle = \langle \mathbf{e}_j, \mathbf{e}_i \rangle$, the matrix A is symmetric.

Furthermore, we have $\mathbf{v}^T A \mathbf{v} = \langle \mathbf{v}, \mathbf{v} \rangle \geq 0$ for every vector **v** in R^n, and $\mathbf{v}^T A \mathbf{v} = \langle \mathbf{v}, \mathbf{v} \rangle = 0$ if and only if $\mathbf{v} = \mathbf{0}$. Thus A is positive definite.

P5. Since M is positive definite, there is (by Theorem 8.4.6) a positive definite symmetric matrix P such that $P^2 = M$. The given inner product $\langle , \rangle$ can be expressed in terms of the dot product as

$$\langle u, v \rangle = \mathbf{u}^T M \mathbf{v} = \mathbf{u}^T P^2 \mathbf{v} = (P\mathbf{u})^T (P\mathbf{v}) = (P\mathbf{u}) \cdot (P\mathbf{v})$$

and so the norm $\| \, \|_M$ corresponding to $\langle , \rangle$ is related the Euclidean norm by $\|\mathbf{u}\|_M = \|P\mathbf{u}\|$. From this it follows that the least squares solution of $A\mathbf{x} = \mathbf{b}$ with respect to $\| \, \|_M$ is the same as the least squares solution of $PA\mathbf{x} = P\mathbf{b}$ with respect to the Euclidean norm. Finally, from Theorem 7.8.3, the solution of the latter coincides with the exact solution of the normal system $(PA)^T PA\mathbf{x} = (PA)^T P\mathbf{b}$ which, in terms of the matrix M, is $A^T MA\mathbf{x} = A^T M\mathbf{b}$.

EXERCISE SET 9.3

1. **(a)** $T(2\mathbf{u} + 4\mathbf{v}) = 2T(\mathbf{u}) + 4T(\mathbf{v}) = 2(1, 2) + 4(-1, 3) = (-2, 16)$.

   (b) $\mathbf{u} = \frac{1}{2}(\mathbf{u} + \mathbf{v}) + \frac{1}{2}(\mathbf{u} - \mathbf{v})$ and $\mathbf{v} = \frac{1}{2}(\mathbf{u} + \mathbf{v}) - \frac{1}{2}(\mathbf{u} - \mathbf{v})$; thus

   $$T(\mathbf{u}) = \frac{1}{2}T(\mathbf{u} + \mathbf{v}) + \frac{1}{2}T(\mathbf{u} - \mathbf{v}) = \frac{1}{2}(2, 4) + \frac{1}{2}(3, 5) = \left(-\frac{1}{2}, \frac{9}{2}\right)$$

   $$T(\mathbf{v}) = \frac{1}{2}T(\mathbf{u} + \mathbf{v}) - \frac{1}{2}T(\mathbf{u} - \mathbf{v}) = \frac{1}{2}(2, 4) - \frac{1}{2}(3, 5) = \left(\frac{5}{2}, -\frac{1}{2}\right).$$

3. If $T(p) = xp(x)$, then $T(cp) = x(cp) = c(xp) = cT(p)$ and $T(p + q) = x(p + q) = xp + xq = T(p) + T(q)$; thus T is linear.

5. $T(cA) = (cA)^T = cA^T = cT(A)$ and $T(A + B) = (A + B)^T = A^T + B^T = T(A) + T(B)$; thus T is linear.

7. $T(cf) = \int_a^b cf(x)dx = c\int_a^b f(x)dx = cT(f)$ and $T(f + g) = \int_a^b [f(x) + g(x)]dx = \int_a^b f(x)dx + \int_a^b g(x)dx = T(f) + T(g)$; thus T is linear.

9. $T(cX) = A_0(cX) = cA_0X = cT(X)$ and $T(X + Y) = A_0(X + Y) = A_0X + A_0Y = T(X) + T(Y)$.

11. $T(-f) = x^2(-f(x))^2 = x^2 f^2(x) \neq -T(f)$; thus T is not linear.

13. $T(\mathbf{0}) = \mathbf{0} + \mathbf{x}_0 \neq \mathbf{0}$; thus T is not linear.

15. $T(c\mathbf{x}) = c_0(c\mathbf{x}) + (c\mathbf{x}) = c(c_0\mathbf{x} + \mathbf{x}) = cT(\mathbf{x})$ and $T(\mathbf{x} + \mathbf{y}) = c_0(\mathbf{x} + \mathbf{y}) + (\mathbf{x} + \mathbf{y}) = (c_0\mathbf{x} + \mathbf{x}) + (c_0\mathbf{y} + \mathbf{y}) = T(\mathbf{x}) + T(\mathbf{y})$; thus T is linear.

17. $T(cf) = (cf(0))(cf(1)) = c^2 f(0)f(1) = c^2 T(f)$; thus T is not linear.

19. **(a)** The polynomials $q_2(x) = x + 5x^2 = x(1 + 5x) = T(1 + 5x)$ and $q_3(x) = 0 = x(0) = T(0)$ are in the range of T.

(b) A polynomial p is in the kernel of T is and only if $p(-1) = p(1) = 0$; thus $q_1(x) = x^2 - 1$ and $q_3(x) = 0$ are in the kernel of T.

21. We have $T\left(k\begin{bmatrix} a & b \\ c & d \end{bmatrix}\right) = T\left(\begin{bmatrix} ka & kb \\ kc & kd \end{bmatrix}\right) = \begin{bmatrix} ka & 0 \\ 0 & ka \end{bmatrix} = k\begin{bmatrix} a & 0 \\ 0 & a \end{bmatrix} = kT\left(\begin{bmatrix} a & b \\ c & d \end{bmatrix}\right)$ and

$$T\left(\begin{bmatrix} a_1 & b_1 \\ c_1 & d_1 \end{bmatrix} + \begin{bmatrix} a_2 & b_2 \\ c_2 & d_2 \end{bmatrix}\right) = T\left(\begin{bmatrix} a_1 + a_2 & b_1 + b_2 \\ c_1 + c_2 & d_1 + d_2 \end{bmatrix}\right) = \begin{bmatrix} a_1 + a_2 & 0 \\ 0 & a_1 + a_2 \end{bmatrix} = \begin{bmatrix} a_1 & 0 \\ 0 & a_1 \end{bmatrix} + \begin{bmatrix} a_2 & 0 \\ 0 & a_2 \end{bmatrix}$$

$$= T\left(\begin{bmatrix} a_1 & b_1 \\ c_1 & d_1 \end{bmatrix}\right) + T\left(\begin{bmatrix} a_2 & b_2 \\ c_2 & d_2 \end{bmatrix}\right)$$

thus T is linear. The kernel of T is the set of matrices of the form

$$\begin{bmatrix} 0 & b \\ c & d \end{bmatrix} = b\begin{bmatrix} 0 & 1 \\ 0 & 0 \end{bmatrix} + c\begin{bmatrix} 0 & 0 \\ 1 & 0 \end{bmatrix} + d\begin{bmatrix} 0 & 0 \\ 0 & 1 \end{bmatrix}$$

and so $\begin{bmatrix} 0 & 1 \\ 0 & 0 \end{bmatrix}, \begin{bmatrix} 0 & 0 \\ 1 & 0 \end{bmatrix}$, and $\begin{bmatrix} 0 & 0 \\ 0 & 1 \end{bmatrix}$ form a basis for the kernel of T. The range of T consists of matrices

of the form $\begin{bmatrix} a & 0 \\ 0 & a \end{bmatrix} = a\begin{bmatrix} 1 & 0 \\ 0 & 1 \end{bmatrix}$; thus $\begin{bmatrix} 1 & 0 \\ 0 & 1 \end{bmatrix}$ forms a basis for the range of T.

23. If $T(X) = T(Y)$ then $A_0 X = A_0 Y$ and so $X = A_0^{-1}(A_0 X) = A_0^{-1}(A_0 Y) = Y$; thus T is one-to-one. Furthermore, if M is an arbitrary matrix in M_{nn} then $M = A_0(X) = T(X)$ where $X = A_0^{-1}M$; thus T is onto.

25. If $T(p) = T(q)$ then $xp(x) = xq(x)$ and so $p(x) = q(x)$; thus T is one-to-one. On the other hand, nonzero constant polynomials (polynomials of degree 0) are not in the range of T; thus T is not onto.

27. **(a)** If $y''(x) = 0$, then $y'(x)$ is constant and $y(x)$ is a linear polynomial, i.e., $y(x) = a_0 + a_1 x$ for some scalars a_0 and a_1. Thus the functions $y_1(x) = 1$ and $y_2(x) = x$ form a basis for the kernel of D.

(b) A general solution of $y'' = 0$ is $y = a_0 + a_1 x$.

29. **(a)** $D(y_1) = D(e^{-\omega x}) = \omega^2 e^{-\omega x} - \omega^2 e^{-\omega x} = 0$ and $D(y_2) = D(e^{\omega x}) = \omega^2 e^{\omega x} - \omega^2 e^{\omega x} = 0$. Furthermore, the functions $y_1 = e^{-\omega x}$ and $y_2 = e^{\omega x}$ are linearly independent since neither is a scalar multiple of the other. Thus y_1 and y_2 form a basis for $\ker(D)$.

(b) A general solution of $y'' - \omega^2 y = 0$ is $y = C_1 e^{-\omega x} + C_2 e^{\omega x}$.

31. The Lagrange interpolating polynomials corresponding to $x_1 = 0$, $x_2 = 1$, $x_3 = 2$ are

$$p_1(x) = \frac{(x-1)(x-2)}{(0-1)(0-2)} = \frac{1}{2}x^2 - \frac{3}{2}x + 1$$

$$p_2(x) = \frac{(x-0)(x-2)}{(1-0)(1-2)} = -x^2 + 2x$$

$$p_3(x) = \frac{(x-0)(x-1)}{(2-0)(2-1)} = \frac{1}{2}x^2 - \frac{1}{2}x$$

The polynomial p for which $T(p) = \mathbf{v} = (1, 3, 7)$ is $p(x) = 1p_1(x) + 3p_2(x) + 7p_3(x) = x^2 + x + 1$.

33. We have $T(c\mathbf{v}) = T(cv_1, cv_2, cv_3, \ldots) = (cv_2, cv_3, cv_4, \ldots) = c(v_2, v_3, v_4, \ldots) = cT(\mathbf{v})$ and

$$T(\mathbf{v} + \mathbf{w}) = T(v_1 + w_1, v_2 + w_2, v_3 + w_3, \ldots, v_n + w_n, \ldots) = (v_2 + w_2, v_3 + w_3, v_4 + w_4, \ldots)$$

$$= (v_2, v_3, v_4, \ldots) + (w_2, w_3, w_4, \ldots) = T(\mathbf{v}) + T(\mathbf{w})$$

thus T is linear. The kernel of T consists of the scalar multiples of $(1, 0, 0, 0, \ldots)$, and the range of T is R^∞.

35. If $\mathbf{v}$ is in V then, since $\{\mathbf{v}_1, \mathbf{v}_2, \ldots, \mathbf{v}_n\}$ is a basis for V, there are scalars $a_1, a_2, \ldots, a_n$ such that $\mathbf{v} = a_1\mathbf{v}_1 + a_2\mathbf{v}_2 + \cdots + a_n\mathbf{v}_n$. It follows that if T is a linear transformation for which $T(\mathbf{v}_1) = T(\mathbf{v}_2) = \cdots = T(\mathbf{v}_n) = \mathbf{0}$ then

$$T(\mathbf{v}) = a_1T(\mathbf{v}_1) + a_2T(\mathbf{v}_2) + \cdots + a_nT(\mathbf{v}_n) = \mathbf{0}$$

for every $\mathbf{v}$ in V.

37. (a) The vector $\mathbf{v} = (-1, 2, 0, 3)$ corresponds to the matrix $\begin{bmatrix} -1 & 2 \\ 0 & 3 \end{bmatrix}$.

(b) A matrix is symmetric and has trace zero if and only if it is of the form

$$\begin{bmatrix} a & b \\ b & -a \end{bmatrix} = a\begin{bmatrix} 1 & 0 \\ 0 & -1 \end{bmatrix} + b\begin{bmatrix} 0 & 1 \\ 1 & 0 \end{bmatrix}$$

thus the matrices $\begin{bmatrix} 1 & 0 \\ 0 & -1 \end{bmatrix}$ and $\begin{bmatrix} 0 & 1 \\ 1 & 0 \end{bmatrix}$ form a basis for this subspace. A basis for the corresponding subspace of R^4 consists of the vectors $(1, 0, 0, -1)$ and $(0, 1, 1, 0)$.

(c) If $A = \begin{bmatrix} a & b \\ c & d \end{bmatrix}$ then $A^T = \begin{bmatrix} a & c \\ b & d \end{bmatrix}$; thus the transformation on R^4 corresponding to $A \to A^T$ is

$(a, b, c, d) \xrightarrow{T} (a, c, b, d)$ and the standard matrix of this transformation is $[T] = \begin{bmatrix} 1 & 0 & 0 & 0 \\ 0 & 0 & 1 & 0 \\ 0 & 1 & 0 & 0 \\ 0 & 0 & 0 & 1 \end{bmatrix}$.

39. If $p = a_0 + a_1x + a_2x^2$ then $J(p) = \int_0^x (a_0 + a_1t + a_2t^2)dt = a_0x + \frac{1}{2}a_1x^2 + \frac{1}{3}a_2x^3$; thus the transformation $p \to J(p)$ corresponds to the transformation $(a_0, a_1, a_2) \xrightarrow{T} (0, a_0, \frac{1}{2}a_1, \frac{1}{3}a_2)$ from R^3

to R^4, and the standard matrix of the latter is $[T] = \begin{bmatrix} 0 & 0 & 0 \\ 1 & 0 & 0 \\ 0 & \frac{1}{2} & 0 \\ 0 & 0 & \frac{1}{3} \end{bmatrix}$. Integration of the polynomial

$p(x) = 1 + x + x^2$ corresponds to the matrix product

$$\begin{bmatrix} 0 & 0 & 0 \\ 1 & 0 & 0 \\ 0 & \frac{1}{2} & 0 \\ 0 & 0 & \frac{1}{3} \end{bmatrix} \begin{bmatrix} 1 \\ 1 \\ 1 \end{bmatrix} = \begin{bmatrix} 0 \\ 1 \\ \frac{1}{2} \\ \frac{1}{3} \end{bmatrix}$$

resulting in $J(p) = x + \frac{1}{2}x^2 + \frac{1}{3}x^3$.

DISCUSSION AND DISCOVERY

D1. $T(2 + 4x - x^2) = 2T(1) + 2T(2x) - \frac{1}{3}T(3x)^2 = 2(1 + x) + 2(1 - 2x) - \frac{1}{3}(1 + 3x) = \frac{11}{3} - 3x$

D2. **(a)** True. Taking $c = 1$ and $\mathbf{u} = \mathbf{v} = \mathbf{0}$, we have $T(\mathbf{0}) = T(\mathbf{0} + \mathbf{0}) = T(\mathbf{0}) + T(\mathbf{0}) = 2T(\mathbf{0})$ and so $T(\mathbf{0}) = \mathbf{0}$. It then follows that $T(c\mathbf{u}) = T(c\mathbf{u} + \mathbf{0}) = cT(\mathbf{u}) + T(\mathbf{0}) = cT(\mathbf{u})$ and $T(\mathbf{u} + \mathbf{v}) = T(\mathbf{u}) + T(\mathbf{v})$; thus T is linear.

(b) True. The transformation T is inverse of the natural isomorphism from P_2 onto R^3 (see Example 18).

(c) True. The transformation T is the sum of two transformations, both of which are linear (see Examples 6 and 7).

(d) True. For each $i, j = 1, 2, \ldots, n$, let $E_{i,j}$ be the $m \times n$ matrix that has 1 as the entry in the ith row and jth column and 0 as the entry in all other positions. These matrices form a basis for M_{mn} and there are mn of them; thus $\dim(M_{mn}) = mn$.

(e) False. There is no finite set in P^∞ that spans P^∞, since the polynomials in any finite set would have some maximum degree, say m, and so it would not be possible to express a polynomial of degree greater than m as a linear combination of the polynomials in the finite set.

D3. **(a)** We have $T(cA) = (cA) - (cA)^T = cA - cA^T = c(A - A^T) = cT(A)$ and

$$T(A + B) = (A + B) - (A + B)^T = (A + B) - (A^T + B^T) = (A - A^T) + (B - B^T) = T(A) + T(B)$$

thus T is linear.

(b) We have $T(A) = A - A^T = 0$ if and only if $A = A^T$; thus the kernel of T consists of the set of all symmetric matrices.

(c) If $B = T(A) = A - A^T$, then $B^T = (A - A^T)^T = A^T - A = -B$, i.e., B is skew-symmetric. On the other hand, if B is skew symmetric then $B = \frac{1}{2}B + \frac{1}{2}B = \frac{1}{2}B - \frac{1}{2}B^T = T(A)$ where $A = \frac{1}{2}B$. Thus the range of T is the set of all skew-symmetric matrices.

(d) If $T(A) = A + A^T$, then the kernel of T is the set of all skew-symmetric matrices and the range of T is the set of all symmetric matrices.

D4. The linear transformation defined by $T(f) = f''''$ has the property that $T(f) = 0$ if and only if f is a polynomial of degree three or less.

D5. If $p(x) = a + bx$ then $T(p) = \int_{-1}^{1}(a + bx)dx = 2a$; thus the kernel of T consists of the set of first degree polynomials whose constant term is zero.

WORKING WITH PROOFS

P1. We have $(kT)(c\mathbf{x}) = kT(c\mathbf{x}) = kcT(\mathbf{x}) = ckT(\mathbf{x}) = c(kT)(\mathbf{x})$ and $(kT)(\mathbf{x}_1 + \mathbf{x}_2) = kT(\mathbf{x}_1 + \mathbf{x}_2) = k(T(\mathbf{x}_1) + T(\mathbf{x}_2)) = kT(\mathbf{x}_1) + kT(\mathbf{x}_2) = (kT)(\mathbf{x}_1) + (kT)(\mathbf{x}_2)$; thus kT is linear. Similarly, we have $(T_1 + T_2)(c\mathbf{x}) = T_1(c\mathbf{x}) + T_2(c\mathbf{x}) = cT_1(\mathbf{x}) + cT_2(\mathbf{x}) = c(T_1(\mathbf{x}) + T_2(\mathbf{x})) = c(T_1 + T_2)(\mathbf{x})$ and

$$(T_1 + T_2)(\mathbf{x}_1 + \mathbf{x}_2) = T_1(\mathbf{x}_1 + \mathbf{x}_2) + T_2(\mathbf{x}_1 + \mathbf{x}_2) = T_1(\mathbf{x}_1) + T_1(\mathbf{x}_2) + T_2(\mathbf{x}_1) + T_2(\mathbf{x}_2)$$
$$= (T_1(\mathbf{x}_1) + T_2(\mathbf{x}_1)) + (T_1(\mathbf{x}_2) + T_2(\mathbf{x}_2)) = (T_1 + T_2)(\mathbf{x}_1) + (T_1 + T_2)(\mathbf{x}_2)$$

thus $T_1 + T_2$ is linear.

P2. **(a)** $0 + T(\mathbf{0}) = T(\mathbf{0}) = T(\mathbf{0} + \mathbf{0}) = T(\mathbf{0}) + T(\mathbf{0})$, thus $T(\mathbf{0}) = \mathbf{0}$.

   (b) $T(-\mathbf{u}) + T(\mathbf{u}) = T(-\mathbf{u} + \mathbf{u}) = T(\mathbf{0}) = \mathbf{0}$; thus $T(-\mathbf{u}) = -T(\mathbf{u})$.

   (c) Using (b), it follows that $T(\mathbf{u} - \mathbf{v}) = T(\mathbf{u} + (-\mathbf{v})) = T(\mathbf{u}) + T(-\mathbf{v}) = T(\mathbf{u}) - T(\mathbf{v})$.

P3. If $T^{-1}(\mathbf{x}) = \mathbf{v}$, then $T(\mathbf{v}) = \mathbf{x}$ and so, by linearity of T, $T(c\mathbf{v}) = c\mathbf{x}$ and $T^{-1}(c\mathbf{x}) = c\mathbf{v} = cT^{-1}(\mathbf{x})$. Similarly, if $T^{-1}(\mathbf{x}_1) = \mathbf{v}_1$ and $T^{-1}(\mathbf{x}_2) = \mathbf{v}_2$, then $T(\mathbf{v}_1) = \mathbf{x}_1$ and $T(\mathbf{v}_2) = \mathbf{x}_2$ and so, by linearity of T, we have $T(\mathbf{v}_1 + \mathbf{v}_2) = \mathbf{x}_1 + \mathbf{x}_2$ and $T^{-1}(\mathbf{x}_1 + \mathbf{x}_2) = \mathbf{v}_1 + \mathbf{v}_2 = T^{-1}(\mathbf{x}_1) + T^{-1}(\mathbf{x}_2)$. This shows that T^{-1} is linear.

P4. For $\mathbf{u}$ and $\mathbf{v}$ in V, let $\langle \mathbf{u}, \mathbf{v} \rangle = T\mathbf{u} \cdot T\mathbf{v}$. Then $\langle \mathbf{u}, \mathbf{v} \rangle = T\mathbf{u} \cdot T\mathbf{v} = T\mathbf{v} \cdot T\mathbf{u} = \langle \mathbf{v}, \mathbf{u} \rangle$,

$$\langle \mathbf{u} + \mathbf{v}, \mathbf{w} \rangle = T(\mathbf{u} + \mathbf{v}) \cdot T(\mathbf{w}) = (T(\mathbf{u}) + T(\mathbf{v})) \cdot T(\mathbf{w}) = T(\mathbf{u}) \cdot T(\mathbf{w}) + T(\mathbf{v}) \cdot T(\mathbf{w}) = \langle \mathbf{u}, \mathbf{w} \rangle + \langle \mathbf{v}, \mathbf{w} \rangle$$

and $\langle k\mathbf{u}, \mathbf{v} \rangle = T(k\mathbf{u}) \cdot T(\mathbf{v}) = kT(\mathbf{u}) \cdot T(\mathbf{v}) = k \langle \mathbf{u}, \mathbf{v} \rangle$. Furthermore, $\langle \mathbf{v}, \mathbf{v} \rangle = T\mathbf{v} \cdot T\mathbf{v} = \|T\mathbf{v}\|^2 \geq 0$ and, since T is an isomorphism, $\langle \mathbf{v}, \mathbf{v} \rangle = \|T\mathbf{v}\|^2 = \mathbf{0}$ if and only if $\mathbf{v} = \mathbf{0}$. Thus shows that $\langle \mathbf{u}, \mathbf{v} \rangle = T\mathbf{u} \cdot T\mathbf{v}$ defines an inner product on V.